52 Springer Series in Chemical Physics

Edited by J. Peter Toennies

Springer Series in Chemical Physics

Editors: Vitalii I. Goldanskii Fritz P. Schäfer J. Peter Toennies

Managing Editor: H.K.V. Lotsch

Volume 40 **High-Resolution Spectroscopy of Transient Molecules**
By E. Hirota

Volume 41 **High Resolution Spectral Atlas of Nitrogen Dioxide 559–597 nm**
By K. Uehara and H. Sasada

Volume 42 **Antennas and Reaction Centers of Photosynthetic Bacteria**
Structure, Interactions, and Dynamics
Editor: M.E. Michel-Beyerle

Volume 43 **The Atom-Atom Potential Method.** Applications to Organic Molecular Solids
By A.J. Pertsin and A.I. Kitaigorodsky

Volume 44 **Secondary Ion Mass Spectrometry SIMS V**
Editors: A. Benninghoven, R.J. Colton, D.S. Simons, and H.W. Werner

Volume 45 **Thermotropic Liquid Crystals, Fundamentals**
By G. Vertogen and W.H. de Jeu

Volume 46 **Ultrafast Phenomena V**
Editors: G.R. Fleming and A.E. Siegman

Volume 47 **Complex Chemical Reaction Systems**
Mathematical Modelling and Simulation
Editors: J. Warnatz and W. Jäger

Volume 48 **Ultrafast Phenomena VI**
Editors: T. Yajima, K. Yoshihara, C.B. Harris, and S. Shionoya

Volume 49 **Vibronic Interactions in Molecules and Crystals**
By I.B. Bersuker and V.Z. Polinger

Volume 50 **Molecular and Laser Spectroscopy**
By Zu-Geng Wang and Hui-Rong Xia

Volume 51 **Space-Time Organization in Marcomolecular Fluids**
Editors: F. Tanaka, M. Doi, and T. Ohta

Volume 52 **Clusters of Atoms and Molecules.** Theory, Experiment, and Clusters of Atoms
Editor: H. Haberland

Volume 53 **Ultrafast Phenomena VII**
Editors: C.B. Harris, E.P. Ippen, G.A. Mourou, and A.H. Zewail

Volume 54 **Physics of Ion Impact Phenomena**
Editor: D. Mathur

Volume 55 **Ultrafast Phenomena VIII**
Editors: J.-L. Martin, A. Migus, G.A. Mourou, and A.H. Zewail

Volume 56 **Clusters of Atoms and Molecules.** Salvation and Chemistry of Free Clusters, and Embedded, Supported and Compressed Clusters
Editor: H. Haberland

Volume 1–39 are listed on the back inside cover

Hellmut Haberland (Ed.)

Clusters of Atoms and Molecules

Theory, Experiment, and Clusters of Atoms

With 197 Figures and 14 Tables

Springer-Verlag
Berlin Heidelberg New York
London Paris Tokyo
Hong Kong Barcelona
Budapest

Professor Dr. Hellmut Haberland
Albert-Ludwigs-Universität
Fakultät für Physik
Hermann-Herder Strasse 3
D-79104 Freiburg, Germany

Series Editors

Professor Dr. Fritz Peter Schäfer
Max-Planck-Institut
für Biophysikalische Chemie
D-37073 Göttingen-Nikolausberg, Germany

Professor Vitalii I. Goldanskii
Institute of Chemical Physics
Academy of Sciences
Kosygin Street 4
Moscow, 117334, USSR

Professor Dr. J. Peter Toennies
Max-Planck-Institut
für Strömungsforschung
Böttingerstrasse 6–8
D-37073 Göttingen, Germany

Managing Editor: Dr. Helmut K.V. Lotsch
Springer-Verlag, Tiergartenstrasse 17,
D-69121 Heidelberg, Germany

ISBN-13: 978-3-642-84331-0   e-ISBN-13: 978-3-642-84329-7
DOI: 10.1007/978-3-642-84329-7

Library of Congress Cataloging-in-Publication Data. Clusters of atoms and molecules: theory, experiment, and clusters of atoms/Hellmut Haberland. ed. p. cm. – (Springer series in chemical physics; v. 52). Includes bibliographical references and index. ISBN 3-540-53332-X (Berlin: alk. paper). – ISBN 0-387-53332-X (New York: alk. paper) 1. Microclusters. 2. Metal crystals. 3. Chemistry, Physical and theoretical. I. Haberland, Hellmut, 1939– . II. Series. QC173.4.M48C55 1993 546.3 – dc20 93-28934

Softcover reprint of the hardcover 1st edition 1994

Typesetting: Macmillan India Ltd., Bangalore 25
54/3140/SPS–5 4 3 2 1 0 – Printed on acid-free paper

Preface

This book was written to fill a gap. Physics, chemistry, and technology of atomic and molecular clusters have progressed tremendously in recent years, but no general text book is available on cluster science. Although several specialised reviews have appeared, a book trying to give an overview for the non-specialist is missing.

The gap is felt even more acutely as people with very different background have entered the field. It is not uncommon that, e.g., somebody from solid-state or nuclear theory has to communicate with the experimental physical chemist; or scientists working in catalysis have to familiarize themselves with the problems of cluster beams in vacuum. It is this interdisciplinary approach which makes cluster science so lively, fruitful and challenging.

But the differences in background and language have also led to many misunderstandings and a lot of confusion. One of the aims of the present book is to reduce these problems, thus leading to a better appreciation of the different disciplines.

The text is split into two self-contained volumes. This first volume gives a general introduction into the field. The book starts with a chapter on *theoretical concepts.* The first five chapters describe methods to calculate cluster properties at temperature $T = 0$. The shell structure common to atoms, nuclei, and metal clusters is discussed next. The last two articles of the theory chapter address various properties related to internal excitation or temperature. *Experimental techniques* are examined in Chap. 3. First the various cluster sources are reviewed in detail. They are an essential part of every experiment and still present a major problem. Then common experimental techniques and problems are dealt with. In Chap. 4 the change of cluster properties with chemical bonding is the major theme: alkali clusters, those of a s^2 electronic structure, semiconductors, transition metals, carbon, oxides and halides of the alkali metals, rare gases, and of neutral molecules are treated.

The second volume reviews solvation, chemistry, as well as the decay and charging of free clusters. Cluster are discussed then, which are supported on a surface or embedded in a gas, liquid or solid. Technological applications can be expected to come from this field, and not so much from the study of cluster beams in vacuum. There exists an intense cross fertilisation between the fields of "free" versus "embedded" clusters. The most spectacular story so far is that of C_{60}, whose exceptional stability was discovered for free clusters, and which becomes superconducting once it is embedded into the right amount of alkali metal.

Each chapter has been written by an expert in the field. It has been the intention of the authors to start as simply as possible, then to progress to more difficult problems, and to end with the latest results of the field. The reader is left to judge if this difficult task has been achieved. From the reaction of my own research group I can tell that this goal has been fulfilled for some articles at least.

It has taken time to collect the contributions from so many authors. Very few manuscripts have never arrived, so that there are some points that have not been covered. In one case an author jumped in, when another one could not finish his article. My sincere thanks go to all authors, their patience and perseverance.

Freiburg, January 1994 Hellmut Haberland

Contents

Contributors

K.H. Bennemann
Institut fur Theoretische Physik der Freien Universität Berlin, Arnimallee 14, D-14195 Berlin, Germany

R.S. Berry
The University of Chicago, Department of Chemistry, Chicago, IL 60637, U.S.A.

S. Bjørnholm
Niels Bohr Institute, Begdamsvej 17, DK-2100 Copenhagen, Denmark

V. Bonačič-Koutecký
Freie Universität Berlin, Institut fur Physikalische und Theoretische Chemie, Takustr. 5, D-14195 Berlin, Germany

C. Bréchignac
Laboratoire Aimé Cotton – Bat. 505 Campus d'Orsay, F-91405 Orsay Cedex, France

U. Buck
Max-Planck-Institut fur Strömungsforschung, Bunsenstr. 10, D-37073 Göttingen, Germany

E.E.B. Campbell
Max-Born-Institut, Rudower Chausee Geb. 19.29, D-12474 Berlin, Germany

P. Fantucci
Dipartimento di Chimica Inorganica e Mettallorganica, Universita di Milano, Centro CNR, I-21033 Milano, Italy

H.-G. Fritsche
Department of Physical Chemistry, Friedrich-Schiller-University, D-07743 Jena, Germany

T.F. George
Department of Chemistry and Physics, Washington State University, Pullman, WA 99164-1046, U.S.A

H. Haberland
Albert-Ludwigs-Universität, Fakultät für Physik,
Hermann-Herder-Str. 3, D-79104 Freiburg, Germany

M.F. Jarrold
Department of Chemistry, Northwestern University, 2145 Sheridan Road,
Evanston, IL 60048, USA

D.A. Jelski
Department of Chemistry, State University of New York, College
at Fredonia, Fredonia, NY 14063, U.S.A

R.O. Jones
Institut fur Festkörperforschung, Forschungszentrum Jülich,
D-52428 Jülich, Germany

J. Koutecký
Freie Universität Berlin, Institut fur Physikalische und Theoretische
Chemie, Takustr. 5, D-14195 Berlin, Germany

T.P. Martin
Max-Planck-Institut für Festkörperforschung, Heisenbergstr. 1,
D-70569 Stuttgart, Germany

H. Müller
Department of Physical Chemistry, Friedrich-Schiller-University,
D-07743 Jena, Germany

G.M. Pastor
Institüt für Theoretische Physik der Universität zu Köln, Zülpicher
Strasse 77, D-50937 Köln, Germany

G. Scoles
Princeton University, Department of Chemistry, Princeton,
NJ 08544-1009, USA

L. Skala
Department of Chemical Physics, Charles University, Prague,
Czechoslovakia

J.M. Vienneau
Department of Chemistry, State University of New York, College
at Fredonia, Fredonia, NY 14063, U.S.A.

1 Introduction

R.S. Berry and *H. Haberland*

1.1 What are Clusters?

From the time of *John Dalton*, when the atomic theory became generally accepted, the study of how matter behaves has divided into two streams. One has been reductionist, concentrating on the properties of individual atoms and molecules. In the 1930's, this line led to nuclear physics and then to particle physics. The other stream has emphasized the properties of very, very many atoms or molecules together, so many that the aggregates they make can be treated as infinite. The realm between these limits concerned only a few independent souls until about the late 1970's. Since then, interest has grown almost explosively in the study of what have come to be called clusters.

Clusters are aggregates of atoms or molecules, generally intermediate in size between individual atoms and aggregates large enough to be called bulk matter. People sometimes argue (but not very seriously) whether three particles are enough to constitute a cluster. And a mole – $6 \cdot 10^{23}$ particles – of anything is far too many particles to be a cluster; that much stuff is certainly bulk matter. Even a hundred billion particles stuck together, far, far fewer than a mole, behave in most ways like bulk matter. Now, when we talk of small clusters, we mean something containing no more than a few hundred or possibly a thousand particles, and a large cluster implies to us something containing thousands of particles. A useful distinction classifies clusters as small, medium-sized or large. Small clusters are those whose properties vary with size and shape so much that no simple, smooth dependence can be given for their dependence on number of component particles. If the cluster's properties are smoothly varying functions of the number of component particles, the cluster is medium-sized or large. If those properties vary smoothly but still reflect the small size of the cluster and the implications of that size, the cluster is medium-sized. If those properties approach those of the corresponding bulk material, the cluster is large. Large clusters typically have dimensions of order a few nanometers, they are spheres with radii between 1 and perhaps 50 nm, or microcrystals with sides of these dimensions. This puts large clusters into the lower reaches of the "nanoscale" materials, and indeed they are.

Aggregates in the small and middle range differ considerably from bulk matter: they differ in at least two important ways. First, a large fraction of their component particles are on the surface. For example in a cluster of 55 atoms of argon or sodium, at least 32 are in some sense on its surface. But small clusters also differ from bulk matter because the quantum states of bulk matter occur in bands of states so close together that we can disregard the energy gaps between those states. (We cannot, of course, neglect the gaps between the bands, such as that between filled and empty levels in an insulator or semiconductor, the very property that makes the substance an insulator and not a metal-like conductor.) In a molecule or a small cluster, the spacings between the individual levels cannot be neglected. The spacings of the energy levels depend on the size of the object; the gaps between energy states of small clusters are quite like those of small and medium-sized molecules. Even in large clusters and nanoscale materials, the small dimensions of the particles give them some properties strongly characterized by quantum effects, properties whose counterparts seem quite classical in larger systems.

What are clusters, and how do they differ from molecules? If we try, we can find borderline cases that are difficult to classify unambiguously, but before we come to these, let us deal with the large majority of clusters. Molecules are characterized by having definite compositions and, in most cases, definite structures. A cluster of, for example, silicon atoms, may contain 3 or 10 or 100 atoms or any number between or greater. Its properties depend on the number of atoms in the cluster and so does the most stable structure. But note that phrase, "the most stable structure". While one structure may be more stable than any of several others, most clusters may assume any of a number of structures. Figure 1 illustrates this for a very simple cluster, Ar_7, one comprised of seven argon atoms. At very low temperatures all the clusters of Ar_7 assume the geometry of lowest energy, the pentagonal bipyramid of Fig. 1a. At higher temperatures, some of the clusters would be found in or near that structure but some could exhibit others of the stable structures, or be near one of them.

How near, and how this comes about, will be one of the main topics of this discussion. But to summarize, clusters differ from conventional molecules because of *composition* and *structure*. For the most part, molecules have definite, highly restricted numbers of atoms and specific compositions. Furthermore they almost always have unique structures. Clusters may be composed of any number N of component particles, and, for most kinds of clusters, as the number of particles of the cluster becomes larger, the number of (locally) stable structures available to the cluster grows rapidly. This latter property has its counterpart in the rapid increase, with number of component atoms, in the number of chemical isomers that exist for a specific composition. However we generally identify each of these isomers as a distinct, isolable chemical species. Clusters of a given composition may exhibit a variety of structures but we generally do not distinguish them as different chemical species. There are, however, a number of exceptions to this, and it sometimes is important to distinguish isomers among clusters of the same composition.

Clusters may be homogeneous, that is, composed of only one kind of atom or molecule, or heterogeneous, made of more than one kind of atom or

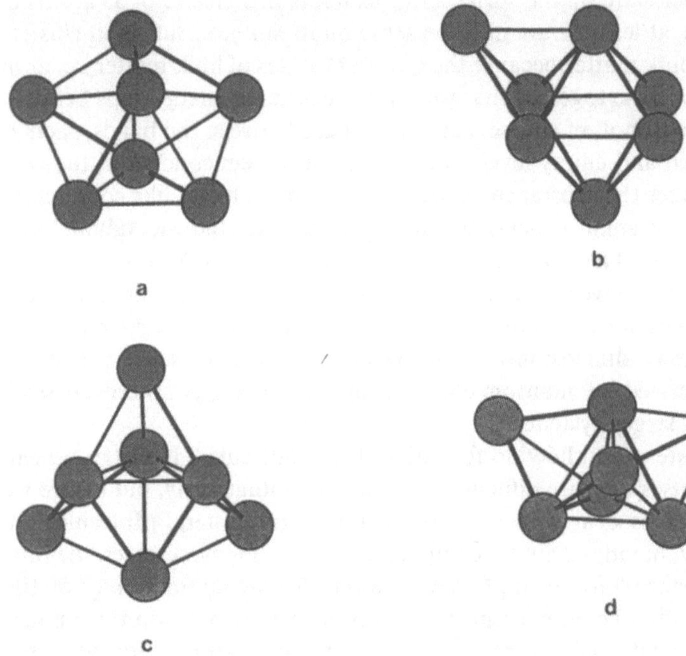

Fig. 1. The four stable forms of the Ar_7 and some other seven-atom clusters. **(a)** The pentagonal bipyramid, the lowest-energy structure; **(b)** the next-to-lowest-energy structure, an octahedron with one capped face; **(c)** the third structure, in order of increasing energy, a tetrahedron with three capped faces; **(d)** the highest-energy structure of the four, which can be looked at as a bicapped trigonal bipyramid

molecule. Clusters of only silicon atoms or of only argon atoms are homogeneous; a cluster of argon atoms around a molecule of benzene is heterogeneous. So is a cluster of three sodium and two potassium atoms; we could also make a cluster of three sodiums and three potassiums, or two sodiums and three potassiums, and so forth. But this terminology loses its usefulness when we come to clusters of some kinds of molecules, notably salt-like molecules of such substances as sodium chloride. We can make clusters of any number of NaCl molecules which we can designate, in general, as $(NaCl)_n$; an example of this kind of species is shown in Fig. 2. We can also make clusters from NaCl or of other alkali halides with one or more excess alkali atoms or halogen atoms, e.g. Na_nCl_{n+y} or $Na_{n+y}Cl_n$. Clusters may be neutral, as in all the examples thus far, or charged: $Na_{n+1}Cl_n$ is more likely, under many laboratory conditions, to occur with a positive charge, i.e. as $(Na_{n+1}Cl_n)^+$, than as the neutral and the cluster with one excess chlorine atom is often found with a negative charge. These are of course not surprising when we recall that the bulk material sodium chloride is best thought of as composed of positive sodium ions and negative chloride ions. But many other kinds of clusters, neutral and charged, have now been observed and studied.

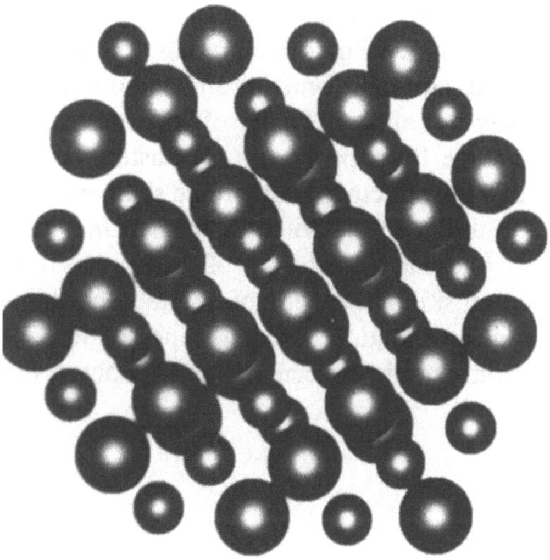

Fig. 2. The $(KCl)_{32}$ cluster in its state of lowest energy. This is the same structure, the rocksalt lattice, as that of bulk KCl crystals

Some clusters are held together by the strong forces of attraction between oppositely charged ions; clusters such as $(NaCl)_n$ are of this kind. Others are held together by the same kinds of strong forces that make covalent chemical bonds; clusters of silicon atoms or of carbon, such as the now-famous C_{60}, are of this kind. Still others are held together by the kind of bonding that holds the atoms of a bulk metal together; we see this kind in the large clusters of metal atoms, large clusters of sodium or copper or iron. Very small clusters of "metal" atoms may be held together by forces more like those of covalent bonds than like the forces exerted by the nearly free-swimming electrons that glue atoms of a metal together. Clusters of the rare gases, and of other closed shell atoms (like very small clusters of beryllium or mercury) belong to the class loosely called van der Waals clusters, held together by the weakest forces that still have their origins in electrical interactions. The last class of clusters we shall single out is that held together by hydrogen bonds. Clusters of water molecules are bound this way; the hydrogens of one water molecule are attracted to specific oxygen atoms of neighboring water molecules. Hydrogen-bonded clusters are held more tightly than van der Waals clusters but less tightly than typical covalent, metallic or ionic clusters.

1.2 What Makes Clusters Interesting?

Clusters have drawn interest for several reasons. Some are matters of technique. We have powerful ways to study clusters, both experimentally and theoretically, which simply did not exist twenty years ago. Others are reasons of purpose. Clusters may offer ways to make altogether new kinds of materials, to carry out chemical reactions in new ways, and to gain new kinds of understanding, not only of this intermediate form of matter but of the behavior of bulk matter too, by learning how bulk properties emerge from properties of clusters, as the clusters grow larger and larger. In this introductory chapter, we shall examine briefly how people study clusters, and touch on the potential for new materials and new chemical reactions, and how studying the behavior of clusters has been giving us new insights into phase transitions, growth of crystals, chemical catalysis, a new method to grow thin films, high T_c organic superconductors, and above all – the slow transition from atomic/molecular/chemical physics to the science of condensed matter.

1.3 How Does One Make Clusters?

Clusters can be made in the laboratory by making a vapor of the elementary component particles and letting them aggregate, or by knocking them directly out of a solid. They can also, in principle, be made in solution, e.g. as small colloidal particles formed during precipitation or in submerged electric arcs. However most of the work during the explosion of interest in this field since the late 1960's has dealt with gaseous clusters and it is these to which we shall direct our attention. Sources of beams of free clusters are typically one of these: i) the most popular and the best understood source, is the "supersonic jet" with its many variants. A gas or vapor is expanded from a high pressure (typically 10^4 to 10^7 Pa) through a small nozzle (typically 0.03 to 1 mm) into vacuum. The mean gas velocity is increasing during this adiabatic and isenthalpic expansion, but the random, thermal motion of the expanding gas is reduced dramatically. This leads to a reduction of the gas temperature and to effective cluster formation. Some of the popular variants are: a) the source can be run continuous or pulsed, b) a pure gas, say Ar, or a mixture, say He with 1% I_2, can be used, c) the "mixture" to be expanded can be generated by a powerful pulsed laser or a strong electric discharge, d) the cluster beam can run, after its formation, through a "cloud" of atoms, molecules, electrons, or ions, picking-up some of them, e) an electric discharge can be ignited on the high or low pressure side, f) . . . Many different combinations have been tried and new sources are constantly developed. ii) Another effective means to generate clusters is the "gas aggregation source". Some material is evaporated or sputtered into a slow flow of argon or helium at a pressure of typically 50 to 500 Pa. The atoms are slowed down by collisions with the rare gas, and start aggregating. This process is similar to cloud formation in the atmosphere. In this case water molecules are evaporated. They do not aggregate, if their density is low enough and their temperature is high enough. Only if these conditions are not fulfilled, aggregation (in common day language cloud formation) starts. Other less common sources are iii) sputtering, or iv) direct laser ablation, without cooling gas.

Once clusters have been made and are in the form of gaseous particles, it is frequently desirable to make them into some kind of controllable beam or stream that can be studied or captured. To observe clusters in a beam, one can probe them while they are free or trap them in a matrix, liquid, glassy or crystalline. Trapped clusters are most easily studied by infrared, visible and ultraviolet spectroscopy. These methods are also used to study free clusters in vacuum. Only for charged clusters has one effective and general means of mass selection and as a result, there are several areas in which more is known about charged clusters than about the neutrals.

One may give clusters positive charge by knocking out one or more electrons, with a beam of electrons or energetic light quanta doing the knock-out. One can also prepare clusters with a negative charge, that is, with an extra electron. These may be prepared by electron transfer from an easily ionized donor such as gaseous rubidium atoms, by capture of electrons from swarms of slow electrons (followed by collision with inert carrier gas atoms or molecules, to carry off the binding energy released by the electron capture and thereby to stabilize the negatively charged clusters), or in the primary process in which the cluster is generated, for example in ablation of material from a solid when it is struck by a high-intensity laser burst or high energy projectile. An effective way to produce intense beams of charged clusters is to generate a plasma in or after the condensation zone. This can be done either by a laser or – less expensively – by an electric discharge.

1.4 Experiments with Clusters

One can eject electrons from neutral or charged clusters. The electrons emitted this way may be collected and studied; so may the newly-produced ions. If one collects electrons, one wants normally to measure their energies. If one collects ions, one may wish to measure their energies and their charge and mass, but Maxwell's laws allows one only to measure the ratio of their charge to their mass.

It is possible to excite clusters, neutral or charged, with light or electron impact and study the radiation they then emit or the particles they lose. The most frequently seen mode for a cluster to lose excess energy is ejection of atoms or larger fragments. It seems that the dominant decay channel is always that having the lowest barrier. If the binding energy of the dimers or trimers is small, atoms are ejected. If the binding energy of the dimer or trimer becomes larger, the ejection of dimers like Na_2 or trimers like C_3 can become the dominant decay channel.

Loss of energy by emission of atoms and of radiation occurs for those clusters which have large band gaps in the bulk, like the rare gases. As there are no potential curve crossings in the rare gases between the ground and electronically excited states, the electronic energy cannot be converted effectively into heavy particle motion. Thus electronic excitation leads to photoemission. Fission into large fragments is observed only for multiply charged clusters. Loss of an electron can of course occur if the excitation energy is higher than the ionization energy.

Charged clusters have the special attraction that they can be sorted easily by electrostatic, magnetic or time-of-flight mass analysis to yield mass spectra and beams of clusters of a single, known size. The numbers of particles in these beams are very small in terms of macroscopic samples, but are sufficient for many experiments. Among the first studied characteristics of charged clusters were their size distributions, which led to the understanding that some sizes and shapes of clusters are particularly stable, in a manner analogous to the stable shells seen in atoms and in nuclei. The numbers N of particles in these particularly stable clusters are sometimes called "magic numbers". As there is nothing magic about these numbers, some people prefer to use the more appropriate term of "intensity anomalies".

One can study the wavelengths of light that clusters absorb. For some elements – rare gases, alkalis, and mercury – the optical absorption spectrum from the atom and the cluster to the bulk could thus be studied. The first femtosecond experiments on free clusters are just appearing. Some of the other properties of charged clusters that have been studied include their modes of fragmentation, their chemistry, and the effects of their collisions with surfaces.

1.5 Experiments Not Possible Today

Preparing vapours or beams consisting mostly of neutral clusters of only one or a few sizes is still a major unsolved experimental challenge. It can be done only for small clusters with less than 8 atoms if a good supersonic beam of the clusters is available.

A very difficult parameter to control or to determine in all experiments is the temperature of a cluster. Temperature for a finite system may be defined as the mean internal kinetic energy of the heavy particles divided by the number of degrees of freedom. Clusters from neat supersonic beams are hot, boiling hot when they leave the condensation zone. They cool by evaporative cooling, i.e. ejection of atoms or larger fragments. Lower temperatures can be obtained in seeded supersonic beams or in cooled gas aggregation sources. But so far no "thermometer" to measure a cluster's temperature is available. Worse, there does not even exist a suggestion how to construct one. Only if the cluster has enough internal energy to evaporate atoms, metastable decay can be used to monitor the temperature.

Magnetic effects have been studied only for clusters which become ferromagnetic in the bulk. No infrared spectroscopy below 10 μm or radiofrequency spectroscopy has been performed on clusters. Inner shell excitation of mass selected clusters has not been performed so far. Other unsolved problems are: can a He cluster become superfluid, or can a free Pb cluster become superconductive? How could one detect this experimentally? How could one do, and what could one learn from an e,2e experiment? Looking on the precision and wide range of experiments in atomic, molecular, and solid state science one can say that the science of clusters has made good progress in recent years, but is still very far from being a mature field.

1.6 Clusters, Tantalizing Subjects for Theoretical Studies

Clusters are of course tantalizing subjects for theoretical studies. It is possible to apply to them the same theoretical tools that are applied to studying molecules, the quantum mechanical methods ranging from ab initio variational calculations of the entire electronic structure, through more approximate methods such as the local approximations to density functionals which approximate the consequences of electron exchange interactions with effective potentials, to efficient but highly oversimplified, effective potentials to describe the motions of the atoms, molecules or ions comprising the cluster. For clusters of argon, krypton and xenon, Lennard-Jones potentials account for the dynamics nearly as well as more accurate effective potentials; likewise, for alkali halides, Born–Mayer potentials of Coulombic long-range interactions and exponential short-range potentials, represent the dynamics well. For other clusters, more subtle approximations are necessary; for example, embedded-atom potentials are one type of approximate, effective potential that has been used to represent metal clusters. If one can find reliable potentials of the last sort, it is now straightforward to use a computer to solve the simultaneous equations of motion of all the atoms of a cluster, or to explore by statistical sampling the accessible phase space of configurations and momenta of the particles, and thereby carry out computer simulations of the behavior of clusters. Solving the equations of motion is called "Molecular Dynamics" (MD); the statistical sampling methods are called "Monte Carlo" (MC) methods.

Both of these have classical and quantum mechanical forms. Furthermore whenever one has in hand a reasonably efficient representation of the potential, for example a set of pairwise interactions between the component particles, one can carry out various examinations of the surface, such as searches for local minima on the surface or for saddle points, and from these, construct reaction paths for particle motion.

One aspect of the theory of the structure of clusters has brought the subject into close contact with nuclear physics. This is the interpretation of structures of clusters in terms of shell models. This is the idea that clusters of particular sizes have structures that correspond to closed shells of component particles and therefore are relatively more stable than clusters lacking one or a few particles that would fill out a closed shell, or than clusters with one or a few particles more than the number for a closed shell. The numbers corresponding to completely filled or closed shells are often called magic numbers; examples are 13, 55 and 147 for clusters that take on icosahedral geometries. The shells here refer to shells of atoms. In metallic clusters, particularly the clusters of alkali metal atoms, the determinant of the shell structure is the electron shell structure. A widely used model for these clusters is the Jellium model, which puts the electrons into a continuous, spherical sea of positive background charge. This model has been refined and made self-consistent, and extended to distorted, nonspherical geometries. However it seems that electronic and atomic forces compete for the dominating role in fixing what forces establish the shell structure and stable geometries of clusters. Small, medium and even moderately large alkali metal clusters have structures determined by their electronic shell structure; quite large clusters of alkali metal atoms have structures that are determined by the shells of atoms that can make them up.

Clusters may exhibit different structures under various external conditions of temperature and pressure. They can show many of the characteristics of solid-like and liquid-like behavior and of the equilibrium and passage between these forms. Some clusters can, under the right conditions, behave like soft solids or like slush. Here is one of the areas in which new phenomena have appeared and new insights have arisen concerning the dynamics of small, medium, and large clusters, and into the nature of phase transitions. Clusters may take on a variety of structural forms, some of them identifiable with particular geometric shapes, others with no specific shape but nevertheless characterizable by some degree of floppiness which, in the extreme, is the same as that of a liquid.

1.7 Clusters Make New Kinds of Materials

Clusters offer several ways to make new kinds of materials, all of them still at the frontiers of research. One is based on the differences between the structures of small clusters and the structures of bulk matter. Many kinds of small clusters have for their most stable forms structures based on polyhedra – the cube, the octahedron and especially the 20-faced, 12-vertex icosahedron. One can build an infinite, repeating array of atoms arranged in cubes or octahedra, but not in icosahedra. One can make infinite arrays from icosahedral structures called quasicrystals, but they do not have the regular, repeating structures we associate with crystals. Instead, while they may show order at short distances, they have a degree of randomness at long distances, if the arrays are allowed to grow large enough. However if one can make regular icosahedral clusters of silicon, for example, each containing 13 silicon atoms – one in the center and 12 others on each of the 12 equivalent vertices of the icosahedron, then one can condense these clusters to form a bulk crystal with regular, repeating units, each a 13-atom cluster. Or one could use icosahedral clusters of 55 atoms each, or of 147 or 309 or 561 atoms, or, in principle, the higher numbers that just complete a regular icosahedron. These are called the Mackay icosahedra, after A. Mackay who pointed out their potential importance in 1962. The properties of a crystalline solid built from regular, repeating icosahedral units will surely be different from those of arrays of individual atoms, but in what ways we do not yet know.

A related solid is the form of carbon built from a regular array of the nearly-spherical balls of 60 carbon atoms, buckminsterfullerene, each molecule of which has precisely the shape of a football (called a soccer ball in the US) with an atom at every corner. This material also forms salt-like compounds with alkalis, up to K_3C_{60} for example, which becomes a superconductor at low temperatures. The C_{60} molecule is an array of 12 five-membered rings and 20 six-membered rings that can be thought of as an icosahedron with planes cutting off all the original 12 vertices, leaving pentagonal faces in their place. Other network-shell structures can be made with 12 five-membered rings and other numbers of six-membered rings; some of these, such as C_{70} and C_{76}, are like ellipsoidal balls, not spherical. In fact C_{70} and C_{76} illustrate a point made earlier: both of these exhibit distinct isomers that can be separated and kept, and there is a school that prefers to call these molecules rather than clusters because of their isolability and distinguishability. Others have been reported, with far more carbon atoms, which are long rods with ends that are presumed to be C_{30} caps like half of buckminsterfullerene molecules. These rods may themselves become or be part of new solid materials.

Other possible new kinds of materials would come from embedding micro-clusters of one material in a host of a different but related substances. For example some investigators have been studying how to introduce unique electrical properties into a material this way. A still more dramatic possibility would be the superatom cluster, a cluster whose core is strongly electron-donating and whose outer part or "rind" is strongly electron-attracting. This structure would have a deficiency of electrons in its core and a surfeit in its rind, just as an atom does. However this superatom could have a core of 55 atoms, for example, not of a single atomic nucleus. Its properties would be different from those of any atoms we now know.

1.8 New Chemistry

Still another facet of cluster science treats the chemical reactions that occur on or within clusters. For example hydrogen molecules striking a cluster of nickel atoms may dissociate into hydrogen atoms that become strongly bound to the nickel cluster. Both hydrogen and ammonia molecules react when they strike clusters of iron atoms. These are examples of reactions that have important counterparts in the chemistry of catalysts. But clusters of metal atoms do not react precisely as the corresponding bulk metals do, even if the metal is in finely divided form, such as micron-sized particles. Even clusters of the same size and composition but different geometric structure may have differing chemical reactivities. Other kinds of reactions of or in clusters that have been studied are such seemingly-simple processes as the recombination of the atoms of a diatomic molecule imbedded in a cluster but dissociated by the absorption of a sufficiently energetic quantum. Other kinds of photochemistry, yielding products different from the reactants, have also been studied in clusters and specific size-dependent effects have been found.

1.9 Outlook

Clusters can be studied by experiment, theory and simulation with the precision and lack of ambiguity we associate with the study of small molecules. Yet we can control the size of the clusters and conditions under which we observe them so that we can extrapolate their properties from those of small clusters through intermediate sizes all the way to bulk matter. This means that clusters have given us a new way to approach the study of matter, a way that allows us to apply the most rigorous, least approximate tools of physics that we have. Cluster science is not only a fascinating field in its own right; it enriches the fields around it and offers tantalizing possibilities for new materials and processes.

2 Theoretical Concepts

2.1 Quantum Chemistry of Clusters

V. Bonačič-Koutecký, P. Fantucci, and *J. Koutecký*

2.1.1 Introduction

Elemental clusters can be roughly defined as agglomerates of atoms subjected to such weak chemical forces that they are unstable under customary experimental conditions. Their existence can not be understood and explained using conventional valence concepts. Clusters are characterized by a large number of surface atoms and this represents a substantial difference with respect to crystals. Consequently, clusters are different from both molecules and solids, and for this reason can have very specific properties [1]. Due to the inherent instability of clusters general information about their fundamental properties have become available only recently using new experimental techniques which are in an expanding development. Therefore, the concepts and models useful in more classical fields of science like conventional chemistry of customary compounds or physics of solids have not such an evident validity in cluster science. Since a sufficient fundament of empirical knowledge about clusters is lacking, it is risky to use semiempirical or empirical theoretical methods which might be successful in organic and inorganic chemistry or in solid state theory. However, it is evident that these methods are very desirable, since the size of the interesting clusters is too large to be easily investigated using more sophisticated first principle quantum molecular theory. Also, the finite dimension of clusters represents a drawback for the applicability of typical solid state theories. Similarly, it is dangerous to take over simplified ideas from other fields of science in order to devise useful models in cluster science.

Nevertheless, there are two features of the theory which seem to be promising in solving this dilemma. First, basic rules determining fundamental cluster properties are consequences of general physical laws and therefore, they are independent of the details of the treatment used. Second, if some laws of cluster physics have general validity, they can be revealed by adequately chosen theoretical methods which provide a deep insight into the structure and characteristic properties of clusters and by careful selection of appropriately simple objects for theoretical investigation. The methodological tools of the theory can be simplified but all the essential ingredients necessary for description of the physical reality must be retained. Moreover, due to the efficiency of modern computers, quantitative checks of general theoretical ideas on small clusters are manageable nowadays.

2.1.2 Quantum Mechanical Background

The electronic structure and properties of a many-electron system like a molecule or a cluster composed of N electrons and n nuclei of charge Z_a ($a = 1, \ldots, n$) may be obtained by solving the Schrödinger equation:

$$\hat{H}|\psi(\mathbf{x}_1, \mathbf{x}_2, \ldots, \mathbf{x}_N, t)\rangle = i\frac{\partial}{\partial t}|\psi(\mathbf{x}_1, \mathbf{x}_2, \ldots, \mathbf{x}_N, t)\rangle \ , \tag{2.1}$$

where $\mathbf{x}_i = (\mathbf{r}_i, \eta_i)$ is the space-spin coordinate of the i-the electron while $\hat{H}$ depends on the space coordinates $\mathbf{r}_i$ and time.

For stationary states Eq. (2.1) reduces to

$$\left(\hat{H}_0 - i\frac{\partial}{\partial t}\right)|\psi_j\rangle e^{-iE_jt} = e^{-iE_jt}(\hat{H}_0 - E_j)|\psi_j\rangle = 0 \ , \tag{2.2}$$

where $\hat{H}_0$ is the time independent Hamiltonian, which in the absence of interaction with external magnetic or electric fields takes on the form:

$$\begin{aligned} \hat{H}_0 &= -\sum_i^N \frac{1}{2}\nabla_i^2 - \sum_a^n \sum_i^N Z_a/|\mathbf{r}_i - \mathbf{R}_a| + \frac{1}{2}\sum_{i \neq j}^N 1/|\mathbf{r}_i - \mathbf{r}_j| \\ &= \sum_i \hat{h}(i) + \frac{1}{2}\sum_{i \neq j} \hat{g}(i,j) \ . \end{aligned} \tag{2.3}$$

In the above equations, atomic units (au) have been adopted ($h = 1, e = 1, m = 1$). The expectation value $E_{\mathrm{el}} = \langle\psi|\hat{H}_0|\psi\rangle$ of the normalized wavefunction (wf) $|\psi\rangle$, together with the nuclear-nuclear interaction energy $E_{\mathrm{nuc}} = \sum_{a \neq b} Z_a Z_b/|\mathbf{R}_a - \mathbf{R}_b|$ gives the total energy $E_T = E_{\mathrm{el}} + E_{\mathrm{nuc}}$ which is a parametric function of the Cartesian coordinates of the nuclei considered as fixed (Born–Oppenheimer (BO) approximation).

2.1.2.1 One-Electron and Many-Electron Theories in Quantum Chemistry

The many-electron wf must obey the Pauli exclusion principle. This requirement is already fulfilled by an antisymmetrized product of one-electron functions $|\varphi_i(\mathbf{x}_i)\rangle = |\bar{\varphi}_i(\mathbf{r}_i)f(\eta_i)\rangle$ ($f(\eta_i)$ equal to α or β) called spin orbitals:

$$|\Phi\rangle = N^{-1/2}\sum_p (-1)^p \hat{P}|\varphi_1(\mathbf{x}_1)\varphi_2(\mathbf{x}_2)\ldots\varphi_N(\mathbf{x}_N)\rangle \ , \tag{2.4}$$

where $\hat{P}$ is a permutation operator assigned to the permutation p. The wf $|\Phi\rangle$ is known also as a "Slater determinant" and optimal spin orbitals can be obtained by minimizing the expectation value $E_{\mathrm{el}} = \langle\Phi|\hat{H}_0|\Phi\rangle$ (Hartree–Fock procedure). Since $\hat{H}_0$ commutes with the spin operators $\hat{S}^2 = \sum_i \hat{S}^2(i)$ and

$\hat{S}_z = \sum_i \hat{S}_z(i)$ the exact wf $|\psi\rangle$ is required to be also an eigenfunction of $\hat{S}^2$ and $\hat{S}_z$. For a given spin quantum number S, the relation $\hat{S}_z|\Phi\rangle = M_S|\Phi\rangle$ is generally satisfied, while the relation $\hat{S}^2|\Phi\rangle = S(S+1)|\Phi\rangle$ is true only for a "closed-shell" wf with all the spatial orbitals $\bar{\varphi}$ doubly occupied or for an "open shell" wf with $M_S = S$. In the latter case all the $M_S/2$ unpaired electrons have the same spin and occupy different spatial orbitals. In all other cases the wf must be expressed as a linear combination of Slater determinants $|\Phi_k\rangle$ with coefficients fixed by spin symmetry requirements. Slater determinants $|\Phi_k\rangle$ enter also the definition of the most general multi-electron wf:

$$|\psi\rangle = \sum_k t_k|\Phi_k\rangle \ , \tag{2.5}$$

where the optimum coefficients minimize the expectation value $E_{el} = \langle\psi|\hat{H}_0|\psi\rangle$ (configuration interaction procedure (CI)).

For a single-determinant wf $|\Phi\rangle$, the energy expectation value in terms of quantities

$$h_{ij} = \langle\varphi_i(1)|\hat{h}(1)|\varphi_j(1)\rangle \tag{2.6}$$

and

$$[ij|kl] = \langle\varphi_i(1)\,\varphi_j(2)|\hat{g}(1,2)|\varphi_k(1)\,\varphi_l(2)\rangle \tag{2.7}$$

involving the occupied spin orbitals $|\varphi_i\rangle \equiv |i\rangle$ is:

$$E_{el} = \sum_i \left\{ h_{ii} + \frac{1}{2}\sum_j ([ij|ij] - [ij|ji]) \right\} . \tag{2.8}$$

Since, according to the definition (2.7) the following relations hold:

$$J_{ij} = [ij|ij] = \langle i(1)|\hat{J}_j(1)|i(1)\rangle \ , \tag{2.9}$$

$$K_{ij} = [ij|ji] = \langle i(1)|\hat{K}_j(1)|i(1)\rangle \ , \tag{2.10}$$

the energy expectation value for $|\Phi\rangle$ can be written in terms of "effective one-electron" contributions:

$$E_{el} = \sum_{i_{occ}} \langle i|\hat{h} + \frac{1}{2}\sum_{j_{occ}} (\hat{J}_j - \hat{K}_j)|i\rangle = \frac{1}{2}\sum_{i_{occ}} \langle i|\hat{h} + \hat{F}|i\rangle \ , \tag{2.11}$$

where

$$\hat{F} = \hat{h} + \sum_{j_{occ}} (\hat{J}_j - \hat{K}_j) \tag{2.12}$$

is the so called Hartree–Fock operator. In (2.11) and (2.12) i_{occ} and j_{occ} run over the set of the occupied spin orbitals.

If the spin orbitals forming the Slater determinant $|\Phi\rangle$ are eigenfunctions of $\hat{F}$, that is

$$\hat{F}|k\rangle = \varepsilon_k|k\rangle \tag{2.13}$$

then, the energy expectation value

$$E_{\mathrm{el}} = \frac{1}{2} \sum_{k_{\mathrm{occ}}} (\varepsilon_k + h_{kk}) \tag{2.14}$$

can be easily derived. In the framework of the so-called Restricted HF (RHF) procedure the optimal energy is obtained under the constraint that the spatial part of the α and β spin orbitals belonging to a closed shell should be pairwise identical. If this constraint is removed the Unrestricted HF (UHF) procedure results. Spin unrestricted wf's, which are not eigenfunctions of the $\hat{S}^2$ operator but a mixture of states corresponding to S, $S+1$, $S+2, \ldots$ spin quantum numbers, are frequently used in cluster theory especially in connection with density functional methods.

Each Slater determinant $|\Phi_k\rangle$ can be uniquely described by a given set of integer occupation numbers of the spin orbitals $|i\rangle$: the symbol $|N_1 N_2 N_3 \ldots\rangle$ indicates that $N_1, N_2, N_3, \ldots$ electrons ($N_i = 0, 1$) occupy the spin orbitals $|1\rangle$, $|2\rangle$, $|3\rangle, \ldots$, respectively. Time independent state vectors $|N_1 N_2 N_3 \ldots\rangle$ can be generated by an ordered product of creation operators $\hat{a}_i^\dagger$ [2] acting on the vacuum state $|\mathrm{vac}\rangle = |0\,0\,0 \ldots\rangle$

$$|\Phi\rangle = |\varphi_1 \varphi_2 \varphi_3 \ldots \varphi_N\rangle = \hat{a}_N^\dagger \ldots \hat{a}_2^\dagger \hat{a}_1^\dagger |\mathrm{vac}\rangle \;. \tag{2.15}$$

Creation ($\hat{a}_i^\dagger$) and annihilation ($\hat{a}_i$) operators satisfy the anticommutation relation $\hat{a}_i^\dagger \hat{a}_j + \hat{a}_j a_i^\dagger = \delta_{ij}$. Related operators are the excitation operators $\hat{E}_{ij} = \hat{a}_i^\dagger \hat{a}_j$ (known as Lie algebra generators) and the electron number operators $\hat{N}_i = \hat{E}_{ii}$. Creation and annihilation operators can be used to express any one- and two-electron operator in compact form:

$$\sum_r \hat{h}(r) = \sum_{ij} h_{ij} \hat{a}_i^\dagger \hat{a}_j = \sum_{ij} h_{ij} \hat{E}_{ij} \tag{2.16}$$

$$\sum_{r \neq s} \hat{g}(r, s) = \sum_{ijkl} [ij|kl] \hat{a}_i^\dagger a_j^\dagger \hat{a}_l \hat{a}_k = \sum_{ijkl} [ij|kl] (\hat{E}_{ik} \hat{E}_{jl} - \delta_{jk} \hat{E}_{il}) \;. \tag{2.17}$$

where r and s label electrons.

Using the above expressions, the expectation value for multi- or one-determinant wf $|\psi\rangle$ takes the form:

$$\begin{aligned} E_{\mathrm{el}} &= \langle \psi | \hat{H} | \psi \rangle \\ &= \sum_{ij} h_{ij} \langle \psi | \hat{E}_{ij} | \psi \rangle + 1/2 \sum_{ijkl} \{ [ij|kl] \langle \psi | \hat{E}_{ik} \hat{E}_{jl} - \delta_{jk} \hat{E}_{il} | \psi \rangle \} \\ &= \sum_{ij} h_{ij} \Gamma_{ij} + 1/2 \sum_{ijkl} [ij|kl] \Gamma_{ijkl} \;, \end{aligned} \tag{2.18}$$

where Γ_{ij} and Γ_{ijkl} are elements of the one- and two-particle density matrices. Eigenfunctions of the one-particle density operator are called "natural orbitals" (NO), and their associated eigenvalues "occupation numbers" (N_i). For a single-determinant wf the one-particle density matrix is diagonal when expressed in the basis of the spin orbitals $|i\rangle$. After summation over the spins, one finds that the occupation numbers for the single determinant spin orbitals are integers: $N_i = 2, 1$ or 0 for closed-shell, open-shell and empty orbitals (MO), respectively. For the general multi-determinant wf the occupation numbers $N_i = \langle\psi|\hat{E}_{ii}|\psi\rangle$ are non-integer numbers, $0 \leq N_i \leq 2$. However, it is often possible to identify natural orbitals with $N_i \cong 2$, or $N_i \cong 1$ (strongly occupied) and other NO's with $N_i \ll 1$ (weakly occupied).

LCAO Type Approximation. Since the operators $\hat{J}_j$ and $\hat{K}_j$ of Eqs. (2.9) and (2.10) depend on the spin orbitals, the form of which is, in principle, unknown, the minimization of E_{el}, (2.14), must be carried out in an iterative way (Self Consistent Field, SCF) [2]. The most widely adopted algorithm is based on the expansion of the spatial functions, $\bar{\varphi}_i(\mathbf{r}_1)$ into a set of fixed functions $\{\chi_\lambda\}$ $(\lambda = 1, \ldots, m)$ according to

$$\bar{\varphi}_i(\mathbf{r}_1) = \sum_\lambda C_{\lambda i}\chi_\lambda(\mathbf{r}_1) \ . \tag{2.19}$$

The auxiliary functions χ_λ (usually chosen as centered at nuclei) are assumed to be Gaussian type (GTO) or Slater type orbitals (STO) and form a nonorthogonal set: $\langle\boldsymbol{\chi}|\boldsymbol{\chi}\rangle = \mathbf{S}$, where $|\boldsymbol{\chi}\rangle = |\chi_1\chi_2\ldots\rangle$ and $\langle\boldsymbol{\chi}| = \langle\chi_1\chi_2\ldots|$ are row and column vectors, respectively. An exact expansion of the single-determinant wf would require a complete basis $\{\chi_\lambda\}$. Such a condition is never satisfied in practice and the results obtained with "saturated" (but still incomplete) basis sets are called "Hartree–Fock limits". Large basis sets $(m > N)$ give enough flexibility to determine spin orbitals which are occupied in the HF configuration, and those which are unoccupied (or virtual). The latter are essential for the description of excited electronic states.

As already mentioned, for a single-determinant wf $\Gamma_{ij} = \langle\psi|\hat{E}_{ij}|\psi\rangle = \langle\psi|\hat{E}_{ii}|\psi\rangle\delta_{ij} = N_i$. In addition, the elements of the two-particle density matrix Γ_{ijkl} can be expressed as simple products of Γ_{ij} elements. Equation (2.18) simplifies to:

$$E_{\text{el}} = \sum_i \Gamma_{ii}\left\{h_{ii} + \frac{1}{2}\sum_j \Gamma_{jj}([ij|ij] - [ij|ji])\right\} \tag{2.20}$$

which is in all equivalent to (2.11). For the simple case of a closed shell system, taking into account the LCAO expansion (2.19), the Eq. (2.20) takes the form:

$$E_{\text{el}} = \sum_{\lambda\mu} R_{\lambda\mu}\left\{h_{\lambda\mu} + \frac{1}{2}\sum_{\nu\sigma} R_{\nu\sigma}\left([\lambda\nu|\mu\sigma] - \frac{1}{2}[\lambda\nu|\sigma\mu]\right)\right\} \tag{2.21}$$

where

$$R_{\lambda\mu} = \sum_{i_{\text{occ}}} N_i C^0_{\lambda i} C^0_{\mu i} \ , \tag{2.22}$$

$(N_i = 2, \forall i)$. $R_{\lambda\mu}$ are elements of the one-particle density matrix expressed in the basis $\{\chi_\lambda\}$ and are related to the local value of the total electron density at the point $\mathbf{r}$, according to:

$$\rho(\mathbf{r}) = \sum_{\lambda,\mu} \chi_\lambda(\mathbf{r})\chi_\mu(\mathbf{r})R_{\lambda\mu} \ . \tag{2.23}$$

Notice that the HF energy (2.21) cannot be simply expressed in terms of $\rho(\mathbf{r})$ only, because of the exchange-energy part. In contrast, the "Local Density Functional" (LDF) theories [3] are based on the theorem which states that the electronic ground state energy is uniquely determined as a functional of density $\rho(\mathbf{r})$. However, the form of this functional is unknown.

2.1.2.2 All-Electron and Valence-Electron-Only Methods

In complex molecular systems it is often possible to identify groups of electrons which play a minor role in determining a given physical observable, e.g. the core electrons of atoms with high atomic number when stability or geometry of the molecule is considered. The "all-electron" (AE) theoretical treatments can be, under given conditions, replaced by the "valence electron only" methods. The latter are all based on the assumption that the atomic cores are rigid (unpolarizable) shells and that the strong orthogonality condition $\int \Phi^c(x_1, x_2, \ldots, x_i \ldots)\Phi^v(x_1, x_2, \ldots, x_i \ldots)\, dx_i = 0,\ \forall i$, holds for Φ^c and Φ^v which are the wf's for N_c core and N_v valence electrons. Consequently, the total electronic energy can be exactly decomposed into contributions $E_{\mathrm{el}} = E^c + E^v + E^{cv}$. In the "frozen-core" approximation, E^c is constant and E^{cv} depends on a fixed potential provided by the core electrons. The corresponding Effective Core Potential (ECP) operators [4] are usually derived from atomic calculations and since the nuclear charge Z_a is substituted by an effective one $Z_a^{\mathrm{eff}} = N_v$ they are required to mimic i) the incomplete screening of the nuclear charge, ii) the core-valence orthogonality, iii) the Coulomb and iv) the exchange core-valence potential. Potentials containing only radial parts would approximately represent the effects i), ii) and iii), while the exchange potential must necessarily be an angular momentum dependent operator. All these features are present in ECP operators of the form:

$$\hat{W}(i) = \hat{W}_0(r_i) + \sum_l \hat{W}_l(r_i)\left(\sum_{m=-l}^{+l} |lm\rangle \langle lm| \right)$$

$$\hat{W}_l(r_i) = \sum_k r^{n_{kl}} C_{kl} \exp(-\alpha_{kl} r_i^2) \ . \tag{2.24}$$

$\hat{W}_0(r_i)$ is a local operator and the sum of operators $|lm\rangle\langle lm|$ is the projector on the $(2l+1)$-degenerate manifold of angular momentum l. The constants n_{kl}, C_{kl} and α_{kl} can be derived by theoretical atomic calculations requiring the best agreement among AE and ECP one-electron functions and energies [4]. In addition to the ECP operator other "effective" operators can be incorporated into the one-electron part of HF Hamiltonian in order to introduce corrections for relativistic effects and core-valence correlation [4, 5]. This leads to an approximate but very powerful theoretical approach especially for systems with heavy atoms.

2.1.2.3 Configuration Interaction (CI) Treatment of Electron Correlation

Considering the form of the operator (2.12) and the energy expression (2.14) it is apparent that in the HF picture each electron moves in an averaged potential generated by all other $N-1$ electrons. Electrons with equal spins are correlated in the sense that the exclusion principle does not permit them to be at the same point. By contrast, electrons with opposite spins are uncorrelated and the HF value E_{el} is always higher than the experimental (non relativistic) one, just because it contains a too high repulsion [2a, b]. A rigorous estimate of the correlation energy $E_c = E_{\mathrm{HF}} - E_{\mathrm{exact}}$ can be made according to the Configuration Interaction (CI) method, or according to any other approach in which the multi-electron effects are explicitly taken into account.

As mentioned in Section 2.1, the second-quantization operators facilitate the definition of a complete set of configurations $\Phi_0, \Phi_1, \ldots, \Phi_\infty$ in the Hartree Fock space, where Φ_0 represents the HF (lowest energy) configuration. All other configurations can be in principle obtained by exciting one or more electrons from orbitals occupied in Φ_0 to virtual orbitals. A wf expanded as in Eq. (2.5) has an energy expectation value:

$$E_{\mathrm{el}} = \langle \psi | \hat{H} | \psi \rangle = \sum_{KL} t_K^\dagger \langle \Phi_K | \hat{H} | \Phi_L \rangle t_L \ . \tag{2.25}$$

Using the excitation operators of Eqs. (2.16) and (2.17) the matrix elements between configurations can be expressed as:

$$H_{KL} = \sum_{ij} h_{ij} \langle \Phi_K | \hat{E}_{ij} | \Phi_L \rangle + \frac{1}{2} \sum_{ijkl} [ij|kl] \left\{ \sum_J \langle \Phi_K | \hat{E}_{ik} | \Phi_J \rangle \langle \Phi_J | \hat{E}_{jl} | \Phi_L \rangle - \delta_{jk} \langle \Phi_K | \hat{E}_{il} | \Phi_L \rangle \right\} \ . \tag{2.26}$$

The configurations Φ_K are usually obtained from HF calculations carried out in a limited basis $\{\chi_\lambda\}$. In addition, several configurations are excluded from the secular problem (2.25) because only low order excitations (single S; double D; Triple T; Quadruple Q . . .) from the ground state configuration Φ_0 or from a chosen set of reference configurations are considered. A truncated $\{\Phi_K\}$ basis leads to energy values far from the "full CI" limit and the computed E_c is in general a basis-set-dependent quantity. Another difficulty met when working with truncated CI expansions is the "size inconsistency" [2a, b]. Size consistent methods give E_c values proportional to the number of electrons; in particular,

$\lim_{N\to\infty} E_c/N = k$ ($k > 0$, constant). In contrast, it can be shown that E_c obtained by truncated CI (e.g. SD-CI) obeys $\lim_{N\to\infty} E_c(\mathrm{SD})/N = 0$. For instance, in the case of two interacting fragments A, a truncated CI gives better estimate of E_c for the separated units A than for the composite system A_2 leading to a biased estimate of the binding (or interaction) energy $\Delta E = E_T(A_2) - 2E_T(A)$ (size-inconsistency).

A way to overcome (or to minimize) the size-inconsistency is offered (among others) by the Multi Reference D-CI method (MRD-CI) [6] which is based on the definition of a subspace of reference configurations $\{\Phi_k\}$ composed of Φ_0 (e.g. the HF configuration) and of some selected configurations (e.g. singly and doubly) excited with respect to Φ_0. Single and double excitations allowed from all the reference configurations Φ_k generate a CI space which may include the most important configurations up to quadruply excited with respect to Φ_0. For an increasing number of reference configurations the method can give an important fraction of the correlation energy since it approaches a full SDTQ-CI and minimizes the size-inconsistency problem. However, size-inconsistency and large computational demand make the CI methods efficiently applicable only to systems with relatively small number of electrons (about 20) [6, 7].

2.1.2.4 Theoretical Determination of the Molecular Geometry

The $3n$ Cartesian nuclear coordinates $\mathbf{x}$ can be transformed by a suitable transformation $\mathbf{B}$ into a set of non redundant $3n-6$ ($3n-5$ for linear systems) internal coordinates $\mathbf{q}$, $\mathbf{q} = \mathbf{xB}$. Correspondingly, the parametric dependence of $E_T = E_{\mathrm{el}} + E_{\mathrm{nuc}}$ from $\mathbf{x}$ can be easily translated into $E = E_T(\mathbf{q})$, the minimum points of which obeys the relations:

$$[\partial E_T(\mathbf{q})/\partial q_k]_{q_k = q_k^e} = 0, \qquad [\partial^2 E_T(\mathbf{q})/\partial^2 q_k]_{q_k = q_k^e} > 0 \quad (k = 1, \ldots, 3n-6) \tag{2.28}$$

where q_k^e are the equilibrium values of the k-th internal coordinate.

Small deviations from the optimum q_k^e values can be described in terms of harmonic vibrational modes. Easy solutions of (2.28) can be found only if the analytical expressions for $E_T(\mathbf{q})$ and its derivatives are known [8]. This is true for the case of HF-type Hamiltonians but only for a few cases of multiconfiguration wf.

The Hellmann–Feynman theorem [2a, b] holding for an exact wf ψ yields a simple relation when nuclei displacements are considered:

$$\frac{\partial E_{\mathrm{el}}}{\partial q} = \langle \psi | \frac{\partial \hat{H}}{\partial q} | \psi \rangle = -\sum_i \sum_a \langle \psi | \frac{\partial}{\partial q} (Z_a / |r_i - R_a|) | \psi \rangle . \tag{2.29}$$

Also a HF wf expanded in a complete basis obeys (2.29), while HF or HF-CI wf's obtained with truncated basis violate the Hellmann–Feynman theorem. In this case the "orbital (or wavefunction) forces" also contribute to the variation of the energy functional

$$\partial E_{el} = \langle \Phi | \partial \hat{H}_0 | \Phi \rangle + \langle \partial \Phi | \hat{H}_0 | \Phi \rangle + \langle \Phi | \hat{H}_0 | \partial \Phi \rangle \tag{2.30}$$

and more complex expressions have to be worked out leading to, for example, the analytical gradient formula for the HF energy [8].

2.1.2.5 Jahn–Teller and Pseudo-Jahn–Teller Effects

The Jahn–Teller theorem [2c] states that for a non-linear system in a spatially degenerate state with a given spin multiplicity and in highly symmetric nuclear configuration, a vibrational mode exists that leads to a geometric deformation removing the degeneracy. If the degeneracy is only approximate, similar behavior can be expected and the so called pseudo-Jahn–Teller effect can arise. In highly symmetric systems the highest occupied MO's would be exactly degenerate only if a sufficient number of electrons make their uniform occupancy possible. Symmetric clusters with incomplete electronic shells must therefore have a tendency to be stabilized by a geometry deformation lowering their symmetry (see Section 3). Roughly speaking, the geometrical distortion is due to the perturbation of the symmetric nuclear framework by the less symmetric electron cloud.

Electron Shell Model. The kinetic energy is an essential term in the energy expectation value (2.11) and (2.18) and since its value increases with the number of nodes of the one-electron functions, it is obvious that MO's with the smallest number of nodal planes are preferentially occupied. Similarly, the NO's with the largest occupation numbers are those with the smallest number of nodes. In general, the ordering of the MO's according to increasing energy or of the NO's according to decreasing occupation numbers is parallel to the ordering predicted by the number of nodal planes. The "shell closing" is a phenomenon which occurs if the number of available electrons is just sufficient to doubly occupy (in the one-electron approximation) the shells of MO's characterized by a small number of nodal planes, leaving the other shells, with larger number of nodal planes, empty. The step from a shell characterized by the number p of nodal surfaces to the next shell with $p + 1$ nodal surfaces can easily result in a considerable increase of the total energy, due to contributions of orbitals with different nodal properties. For example, if the number n_p of the one-electron functions in the p-th shell is assumed to be equal to the number of distinct p-th order products of space variables $(x^k y^l z^m;\ k + l + m = p)$, then the number of available one-electron levels in a shell is $n_0 = 1, n_1 = 3, n_2 = 6, n_3 = 10$. Correspondingly, the "magic numbers" of electrons are 2, 8, 20, 40. Indeed, these are the numbers of electrons which are considered as magic in the cluster science of the simple metals. In two dimensions, the magic numbers are 2, 6, 14, 22. Deviations from the "magic numbers" 2, 8, 20, 40, ... can be expected when realistic potentials are considered, but the large stability of symmetric clusters with "magic number" of electrons is a general tendency which can be well understood from a qualitative point of view. The "magic numbers" of valence electrons [9] which cause a special stability of metal clusters are consequently not at all magic but a natural consequence of general laws of quantum theory.

2.1.2.6 Interaction with Electromagnetic Field

The response of a system (e.g. a cluster) to a perturbing electromagnetic field can give very basic information about the electronic and geometrical structure of the investigated system.

Two different conceptional approaches can be chosen to formulate the quantum theory for this response. Either a time-dependent state of the perturbed system is considered, or the probabilities of jumps in the "excited" states of the unperturbed system are determined.

If the wavelengths of the electromagnetic waves interacting with the system are large, it is sufficient to write the perturbation in the form $\hat{V}(t) = -\sum_i \hat{\mu}_i E(t) + \cdots$ where $E(t)$ is a time-dependent electric field and $\hat{\mu}_i$ is the dipole moment operator. Usually the perturbation caused by the electromagnetic field is assumed to be small, so that the time dependent perturbation theory is an appropriate methodological tool. Very often first order perturbation theory (or linear response approach) yields results of sufficient accuracy.

2.1.2.6a) Time-Dependent Perturbation Theory and Dynamic Polarizability

The time-dependent ket $|\psi(t)\rangle$ which satisfies Eq. (2.1) with $\hat{H} = \hat{H}_0 + \hat{V}(t)$ can be expanded in a series of the eigenfunctions $|\psi_j^0\rangle$ of the time-independent Hamiltonian $\hat{H}_0$, $\hat{H}_0|\psi_j^0\rangle = E_j^0|\psi_j^0\rangle$,

$$|\psi(t)\rangle = \sum_j e^{-i\hat{H}_0 t}|\psi_j^0\rangle a_j(t) . \tag{2.31}$$

Let us assume that the periodic perturbation caused by the electric field is:

$$\hat{V}(t) = -2\hat{\mu}_x E_x(1 - e^{t/\tau})\cos(\omega + i\eta)t , \tag{2.32}$$

where τ is a switching time of the perturbation. $\hat{V}(t)$ has matrix elements defined as $V_{0j}(t) = \langle\psi_0^0|\hat{V}(t)|\psi_j^0\rangle$. If it is assumed that all expansion coefficients remain small during the whole process with exception of the coefficient a_0 of the initial state $|\psi_0^0\rangle$ of the system, the following simple relation results:

$$a_j(t) = -i\int_0^t V_{j0}(t)e^{i\omega_{j0}t}\,dt , \tag{2.33}$$

with $\omega_{jo} = E_j^0 - E_0^0$. From Eqs. (2.32) and (2.33) follows for $t \gg \tau$:

$$a_j(t) = -E_x \mu_{jo}^x \left\{ \frac{e^{i(\omega_{jo}+\omega)t}}{\omega_{jo}+\omega+i\eta} - \frac{e^{i(\omega_{jo}-\omega)t}}{\omega_{jo}-\omega-i\eta} \right\}, \tag{2.34}$$

$$\mu_{jo}^x = \langle \psi_j^0(0) | \hat{\mu}_x | \psi_0^0(0) \rangle . \tag{2.35}$$

If the expansion (2.31) with Eq. (2.33) for $a_j(t)$ is substituted into the definition of the dipole moment of the system $\langle \psi | \hat{\mu}_x | \psi \rangle$ the following relation showing the time dependence of the dipole moment to the first order in E_x is obtained:

$$\langle \psi | \hat{\mu}_x | \psi \rangle = \mu_{00}^x + 4 \sum_j \left\{ (\omega_{jo}+\omega)^{-1} + (\omega_{jo}-\omega)^{-1} \right\} (\mu_{jo}^x)^2 E_x \cos \omega t . \tag{2.36}$$

Consequently, the polarizability α_{xx} is expressed as

$$\begin{aligned} \alpha_{xx}(\omega) &= 2 \sum_j \frac{\omega_{jo}}{\omega_{jo}^2 - \omega^2} (\mu_{jo}^x)^2 \\ &= 2 \langle \psi_0^0 | \hat{\mu}_x \{ \hat{G}(E_0^0 - \omega) + \hat{G}(E_0^0 + \omega) \} \hat{\mu}_x | \psi_0^0 \rangle , \end{aligned} \tag{2.37}$$

where the Green operator is defined as: $\hat{G}(z) = \hat{Q}_0 (\hat{H} - z \hat{I})^{-1} \hat{Q}_0$, with $\hat{Q}_0 = I - |\psi_0^0\rangle \langle \psi_0^0|$.

The last expression on the right-hand side of Eq. (2.37) shows the connection with Green functions and linear response theory which avoids the "spectral decomposition" in terms of eigenvalues of the unperturbed Hamiltonian which is a typical drawback of the perturbation theory. The mean polarizability defined as:

$$\begin{aligned} \bar{\alpha}(\omega) &= 1/3 [\alpha_{xx}(\omega) + \alpha_{yy}(\omega) + \alpha_{zz}(\omega)] \\ &= 2/3 \sum_j \frac{\omega_{jo}}{\omega_{jo}^2 - \omega^2} |\mu_{jo}|^2 = \sum_j f_{jo}/(\omega_{jo}^2 - \omega^2) \end{aligned} \tag{2.38}$$

is expressed in terms of the square of the total transition moment $|\mu_{jo}|^2$ and, for $\omega \to 0$, reduces to the static polarizability:

$$\bar{\alpha}(0) = 2/3 \sum_{j>0} |\mu_{jo}|^2/\omega_{jo} = \sum_j f_{jo}/\omega_{jo}^2 . \tag{2.39}$$

Equations (2.38) and (2.39) are also written in terms of the oscillator strengths:

$$f_{jo} = (2/3)\, \omega_{jo} |\mu_{jo}|^2 \tag{2.40}$$

which satisfy the Kuhn–Thomas summation rule $\sum_j f_{jo} = N$.

2.1.2.7 Plasmons

Since some characteristic spectroscopic properties of relatively small metal clusters are often interpreted with the help of the plasmon concept it is appropriate to outline shortly the classical and quantum mechanical theory of plasmons.

If a system is forced to move from the geometric configuration in which it has minimal energy it tries to come back to this optimal configuration and, in absence of damping harmonic oscillations can take place (e.g. the classical harmonic oscillator). Numerous phenomena in different branches of science have laws of motion strictly similar, from a mathematical point of view, to the equation of the harmonic oscillator. Consequently, such systems exhibit resonances when periodically perturbed. However, even if simple classical picture seems to reproduce some experimental facts such as for example, appearance of giant resonances in absorption spectra of clusters with given nuclearity, the intrinsic microscopic (quantum) structure of small clusters cannot be ignored if a non-empirical interpretation of a physical phenomenon is attempted.

The volume plasma motion in an infinite metal is predominantly caused by the reaction of the electron density to a periodic perturbation and the driving force for the oscillations of the electron density is the electron-electron interaction. Since the surface is a very important feature of relatively small clusters the notion of surface plasmon has been introduced in analogy to that used in the physics of metals. The relative shift of the negative charge of the more movable electron density relative to the more inert background of the positive charge can lead to "plasma" oscillations. Some similarities to the dynamic polarizability of dielectrics are noticeable since the restoring force here is mainly due to the interaction between negative electron charge and positive charge of the atomic nuclei or atomic cores.

2.1.2.7a) Classical Description of Plasmons in Metals

For the sake of simplicity let us assume that the positively charged background is homogeneous. If the electronic charge density $\rho(\mathbf{r}, t)$ which is a periodic function of time, can be expanded in the plane waves compatible with the periodicity of lattice $\mathbf{q}$ by a Fourier transform [10] then:

$$\rho(\mathbf{r}\, t) = V^{-1/2} \sum_{\mathbf{q}} \rho_{\mathbf{q}}(0)\, e^{i(\omega_{\mathbf{q}} t - \mathbf{q}\mathbf{r})} \; . \tag{2.41}$$

The velocity $\mathbf{v}_i$ of the i-th electron is changed under the influence of the electric field $\mathbf{E}$ according to the Newton law:

$$m \frac{\partial \mathbf{v}_i}{\partial t} = e\, \mathbf{E}(\mathbf{r}_i) \; . \tag{2.42}$$

Using basic laws of the classical electrodynamics for average electric field $\mathbf{E}$ and averaged current density $\mathbf{j}$ an oscillator-like equation can be obtained:

$$\frac{\partial^2 \rho(\mathbf{r}, t)}{\partial t^2} = -\, \omega_p^2 \rho(\mathbf{r}, t) \; , \tag{2.43}$$

with the plasmon frequency (in non-atomic units):

$$\omega_p = (e^2 \bar{\rho}/m)^{1/2} \; . \tag{2.44}$$

Notice that besides universal constants only the average electron density $\bar{\rho} = N/V$ in the metal is needed to define the plasmon frequency. It is easy to establish relations between the plasmon frequency ω_p and the dynamic and static polarizabilities. For $\omega^2 \gg \omega_p$, Eq. (2.38) simplifies to:

$$\bar{\alpha}(\omega) = -N(e^2/m\omega^2) = -(\omega_p/\omega)^2 \ . \tag{2.45}$$

Similarly, the expression

$$\bar{\alpha}(0) = (\omega_p/\bar{\omega})^2 \ , \tag{2.46}$$

can be obtained if the mean excitation energy $\bar{\omega}$ replaces the individual ω_{j0}'s in Eq. (2.39).

Polarizability of a Dielectric. The mathematical expressions used for the description of surface plasmons are similar to those valid for the response of dielectrics on the electromagnetic field [10].

The equation of motion of an electron under the influence of the electric field E_x reads:

$$eE_x e^{i\omega t} = m_c(\ddot{x} + \gamma\dot{x} + \omega_0^2 x) \ , \tag{2.47}$$

where $\omega_0^2 x$ is the restoring force and $\gamma\dot{x}$ is the damping.

Under the assumption that the equation of motion (2.47) is valid for all the electrons, the polarizability $\alpha_{xx}(\omega)$ defined by:

$$P_x = Ne\sum_j x_j = E_x \alpha_{xx} = P_{x0}\, e^{i\omega t} \ , \tag{2.48}$$

takes the form:

$$\alpha_{xx}(\omega) = \omega_p^2/(\omega_0^2 - \omega^2 + i\gamma\omega) \ , \tag{2.49}$$

where ω_p is given by Eq. (2.44). Evidently, the static polarizability is:

$$\alpha_{xx}(0) = (\omega_p/\omega_0)^2 \ . \tag{2.50}$$

The imaginary part of the dynamic polarizability can be expressed in the form:

$$\mathrm{Im}(\alpha(\omega)) = -\frac{\gamma\omega\,\omega_p^2}{(\omega_p^2/\alpha(0) - \omega^2)^2 + \gamma^2\omega^2} \ , \tag{2.51}$$

which is frequently used for the interpretation of the absorption spectra of metal clusters [11].

2.1.2.7b) Quantum Theory of Plasmons

The oscillation of the electron density in a system (supposed to be in a singlet state) can be described by the time dependent quantum mechanical quantity

$\rho(\mathbf{r}, t)$ associated with the time dependent function $|\psi(t)\rangle$:

$$\rho(\mathbf{r}, t) = \sum_{i,j} \bar{\varphi}_j(\mathbf{r}) \bar{\varphi}_i(\mathbf{r}) \langle \psi(t) | \hat{\mathscr{E}}_{ji} | \psi(t) \rangle \; , \tag{2.52}$$

where $\hat{\mathscr{E}}_{ji} = \hat{E}_{ji} + \hat{E}_{\bar{j}\bar{i}}$; $\hat{E}_{ji}$ and $\hat{E}_{\bar{j}\bar{i}}$ act on α- and β-spin electrons, respectively.

Since the formal solution of the Schrödinger equation can be written as $|\psi(t)\rangle = e^{-i\hat{H}t}|\psi(0)\rangle$, for an arbitrary time dependent operator $\hat{A}$ the following expression holds:

$$i\langle \psi(0)| \frac{d\hat{A}}{dt} |\psi(0)\rangle = \langle \psi(0)| [\hat{A}^H, H] |\psi(0)\rangle \; , \tag{2.53}$$

where $\hat{A}^H = e^{-i\hat{H}t} \hat{A} e^{-i\hat{H}t}$ is the operator in the so called Heisenberg picture. In particular

$$i\frac{d[\rho(\mathbf{r}, t)]^H}{dt} = \sum_{i,j} \bar{\varphi}_j(\mathbf{r}) \, \bar{\varphi}_i(\mathbf{r}) \langle \psi(0) | [\hat{\mathscr{E}}^H_{ij}, \hat{H}] | \psi(0) \rangle \; . \tag{2.54}$$

The commutator of an arbitrary $\hat{\mathscr{E}}_{ji}$ with $\hat{H}$ has the form

$$[\hat{\mathscr{E}}_{ji}, \hat{H}] = (\varepsilon_i - \varepsilon_j)\hat{\mathscr{E}}_{ji} + (N_j - N_i) \sum_{l \neq k} \hat{\mathscr{E}}_{lk}\left([il|kj] - \frac{1}{2}[il|kj] \right) , \tag{2.55}$$

where the higher order terms in $\hat{\mathscr{E}}_{ij}$ are neglected as is also the case in the so called random phase approximation (RPA) [12 and references therein]. In (2.55) the number operators are replaced by their expectation values N_k and ε_k are HF MO energies. In this approximation the Fourier transform of $\langle \psi(0) | \hat{\mathscr{E}}^H_{ji} | \psi(0) \rangle$ is:

$$\hat{\mathscr{E}}_{ij}(\omega) = \frac{(N_i - N_j)}{\varepsilon_j - \varepsilon_i - \omega} \sum_{l \neq k} \hat{\mathscr{E}}_{lk}(\omega) \left([il|jk] - \frac{1}{2}[il|kj] \right) . \tag{2.56}$$

Consequently,

$$\rho(\mathbf{r}, \omega) = \sum_{i,j} \frac{N_j - N_i}{(\varepsilon_j - \varepsilon_i - \omega)} \bar{\varphi}_j(\mathbf{r}) \, \bar{\varphi}_i(\mathbf{r}) \langle \psi(0) | \times \sum_{l \neq k} \hat{\mathscr{E}}_{kl}(\omega) \left([il|jk] - \frac{1}{2}[il|kj] \right) | \psi(0) \rangle \; . \tag{2.57}$$

If $\bar{\varphi}_k(\mathbf{r}) \, \bar{\varphi}_l(\mathbf{r}) = \bar{\varphi}_j(\mathbf{r}) \, \bar{\varphi}_i(\mathbf{r})$ and if all integrals $[il|jk] - \frac{1}{2}[il|kj]$ have the same value v, then

$$\rho(\mathbf{r}, \omega) = v \sum_{ij} \frac{N_j - N_i}{(\varepsilon_j - \varepsilon_i - \omega - i\delta)} \rho(\mathbf{r}, \omega) \tag{2.58}$$

and the resonance frequencies ω are solutions of the simple relation:

$$1 = v \sum_{i,j} \frac{N_j - N_i}{(\varepsilon_j - \varepsilon_i - \omega - i\delta)} \; , \tag{2.59}$$

where δ is the damping factor.

A graphical analysis reveals that nearly all frequencies ω which satisfy (2.59) are close to one-electron "HF excitation energies" $\varepsilon_j - \varepsilon_i$, but one value ω_p lies outside this range and is therefore said to belong to a collective excitation or plasmon. This clearly is a definition using the concepts of a one-electron picture which is not adequate, in general, for description of the excited states of clusters, as will be shown in Section 3.

The conditions under which Eq. (2.58) has been derived are fulfilled in the infinite crystal for free electrons due to translation symmetry with $i = \mathbf{k}$, $j = i + \mathbf{q} = \mathbf{k} + \mathbf{q}$, $v = v_\mathbf{q}$ and $\Sigma_{ij} = \Sigma_{k'q}$. Since the interaction among one-electron (particle-hole) transitions: $\mathbf{k}' \to \mathbf{k} + \mathbf{q}$ and $\mathbf{k}'' = \mathbf{k} + \mathbf{q}'$ for $\mathbf{q} \neq \mathbf{q}'$ is not allowed, the electron density $\rho(\mathbf{r}, \omega)$ can be decomposed into the components $\rho_\mathbf{q}(\mathbf{r}, \omega)$. Analogously, the electron density $\rho(\mathbf{r}, \omega)$ have independent components $\rho_\Gamma(\mathbf{r}, \omega)$ for each irreducible representation Γ of the symmetry group of the cluster.

In the random phase approximation the relation (2.59) for the free-electron can be written as:

$$1 \approx v_\mathbf{q} \sum_\mathbf{k} N_\mathbf{k} \left\{ \frac{1}{[\omega - (\varepsilon_\mathbf{k} - \varepsilon_{\mathbf{k}-\mathbf{q}}) + i\eta]} - \frac{1}{[\omega - (\varepsilon_{\mathbf{k}+\mathbf{q}} - \varepsilon_\mathbf{k}) + i\eta]} \right\}$$

$$= v_\mathbf{q} \mathbf{q}^2 \sum_\mathbf{k} \frac{N_\mathbf{k}}{(\omega_p - \mathbf{qk} + i\eta)^2 - \mathbf{q}^4/4} . \tag{2.60}$$

Since in the free-electron model the integral $v_\mathbf{q}$ (in the atomic units) is

$$v = v_\mathbf{q} = \frac{1}{\mathbf{q}^2 V} , \tag{2.61}$$

the simple relation for the plasmon frequencies ω_p follows in the limit $\mathbf{q} \to 0$ (cf. Eq. (2.44)):

$$\frac{\sum_\mathbf{k} N_\mathbf{k}}{V\omega_p^2} = \frac{\rho_0}{\omega_p^2} = 1 , \tag{2.62}$$

where V is the volume and ρ_0 is the average electron density.

The evaluation of the commutator $[\hat{\mathscr{E}}_{ij}, \hat{H}]$ assuming a ground atate HF wf and keeping only first order terms in $\hat{\mathscr{E}}_{ji}$ leads to a theory for the time-dependency of $\rho(\mathbf{r}, t)$ in which only the interaction between the single one-electron excitations is considered. An equivalent approach leading to (2.57) is the CI including only the singly excited configurations $|i \to j\rangle$ (S-CI or Tamm–Dancoff method).

The matrix elements between single excitations $A_{ij} = \langle i \to j | \hat{H}_0 | k \to l \rangle$ are the dominant contributions to the time-dependent Hartree–Fock or RPA equations:

$$\begin{pmatrix} \mathbf{A} & \mathbf{B} \\ -\mathbf{B} & -\mathbf{A} \end{pmatrix} \begin{pmatrix} \mathbf{y} \\ \mathbf{z} \end{pmatrix} = \omega \begin{pmatrix} \mathbf{y} \\ \mathbf{z} \end{pmatrix} . \tag{2.63}$$

In addition, this kind of interaction is coupled to the interaction between the HF ground state and the doubly excited configurations $|i \rightarrow j, k \rightarrow l\rangle$ via the matrix B. A simple calculation shows that the assumption of a constant interaction integral v again leads to Eq. (2.59) for the eigenvalues ω.

2.1.3 Ground State Properties of Metal Clusters

Let us specify several fundamental questions in cluster research which will be addressed using ab initio CI quantum chemical methods outlined in Section 2: i) How is the stability and how are other properties like ionization potentials (IP) electron affinities (EA) and static polarizabilities ($\bar{\alpha}$) of the neutral and charged M_n, M_n^+ and M_n^- clusters dependent on the cluster size and the cluster geometry? ii) Are the forces responsible for the stability of clusters and solids of similar nature and do clusters have specific electronic features different from those of molecules and solids? iii) Are the measured abundances [9, 13–15] of ionized clusters related only to the stability of the parent species or are they also connected with fragmentation processes?

2.1.3.1 Stability and Structure of Neutral and Charged Alkali Metal and Mixed Ia–IIa Clusters

Since Ia metal atoms have only one valence electron, the rules determining the electronic structure of clusters are expected to be relatively simple. In addition to the total energy, possible measures of cluster stability are the binding energy per atom and the second differences of the cluster energies defined as:

$$E_b/n = (nE_1 - E_n)/n; \quad E_b^{\pm}/n = [(n-1)E_1 + E_1^{\pm} - E_n^{\pm}]/n \tag{3.1}$$

and

$$\Delta^2 E_n = E_{n+1} + E_{n-1} - 2E_n; \quad \Delta^2 E_n^{\pm} = E_{n+1}^{\pm} + E_{n-1}^{\pm} - 2E_n^{\pm} \tag{3.2}$$

for neutral and charged clusters, respectively. E_n and $E_n^{\pm}$ are energies of the neutral and the charged clusters with n atoms, respectively. The quantity $\Delta^2 E_n$ can be interpreted as the difference between the energies of the two fragmentation channels: $M_n \rightarrow M_{n-1} + M_1$ and $M_{n+1} \rightarrow M_n + M_1$. Positive values of $\Delta^2 E_n$ indicate that the dissociation of M_{n+1} into M_n is a more favorable process than the fragmentation of M_n into M_{n-1}, indicating a special stability of the M_n clusters. In order to obtain E_b/n and $\Delta^2 E_n$ the determination of the cluster structure is a prerequisite. Therefore, the first step is the geometry optimization at the HF level (cf. Sections 2.1.2.1, 2.1.2.4). The BO energy potential surfaces of alkali metal clusters are relatively shallow indicating a relatively large mobility of atoms in comparison with normal stable molecules. Moreover, several local minima with comparable energies might be present, the number of which increases with n. This requires a careful choice of the starting points for the geometry optimization for which the cluster symmetry, compactness and possible Jahn–Teller distortions might be useful. Due to the complicated nature of the BO surfaces the application of the energy gradient method without approximations which neglect orbital forces (cf. Section 2.4), is required. Although the details of the cluster geometry should not be overestimated, the interplay between the cluster geometry and characteristic electronic features is essential for an understanding of their specific properties.

The topologies of the HF optimized geometries for Li_n, $Li_n^{\pm}$, Na_n and Na_n^+ clusters next to the CI binding energy per atom E_b/n are drawn in Fig. 1 [17–20]. All five curves show a tendency to increase as a function of the cluster size. The oscillatory behavior for even-odd number of valence electrons N is particularily pronounced for the $\Delta^2 E_n$ and $\Delta^2 E_n^+$ quantities (cf. window of Fig. 1b) indicating larger stability towards fragmentation for clusters with even numbers of valence electrons. This includes a particular stability of clusters corresponding to so-called closing of shells with $N = 2$ and 8 (magic numbers) (cf. Ref. [9] and Section 2.5) and predicts in addition the maxima for $N = 4$ and 6. Note that the $\Delta^2 E_n$ and $\Delta^2 E_n^+$ are particularly suitable for interpretation of the experiments using hot cluster beams, because the measured abundances correspond to clusters which have survived the fragmentation. Very recent experiments using cold beams [21] show no pronounced maxima of the abundances which increase monotonically with the cluster size. Therefore the binding energy per atom E_b/n is more suitable as a measure of stability. Three important points should be emphasized: i) the described behavior of the stability measures E_b/n and $\Delta^2 E_n$ have been obtained only for the optimized cluster geometries which for small n are deformed sections of the crystal lattices and for larger n are composed of tetrahedral or (distorted) pentagonal bipyramidal units. ii) The inclusion of correlation effects is an appropriate manner plays a decisive role, since $\sim$ 50% of the binding energy is due to correlation effects (cf. Section 2.3). Moreover, the correlation energy can differ substantially for states of different multiplicity and its contribution is essential for determining the energy ordering of isomers with comparable stability (e.g. planar and three dimensional Li_5, Li_6 or Na_5, Na_6 clusters). iii) Relatively small AO basis sets can be chosen but they have to account for an appropriate s–p hybridization. For example, a relatively large AO basis set very close to the so called HF limit (cf. Section 2.1) without p functions does not yield appropriate stability of Li_4 and Li_8 clusters towards the atomization. But, a minimal basis set of s-orbitals augmented by one p-function is already appropriate for qualitative predictions. This is an important finding since p functions are not occupied in the ground state of alkali metal atoms, but are essential for metallic bulk properties. This is also the case for IIa and IIb elements. The degree of the s–p hybridization can be used with caution only as one of possible measures of the metallic features of the cluster.

The most stable neutral and cationic Li_n and Na_n clusters have planar forms for $n \leq 6$ which are deformed sections of the (111) plane of the fcc (or hcp) lattice and for $n \geq 7$ they change into three dimensional structures with pronounced characteristics of the compact tetrahedral growth (cf. Fig. 1). Notice that for the

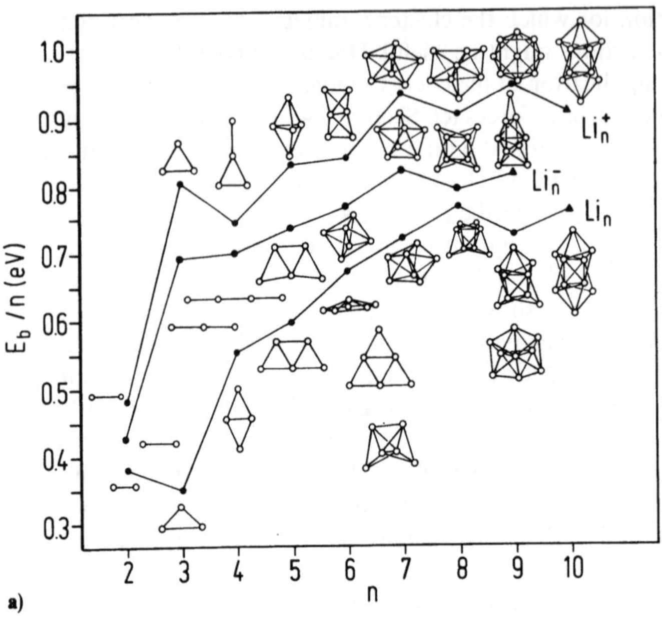

a)

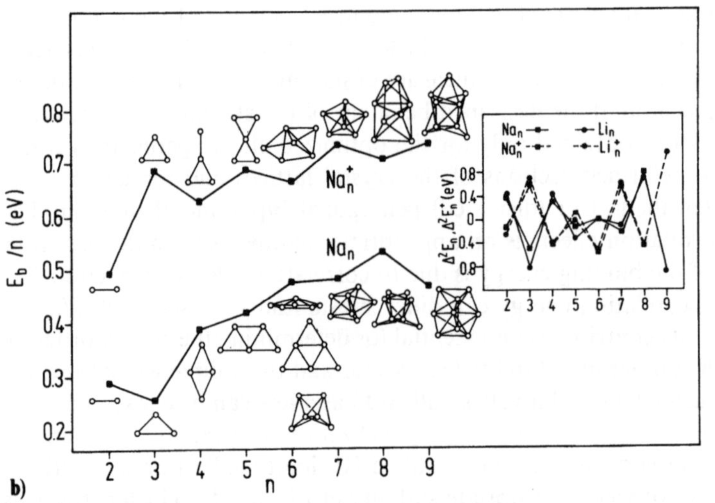

b)

Fig. 1. The CI binding energy per atom E_b/n as a function of the nuclearity n for **a)** neutral, anionic and cationic Li_n clusters and for **b)** neutral, and cationic Na_n clusters. The topologies of the H-F optimized geometries are shown. The measures of the stabilities Δ^2E_n and $\Delta^2E_n^+$ for neutral (—) and cationic (– – –) Li_n and Na_n clusters are drawn in the window. Small, but appropriate AO basis sets are used. For details cf. Refs. [17–20]

cationic pentamer, the planar structure and the trigonal bipyramid are energetically close lying, and for neutral hexamers there is a strong competition between the planar and the two three-dimensional structures. This is easy to understand, since up to $n = 6$ there are not enough valence electrons to occupy degenerate one-electron levels of the highly symmetrical species and therefore Jahn–Teller deformation takes place (cf. Section 2.5). For example, the energy stabilization from the tetrahedral to the rhombic structure occurs for tetramers [22] (Fig. 2). On the contrary, for $n = 8$, Li_8 and Na_8 tetrahedral (T_d) clusters exhibit a particularly large stability. They are formed by a central tetrahedron with one atom capping each face and building additional tetrahedral subunits. The eight valence electrons can fill up a nondegenerate *s*-type MO and triply degenerate *p*-type MO's (cf. "closing of shell" Section 2.5). Similarily, the highly symmetrical structure of Li_{20} (cf. Fig. 2) which allows the 20 valence electrons to occupy "*s, p, s, d*" type one-electron functions seems to be more stable than other less symmetrical structures representing different sections of the fcc lattice (cf. Section 2.5 and Ref. [23]).

The interatomic distances in optimized structures do not fluctuate substantially with *n* (for details compare Refs. [17–20]) and their average does not differ considerably from the interatomic distances found for the corresponding metal bulks. For example, two calculated distances in the T_d structure of Li_8 are 3.09 and 3.11 Å compared to 3.1 Å in the fcc Li crystal. Similarily, in the T_d structure

Fig. 2. Schematic one electron functions with occupation numbers for different structures of Li_4, Li_8 and Li_{20} (cf. Ref. [23])

of Na_8 the two independent distances are 3.75 Å and 3.81 Å, compared to $r = 3.7$ Å in the crystal.

The geometries of cationic and mainly of anionic clusters (Li_n^-) differ considerably from the structures of the most stable neutral species. These differences are due not only to the different numbers of valence electrons but also to the fact that charged clusters prefer such geometries for which the polarization of the electronic cloud optimally compensates the unfavorable contribution due to the Coulomb repulsion. In fact, the linear geometries of Li_n^- and Na_n^- small clusters are the most stable ones since negative charges are located at the peripherical atoms (cf. Refs. [18] and [24]). In the case of cationic clusters the positive charge is distributed over the whole cluster, although the difference between the positive charges at the nonequivalent centers remains fairly large even for the largest clusters considered (cf. Ref. [20]).

The geometries presented here will be used in Sections 2.1.3.2 and 2.1.4 for structural assignments to measured quantities such as ionization potentials (IP) electron affinities (EA) and optical response properties.

2.1.3.2 One Electron Densities, Ionization Potentials, Electron Affinities, Fragmentation Channels and Polarizability

The difference between the one-electron density of a cluster (obtained from the CI correlated wavefunction) and the superposition of one-electron densities of non-interacting atoms (DOD) can be useful for characterization of the type of bonding which takes place in clusters. Regions with positive and negative values of DOD can be interpreted as bonding and antibonding regions without great ambiguity under the condition that no important rehybridization takes place in cluster formation. The DOD's of clusters show clearly that the conventional two center bond is not present in Li_n clusters. Planar clusters (Li_4, Li_5, Li_6) exhibit enhanced positive DOD values in the triangular regions indicating three center delocalized bonds (Fig. 3). For three dimensional clusters Li_7 and Li_8 delocalization of the electronic cloud is present but the electronic distribution is not homogeneous. The spherical shell with large and positive values of DOD for Li_8 (T_d) has four windows opposite to the Li atoms of the inner tetrahedron (Fig. 3). This illustrates non-conventional features of bonding in clusters as indicated in the Introduction (see also Ref. [25]).

The vertical and adiabatic ionization potentials (IP_v, IP_a) for Li_n and Na_n clusters are reported as functions of cluster size in Fig. 4. All four curves show i) decreasing trend in IP with the increasing n in a qualitative agreement with the classical spherical droplet model, ii) the even-odd oscillations with maxima for clusters with even n indicating their stability and iii) a large decrease in IP from $n = 2$ to 3 and from $n = 8$ to 9 supporting the argument for special stability of dimers and octamers. The IP curves for Li_n and Na_n clusters exhibit parallel behavior, the latter having lower values similar to the behavior of E_b/n which indicates smaller stability of Na_n mostly probably due to the lower degree of the

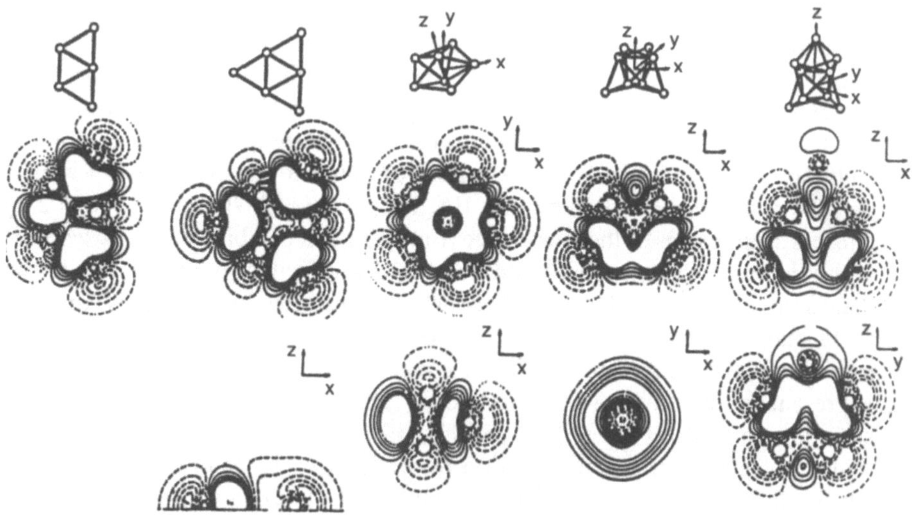

Fig. 3. Difference one-electron (DOD) density maps for Li_n ($n = 5-9$). The coordinate axes define the planes of cuts displayed. The full and broken lines represent constant positive and negative values of DOD, respectively (cf. Ref. [17])

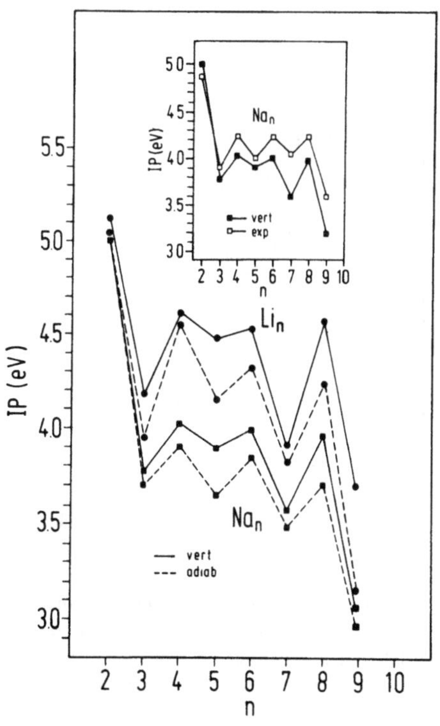

Fig. 4. Vertical and adiabatic ionization potentials (IP_v and IP_a) for Li_n and Na_n clusters defined as $E_{CI}(M_n^+) - E_{CI}(M_n)$ as functions of the nuclearity n. A comparison between calculated IP_v of Na_n and measured IP [28] is in the window. For details cf. Refs. [17, 19 and 20]. The vertical process is assumed to take place at the fixed geometry of the neutral species and in the adiabatic process the cationic geometry is allowed to relax

s–p hydridization than in the case of Li_n clusters. A comparison between calculated IP_v and measured appearance potentials for Na_n [26] (window of Fig. 4) shows a parallel behavior. The calculated values are slightly lower than the experimental ones, since core valence correlation has not been included in this treatment (cf. Section 2.2). The introduction of core valence polarization [5] shifts the values of ionization potentials even closer to the experimental values.

The adiabatic electron affinities EA_a for Li_n^- ($n = 2$–9) and Na_n^- ($n = 2$–5) clusters as functions of the cluster size (Fig. 5a) again exhibit oscillatory features with maxima for clusters with odd number of atoms in striking analogy to the measured EA for Cu_n^- clusters [27] which demonstrates some features common to Ia and Ib elemental clusters. However, the analogies should not be overestimated since the structural difference among small Ia and Ib anionic clusters is present in some cases (for example Cu_4^- [28] and Ag_4^- [29] assume rhombic structures in contrast to Li_4^- [18] and Na_4^- [24] which prefer linear forms). Note that for a quantitative determination of the EA the choice of an AO basis set which allows for flexible enough distribution of the extra electronic charge is particularly important. Similarly, the calculation of static dipole polarizabilities (Fig. 5b), which are connected with the charge of the electronic cloud under the influence of an external electric field, requires a careful choice of the AO basis set and an adequate electron correlation treatment (cf. Section 2.1.2.3). The one-electron methods yield too "compact" electron distributions, while correlation effects, keeping the electrons apart, produce more diffuse and polarizable electron densities. As a consequence the average polarizability $\bar{\alpha}$ computed with CI

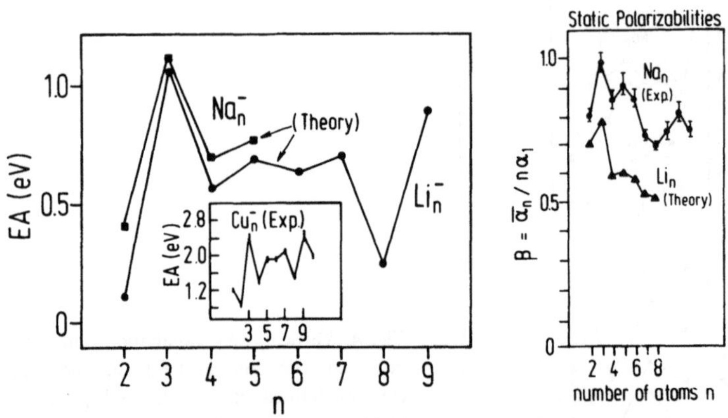

Fig. 5. a) The adiabatic electron affinities $EA_a = E_{CI}(M_n) - E_{CI}(M_n^-)$ of Li_n and Na_n clusters [18, 24] as functions of the nuclearity n (for the best geometry of the neutral and the anionic cluster, respectively). The experimental EA for Cu_n clusters [29] are shown in the window. **b)** A comparison between calculated static polarizabilities for Li_n as a function of the cluster size [30] and the measured polarizabilities for Na_n [31]. The CI values for the static polarizabilities α are obtained using the Hamiltonian in which homogeneous external field $\underline{E}$ is considered, from the expression $E(\underline{E}) = E_0 - \underline{\mu}\underline{E} - \frac{1}{2}\underline{E}\alpha\underline{E}$ where E_0 is the energy of the unperturbed system and dipole moment $\underline{\mu}(\underline{E}) = \underline{\mu}_0 + \alpha\underline{E}$

is usually considerably higher than that given by independent-particle models. The average static polarizabilities per atom β calculated for Li_n clusters are decreasing functions of the cluster size (cf. Fig. 5b) particularly from $n = 3$ to 4 and from $n = 6$ to 8 [30] which is parallel to the experimental finding for Na_n clusters [31]. Note that the values of all three properties IP, EA and β are strongly dependent on the cluster geometry. By comparing measured and predicted IP and EA an indirect assignment of geometries for neutral and anionic clusters, respectively can be made.

Another property such as fragmentation energy, for example for the two channels $Na_n^+ \rightarrow Na_{n-1}^+ + Na_1$ and $Na_n^+ \rightarrow Na_{n-2}^+ + Na_2$ also needs careful consideration. The fragmentation involving atom loss is given by energy differences between an odd and an even Na_n^+ cluster and the fragmentation with dimer loss depends on the energy difference between two clusters with even or two clusters with odd n. Consequently, it is not surprising that Na_5^+, Na_7^+ and Na_9^+ clusters with an even number of electrons favor a fragmentation with loss of a dimer [20] which has been confirmed experimentally [32].

2.1.4 Excited States of Alkali Metal Clusters and their Spectroscopical Properties

The theoretical study of excited states of alkali metal clusters [24, 33–36] is of particular interest since it offers an excellent opportunity to gain knowledge about the development of specific structural and electronic properties as a function of cluster size. The question of which extent clusters exhibit characteristic features different from properties of normal molecules and of solid bulk and how this difference is connected with the size and the shape of a cluster, can be very precisely addressed by investigating the properties of the excited state. For this purpose, a comparison between the results obtained from the ab initio CI method (cf. Sections 2.1.2.1, 2.1.2.3) and the spectra recorded by different experimental techniques such as photodetachment [37, 38], resonant-two photon ionization [39–41], and photodepletion [11, 42–48] is particularly valuable.

2.1.4.1 Quantum Chemical Interpretation of Photodetachment Spectra of Anionic Clusters

Electron photodetachment spectra of small alkali metal and transition metal clusters are highly structured and have a very regular pattern as a function of cluster size [37, 38]. The individual peaks arise due to photodetachment transitions between the ground state of the anion and the ground and the excited states of the neutral clusters. In this manner the "dark" *and* optically allowed excited states of the neutral species are accessible. The recorded photoelectron intensities are functions of the electron binding energy which is equal to the difference between the photon energy and the measured electron kinetic energy. For the complete interpretation of the spectrum, the geometries of anions have to be determined first and then the energies of the ground and excited states with appropriate spin multiplicities of the neutral species have to be calculated at the anion geometries.

A comparison of calculated [24, 36] and recorded [38] spectra for Na_{2-5}^- which permits an assignment of the cluster geometry is given in Fig. 6. The calculated vertical detachment energies VDE (defined as the energy difference between the anionic and the neutral ground states at the best geometry of anion)

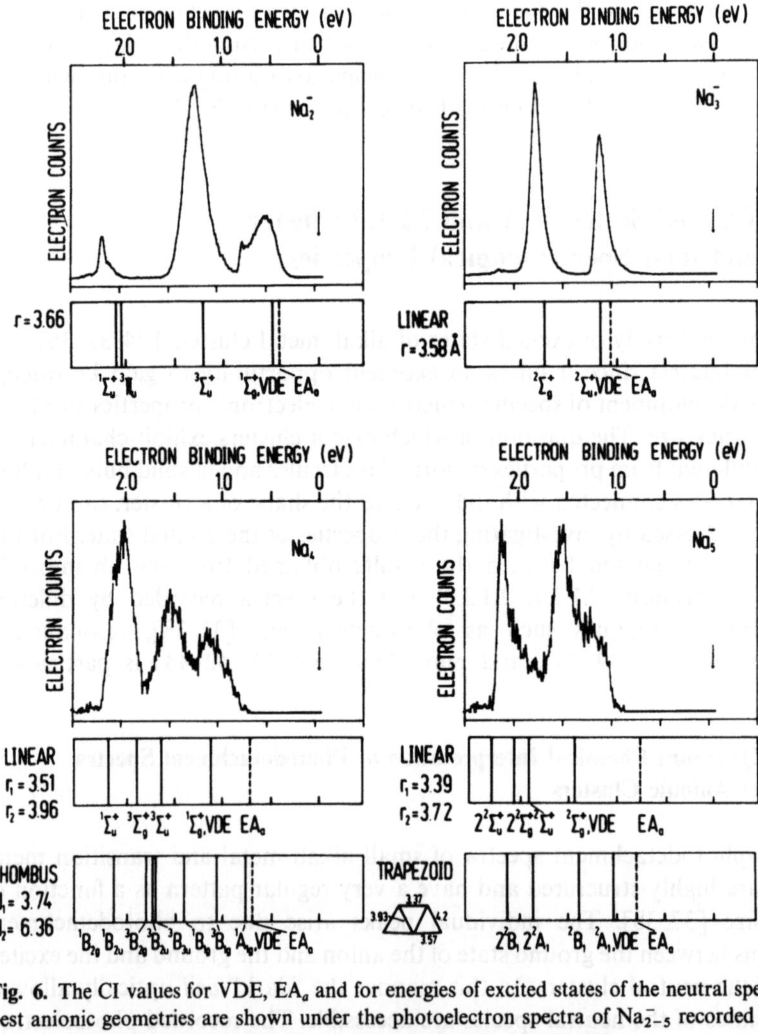

Fig. 6. The CI values for VDE, EA_a and for energies of excited states of the neutral species at the best anionic geometries are shown under the photoelectron spectra of Na_{2-5}^- recorded by Bowen and co-workers with 2.54 eV photons. For details cf. Refs. [24, 36, 38]

for Na_2^- and Na_3^- at their optimal *linear* geometries are in good agreement with maxima of the bands with the lowest binding energy and the EA_a's correspond to the signal onsets. The assignment of the lowest excited states of neutral Na_2 and Na_3 at linear geometries of their corresponding anions to the other bands is straightforward.

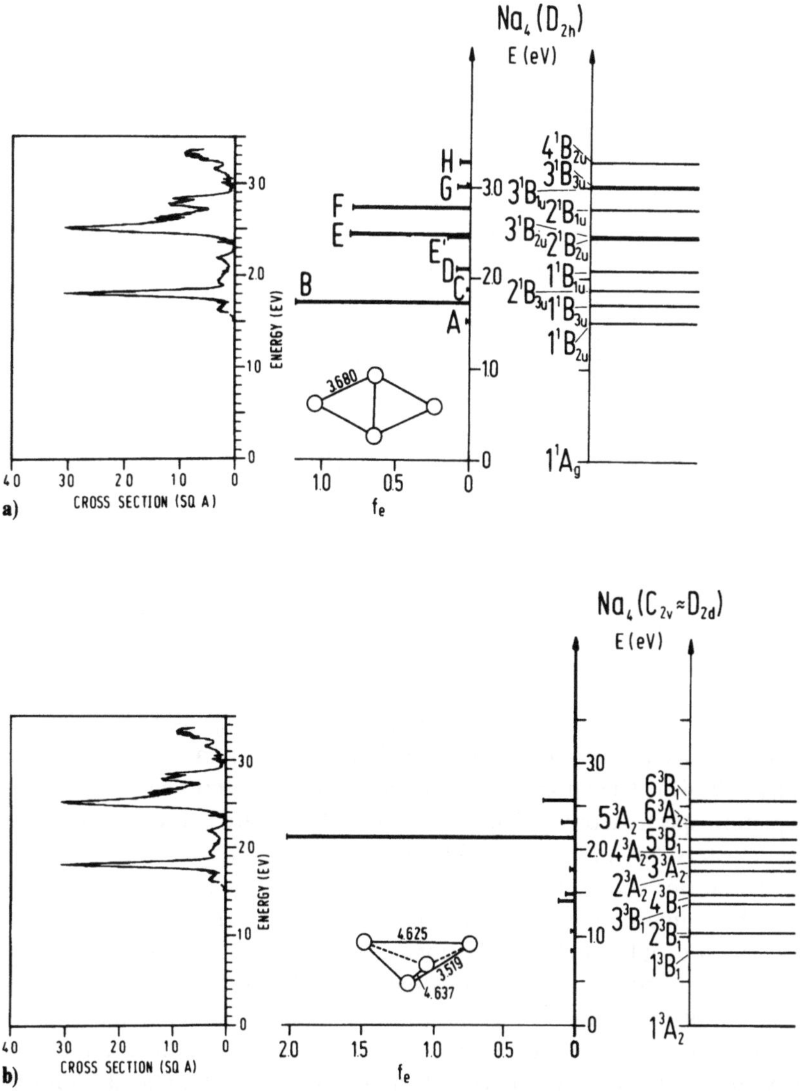

Fig. 7. A comparison between the depletion spectrum of Na_4 [42, 43] and the CI energies of optically allowed states and the oscillator strengths f_e for **a)** the best neutral rhombus (singlet state) and **b)** the deformed tetrahedron (best triplet state) structures. For details cf. Refs. [34] and [36]

The two Na_4^- isomers with linear (L) and rhombic (Rh) geometries have almost equal ground state energies, the former being slightly lower than the latter. The calculated VDE_L for the best of linear geometry of Na_4^- agrees well with the experimental value and the energies of the lowest excited states $^2\Sigma_u^+$ and $^{1,3}\Sigma_g^+$ of Na_4 coincide with the high intensities of the other two observed peaks (Fig. 7). The spectrum computed for Na_4 at the optimized rhombic geometry of Na_4^- agrees less well with the observed one since VDE_{Rh} is not apparent in the spectrum. Therefore, the conclusion can be made that the linear structure of Na_4^- is responsible for the observed photodetachment spectrum while the contribution of the rhombic structure is not substantial.

The ground state energy difference of trapezoidal (Tr) and linear (L) Na_5^- isomers is also very small although the former is slightly favored, independently on the details of the treatment. The calculated VDE_{Tr} for the trapezoidal structure coincides with the onset of the spectral profile, the first excited state 1^2B_1 lies in the region of the observed shoulder and the next two excited states 2^2A_1 and 2^2B_1 fit into the second band. For the linear Na_5^-, VDE_L coincides with the energy interval of the first broad band, and the two states $1^2\Sigma_u^+$ and $2^2\Sigma_g^+$ lie in the interval of the second band (Fig. 7). Due to the broad feature of the spectrum both anionic isomers (trapezoidal and linear) seem to be responsible for the photodetachment transitions in Na_5^-.

These examples illustrate that the ab initio CI study of Na_{2-5}^- and Na_{2-5} yields a complete and precise interpretation of the photoelectron detachment spectra. It is worth mentioning that the complete analogy has been found for K_{2-5}^- [49], but not for Ag_{2-5}^- [29b].

2.1.4.2 Quantum Chemical Interpretation of Absorption Spectra of Clusters

The absorption spectra of alkali trimers and tetramers exhibit rich structures. The spectra of trimers which were measured several years ago using the resonant two photon scheme have been interpreted as a typical case of Jahn–Teller instability [39–41]. The spectra of tetramers have been recorded very recently with depletion spectroscopy [42, 43, 45]. The first measured absorption spectra for larger clusters: Na_n ($n = 8, 9, 10, 12$ and 20) [11] and K_n^+ ($n = 9$ and 21) [47] contained one (for $n = 8, 20$) or two (for $n = 10, 12$) intense transitions and they have been interpreted in terms of classical surface plasma oscillations in spherical metal droplets [11] (cf. Sections 2.1.2.6 and 2.1.2.7). However, in the new experiments additional transitions have been found [42–44, 48]. For the interpretation of these absorption spectra as well as for understanding of the physical reasons for appearance of their characteristic pattern, the ground state geometry of the neutral cluster as well as energies of the optically allowed excited states for the given spin multiplicity and the corresponding transition intensities (cf. Section 2.1.2.6) have to be determined using a reliable quantum molecular theory.

The structured rich pattern of the optical response spectrum of Na_4 obtained by *Kappes* and *coworkers* [42, 43] presents a good test for a molecular interpretation. The energies, the oscillator strengths and the lifetimes of ten optically allowed excited states with excitation energies up to 3.3 eV have been determined for the best neutral rhombus geometry. A comparison between calculated [34, 36] and recorded spectrum [42, 43] of Na_4 shown in Fig. 7a allows for the complete assignment of eight bands to the rhombic Na_4 structure. Among several transitions to electronic states belonging to the same irreducible representation of D_{2h} point group only one is characterized by high intensity: B, E and F corresponding to 1^1B_{3u},3^1B_{2u} and 2^1B_{1u}, respectively. The analysis of the CI wavefunctions for each state shows that linear combinations of the same dominant one-electron or two-electron excitations can lead either to very intense or very weak transitions. Moreover, the wavefunctions of $^1B_{3u}$ states are dominated by one single electron excitation, while those of $^1B_{1u}$ and $^1B_{2u}$ states have large contributions of two singly excited and one doubly excited configurations [34, 36]. Note, however, that the contribution of a large number of configurations included in the large scale CI (in addition to the leading ones) is essential for reliable prediction of the energies and oscillator strengths of the transitions.

In order to find out whether other isomers can contribute to the spectrum and whether a different cluster geometry gives rise to a substantially different spectroscopic pattern, vertical optically allowed transitions have been also determined for the deformed tetrahedral structure of C_{2v} symmetry, but very close to a D_{2d} shape, representing the best geometry for the triplet ground state. Its energy is 0.256 eV higher than the ground state energy of the rhombus. A single intense transition into the 5^3B_1 dominates the spectrum of the deformed tetrahedron (Fig, 7b). Its location does not correspond to any recorded intense transition of Na_4. This is an important finding for two reasons: it confirms that solely the rhombic structure is responsible for the absorption spectrum of Na_4, and it illustrates that the interplay between the nuclear framework and the electronic excitation can give rise to a single intense transition resembling a giant resonance also in a cluster as small as a tetramer.

A comparison of the recently recorded depletion spectrum of Li_4 [45] and the quantum chemical predictions for the location of transitions and their oscillator strengths for the optimal rhombic ground state structure permitted again the full assignment of the spectrum [33, 45]. The similarity between absorption spectra of Li_4 and Na_4 is striking.

In order to explain the depletion spectrum of Na_8 which is characterized by a single dominant intense transition located at 494 nm and of fine structure shifted to the red with low intensity [42, 43], the optically allowed excited states up to 3.0 eV of excitation energy, their oscillator strengths and lifetimes have been determined for three structures of T_d, D_{2d} and D_{4d} symmetry with energetically close lying ground states [35, 36]. However, only the highly symmetrical tetrahedral T_d form represents a local minimum on the SCF energy surface, while both the closely related D_{2d} structure (which is deformed section of the fcc

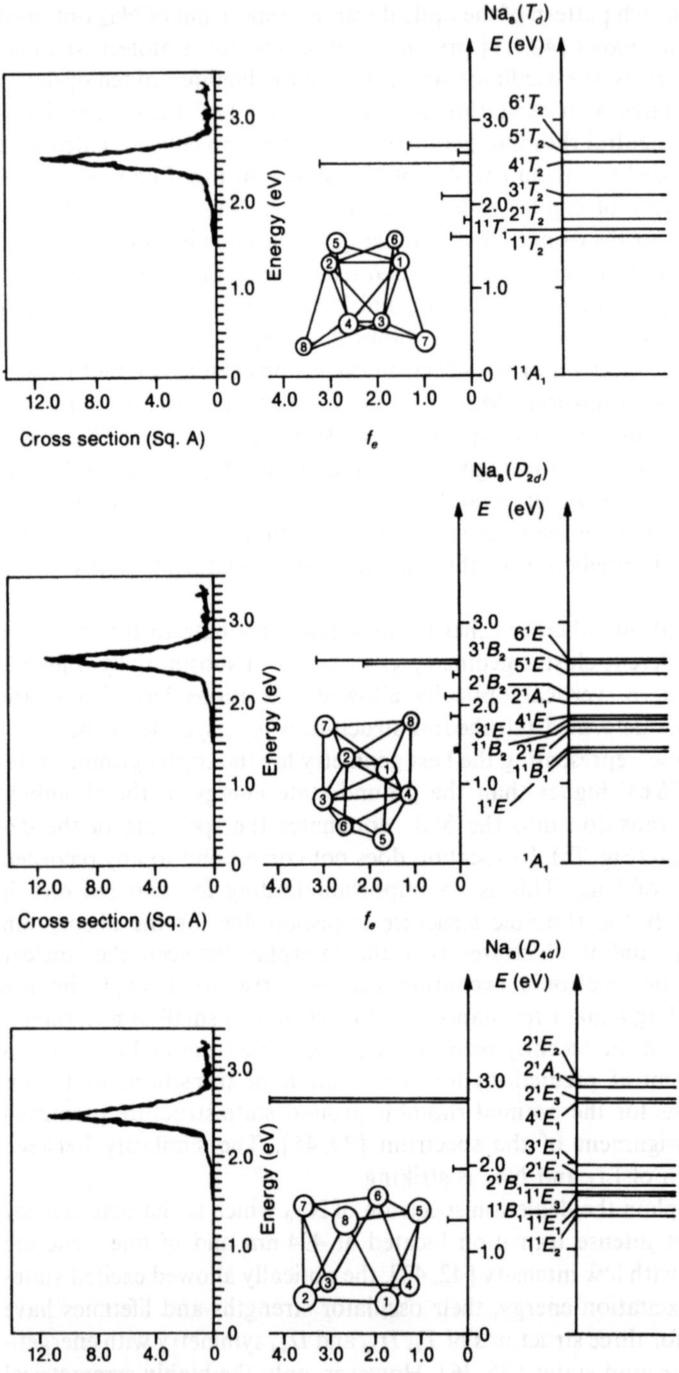

$Na_8(T_d)$
E (eV)
Energy (eV)
3.0
2.0
1.0
0
6^1T_2
5^1T_2
4^1T_2
3^1T_2
2^1T_2
1^1T_1
1^1T_2
1^1A_1
12.0 8.0 4.0 0
Cross section (Sq. A)
4.0 3.0 2.0 1.0 0
f_e
$Na_8(D_{2d})$
6^1E
3^1B_2
5^1E
2^1B_2
2^1A_1
4^1E
3^1E
1^1B_2
2^1E
1^1B_1
1^1E
$^1A_1$
$Na_8(D_{4d})$
2^1E_2
2^1A_1
2^1E_3
4^1E_1
3^1E_1
2^1E_1
2^1B_1
1^1E_3
1^1B_1
1^1E_1
1^1E_2

Fig. 8

lattice [50]) and the highly compact D_{4d} antiprism transform into the T_d geometry after relaxing the symmetry constraints without passing through barriers. A comparison between the measured absorption spectrum and theoretical predictions for all three T_d, D_{2d} and D_{4d} structures is presented in Fig. 8. For the T_d geometry, the transition to the 4^1T_2 state at 494 nm has a dominant oscillator strength corresponding to the maximum of the observed intense band. The CI wavefunction of the 4^1T_2 state is dominated by three configurations arising by single excitations from $1t_2$ to $3t_2$ and to $1e$ one-electron levels from the ground state configuration [35, 36]. This represents a many-electron effect which does not, however, necessarily coincides with a collective excitation in the continuous spectrum like that described by the classical or quantum mechanical description of a plasmon (cf. Section 2.7). One transition with appreciable intensities is slightly blue-shifted with respect to the dominant one and lies within the energy interval of the measured intense band. The two red-shifted weak transitions can be assigned to the fine structure recently observed [42, 43]. Both related structures T_d and D_{2d} give rise to a very similar spectroscopic pattern which is in complete agreement with the observed absorption spectrum of Na_8 [35, 36]. The calculated spectrum of the antiprism (D_{4d}) is entirely dominated by a single intense transition with a location slightly blue-shifted with respect to the measured one.

Again, the observed absorption spectrum of Li_8 and the quantum molecular assignment [46] show striking similarities to the described absorption features of Na_8: One dominant intense transition, additional energetically close lying blue shifted less intense transition and the fine structure shifted to the red have almost same locations in both cases. The appearance of relatively small number of intense transitions and a large number of weak ones is due to molecular interference phenomena. A small number of leading excitations (two–four) determine $\sim$ 80% of the wavefunctions of excited states for which large values of oscillator strength f_e have been obtained. Same configurations with different coefficients and signs enter the wavefunctions of excited states to which transitions with small f_e have been calculated. The molecular interpretation has been recently confirmed by femtosecond time resolved spectroscopy which yields two energetically close lying excited states located at $\sim$ 2.5 eV with different lifetimes 0.4–0.5 and 4.0–5.0 ps [51]. Moreover, if the appearance of a giant resonance is taken as the sign of collective type behavior due to plasmon excitation, one would not expect very similar absorption spectra for Li_8 and Na_8, since the polarizabilities of both clusters as well as of both bulks are very different. In a fact, an estimate of plasmon frequencies for Li_n clusters using static polarizabilities (cf. Section 2.1.2.7) yields considerably poorer agreement with experiment than in the case of Na_n clusters [11].

Fig. 8. A comparison between the depletion spectum of Na_8 (cf. Refs. [42, 43]) and the CI energies of optically allowed states and the oscillator strengths f_e for the T_d, D_{2d} and D_{4d} structures. The ECP-CPP-CI calculations, for details see Refs. [35, 36]

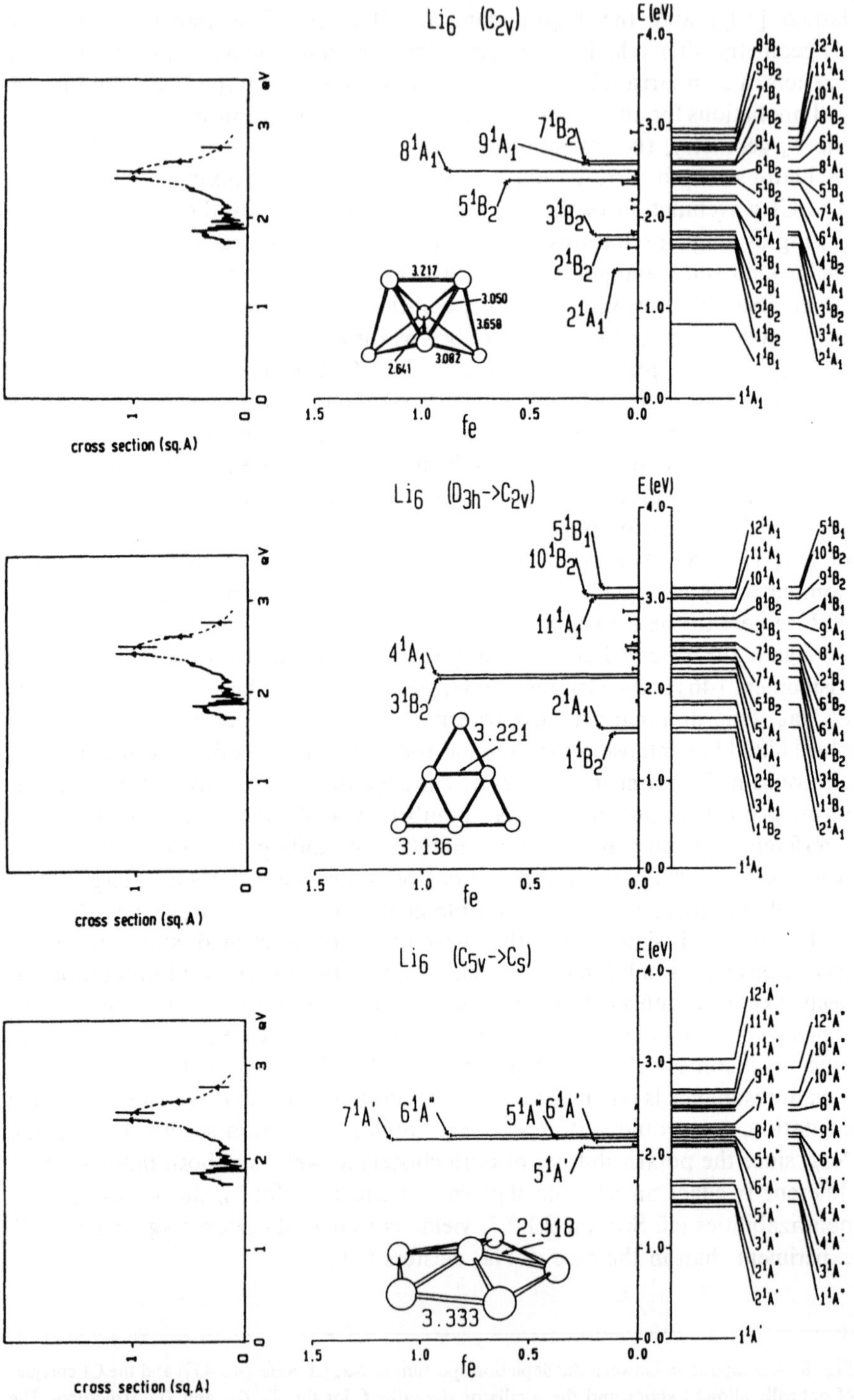
Li6 (C2v)
E(eV)
cross section (sq.A)
eV
fe
3.217
3.050
3.658
2.641
3.082
8¹A₁
9¹A₁
7¹B₂
5¹B₂
3¹B₂
2¹B₂
2¹A₁
1¹A₁
Li6 (D3h->C2v)
5¹B₁
10¹B₂
11¹A₁
4¹A₁
3¹B₂
2¹A₁
1¹B₂
3.221
3.136
Li6 (C5v->Cs)
7¹A'
6¹A"
5¹A"
6¹A'
5¹A'
1¹A'
2.918
3.333

Fig. 9(a)

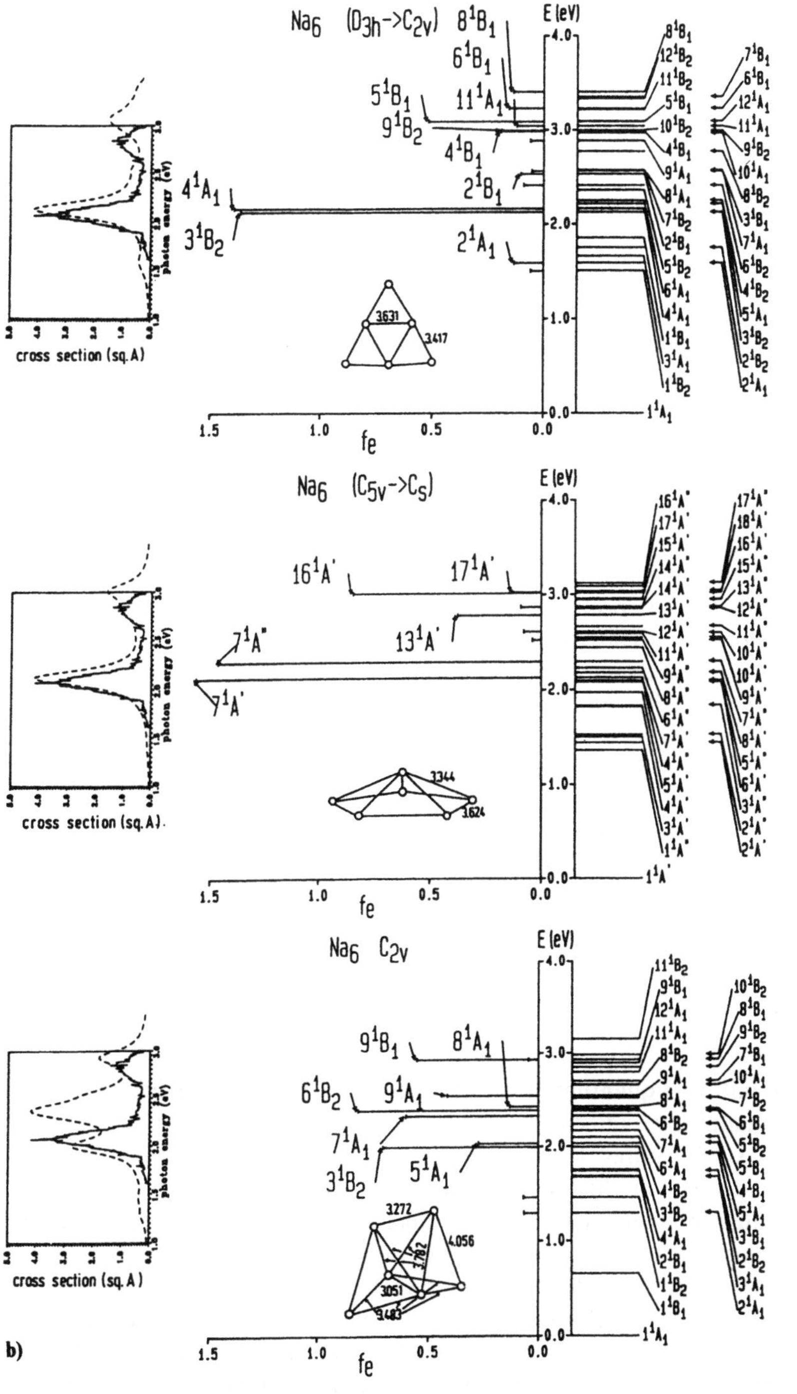

Fig. 9a, b. Comparison of predicted CI transition energies and oscillator strengths for three energetically close lying structures: planar (D_{3h}), flat pentagonal pyramid (C_{5v}) and $3D$-C_{2v} form of **a**) Li_6 [52] and **b**) Na_6 [54] with depletion spectra [52] and [53]

Transitions to the lowest optically allowed states have low oscillator strengths, while the transitions with dominant intensities occur to higher lying excited states with equivalent leading features for both Li_8 and Na_8 T_d and D_{2d} structures. This might be one of the reasons for the similar location ($\sim$ 500 nm) of the intense bands found for both Na_8 and Li_8 clusters.

The question can be raised at which cluster size the transition from two- to three-dimensional structures occur ($n > 6$) (cf. Fig. 1). A comparison of depletion spectra for Li_6 [52] and Na_6 [53] with predicted transition energies and oscillator strengths [52, 54] for all three energitically close lying structures shown on Figs. 9a and 9b demonstrates clearly that Na_6 is still two-dimensional or close to planarity, but Li_6 assumes already $3D$–C_{2v} structure built from the tetrahedral units. Moreover, although the experiments are not carried out at $T = 0$, the structural assignment based on quantum chemical predictions can be made. If all three structures of Li_6 or Na_6 would contribute with the same weights, the recorded spectra would have intense bands located at $\sim$ 2.2, 2.5 and $\sim$ 3.0. On the contrary, the intense transition for Na_6 have been recorded at $\sim$ 2.2 and a considerably weaker ones at $\sim$ 3.0 eV corresponding to the predictions for the most stable planar form of Na_6. The flat pentagonal pyramid of Na_6 yields similar spectroscopic pattern (as the planar one) and although it does not represent a Hartree–Fock local minimum it is difficult to rule out its contribution to spectrum, since the nature of the minima on very flat energy surfaces can be easily changed by inclusion of correlation effects. For Li_6 the agreement between predicted spectrum for the $3D$–C_{2v} structure (which is the most stable one) with the recorded spectrum is excellent and the contributions from the other two structures can be excluded. Figure 9 illustrates a fine structural difference between Li_6 and Na_6 which might arise due to a large ability of Li atom undergo *s*–*p* hybridization. For Na_n the transition to three-dimensional structures occurs for Na_7 [54].

2.1.5 Conclusions

1. The ab initio-CI quantum molecular theory is capable of predicting observables in quantitative agreement with experiment and of extracting the basic leading physical features. If simplified theories contain these features they can also yield adequate qualitative explanations of the studied phenomena.

2. Because of their simplicity the alkali metal clusters are particularly suitable objects for study of general cluster properties. The theoretical investigation shows that the character of the interatomic interaction remains overall the same for small and large clusters as well as for metals. This fact is clearly testified by very small changes in the interatomic distances in clusters of different sizes. The degree of electron promotion into *p*-type atomic orbitals increases with the number of nearest neighbors in clusters and can be considered as a sign of increasing degree of metallic character.

3. Special stability and the high symmetry of Ia clusters with 2, 8 and 20 valence electrons is a general consequence of the nodal properties of the relevant one-electron functions which determine their occupation numbers. The lowering of the symmetry of clusters with "incomplete electron shells" is interpreted in terms of the Jahn–Teller or pseudo Jahn–Teller effect. These rules are in general independent of the theoretical model used. Consequently, the jellium model does not represent a prerequisite for the validity of rules concerning shell closings.

4. Stabilities and other measurable quantities such as ionization potentials, electron affinities and polarizabilities are dependent on the geometry and the size of the cluster. They can be quantitatively determined and their behavior as functions of the cluster size can be qualitatively understood. Again, the applicability and limitations of simpler theoretical approaches can be explained from the point of view of a more general theory.

5. The interplay between molecular quantum theory and two types of spectrocopies recently available, depletion- and photodetachment-, makes possible a comparison of several calculated and measured properties. Moreover, a comparison between theory and experiment also supplies additional information which cannot be gained directly from experiment or theory alone. For instance, the ab initio CI calculations of ground states of Na_{2-5}^- and Na_{2-5} and the excited states of the latter account for the observed spectroscopic patterns, reproduce in a quantitative way the observed excitation energies and allow for an assignment of cluster geometries. The quantum molecular CI investigation of the Na_{4-8} and Li_4–Li_8 excited states yields the complete assignments of the experimentally found bands, i.e. quantitative predictions of their positions and right order of magnitude of their oscillator strengths for planar 2D geometries of Na_{4-6}, Li_{4-5} and 3D-structures of Li_{6-8} and Na_{6-8}. The agreement between experiment and theory is equally good for octamers characterized by one or two close lying intense bands as well as for tetramers and trimers which exhibit considerably richer spectroscopic patterns.

6. The qualitative discussion of the excited states wavefunctions for optically allowed states clearly shows that several single as well as double excitations determine the nature of the transitions which means that the excited states of clusters cannot be described in a one-electron picture by particle-hole excitations. Therefore, the quantum theory of plasmon is not sufficiently general. The nature of excitation in small clusters is just a many-electron effect which is adequately described by methods respecting electronic correlation and should not be confused with collective excitation represented by a surface plasmon (cf. Section 2.1.2.7).

7. The plasmon theory can yield a very rough estimate of the energy interval in which the absorption frequencies of intense bands are measured. There predictions might agree well with the observed spectra in some cases and not in others if the spectra do not exhibit a rich pattern. Therefore, it is not surprising that the presence of additional bands in small clusters due to quantum effects have been recently reported and that a larger deviation from the approximate

estimate of the location of intense transitions in depletion spectra or these obtained by femtosecond spectroscopy can occur.

Acknowledgement. This work has been supported by the Deutsche Forschungsgemeinschaft (Sfb 337, Energy transfer in molecular aggregates) and Consiglio Nazionale delle Ricerche (CNR).

References

1. a) V. Bonačič-Koutecký, P. Fantucci, J. Koutecký: Chem. Rev. and references therein, **91**, 1035 (1991); b) J. Koutecký, P. Fantucci: Chem. Rev. **86**, 539 (1986); c) A.W. Castleman Jr., R.G. Keesee: Chem. Rev. **86**, 589 (1986); d) see also: *Elemental and Molecular Clusters*, eds. G. Benedek, T.P. Partin, G. Pacchioni, Springer Series in Material Science, **6**, 1987, Springer-Verlag, Heidelberg and references therein; e) E. Schumacher: Chimia **42**, 357 (1988); f) M.M. Kappes: Chem. Rev. **88**, 369 (1988)
2. a) A. Szabo, N.S. Ostlund: *Modern Quantum Chemistry*, McMillan, New York, 1982; b) R. McWeeny, *Methods of Molecular Quantum Mechanics*, Academic Press, London, 1989; c) P.W. Atkins, Molecular Quantum Mechanics, Oxford University Press, Oxford, 1983
3. R.G. Parr, W. Yang: *Density-Functional Theory of Atoms and Molecules*, Oxford University Press, New York, 1989
4. L. Szasz: *Pseudopotential Theory of Atoms and Molecules*, John Wiley, New York, 1985
5. P. Fantucci, S. Polezzo, V. Bonačič-Koutecký, J. Koutecký: Z. Phys. D, Atoms, Molecules and Clusters **13**, 355 (1989)
6. R.J. Buenker, S.D. Peyerimhoff: Theor. Chim. Acta **35**, 33 (1974); R.J. Buenker, S.D. Peyrimhoff, W. Butscher: Mol. Phys. **35**, 771 (1978); R.J. Buenker, in *Current Aspects of Quantum Chemistry*, ed. by R.T. Carbo, Studies in Physical and Theoretical Chemistry, Vol. **21**, p. 17 (Elsevier, Amsterdam, 1982)
7. P.E. Siegbahn: J. Chem. Phys. **72**, 1647 (1980); V.R. Saunders, J.H. van Leuthe: Mol. Phys. **48**, 923 (1983)
8. P. Pulay: Mol. Phys. **17**, 197 (1969); R. Fletcher: Mol. Phys. **19**, 55 (1970); J.A. Pople, S. Krishnan, H.B. Schlegel, J.S. Brinkley: Int. J. Quant. Chem. **S 13**, 225 (1979)
9. W.D. Knight, K. Clemenger, W.A. de Heer, W.A. Saunders, M.Y. Chou, M.L. Cohen: Phys. Rev. Lett. **52**, 2141 (1984)
10. C. Kittel: *Introduction to Solid State Physics*, John Wiely, New York, 1967
11. a) W.A. de Heer, K. Selby, V. Kresin, J. Masui, M. Vollmer, A. Chatelain, W.D. Knight: Phys. Rev. Lett. **59**, 1805 (1987); b) K. Selby, M. Vollmer, J. Masui, V. Kresin, W.A. de Heer, W.D. Knight: Phys. Rev. **B 40**, 5417 (1989)
12. J. Koutecky, C. Scheuch: Int. Quant. Chem. **XXXVII**, 373 (1990)
13. O. Echt, K. Sattler, E. Recknagel: Phys. Rev. Lett. **47**, 1121 (1981)
14. M.M. Kappes, R.W. Kunz, E. Schumacher: Chem. Phys. Lett. **91**, 413 (1982)
15. M.M. Kappes, M. Schär, P. Radi, E. Schumacher: Chem. Phys. Lett. **119**, 11 (1985)
16. C. Brechignac, Ph. Cahuzac: Z. Phys. D **3**, 121 (1986)
17. I. Boustani, W. Pewestorf, P. Fantucci, V. Bonačič-Koutecký, J. Koutecký: Phys. Rev. **B 35**, 9437 (1987)
18. I. Boustani, J. Koutecký: J. Chem. Phys. **88**, 5657 (1988)
19. V. Bonačič-Koutecký, P. Fantucci, J. Koutecký: Phys. Rev. **B 37**, 4369 (1988)
20. V. Bonačič-Koutecký, I. Boustani, M.F. Guest, J. Koutecký: J. Chem. Phys. **89**, 4861 (1988)
21. E.C. Honea, M.L. Homer, J.L. Persson, R.L. Whetten: Chem. Phys. Lett. **171**, 147 (1990)
22. H.-O. Beckmann, J. Koutecký, V. Bonačič-Koutecký: J. Chem. Phys. **73**, 5182 (1980)

23. J. Koutecký, V. Bonačić-Koutecký, I. Boustani, P. Fantucci, W. Pewestorf: in *Large Finite Systems*, ed. by J. Jortner, A. Pullman, B. Pullman (Reidel, Dordrecht, 1987), p. 303
24. V. Bonačić-Koutecký, P. Fantucci, J. Koutecký: J. Chem. Phys. **91**, 3794 (1989)
25. C. Gatti, P. Fantucci, G. Pacchioni: Theoret. Chim. Acta **72**, 433 (1987)
26. M.M. Kappes, M. Schär, U. Röthlisberger, Ch. Yeretzlav, E. Schumacher: Chem. Phys. Left. **143**, 251 (1988)
27. D.G. Leopold, J. Ho, W.C. Lineberger: J. Chem. Phys. **86**, 1715 (1987)
28. H. Akeby, I. Panas, L.G.M. Petterson, P. Siegbahn, U. Wahlgreen: J. Chem. Phys. **94**, 5471 (1990)
29. a) C.W. Bauschlicher, S.R. Langhoff, H. Partridge: J. Chem. Phys. **93**, 8133 (1990); b) V. Bonačić-Koutecký, L. Češpiva, P. Fantucci, J. Koutecký: Z. phys. D **26**, 287 (1993)
30. P. Fantucci, V. Bonačić-Koutecký, S. Polezzo, M.F. Guest, J. Koutecký: to be published
31. W.D. Knight, K. Clemenger, W.A. de Heer, W.A. Saunders: Phys. Rev. B **31**, 2539 (1985)
32. C. Brechignac, Ph. Cahuzac, J.-Ph. Roux, D. Pavolini, F. Spiegelmann: J. Chem. Phys. **87**, 5694 (1987)
33. V. Bonačić-Koutecký, P. Fantucci, J. Koutecký: Chem. Phys. Lett. **146**, 518 (1988)
34. V. Bonačić-Koutecký, P. Fantucci, J. Koutecký: Chem. Phys. Lett. **166**, 32 (1990)
35. V. Bonačić-Koutecký, M.M. Kappes, P. Fantucci, J. Koutecký: Chem. Phys. Lett. **170**, 26 (1990)
36. V. Bonačić-Koutecký, P. Fantucci, J. Koutecký: J. Chem. Phys. **93**, 3802 (1990)
37. S.T. Arnold, J.G. Eaton, D. Patel-Misra, H.W. Sarkas, K.H. Bowen: in *Ion and Cluster Spectroscopy and Structure*, ed. by J. Maier (Elsevier, New York, 1989)
38. K.M. McHugh, J.G. Eaton, G.H. Lee, H.W. Sarkas, L.H. Kidder, J.T. Snodgrass, M.R. Manea, K.H. Bowen: J. Chem. Phys. **91**, 3792 (1989)
39. A. Herrmann, M. Hofmann, S. Leutwyler, E. Schumacher, L. Wöste: Chem. Phys. Lett. **62**, 216 (1979)
40. M. Broyer, G. Delacretaz, P. Labastie, J. Wolf, L. Wöste: Phys. Rev. Lett. **57**, 1851 (1986)
41. M. Broyer, G. Delacretaz, N. Guoquon, J. Wolf, L. Wöste: Chem. Phys. Lett. **145**, 239 (1988)
42. C.R.C. Wang, S. Pollack, M.M. Kappes: Chem. Phys. Lett. **166**, 26 (1990)
43. C.R.C. Wang, S. Pollack, D. Cameron, M.M. Kappes: J. Chem. Phys. **93**, 3787 (1990)
44. S. Pollack, R.C. Wang, M.M. Kappes: J. Chem. Phys. **94**, 2496 (1991)
45. M. Broyer, J. Chevaleyre, Ph. Dugourd, J.P. Wolf, L. Wöste: Phys. Rev. A, **42**, 6954 (1990)
46. J. Blanc, V. Bonačić-Koutecký, M. Broyer, J. Chevaleyre, Ph. Dugourd, J. Koutecký, C. Scheuch, J.P. Wolf, L. Wöste: J. Chem. Phys. **96**, 1793 (1992)
47. C. Brechignac, Ph. Cahuzac, F. Carlier, J. Leygnier: Chem. Phys. Lett. **164**, 433 (1989)
48. C. Brechignac: private communication
49. V. Bonačić-Koutecký, A. Blase, G. Perez-Bravo: to be published
50. J.L. Martins, J. Buttet, R. Car: Phys. Rev. B. **31**, 1804 (1985)
51. G. Gerber, private communication
52. P. Dugourd, J. Blanc, V. Bonačić-Koutecký, M. Broyer, J. Chevaleyre, J. Koutecký, J. Pittner, J.-P. Wolff, L. Wöste: Phys. Rev. Letters **67**, 2638 (1991)
53. C.P.C. Wang, S. Pollack, T.A. Dahlseid, G.M. Koretsky, M.M. Kappes: J. Chem. Phys. **96**, 7931 (1992)
54. V. Bonačić-Koutecký, J. Pittner, C. Scheuch, M.F. Guest, J. Koutecký: J. Chem. Phys. **96**, 7938 (1992)

2.2 Tight-Binding and Hückel Models of Molecular Clusters

D.A. Jelski, T.F. George, and *J.M. Vienneau*

2.2.1 Introduction

We now consider a simplified method for calculating electronic states of clusters. In the next chapter, local density calculations originally developed for the bulk metal are adapted for cluster calculations. This results in a "solid" in which surface effects are important. Similarly, Hartree–Fock methods are frequently used to calculate the structure of small molecules and clusters. In the current chapter, we shall not be concerned with solving the Schrödinger equation from first principles, but shall instead adopt a semiempirical approach. Apart from computational simplicity, the benefits of this approach are several-fold. First, surface effects are automatically taken care of. Secondly, the geometry of the cluster can be given explicitly, and there is no need to assume a spherical or model geometry. Finally, the electronic structure can be probed more accurately, since the electronic states are calculated explicitly.

The limitations of the present method must also be explored. In particular, the semiempirical parameters are usually calculated from the band structure in the bulk. There is no guarantee that this parametrization will continue to be valid for very different cluster systems. In any event, some modification of the method must be undertaken before these methods can be applied to clusters. Secondly, while some information about the electronic structure is inherent in any method which depends on the bulk band gap, how accurate that information is for systems different from the bulk and/or for states not near the band gap remains to be determined. Finally, one is limited to materials for which suitable parameters are known, and for which a variety of approximations are appropriate. Nevertheless, there are many systems for which quantum chemical techniques can yield valuable information, and the purpose of this chapter is to indicate that many of the above problems are soluble.

We should distinguish the present techniques from classical force field methods. These methods attempt to model the potential around an atom using a classical field, and from this to calculate stable structures and heats of formation from molecular dynamics calculations. This is very widely used in cluster research and yields much valuable information, but it is beyond the scope of this chapter. Since it is not a quantum mechanical method, little information about electronic structure can be obtained.

The most straightforward quantum technique is, of course, the ab initio Hartree–Fock calculation. In this case the Schrödinger equation is solved explicitly, and solutions of arbitrary accuracy can be obtained, depending on the size of the computer and the patience of the researcher. Most ab initio calculations are done for small molecules, and the amount of computation involved makes this method inappropriate for larger systems.

Short of ab initio techniques, there exists a large number of semiempirical methods. These begin with the principles of quantum mechanics, but then greatly simplify the calculation by introducing empirical parameters, which are chosen to produce as accurate a result as possible. In many cases the results are very accurate indeed, and it is not always fair to say that semiempirical methods are less exact than ab initio techniques. Semiempirical methods are widely used in studies of organic systems. Pharmaceutical firms make much use of these methods in searching for new drugs.

The methods we are about to describe are semiempirical of the simplest sort, involving only one or a few parameters. Their great advantage is computational and conceptual simplicity, and they are readily adaptable to calculations of large clusters. Specifically, in what follows we shall consider the tight-binding model and the Hückel model. Both of these are so-called nearest-neighbor models, in which the interaction between nearest-neighboring atomic sites is accounted for, and all other interactions are ignored.

We start with a short background of the theory underlying the two methods. We then continue with a description of how the methods have been used in cluster research, concentrating first on application of the tight-binding model to silicon clusters, and then on applications of the Hückel model to metallic clusters. Finally, we close with some comments on the relative merits of the methods, and on possible directions for future research.

2.2.2 Quantum Chemistry Background

In order to define notation, and also to ensure completeness, we review here some elementary quantum chemistry as will be used in what follows. The Hamiltonian for a molecule can be written as

$$H\psi_i = E_i\psi_i \ , \tag{1}$$

where ψ_i refers to a molecular electronic state with energy E_i, which are the quantities we want to determine. In principle, of course, H depends on all electron-nuclear and electron-electron interactions. But the tight-binding (TB) model assumes that the total Hamiltonian can be simplified into a series of one-electron Hamiltonians, i.e., that each electron feels an average field made up of the nuclei and all other electrons. This is the same approximation as made by the Hartree–Fock approach, but the TB method is not a self-consistent approach.

The second characteristic of the TB model is that local, atomic orbitals are used as a basis set. It is from this that the method takes its name, i.e., it is assumed that each electron is well localized around a given nucleus. The opposite standpoint would be to assume plane-wave Bloch functions for the basis set, which is the approach used for many metals. *Bullett* [1] has shown that by including *d*-orbitals, the TB basis set looks more and more like the plane-wave approach, and thus can be used for metals as well. Our discussion here, however, restricts attention to *s* and *p* orbitals, and hence the TB method described here is not applicable to systems such as metals where electrons are nearly free.

We can write the one-electron states as

$$H_\alpha \phi_\alpha = \varepsilon_\alpha \phi_\alpha \ , \tag{2}$$

where α refers to the one-electron, atomic Hamiltonian, energy and orbital. For an isolated atom, this is an exact expression. The ϕ_α constitute our basis set, and so we can express the molecular orbitals in terms of them as

$$\psi_i = \sum_\alpha c_{i\alpha} \phi_\alpha \ . \tag{3}$$

Operating on this with the molecular electronic Hamiltonian, H, we get

$$H\psi_i = H\left(\sum_\alpha c_{i\alpha}\phi_\alpha\right) = E_i \psi_i \ . \tag{4}$$

We can rewrite Eq. (4) using Dirac notation as

$$H|i\rangle = \sum_\alpha c_{i\alpha} H|\alpha\rangle = E_i|i\rangle \ . \tag{5}$$

From this, the energy E_i can be written as

$$E_i = \sum_{\alpha,\beta} c^*_{\beta i} c_{i\alpha} \langle\beta|H|\alpha\rangle = \sum_{\alpha,\beta} c^*_{\beta i} c_{i\alpha} V_{\beta\alpha} \ , \tag{6}$$

where Greek letters refer to atomic states, and where $V_{\alpha\beta}$ is the numerical value of the integral expressed by the brackets.

Equation (5) is a system of n equations with n unknowns, where n is the size of the basis set, in principle infinite. We can rewrite Eq. (5) in determinantal form

$$\det|\mathbf{H} - E_i \mathbf{S}| = 0 \ , \tag{7}$$

where **S** refers to the overlap matrix, and **H** refers to the Hamiltonian matrix given by Eq. (5). For an orthogonal basis set, **S** is simply a unit matrix. For a non-orthogonal basis set, **S** must be calculated explicitly. However, in our derivation we will assume that ϕ_α and ϕ_β are orthogonal, i.e., that $\langle\alpha|\beta\rangle = 0$ for $\alpha \neq \beta$, and thus $S_{\alpha,\beta} = \delta_{\alpha\beta}$. This is obviously true if the orbitals are on the same atom, since hydrogen-like orbitals are orthogonal. It is not true if the orbitals are on different atoms, since it is precisely the overlap which creates the bond.

But since the overlap is generally small, it can be included in the parametrization, and therefore ignored. Simplicity demands such a procedure, and it is the course we follow here. A more detailed account is found in Bullett's review article [1].

The second approximation is to assume that the molecular electronic Hamiltonian acts on the atomic orbital as

$$H|\alpha\rangle = H_\alpha|\alpha\rangle = \varepsilon_\alpha|\alpha\rangle \ . \tag{8}$$

This is the same as saying that the molecular electronic Hamiltonian for an isolated atom is the same as the atomic Hamiltonian from which the basis set is derived. A more mathematically sophisticated way of saying this is to use projection operators, as discussed by Bullett.

The final approximation is to note that

$$\langle\beta|H|\alpha\rangle = V_{\beta\alpha} = 0 \quad \text{for } \alpha, \beta \text{ on same atom, or on distant atoms}$$
$$\neq 0 \quad \text{for } \alpha, \beta \text{ on nearest-neighbor atoms} \ . \tag{9}$$

This is how the interaction between atoms in the molecule is taken into account. In general, of course, $V_{\beta\alpha}$ has to depend on the interatomic distance, and the TB model does that. In the present case, $V_{\alpha\beta}$ is calculated for a fixed interatomic distance, and then is allowed to vary as $1/r^2$.

The parameters for silicon are well known: $V_{ss\sigma} = -1.938$ eV, $V_{sp\sigma} = 1.745$ eV, $V_{pp\sigma} = 3.050$ eV and $V_{pp\pi} = -1.075$ eV [2]. These have been calculated to fit the band structure of the bulk material, and have no other immediate physical significance other than that they work. At large distances, the decay is much faster than $1/r^2$, and therefore a cutoff distance is chosen after which the matrix elements are set to zero, and no bond is said to exist. In addition, we include the diagonal terms of the matrix in Eq. (7), which are adjusted so that an isolated atom in the ground state is at zero energy. These are $E_s = -5.25$ eV and $E_p = 1.20$ eV.

Equation (7) can then be solved. Each bond is projected along the x, y and z coordinates, and then weighted according to the interatomic distance. The matrix is then diagonalized. The eigenvalues correspond to the energy levels, and the eigenvectors are the coefficients, $c_{i\alpha}$. The value $c_{i\alpha}^* c_{\alpha i}$ corresponds to the contribution of atomic orbital i to molecular orbital α, or, the probability of an electron in molecular orbital α being in atomic orbital i. In the model which we have presented, the matrix to be diagonalized is real and symmetric, and therefore all eigenvalues and eigenvectors are real.

Similarly, the sum $\sum_\alpha c_{i\alpha}^* c_{\alpha i} n_\alpha$ is the total charge density in atomic orbital i, where n_α is the occupation number of molecular orbital α. In the case of s and p orbitals, there are four orbitals per atomic site, and thus the total charge on a given atom (labelled k) as a result of an electron in molecular orbital α is

$$q_k = \sum_{i \text{ on } k} \sum_\alpha n_\alpha c_{i\alpha}^* c_{\alpha i} \ . \tag{10}$$

The evaluation of Eq. (10) is known as a Mulliken population analysis, and from it one can gather information not only about the energy of a cluster, but also about the charge distribution. Such data must be handled with care, however, as will be illustrated in the next section.

The Hückel model differs from the general TB model in two respects [3]. First, instead of including *s* and *p* orbitals in the basis set, it only includes *s* orbitals. Hence the assumption is that there is only one valence electron per atom. This means that there are only two parameters, the diagonal terms, known as the Coulomb integral, usually abbreviated as α, and the off-diagonal terms, corresponding to the V_{ss} interaction between nearest neighbors, which is denoted by β and termed the resonance integral. Since the s orbital is spherically symmetric, there is no angle dependence in the model. Further, in the simplest case, there is no length dependence on β, and so the result depends only on topology. However, it is a relatively simple matter to include a distance-dependent term, in which case the Hückel model is extended and more accurate.

2.2.3 Application to Clusters

The parameters given above are fitted to solid silicon in a diamond lattice structure [2]. They are chosen to match the band structure for that system. The self-energy terms, E_s and E_p, are chosen so that the zero-point energy is the Fermi level of the solid. It is not entirely clear that the above model holds for clusters, and in any event, some additional features must be considered.

In the first case, the bulk parameters are chosen for the equilibrium geometry, i.e., a fixed system. This is easy since all bonds and atoms in the lattice are identical. But for clusters, it is necessary to accommodate varying bond lengths. For this, some reasonable potential must be included. *Tománek* and *Schlüter* [4] have done this by using an analytic expression for the silicon dimer, matched to ab initio results. This is used to develop a repulsive potential curve. The attractive part of the potential is fitted by multiplying the TB parameters by $1/r^2$, while continuing to insist that the band structure of the bulk comes out properly. Thus the repulsive part of the potential is assumed to be a pair-wise classical potential, whereas the attractive part comes from the band structure calculation.

The second problem in using the model for clusters comes from the varying coordination numbers. In the bulk, all atoms are tetrahedrally coordinated, and hence it is not necessary to account for energy differences arising from this source. But for a cluster, this can yield important variations. Thus it is necessary to include a parametrization which depends on the ratio between bonds and atoms. Tománek and Schlüter have done this by fitting a quadratic equation to the dimer and two bulk structures, diamond and fcc. This equation is

$$E_{CN} = n[\chi_1(n_b/n)^2 + \chi_2(n_b/n) + \chi_3] , \tag{11}$$

where n refers to the total number of atoms, and n_b to the total number of bonds. The χ_i's are empirically fitted.

The final problem in applying the TB model to clusters concerns charge separation. For the solid lattice this difficulty does not arise (at least for a homonuclear species like silicon), but for clusters, it is necessary to account for the fact that there may be a dipole moment. Thus Coulomb forces must be accounted for explicitly. Tománek and Schlüter have done this by including $U(q_k^2 - q_0^2)$ in the Hamiltonian, where q_k is calculated from Eq. (10), and q_0 is the number of valence electrons on a neutral silicon atom. The constant U is to be empirically determined, but in the absence of data with which to fit it, it is chosen as unity.

Thus the complete Hamiltonian to be applied to clusters is [4]

$$E_{\mathrm{coh}} = \sum_i n_i E_i - \sum_\alpha n_\alpha E_\alpha + \sum_{k \neq j}\sum E_{\mathrm{repulsive}}(j, k) + E_{\mathrm{CN}} + \sum_k U(q_k^2 - q_0^2) \, , \quad (12)$$

where the first term on the rhs is the band energy. The second term sets the zero energy to an isolated atom in the ground state, so that the resulting energy is the cohesion energy of the cluster, E_{coh}. E_{CN} is the bond-number-dependent term from Eq. (11), and the last term is the Coulomb term. The above Hamiltonian is fitted to match the bulk and the dimer exactly, and it is hoped that it will apply to situations in between. Whether or not it does is the subject of the next section.

2.2.4 TB Model Applied to Silicon Clusters

The test of the TB model when applied to silicon clusters ultimately rests on comparison with experiment. Experimental data is more difficult to come by, and so some of our results remain speculative. Nevertheless, there is a sufficiently good match to indicate that the TB model is valid over a range of structures. *Tománek* and *Schlüter* [4], having developed the model described above, were the first to apply it to small silicon clusters, where they compared their results with a local-density approximation calculation. While we refer the reader to the original article for the details, it may safely be said that the results largely agree, at least in terms of the cohesion energy. This constitutes the first evidence that the TB model may be useful. Tománek and Schlüter performed a calculation on two structures of Si_{10}, a tetracapped octahedron (TO, shown in Fig. 1) and an adamantane-like structure, which is a bulk fragment. The TO isomer was found to have a cohesion energy of -3.6 eV/atom, whereas the adamantane form was found to be unstable.

We have performed an extensive study of Si_{10} isomers [5], comparing not only energies, but also whatever other experimental parameters can be culled from the literature. The various structures we found are shown in Fig. 1. We review these results with an eye toward establishing the benefits of the TB model. The most stable structure is a distorted form of a bicapped tetragonal

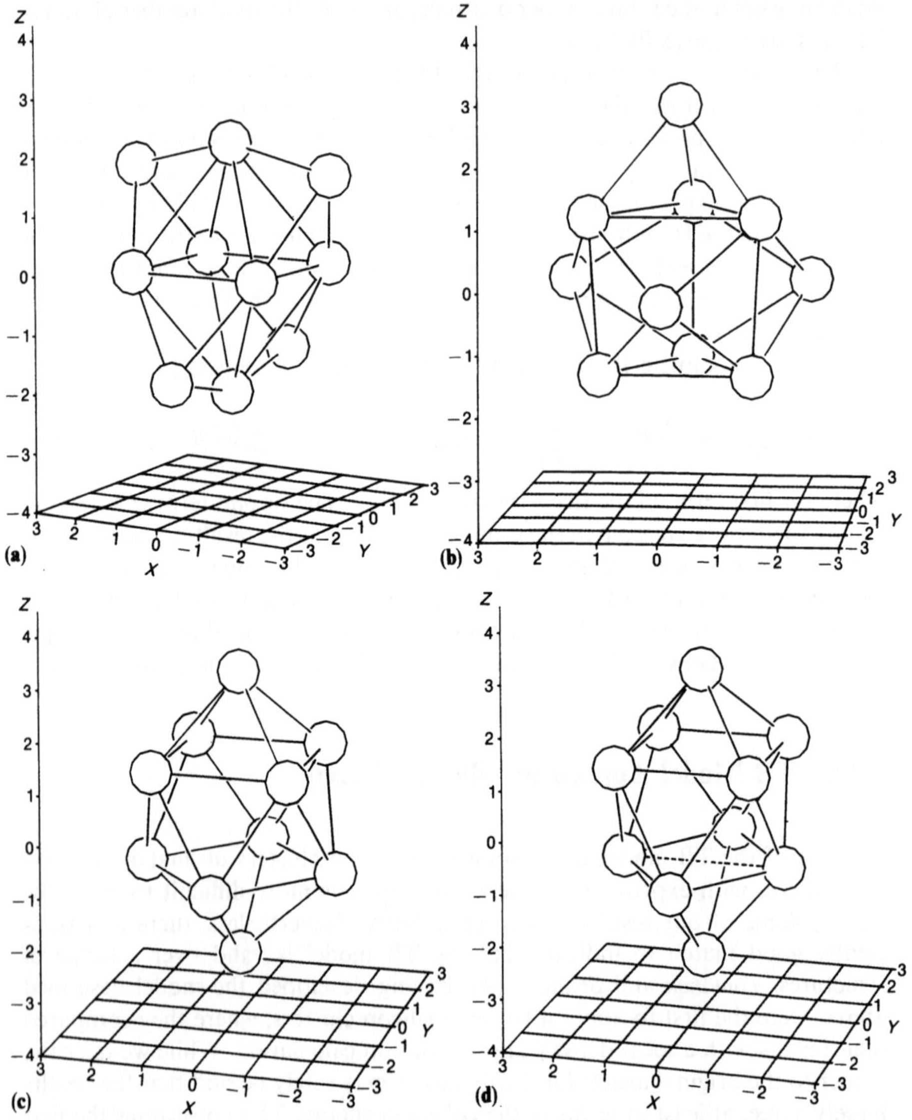

Fig. 1. Structures of various Si_{10} isomers: (**a**) tetracapped octahedron (TO); (**b**) tetracapped trigonal prism (TTP); (**c**) bicapped tetragonal antiprism (BTA); (**d**) distorted BTA-I (DBTA-I), with the possible 25th bond indicated by the dashed line. The unit on each axis is Å

antiprism (DBTA-I). The distortion results from Jahn–Teller effects since the ground state of undistorted BTA is degenerate. As shown in Table 1, the cohesion energy of DBTA-I is -3.98 eV/atom. Equally important, the HOMO-LUMO gap is 1.4 eV. This matches very closely to the experimental value of about 1.2 eV [6].

This result for DBTA dramatically illustrates the advantages of the TB model. In order to derive a distorted structure, global optimization is necessary. Indeed, we find another metastable structure (DBTA-II) which is also a Jahn–Teller distortion of BTA. In this instance, the HOMO-LUMO gap is 0.6 eV. An ab initio calculation can do a global optimization only with great difficulty, and classical force field approaches cannot deal with Jahn–Teller effects. Thus semiempirical techniques are best from this point of view. Ideally, however, the result can be confirmed by an ab initio calculation.

Closer consideration of DBTA-I yields more information about both the TB model and the nature of silicon clusters. Figure 2 contains information about bond lengths for various isomers. A difficulty with the TB model is that it depends on the number of bonds through Eqs. (9) and (11). This problem is usually solved by choosing a cutoff value, i.e., two atoms are considered bonded when separated by less than the cutoff value, and unbonded when further apart. This is another parameter put into the calculation, and is chosen to best match experimental evidence. Tománek and Schlüter have chosen the cutoff value to be 0.255 nm, which is the average of nearest-neighbor and next-nearest-neighbor distances in the bulk. This value is fine as long as coordination numbers and structures are similar to those in the bulk, but it fails for clusters with coordination numbers larger than 4 or 5.

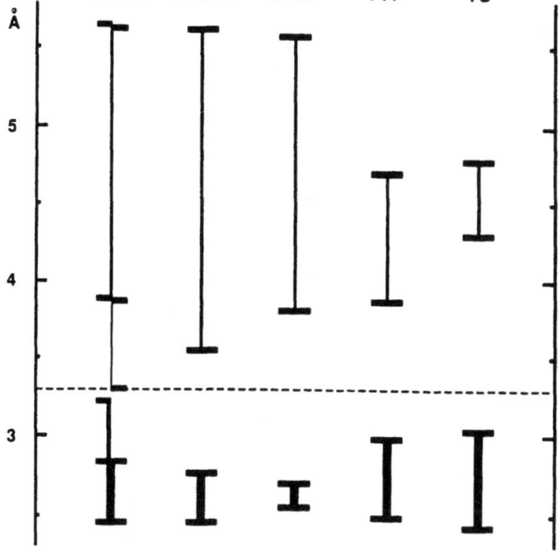

Fig. 2. Range of bond lengths for different isomers. The cutoff parameter chosen (0.33 nm) is shown by the dashed line. The thick bars below illustrate the range of bond lengths in the cluster. The thin bars above indicate the range of non-bonded distances, with the upper limit corresponding to the maximum diameter of the cluster. For DBTA-I two possibilities are indicated. On the left, the 25-bond form is shown, whereas on the right, the 24-bond form is shown. The unit on each axis is Å

The importance of the cutoff value is shown in Fig. 2. This value is illustrated by the dashed line, and as can be seen, for most clusters a choice between 0.3 nm to 0.35 nm makes no difference and is therefore arbitrary. For DBTA-I, however, there are two atoms which are approximately 0.33 nm apart, and this provides a test for the appropriate cutoff distance. We have performed the calculation for DBTA-I under two situations, with the 24-bond version of the isomer and also with the 25-bond version. The results are given in Table 1, and it can be seen that the 25-bond version more closely matches the experimental parameters. In fact, a partial bond is probably a better way of dealing with the situation, and in principle one could use some smooth curve to delineate bonded from unbonded atoms, rather than a cutoff value. Nevertheless, we now have direct evidence for appropriate bond lengths in clusters, and we are thus also able to extend the applicability of the model to larger coordination numbers.

The results in Table 1 are consistent with those calculated elsewhere. A classical force field model yields BTA as the most stable isomer. A calculation by Parinello and co-workers suggests a tie between the TO and TTP isomers [7], and ab initio results vary between TTP and TO [8–10]. None of these calculations include Jahn–Teller effects, and none of them reproduce the experimentally observed band gap. BTA has a degenerate ground state, and TTP and TO have band gaps more than twice that observed. Further, if one assumes that the activation barrier separating the BTA related forms is very small, then two isomers exist in equilibrium, these being DBTA and TTP, with DBTA as the most abundant species. Experimental evidence indicates that at least two isomers of Si_{10} exist with 85% and 15% population distributions [11]. This data is entirely consistent with our result.

We thus conclude that the TB model is a fairly accurate way of dealing with silicon clusters. The original parameters were chosen by Tománek and Schlüter, and we have further investigated the problem of the cutoff distance. We find that a good physical argument can be made for choosing 0.33 nm. Our relative energies are consistent with those of other workers, and our data match the PES

Table 1. Cohesion energies, HOMO-LUMO transition energies and LUMO energies of various Si_{10} structures, shown in Fig. 1. The LUMO one-electron energy, which is a rough indication of the electron affinity, is given relative to the "HOMO" level of the bulk

Species	Cohesion energy (eV/atom)	HOMO-LUMO gap (eV)	LUMO level (eV)
DBTA-I[a]	−3.98 (−3.92)	1.4 (0.9)	+0.18 (+0.13)
DBTA-II	−3.92	0.6	−0.50
BTA	−3.90	0.0	−0.92[b]
TTP	−3.91	2.6	+1.21
TO	−3.61	2.9	+2.00

[a] The values for the 24-bond structure are given in parentheses

[b] Since the ground state is degenerate, this is also the HOMO level

spectra and the thermal distribution data. We thus conclude that the TB model is probably useful for investigating larger clusters.

The first larger system we have investigated is silicon cluster cations ranging from 30 to 45 atoms [12]. This was inspired by experimental evidence indicating a dramatic difference in chemical reactivity for different sized clusters [13]. Ammonia and methanol are found to react quickly with clusters containing 30, 36, 43–44 and 46 atoms, and are found to react only slowly with clusters containing 33, 39 and 45 atoms. There are two orders of magnitude separating the reactivity of Si_{36}^+ from Si_{39}^+. Conversely, there is no significant variation found in reactivity with oxygen or nitrogen oxide.

Several salient points can now be mentioned. First, the periodicity in reactivity is roughly six atoms. This seems to indicate a six-membered ring as a basic unit. Secondly, the reactivity with strong free radicals (such as O_2 and NO) is non-periodic. This seems to indicate that the number of dangling bonds is not the determining factor, since free radicals would be expected to react with dangling bonds. Thus there must be some other feature of the structure which determines the reaction rate. We have used the TB model to investigate this more closely. Since this work was done prior to the work on Si_{10}, the cutoff distance used for this study was 0.255 nm, taken from Tománek and Schlüter.

Six-membered rings of silicon atoms are stacked as shown in Fig. 3. When the number of atoms is not divisible by six, the remaining atoms are arranged as a cap at one end of the cylinder. In the first instance, we are interested in determining the stability and geometry of the structures. We find that all such structures, from $n = 30$ to 45 are metastable. To our surprise, we also find that the rings are quite flat. We also find that while bending increases the stability of

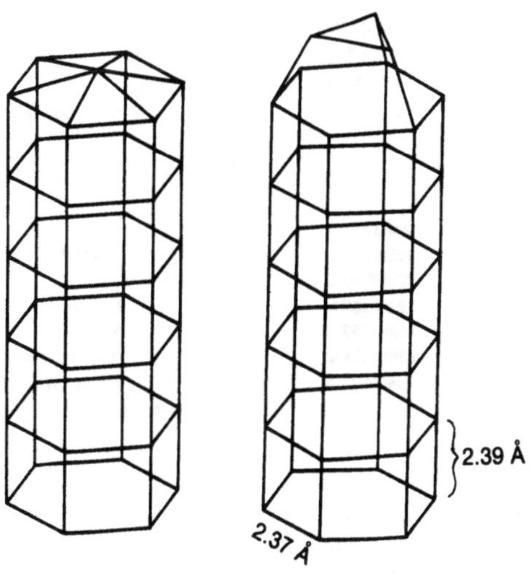

Fig. 3. Geometry of two silicon clusters, Si_{37}^+ and Si_{39}^+, with a silicon atom at each vertex

the ring, it substantially weakens the bonds between the rings, thereby raising the energy of the cluster. The precise geometry of the clusters is given in the original reference.

The problem is to account for the difference in reactivity. Figure 4 depicts the charge distribution as a function of cluster size. Given the large variation in reactivity, we require a large qualitative difference in some property. It is observed that a three-atom cap has a negative total charge, whereas caps of other sizes tend to be positive. Exceptions are for $n = 28$, 29, 40 and 41, which correspond to four or five atom caps. These are also slightly negative, but are

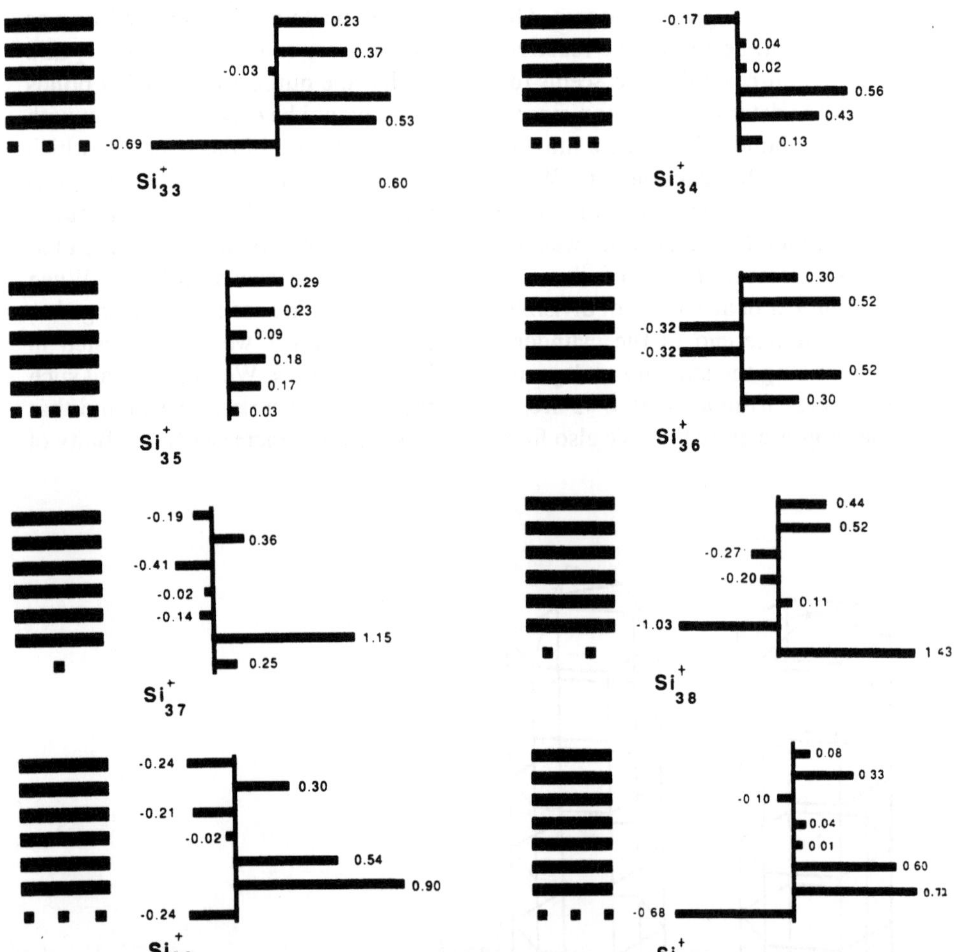

Fig. 4. Total electronic charge per layer for different silicon clusters. The diagrams on the left of each figure represent a schema of the cluster geometry, with the solid bars denoting a six-membered ring, and the individual atoms of the cap are shown explicitly. The bar graphs on the right show the total charge per layer, i.e., the sum of the charges of each atom constituting a single layer

not especially reactive. The Si_{39}^+ cluster is the least reactive, and it is the only species to be negatively charged at both ends.

Thus the charge distribution appears to vary qualitatively as a function of cluster size. If ammonia and methanol are thought of as nucleophiles, then their reactivity with a positively-charged cap is straightforward. Similarly, the reactivity with a negatively charged cap is expected to be greatly diminished. A three-atom cap tends to draw electrons toward it, and thus tends to be negatively charged. This can be considered a resonance effect. Similarly, since the variation in dangling bonds is not qualitatively different as a function of *n*, it follows that free radicals react indiscriminately.

It is interesting to note that these experiments were done for the cation, which implies that a local negative charge is less likely. The prediction, therefore, is that the reactivity pattern for the neutral or anionic species will be much less pronounced since four and five atom caps would draw significant electron density.

The second project is to investigate the structure of Si_{60}^+. Again, the impetus is a remarkably simple experimental result, notably the photofragmentation data [14]. It is found for Si_{60}^+, as well as for a wide range of other sizes of silicon and germanium clusters, that the photofragmentation pattern yields predominantly Si_{10}^+. The experimenters point out that the charge on the original cluster is most likely to migrate to the largest fragment, and therefore the complete absence of fragments larger than 11 or 12 atoms indicates that the Si_{60}^+ ion explodes rather than just breaking into two or several pieces.

There are two approaches which can be used here. One is to suggest that Si_{10}^+ is more stable than other small clusters, and hence ten atoms is the preferred size. The difficulty is that this does not appear to be the case, for it appears that cohesion energy rises as a function of size (for small clusters), and further, that Si_{13} is considerably more stable than the Si_{10} fragment. The second possibility is that silicon clusters fragment into ten-atom pieces by virtue of the construction of the parent species. If this is true, then it imposes a significant condition on the structure of clusters larger than twenty atoms.

We choose to investigate the second possibility by studying the properties of an Si_{60} cluster arranged as shown in Fig. 5, and which we describe as stacked, Si-

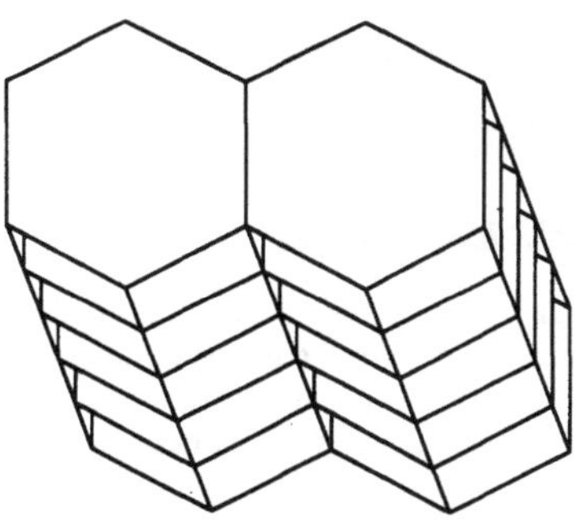

Fig. 5. The proposed structure for the Si_{60} cluster, consisting of six, stacked, ten-atom naphthalene-like rings

naphthalene rings. This is the next logical step beyond stacked six-membered rings. The purpose of this exercise is to see if this constitutes a stable structure, to determine a logical fragmentation pattern, and to estimate the degree of conjugation around the rings. If fragmentation occurs between the planes, then a ten-atom fragment is a likely product.

A larger model can now be considered. For small clusters, $n < 20$, the structure is molecular in form, i.e., close-packed with large coordination numbers. Very large clusters assume a bulk structure. The difficulty arises for intermediate forms. The molecular framework does not work, since then a close-packed solid would result. Similarly, a large surface area precludes a straightforward application of the bulk model. Thus the arrangement in Figs. 3 and 5 seems reasonable. Most atoms are tetra-coordinated, the surface area is minimized, and reactivity and fragmentation patterns can be accommodated. Other models have also been proposed which solve many of the same difficulties [15, 16]. The experimental art does not yet permit resolution between these conflicting theories.

The cohesion energv of the Si_{60} cluster in Fig. 5 is -3.6 eV. This compares to a bulk value of -4.4 eV, and the value for Si_{10} of -4.0 eV. This trend is consistent with the above hypothesis, since both a tightly-packed small cluster or a stable solid would be more stable than the intermediate system. The charge distribution is given in Fig. 6, which indicates that negative charge is concentrated in the center of the molecule.

Of primary interest are the bond strengths of the cluster, shown in Fig. 7. This is calculated using values for overlap integrals given by Mulliken. It can be seen that the weakest bonds are between the planes second from the ends. Thus

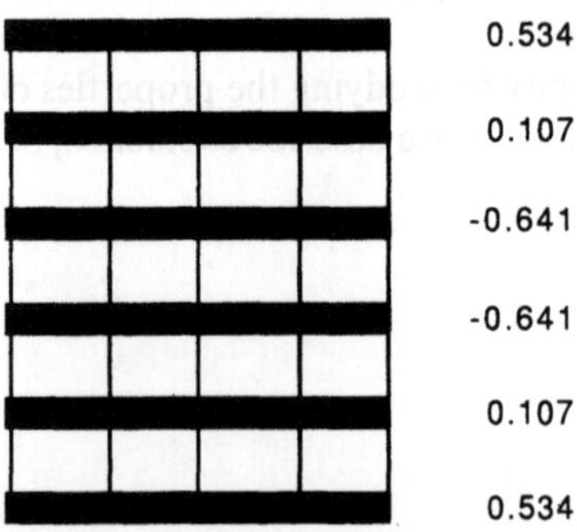

Fig. 6. The charge density distribution in Si_{60}. Each line represents one naphthalene plane, and the number is total change on that plane, in units of electronic charge

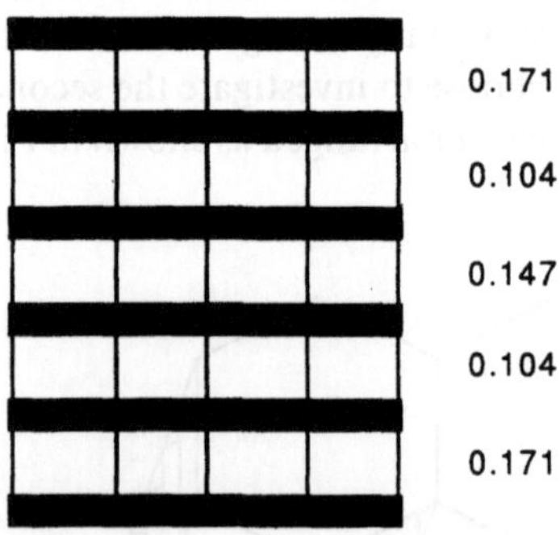

Fig. 7. The average bond strength of the bonds between each of the planes in the proposed Si_{60} cluster. Each number is an average over ten bonds. Numbers are unitless

if the cluster were to fragment, it would most likely fragment along these lines, leaving three, twenty-atom fragments. The Si_{20} fragment has a cohesion energy of -3.2 eV and is probably unstable with respect to dissociation to Si_{10}. Hence the appearance of ten-atom daughter fragments follows.

To conclude this section, we review what we have tried to do. We have used the tight-binding model to determine properties of silicon clusters. The advantages of this are that it is simple enough to allow complete geometric optimization, while still providing data about electronic states. Strong evidence for the suitability of the model is provided by the results for Si_{10}, and also by the work of Tománek and Schlüter on smaller clusters. This model has subsequently been applied to larger clusters with an eye toward explaining the reactivity and photofragmentation data.

2.2.5 Applications of the Hückel Model

We very briefly review some work using the Hückel model [3]. A Hückel calculation is valid primarily for atoms with a spherical valence shell, and thus has been used mainly for alkali metals. It was originally developed, however, to describe conjugated carbon systems, in which π electrons are considered to be valence electrons, and σ electrons are considered to be inner-shell. This approach is very primitive and is valid provided the angular dependence of the Hamiltonian can be neglected. In the simplest case, not even the radial dependence is included, though Salem indicates simply how that can be accomplished.

Our laboratory has used the Hückel model to calculate stable structures of sodium clusters [17, 18]. This simple model was used so as to permit a very rapid calculation of a large number of different isomers, necessary because there are as many as 11-million ways of constructing Na_{10} [17]. Graph theory was used in analyzing this problem. This is possible because the Hückel model depends only on the adjacency matrix, and hence a molecule can be represented topologically. A scheme was developed to eliminate most isomorphic graphs. However, since graph theory does not accommodate bond lengths, a geometric constraint reduces the number of possible isomers dramatically. Thus the energies of all topologically and geometrically allowed structures were calculated using the Hückel method, both for the neutral and cationic species. This yields stable structures, a representative sample of which are shown in Fig. 8.

Some interesting data about the transition from molecule to solid now emerge. The energy spectrum can be calculated using the Hückel model [18]. It is found that this falls into certain bands which can be described by the usual *s*, *p*, and *d* notation, depending upon the orbital angular momenta. The band energies can be modelled by a least squares fit to the function

$$E_l(n) = A_l + B_l(n^{1/3} + 1)^{-2} , \tag{13}$$

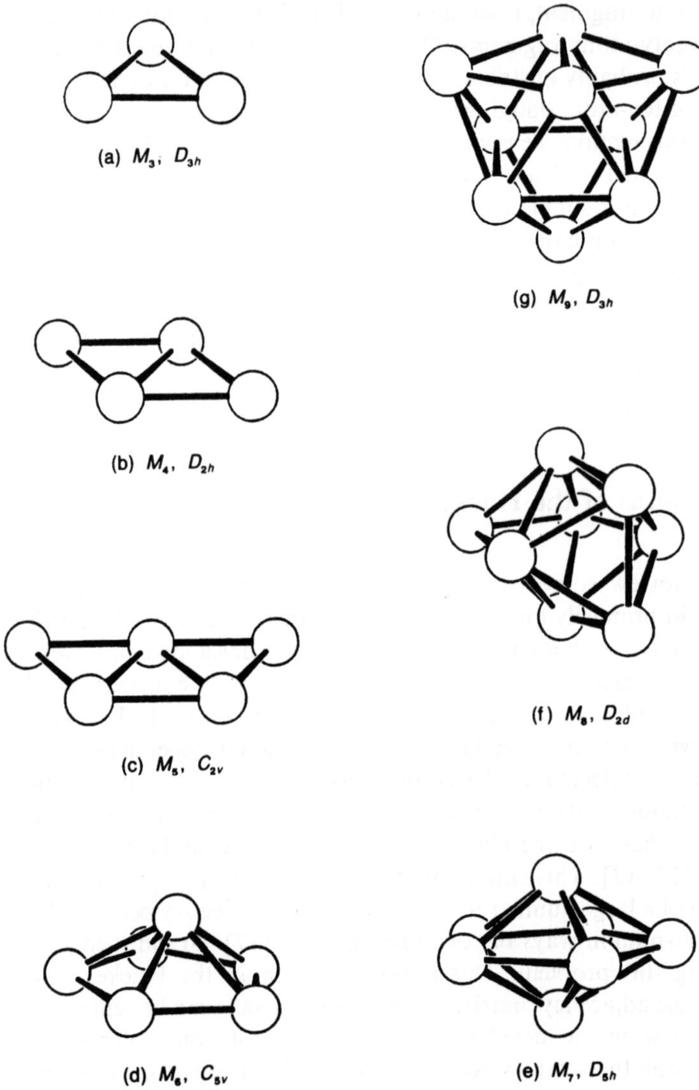

Fig. 8. Most stable structures for neutral metal clusters up to nine atoms

where n is the number of atoms in the cluster, and A_l and B_l are determined from the curve fit. The subscript l distinguishes between orbitals with s, p or d symmetry. This curve matches the energy spectrum very well. However, there are two breaks where the energy of the bands is significantly different. These occur at $n = 7$ and $n = 12$. At $n = 7$ it is observed that the cluster structures change from two-dimensional to three-dimensional. At $n = 12$ it is observed that there are body atoms inside the cluster that appear to be bulk-like. Thus the breaks in the band structure can be directly correlated to the size of the cluster, and it appears that clusters with 12 or more atoms will behave more like the bulk metal.

2.2.6 Conclusions

The last paragraph is a good example of both the advantages and disadvantages of the tight-binding and Hückel model approaches. On the plus side is computational simplicity. This allows the rapid determination of a large number of structures, and also the calculation of general trends in electronic properties. This is also true with silicon clusters, where full geometric optimization is found to be important. State-of-the-art ab initio calculations are as yet unable to accomplish this.

Interest in clusters has been renewed because of their optical properties. Here data about excited states is crucial, and the tight-binding method may be able to provide this. By definition, the TB model is fitted to the band gap in the bulk, and therefore should be expected to yield accurate excited state data near the band gap. Far from the band gap is a different story, and there the accuracy of the method remains to be tested. Nevertheless, the TB method provides an excellent method of estimating optical parameters with relatively little work, and incorporating full geometric optimization.

A second virtue of the TB method is that it is readily applicable to solid state chemistry. In a sense, a cluster can be viewed as a defect in a solid. There has recently been some interesting work on lattice defects using variations on the TB model. This becomes especially interesting since it is supposed that embedded clusters will have special optical properties. The nature of these embedding effects is something that can conceivably be investigated.

Acknowledgments. Acknowledgment is made by DAJ to the donors of the Petroleum Research Fund, administered by the American Chemical Society, for partial support of this research. DAJ also acknowledges that this research was partially supported by a grant from the Research Corporation of the State University of New York, and he thanks the SUNY Research Foundation for an Equipment Matching Grant. TFG acknowledges support by the Office of Naval Research and the National Science Foundation under Grant CHE-8620274.

References

1. D.W. Bullett: in *Solid State Physics*, Vol. **35**, ed. by H. Ehrenreich, F. Seitz, D. Turnbull (Academic, New York, 1980), pp. 129–214
2. D.J. Chadi: Phys. Rev. B **29**, 785 (1984)
3. L. Salem: *Molecular Orbital Theory of Conjugated Systems* (W.A. Benjamin, New York, 1966)

4. D. Tománek, M.A. Schlüter: Phys. Rev. B **36**, 1208 (1987). See also P.B. Allen, J.Q. Broughton, A.K. McMahan: Phys. Rev. B **34**, 859 (1986); F.S. Khan, J.Q. Broughton: Phys. Rev. B **39**, 3688 (1989)
5. T.T. Rantala, D.A. Jelski, T.F. George: J. Cluster Sci, submitted
6. O. Cheshnovsky, S.H. Yang, C.L. Pettiette, M.J. Craycraft, Y. Liu, R.E. Smalley: Chem. Phys. Lett. **138**, 119 (1987)
7. P. Ballone, W. Andreoni, R. Car, M. Parrinello: Phys. Rev. Lett. **60**, 271 (1988)
8. K. Raghavachari, V. Logovinsky: Phys. Rev. Lett. **55**, 2853 (1985)
9. K. Raghavachari: J. Chem. Phys. **83**, 3520 (1985); **84**, 5672 (1986)
10. K. Raghavachari, C.M. Rohlfing: J. Chem. Phys. **89**, 2219 (1988)
11. M.F. Jarrold, J.E. Bower, K. Creegan: J. Chem. Phys. **90**, 3615 (1989)
12. D.A. Jelski, Z.C. Wu, T.F. George: Chem. Phys. Lett. **150**, 447 (1988)
13. Y. Liu, Q.-L. Zhang, F.K. Tittel, R.F. Carl, R. E. Smalley: J. Chem. Phys. **85**, 7434 (1986); J.L. Elkind, J.M. Alford, F.D. Weiss, R.T. Laaksonen, R.E. Smalley: J. Chem. Phys. **87**, 2397 (1987)
14. Q.-L. Zhang, Y. Liu, R.F. Curl, F.K. Tittel, R.E. Smalley: J. Chem. Phys. **88**, 1670 (1988)
15. E. Kaxiras: Chem. Phys. Lett. **163**, 323 (1989); Phys. Rev. Lett. **64**, 551 (1990)
16. H. Kupka, K. Jug: Chem. Phys. **130**, 23 (1989)
17. Y. Wang, T.F. George, D.M. Lindsay, A.C. Beri: J. Chem. Phys. **86**, 3493 (1987)
18. D.M. Lindsay, Y. Wang, T.F. George: J. Chem. Phys. **86**, 3500 (1987)

2.3 Density Functional Calculations for Clusters

R.O. Jones

2.3.1 Introduction

One of the most important properties of any aggregate of atoms – including clusters – is the geometrical arrangement of the constituents. The structure of clusters is the goal of many experimental investigations and is essential input for most calculations, and we find examples of both elsewhere in this volume. The determination of cluster structures from experimental data provides severe challenges, although significant progress has been made recently in both the preparation and identification of small clusters. The aim of the present chapter is to examine the problems facing the theorist who would like to make predictions about the structure of small clusters or molecules. In principle it should be easy to determine the most likely structure of such systems, or even of condensed matter in general. One chooses a structure, determines the total energy E of the system of electrons and ions, and then repeats this calculation for all possible geometries. The ground state structure is then that with the lowest energy. Is this a practicable approach? Can the standard methods of molecular physics (theoretical chemistry) be adapted to this problem, and what are their limitations? How far can we simplify the calculations and still get useful results?

The calculation of the total energy E has been a central goal of molecular physicists for decades. In principle, E and many other quantities of interest can be found from a quantum mechanical calculation of the exact wave function Ψ of the system of electrons and ions. Methods for doing this are discussed in Chapter 2.1, where we can see that theoreticians have also been active. Nevertheless, there are many identifiable clusters that are too large to be handled by these methods. We discuss some of the reasons for this in the following section. Not only is the calculation of the total energy difficult, but the number of local minima in the energy surface increases dramatically with increasing cluster size. Together, they lead to a problem that appears to be insoluble.

I shall discuss here a combination of density functional (DF) and molecular dynamics (MD) methods, which has developed into a general method for calculating structures. While it cannot *solve* the above problems – they would not be insoluble then – the method addresses *both* aspects. First, the density functional formalism, with a local density approximation for the exchange-correlation energy, allows us to carry out efficient, if approximate, calculations of the total energy for a single geometry. Second, we introduce the MD strategy of "simulated annealing" to determine low-lying energy minima, i.e. for finding solutions to the minimization problem that are *almost* optimal. The motion associated with finite temperature MD prevents the system being trapped in unfavorable energy minima. After outlining the important features of the approach, we show that reliable geometries are found for small clusters of sulphur. Large changes in cluster geometries can be followed, as we show for S_7O, and, for heterocyclic clusters of the form Se_mS_n, the method describes accurately the energy differences between different isomers. Calculations for small phosphorus clusters have indicated important low-lying structures that had not been found with other methods. Finally, we discuss the approximations involved in the calculations and ask whether it might be possible to do calculations based on an even simpler electron gas ("jellium") model.

2.3.2 Calculating Structures

2.3.2a) Calculation of the Total Energy, *E*

For an interacting system of electrons and ions, a knowledge of the exact wave function Ψ would allow one to determine many quantities of interest, including the total energy

$$E = \langle \Psi | \hat{\mathscr{H}} | \Psi \rangle / \langle \Psi | \Psi \rangle \ , \tag{1}$$

where the Hamiltonian of the system $\hat{\mathscr{H}}$ includes the kinetic energy and the potential energy contributions due to the electron-electron and electron-nucleus interactions. Calculations of this type can be carried out for systems with very few electrons, the best-known being the hydrogen atom. Even for the helium atom, however, it is not possible to find an analytic expression for the wavefunction and energy, and approximations for Ψ are unavoidable.

Approximate solutions for Ψ are usually based on the variational principle of Rayleigh and Ritz, i.e., if $|\Phi\rangle$ is an approximate wavefunction, then

$$\langle \Phi | \hat{\mathscr{H}} | \Phi \rangle / \langle \Phi | \Phi \rangle \geq E_{GS} \ . \tag{2}$$

The simplest form of the many-particle wavefunction is that of *Hartree* (1928), who represented Φ as a product of single-electron functions:

$$|\Phi\rangle = \phi_1(\mathbf{r}_1)\,\phi_2(\mathbf{r}_2)\ldots\phi_n(\mathbf{r}_n) \ , \tag{3}$$

where the functions ϕ_i each satisfy a Schrödinger equation whose potential term is given by the mean field of the other electrons,

$$\left(-\frac{1}{2}\nabla^2 + V_{\text{ext}} + \varphi\right)\phi_i(\mathbf{r}_i) = \varepsilon_i\phi_i(\mathbf{r}_i) \ . \tag{4}$$

Here V_{ext} is the external field of the nuclei, and the Coulomb potential φ is determined by solving Poisson's equation [we use atomic units, $e = m = \hbar = 1$]. An improved approximation that incorporates Fermi statistics results if we replace the product in (3) by a single (Slater) determinant [also known as a "configuration"]. This leads to an additional potential term (the "exchange" potential) in (4), but does not affect the single-particle picture. This approach – the "Hartree–Fock approximation" – has been the basic method of atomic and molecular physics since its inception 60 years ago. Programs are available from several sources and calculations can be performed in a routine fashion for systems with up to several hundred electrons. The single-particle picture is also very useful in understanding the results.

The Hartree–Fock method is so familiar that it is sometimes overlooked that the use of a wavefunction with a single determinant is an approximation that is often insufficient to obtain satisfactory total energies. In fact, *Coulson* [1] noted thirty years ago that "*it is now perfectly clear that a single configuration wavefunction must inevitably lead to a poor energy*". In principle, the representation of the wavefunction as a linear combination of determinants ("configuration interaction", CI) leads to an exact wavefunction and energy, but the numerical effort required increases explosively with increasing electron number. Even today it is difficult to obtain accurate energies for systems with more than ~ 20 electrons. We can see that the first problem – calculating the total energy for one geometry – is very difficult if we demand exact answers.

2.3.2b) Determination of the Lowest-Lying Minima

These arguments apply to a *single* geometry, i.e. for a particular set of "internal coordinates" of a system. If we have N atoms in a cluster, the number of such coordinates is $3N - 6$, since we obtain the same energy if we rotate or translate the cluster. If we required 10 calculations for each coordinate in a 10-atom cluster, for example, the 24 independent interatomic degrees of freedom would mean that we would need 10^{24} calculations in a cluster where even *one* is difficult! This is not all. Even if we could locate easily the closest minimum to a particular geometry, it would be difficult to find the most stable structure if the number of local minima in the energy surface is very large. We now show that the number of minima increases very rapidly with increasing cluster size, so that this is usually the case.

The enumeration of topologically different isomers consistent with a given chemical formula is one of the oldest problems in theoretical chemistry. In 1874, only a decade after the chemical bond was first denoted by a line joining the nuclei, *Cayley* [2] showed that the number of isomers could grow rapidly with increasing N, and much work in the intervening period has confirmed this. Recently, *Hoare* and *McInnes* [3] studied clusters where the atoms interact with a pairwise Lennard-Jones potential:

$$V_{\text{Lennard-Jones}} = 4\varepsilon_0\left[\left(\frac{\sigma}{r}\right)^{12} - \left(\frac{\sigma}{r}\right)^{6}\right]. \tag{5}$$

Table 1. Number of minima in cluster interacting with a Lennard-Jones potential [Eq. (5)] (*Hoare* and *McInnes* [3])

N	LJ
6	2
7	4
8	8
9	18
10	57
11	145
12	366
13	988

For small clusters it was possible to locate *all* the minima, and the number of minima found for clusters of different sizes is shown in Table 1. This number increases rapidly with increasing N, with signs that the increase is exponential. In fact, for clusters of identical atoms interacting with a pairwise potential, *Wille* and *Vennik* [4] have shown that there is no known algorithm that grows with time as power of N and with which one can determine the ground state energy and structure. Such problems are classified as "*NP-hard*" [5] and are termed "intractable". It is sobering indeed when mathematicians view a problem of widespread interest to be insoluble in practice. We should also ask whether it is *physically* relevant to look for the most stable structure among many of similar energy. Finally, parametrized force laws, such as Eq. (5), are inherently simplified and semi-empirical, and it is difficult to know how the choice of parameters affects the final results. It is well-known, for instance, that *no* simple pairwise force can stabilize the diamond structure, and there are many other examples where three- and four-body potential energy terms are essential.

This discussion shows that there are two distinct problems to be solved, with *multiplicative* difficulties. As we noted above, the combination of density functional formalism and simulated annealing addresses (if not solves) both aspects.

2.3.2c) Energy Calculations Using Density Functional Theory

The basic theorems of the density functional formalism were derived by *Hohenberg* and *Kohn* [6]. They showed that:

(1) Ground state (GS) properties of a system of electrons and ions in an external field, V_{ext}, can be determined by a knowledge of the density *alone*. In the language of the experts, GS properties are *functionals* of the electron density $n(\mathbf{r})$. The total energy E is an example, i.e., $E = E[n]$.
(2) $E[n]$ satisfies the variational principle $E[n] \geq E_{GS}$, and the density for which the equality holds is the ground state density, n_{GS}.

A general proof of these assertions has been provided by *Levy* [7].

The step from formalism to a practical scheme for energy calculations results from the observation by *Kohn* and *Sham* [8] that the minimization of $E[n]$ is simplified if we write:

$$E[n] = T_0[n] + \int d\mathbf{r}\, n(\mathbf{r}) \left(V_{\text{ext}}(\mathbf{r}) + \frac{1}{2}\varphi(\mathbf{r}) \right) + E_{\text{xc}}[n] \ , \tag{6}$$

where T_0 is the kinetic energy that a system with density n would have in the absence of electron-electron interactions, $\varphi(\mathbf{r})$ is the Coulomb potential, and E_{xc} is our definition of the exchange-correlation energy. The variational principle now yields

$$\frac{\delta E[n]}{\delta n(\mathbf{r})} = \frac{\delta T_0}{\delta n(\mathbf{r})} + V_{\text{ext}}(\mathbf{r}) + \varphi(\mathbf{r}) + \frac{\delta E_{\text{xc}}[n]}{\delta n(\mathbf{r})} = \mu \ , \tag{7}$$

where μ is the Lagrange multiplier associated with the requirement of constant particle number.

If we now consider the corresponding equation for a system of *noninteracting* particles,

$$\frac{\delta E[n]}{\delta n(\mathbf{r})} = \frac{\delta T_0}{\delta n(\mathbf{r})} + V(\mathbf{r}) = \mu \ , \tag{8}$$

the mathematical problems (7) and (8) are *identical*, provided that

$$V(\mathbf{r}) = V_{\text{ext}} + \Phi(\mathbf{r}) + \frac{\delta E_{\text{xc}}[n]}{\delta n(\mathbf{r})} \ . \tag{9}$$

The solution of Eq. (8) follows then if we can solve the Schrödinger equation for noninteracting particles,

$$\left(-\frac{1}{2}\nabla^2 + V(\mathbf{r}) \right)\psi_i(\mathbf{r}) = \varepsilon_i \psi_i(\mathbf{r}) \ ,$$

$$n(\mathbf{r}) = \sum_{i=1}^{N} |\psi_i(\mathbf{r})|^2 \ . \tag{10}$$

We must satisfy Eq. (9), and this is usually done by a "self-consistency" cycle. From a starting value of n we compute $V(\mathbf{r})$ and the n' that results, and continue this procedure until $n = n'$.

The equations we must solve are single-particle equations of Hartree form (cf. Eq. (4)) and share a major advantage of the HF method, namely the single-particle interpretation of the results. In contrast to the Hartree–Fock potential, however, the effective potential has a *local* dependence on the density. However, some words of caution are in order before we rejoice that the problem of determining the total energy of a system of electrons and ions has been reduced to such a simple form:

(1) The equations we solve to determine the total energy of the system of interacting electrons – the real, physical system – refer to a fictitious system of noninteracting "electrons". We must therefore be cautious in assigning a physical meaning to the eigenvalues ε_i and eigenfunctions ψ_i in Eq. (10).

(2) While T_0, the electron-ion and electron-electron interaction energies in the energy expression (6) can readily be evaluated, E_{xc} is not known. It is a relatively small but far from negligible contribution to the total energy, at least for atoms and molecules. Furthermore, $V(\mathbf{r})$ involves the functional derivative of E_{xc} with respect to the density, and this poses additional formal problems.

We can use the relationship between the non-interacting and interacting systems to derive an exact expression for E_{xc} [9]:

$$E_{xc} = \frac{1}{2} \int d\mathbf{r}\, n(\mathbf{r}) \int d\mathbf{r}' \frac{1}{|\mathbf{r} - \mathbf{r}'|} n_{xc}(\mathbf{r}, \mathbf{r}' - \mathbf{r}) \ , \tag{11}$$

where

$$n_{xc}(\mathbf{r}, \mathbf{r}' - \mathbf{r}) \equiv n(\mathbf{r}') \int_0^1 d\lambda (g(\mathbf{r}, \mathbf{r}', \lambda) - 1) \ . \tag{12}$$

E_{xc} can then be understood as the interaction between $n(\mathbf{r})$ and the "exchange-correlation hole", $n_{xc}(\mathbf{r})$. The function $g(\mathbf{r}, \mathbf{r}', \lambda)$ is the pair-correlation function – the probability of finding an electron at both $\mathbf{r}$ and $\mathbf{r}'$ – of a system with density $n(\mathbf{r})$ and electron-electron interaction $\lambda\varphi$. The evaluation of such correlation functions generally requires a knowledge of the exact wavefunction, which is precisely a problem we should like to avoid. The experience of the last 20 years indicates that the *exact* exchange-correlation energy functional $E_{xc}[n(\mathbf{r})]$ must be extremely complicated, and there seems to be little prospect of determining it and its functional derivative in forms appropriate for calculation. For practical applications of the DF formalism it is essential to develop and test approximations for E_{xc}, and the results described below can be viewed as tests of one of these.

The most widely used approximation is the local spin density (LSD) approximation

$$E_{xc}^{LSD} = \int d\mathbf{r}\, n(\mathbf{r})\, \varepsilon_{xc}[n_\uparrow(\mathbf{r}), n_\downarrow(\mathbf{r})] \ . \tag{13}$$

Here $\varepsilon_{xc}[n_\uparrow, n_\downarrow]$ is the exchange and correlation energy per particle of a homogeneous, spin-polarized electron gas with spin-up and spin-down densities $n_\uparrow$ and $n_\downarrow$, respectively. In the work described here we use the expression of *Vosko* et al. [10], which was derived from Monte Carlo calculations for an electron gas of various densities. This approximation avoids the use of adjustable parameters, but its application to atoms, molecules and solids cannot be justified by small departures from homogeneity. A detailed discussion is given in the review article by *Jones* and *Gunnarsson* [11].

There have been many applications of the LSD approximation to small molecules, and a survey of some of the results can be found in the review article mentioned above [11]. It is difficult to summarize in a few sentences all the work that has been done in the past 15 years or so, but some features of the results appear to be almost universal:

(1) Equilibrium geometries are given reliably in many families of molecules and extended systems. Furthermore, the energy variations in the neighborhood

of local minima are described well, so that vibration frequencies in molecules and phonon frequencies in solids are generally in satisfactory agreement with experiment.

(2) While bonding trends in families of molecules or solids are given reliably, the calculated energy differences [such as binding, ionization and transfer energies] can show substantial departures from the experimental values.

In the case of ozone (O_3), for example, the dissociation energy (experimentally 1.1 eV) is overestimated by $\sim$2 eV [12]. The wavefunction of ozone, however, is described very poorly by a single determinant, and Hartree–Fock calculations give a ground state with the wrong symmetry. Even in this case, the LD-approximation gives a good description of both the ground state geometry and the energy difference between the ground state and the lowest state with bond angle $\alpha = 60°$. In S_3 [13], there are two states with quite different geometries that are almost degenerate, and this prediction has been confirmed by CI calculations [14].

The practical necessity of approximating E_{xc} leads to an essential difference in perspective between the density functional and configuration interaction approaches. While the CI method seeks an exact numerical solution of the Schrödinger equation, which would yield precise answers for many quantities of interest, even the best solution of the density functional equations can only reflect the accuracy of the approximation for E_{xc}.

2.3.2d) The Molecular Dynamics/Density Functional Approach

In systems where the ground state is unknown or there are many local minima, it is necessary to develop alternative methods for finding solutions that are near to optimal. *Kirkpatrick* et al. [15] noted the connection between statistical physics and the minimization of a function of many variables, and suggested "simulated annealing" based on a Monte Carlo sampling as a means of finding such solutions. Annealing at a finite temperature is Nature's way of finding the most stable structure. If the temperature is chosen appropriately, the system avoids being trapped in unfavorable local minima in the energy surface, and a great range of structures can be probed. On the other hand, the temperature should not be too high, or the system under consideration could dissociate or melt.

Molecular dynamics (MD) provides an alternative to the Monte Carlo approach, and *Car* and *Parrinello* [16] showed that it could be combined with the density functional (DF) scheme for calculating total energies. The result is a method for calculating electronic properties that is free of adjustable parameters and makes no assumptions about ground state geometries. Finite temperature MD techniques allow an efficient sampling of the potential energy surface, and the DF method [11], with the local spin density (LSD) approximation for the exchange-correlation energy, avoids the parametrization of the interatomic forces common in MD schemes [17].

There are *two* minimization problems to be solved: The total energy must be minimized for each geometry, and we must find the geometry with the lowest energy. This is done conveniently by viewing E as a function of the interdependent sets of degrees of freedom, $\{\psi_i\}$ and $\{\mathbf{R}_I\}$, and using standard MD techniques to follow trajectories defined by the Lagrangian

$$\mathscr{L} = \sum_i \mu_i \int_\Omega d\mathbf{r}\,|\dot{\psi}_i^* \dot{\psi}_i| + \sum_I \frac{1}{2} M_I \dot{\mathbf{R}}_I^2 - E[\{\psi_i\},\{\mathbf{R}_I\}] + \sum_{ij} \Lambda_{ij}\left(\int_\Omega d\mathbf{r}\,\psi_i\psi_j^* - \delta_{ij}\right) \quad (14)$$

and the corresponding equations of motion

$$\mu\ddot{\psi}_i(\mathbf{r},t) = -\frac{\delta E}{\delta\psi^*(\mathbf{r},t)} + \sum_k \Lambda_{ik}\psi_k(\mathbf{r},t)\ ,$$
$$M_I\ddot{\mathbf{R}}_I = -\nabla_{\mathbf{R}_I} E\ . \quad (15)$$

Here M_I and $\mathbf{R}_I$ denote the ionic masses and coordinates, μ_i are "masses" associated with the electronic degrees of freedom, and the Lagrange multipliers Λ_{ij} guarantee the orthonormality of the single particle orbitals ψ_i. The artificial Newton's dynamics for the electronic degrees of freedom (with appropriately chosen μ_i) ensure that the electrons remain close to their ground state. From the orbitals and the resultant density $n(\mathbf{r},t) = \sum_i |\psi_i(\mathbf{r},t)|^2$ we use the density functional expression (6, 13) for the total energy E.

2.3.3 Application to Clusters of Group VIa Elements

The group VIa elements provide some of the best characterized of small atomic clusters. They are unique in that many allotropes comprise regular arrays of well-separated ring molecules, and X-ray structure analyses have been performed for S_n, $n = 6$–8, 10–13, 18, 20 [18, 19] and Se_n, $n = 6, 8$ [18, 20]. S_9 has been found as microcrystals, and S_{18} exists in two distinct forms. Several mixed crystals of the form Se_mS_n and a range of sulphur oxides and ions are also known. These clusters are examples of "solid state clusters", a branch that is sometimes overlooked by workers who focus on the gas phase. The preparation of these clusters has been reviewed by *Steudel* [19].

Sulphur and selenium have much in common and it is not surprising that mixed isomers coexist. In addition to the intrinsic interest in Se_mS_n molecules, an understanding of their structures could provide insight into the complex structures of related liquid and amorphous materials, including Se itself. Mixed molecules have been studied in the vapor phase [21, 22], as liquids and as solid solutions [23]. The different species crystallize together, leading to structures with sulphur and selenium atoms distributed over the atomic sites [24]. The possible complexity is evident from the example of eight-membered rings, where

30 different crown-shaped isomers can occur. We have performed MD/DF calculations for some of these systems, using a large unit cell with periodic boundary conditions to guarantee a weak interaction between the clusters. For details of the calculations see the original articles.

2.3.3a) Calculations for Sulphur Clusters, S_n

The results for sulphur clusters S_2 to S_{13} [25] show that it is possible to determine low-lying energy minima even if one starts from a geometry that is quite incorrect. In $S_{3\text{-}6}$ we started from almost linear chains and in $S_{7\text{-}13}$ from nearly planar rings. After heating and cooling the system repeatedly, and ultimately cooling the molecule to $T = 0$, the final structures agreed well with experiment in all cases where X-ray data were available. We show in Fig. 1 the structures of S_9 and S_{12}. The D_{3d} symmetry in S_{12} [Fig. 1(b)] is reproduced very closely, and the structural parameters are in good agreement with measured values. Calculations using a more accurate pseudopotential [26] give even better agreement with measured geometries, an example being the D_{4d} ("crown") structure of S_8, where the calculated structural parameters ($d = 206.4$ pm, $\alpha = 108.5°$, $\gamma = 98.1°$) are almost indistinguishable from the measured values ($d = 206.0 \pm 0.3$ pm, $\alpha = 108.5 \pm 0.7°$, $\gamma = 98.1 \pm 2.1°$).

The accurate reproduction of known structures gives us confidence in the predictions found for clusters where the ground state structures are not yet known. In the case of S_9, the crystallites that can be prepared are too small for X-ray diffraction analysis. We predict that the most stable isomer has C_2 symmetry [Fig. 1(a)], with several local minima with C_s symmetry with energies > 0.2 eV higher. The calculated ground state structure is consistent with measured Raman spectra, which indicate almost equal bond lengths and bond angles, and dihedral angles in the range 70–130°.

2.3.3b) Structural Change in S_7O

There are many cases where structures of interest are separated by large energy barriers. These are particularly difficult tests for simulation methods, since the

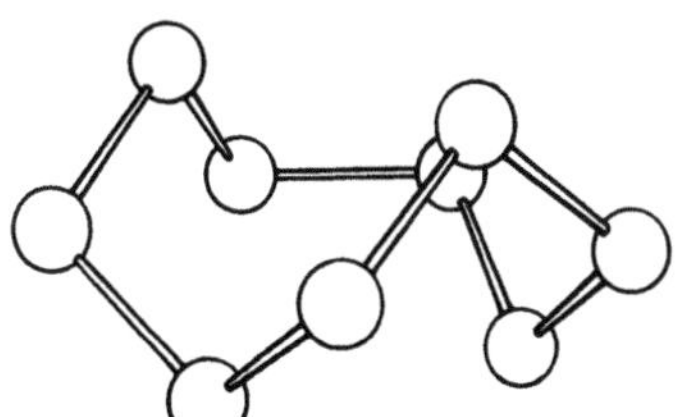

a $S_9(C_2)$

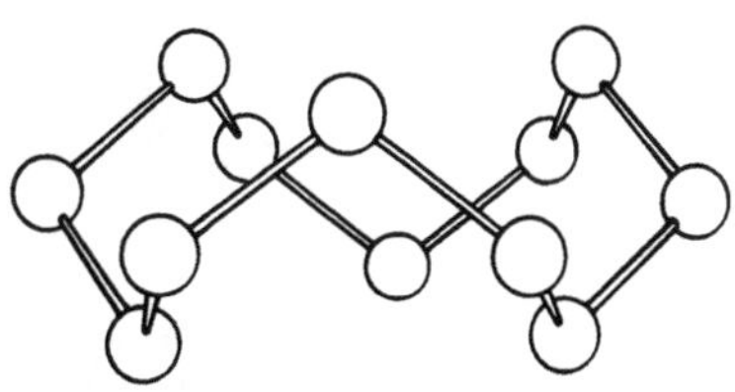

b $S_{12}(D_{3d})$

Fig. 1. Structures of (**a**) S_9 and (**b**) S_{12}

real time required for a system to cross the energy barrier may be many orders of magnitude greater than that available. An interesting example where structural optimization nevertheless proved possible is the S_7O molecule [27]. In the course of the simulation shown in Fig. 2 – 1200 time steps at $T = 2000$ K – the molecule changed from a stable local minimum with a ring structure similar to that in S_8 [Fig. 2(a)] to a structure with different topology, namely an oxygen atom outside the S_7 structure [Fig. 2(i)]. If we reduce the temperature of the system slowly, we find the geometry shown in Fig. 2(j) and structural parameters in good agreement with those of the experimental ground state [27]. The energy difference between ring and ground state structures is only ~ 0.2 eV and the entire structural transformation took place in only ~ 8000 time steps ($< 10^{-12}$ s), although the energy barrier is ~ 5 eV. It appears that the molecular dynamics technique is a particularly efficient generator of geometrical configurations.

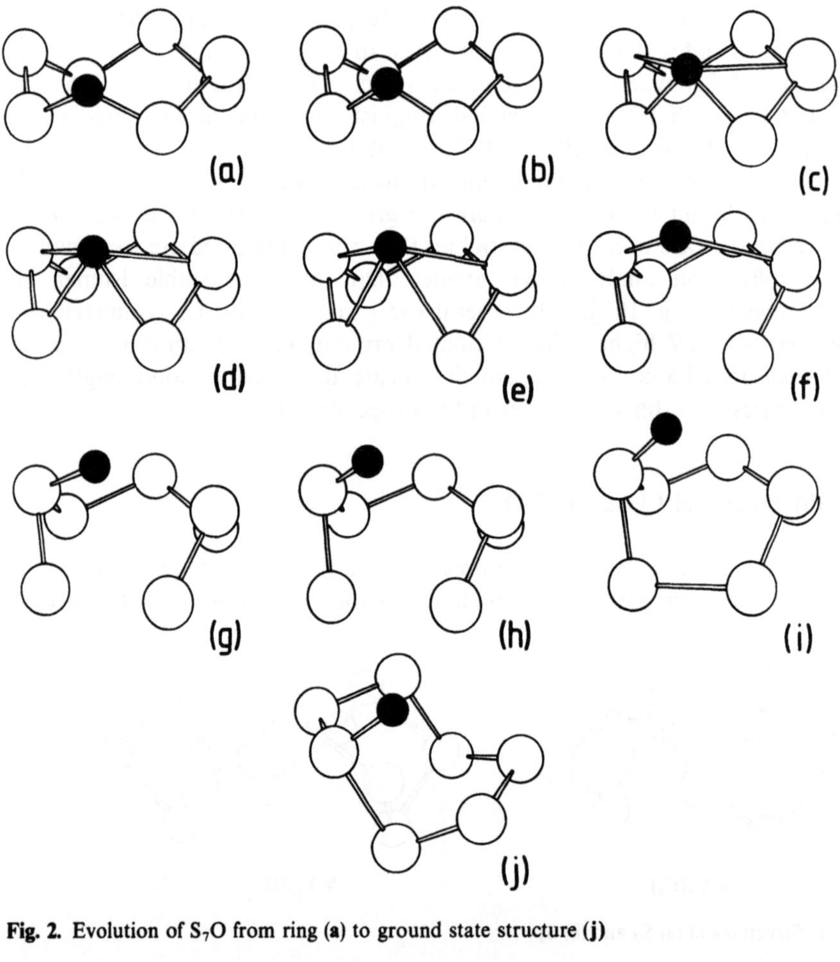

Fig. 2. Evolution of S_7O from ring **(a)** to ground state structure **(j)**

2.3.3c) Isomers of Se_2S_6, Se_6S_2

There has been a considerable amount of work on S_nSe_{8-n} molecules, and Raman and NMR spectroscopies have been used recently to study both these molecules and molten mixtures of Se and S [28]. Theoretical predictions of the relative stability of different isomers are difficult, since the interconversion of various Se_mS_n molecules and ring-chain equilibration processes involve the rupture and formation of S–S, Se–Se and Se–S bonds. Reactions involving the groups

$$\text{-S-S-} \; + \; \text{-Se-Se-} \longrightarrow 2 \; \text{-Se-S-} \tag{16}$$

have been studied recently using Hartree–Fock techniques with a minimal Gaussian basis [29], without reaching the accuracy needed for predicting the small energy differences involved. For example, the reaction (16) is predicted to be exothermic, in contrast to calorimetric measurements in the liquid [30]. For diatomic molecules in the gas phase [31], the measured enthalpy change is $+0.082 \pm 0.069$ eV.

Jones and *Hohl* [28] have performed extensive calculations on seven- and eight-membered rings of the type Se_mS_n, starting in each case from the ground state geometry of S_8 (D_{4d}) and determining the relative energies were determined to better than 1×10^{-3} eV. The cyclic structures possible in Se_2S_6 and Se_6S_2 are shown in Fig. 3. The calculated structures are characterized by small deviations from the crown-shaped [D_{4d}] structures familiar from S_8 and Se_8, and show

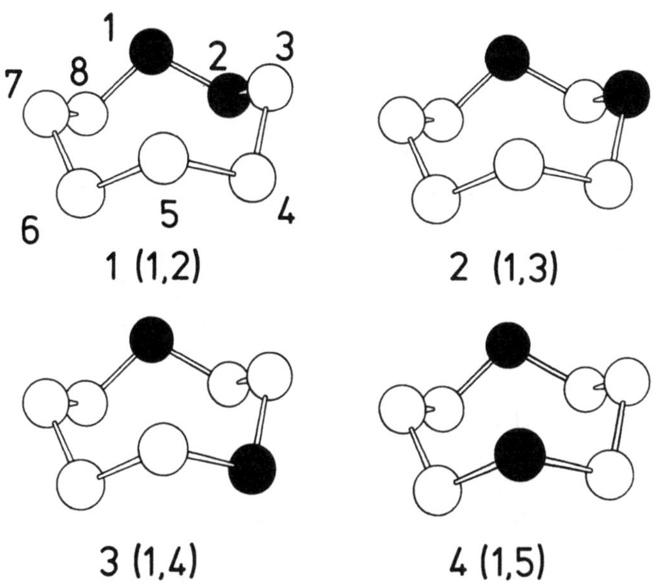

Fig. 3. Structures of Se_2S_6 (and Se_6S_2)

several remarkable features. In all eight structures, the bond lengths ($d_{S\text{-}S}$, $d_{S\text{-}Se}$, $d_{Se\text{-}Se}$) are the same within 1 pm (208 pm, 223 pm, 235 pm), and $d_{S\text{-}S}$ and $d_{Se\text{-}Se}$ are the same as found in the calculations for the ring structures of S_8 and Se_8. The energy calculations show that the (1, 2) structures are the most stable isomers in both Se_2S_6 and Se_6S_2. The energies of the other structures are *degenerate* to within 3 meV and lie 0.08 eV above the ground state in each case. These results can be understood in terms of the very simple model we now describe.

If we associate with each bond length ($d_{S\text{-}S}$, $d_{S\text{-}Se}$, $d_{Se\text{-}Se}$) a corresponding energy contribution ($E_{S\text{-}S}$, $E_{S\text{-}Se}$, and $E_{Se\text{-}Se}$, respectively), we obtain a simple expression for the energies. In the case of the (1, 2) structure in Se_2S_6,

$$E_{12} = E_{Se\text{-}Se} + 2E_{S\text{-}Se} + 5E_{S\text{-}S} \ . \tag{17}$$

and in *all three* remaining structures,

$$E_{13,14,15} = 4E_{S\text{-}Se} + 4E_{S\text{-}S} \ . \tag{18}$$

From this argument we may then expect the energies of the last three structures to be equal, and the energy of the (1, 2) structure to differ by an amount

$$\Delta E = E_{Se\text{-}Se} + E_{S\text{-}S} - 2E_{S\text{-}Se} \ . \tag{19}$$

If we apply the same argument to the Se_6S_2 structures, Eqs. (17–18) still apply, with S and Se interchanged. Again we find the (1, 2) isomer separated from the other three, and the energy difference is precisely that in Eq. (19). We now understand why the energy orderings in Se_2S_6 and Se_6S_2 are the same, and why the most stable (1, 2) isomers are each separated from three structures with almost identical energies. The separation in the present calculations is approximately 0.08 eV, which is consistent with the endothermicity of the reaction (16) and in reasonable agreement with the measured heats of reaction in the gas and liquid phases.

The results of this model agree with trends found in recent measurements, in particular with the relative abundance of structures with adjacent minority atoms. The symmetry between the energy ordering of the isomers of Se-rich and S-rich molecules is in direct contrast to earlier discussions of these systems [21, 32]. This symmetry is perturbed, of course, by differences between the elements such as the size, but the essential features are not changed. The trend to "segregation" of minority components could also occur in the disordered liquid and amorphous states. Segregation is a consequence, in this model, of the sign of ΔE in Eq. (19). Other systems with a small value of ΔE or with one of opposite sign would show a different behavior.

This example shows some of the advantages of the approach. Although the total energies obtained from LSD calculations are not as accurate as those obtainable from calculations based on the wave function, energy differences between similar systems are given surprisingly accurately. In particular, the model described above would be treated with scepticism in the absence of the calculations.

2.3.4 Structure of Phosphorus Clusters, P_2 to P_{11}

The final example we discuss here involves small phosphorus clusters. In elemental form, phosphorus shows a structural variety exceeded only by sulphur, and there have been many studies of the allotropes as well as amorphous forms [18, 33]. Gas phase clusters have been of interest for many years, with particular attention being paid to the occurrence of only P_2, P_3, P_4 and (possibly) P_8 in the vapor phase at high temperature [34]. There have been numerous speculations about the apparent absence of heavier clusters [35], and several calculations indicate that the cubic form of P_8 is, in fact, less stable than two P_4-tetrahedra [36]. Recently, *Martin* [37] was able to detect mass spectroscopically P_n clusters up to $n = 24$ in a supercooled beam. In spite of this, almost nothing is known about the structure of clusters with $n > 4$.

MD/DF calculations have been performed on phosphorus clusters up to P_{11} [38], and the calculated geometries and vibration frequencies agree well with experiment in those cases (P_2, P_4) where measurements are available. In the larger clusters there were interesting and unexpected results: (1) Although the tetrahedral structure is energetically favored in P_4, there is a large "basin of attraction" for a D_{2d} "roof" structure, i.e. this structure is the closest minimum for a large region of configuration space. (2) The "roof" structure is a prominent feature in the low-lying isomers in P_5 to P_8. The calculated ground states in P_5, P_6, and P_7 have a P_4-roof with an additional one, two and three atoms, respectively. In the last case, the trimer is approximately equilateral.

In P_8, we have found two isomers that are more stable than the cubic structure and correspond to local minima in the energy surface. The most stable is shown in Fig. 4(a) and can most easily be understood as a (distorted) cube with one bond rotated through 90°. This "wedge" or "cradle" structure is found as

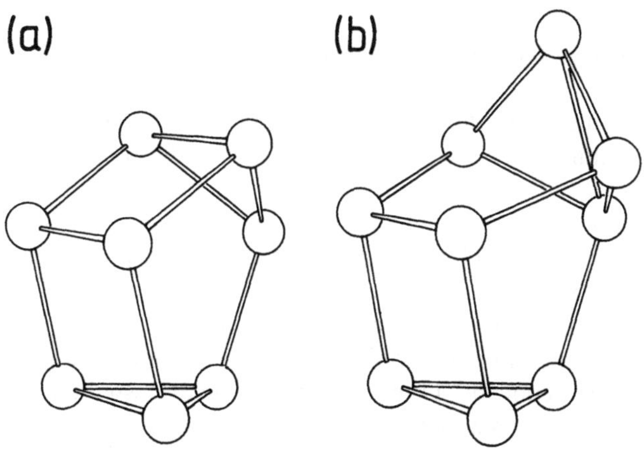

Fig. 4. Most stable isomers of **(a)** P_8 and **(b)** P_9

a structural unit in violet (monoclinic, Hittorf) phosphorus [39]. There is also an interesting analogy between the structures of the P_8-isomers and those of the corresponding isoelectronic hydrocarbons C_8H_8. The cubic form of the latter (cubane) has been prepared by *Eaton* and *Cole* [40], and can be converted catalytically to the wedge shaped form cuneane [41]. A third possible isomer of C_8H_8 was estimated to have a strain energy intermediate between the other two, in agreement with the energetic ordering found in the P_8 isomers [38]. The most stable P_9 isomer is shown in Fig. 4(b). The relationship to the wedge structure of P_8 is clear.

The results for phosphorus clusters demonstrate the usefulness of the simulated annealing approach. The calculations reproduce known structures very well, so that we may have confidence in the predictions for those molecules where the structures are presently unknown. The predictions show very plausible trends, even though several of the structures were quite unexpected.

2.3.5 Models Based on Electron Gas Calculations

The above calculations are all based on the local spin density approximation for E_{xc} [Eq. (13)]. This approximation was originally proposed for use in systems that are close to homogeneity, and *Kohn* and *Sham* [8] noted that "*we do not expect an accurate description of chemical bonding*" with it. The electron density in atoms, molecules and solids is usually far from uniform, and we now ask how it is that this approximation can give such a reasonable description of bonding trends and geometries in so many systems. We then examine some recent work that makes further simplifications based on the uniform electron gas (or "jellium") model.

2.3.5a) Why Should Local Density Approximations "Work"?

Approximate exchange terms have often been used in theories based on the density. *Dirac* [42], for example, included exchange effects in the well-known Thomas-Fermi theory by adding a term derived from the exchange energy density in a homogeneous system. The use of an approximate exchange *potential* in addition to the Hartree term in Eq. (4) was suggested by *Slater* [43], who used a dimensional argument to show that the exchange potential in a system of variable density could be approximated by a term with a local dependence $\sim [n(\mathbf{r})]^{1/3}$ on the density. It is often overlooked that this dependence on the density is a consequence of the concept of the "exchange" hole discussed above, i.e. the region near an electron that is avoided by electrons of the same spin, and not on the exchange potential in a homogeneous system. Shortly afterwards, Gáspár [44] adopted the Dirac approximation to the exchange energy, and used a variational approach to derive an exchange potential two thirds the size of Slater's. Solutions of the resulting equations for the Cu^+ ion reproduced the Hartree-Fock eigenfunctions and eigenvalues values very well [44]. This was an interesting demonstration that an approximation based on results for a homogeneous system could give a satisfactory description, even if the density is very far from uniform.

In spite of its complexity, the exact expression for E_{xc} [Eq. (11)] enables us to make several observations: First, since $g(\mathbf{r}, \mathbf{r}')$ tends to unity as $|\mathbf{r} - \mathbf{r}'| \to \infty$, the above separation into electrostatic and exchange-correlation energies can be viewed as an approximate separation of the consequences of long- and short-range effects, respectively, of the Coulomb interaction. The total interaction energy should then be less sensitive to changes in the density, since the long-range part can be calculated exactly.

The second observation [9] arises from the isotropic nature of the Coulomb interaction, and has important consequences. A variable substitution $\mathbf{R} \equiv \mathbf{r}' - \mathbf{r}$ in Eq. (11) yields

$$E_{xc} = \frac{1}{2} \int d\mathbf{r}\, n(\mathbf{r}) \int_0^\infty dR\, R^2 \frac{1}{R} \int d\Omega\, n_{xc}(\mathbf{r}, \mathbf{R}) \ . \tag{20}$$

This shows that the xc-energy depends only on the spherical average of $n_{xc}(\mathbf{r}, \mathbf{R})$, so that approximations for E_{xc} can give an *exact* value, even if the description of the non-spherical parts of n_{xc} is quite inaccurate. Third, from the definition of the pair-correlation function, there is a sum-rule requiring that the xc-hole contains one electron, i.e., for all $\mathbf{r}$,

$$\int d\mathbf{r}'\, n_{xc}(\mathbf{r}, \mathbf{r}' - \mathbf{r}) = -1 \ . \tag{21}$$

This means that we can consider $-n_{xc}(\mathbf{r}, \mathbf{r}' - \mathbf{r})$ as a normalized weight factor, and define locally the radius of the xc-hole,

$$\left\langle \frac{1}{\mathbf{R}} \right\rangle_{\mathbf{r}} = -\int d\mathbf{r}\, \frac{n_{xc}(\mathbf{r}, \mathbf{R})}{|\mathbf{R}|} \ . \tag{22}$$

This leads to

$$E_{xc} = -\frac{1}{2} \int d\mathbf{r}\, n(\mathbf{r}) \left\langle \frac{1}{\mathbf{R}} \right\rangle_{\mathbf{r}} , \tag{23}$$

showing that, provided the sum-rule (22) is satisfied, the exchange-correlation energy depends only weakly on the details of n_{xc} [9]. In fact, we can say that it is determined by the first moment of a function whose second moment is known exactly.

These arguments are made clearer by the example shown in Fig. 5, where we compare – for one value of $\mathbf{r}$ in the nitrogen atom – the results of local density (exchange only) calculations of the exchange hole with exact results, i.e. those calculated from Hartree–Fock wavefunctions [11]. The approximate and exact holes are qualitatively different, the former being spherically symmetric about

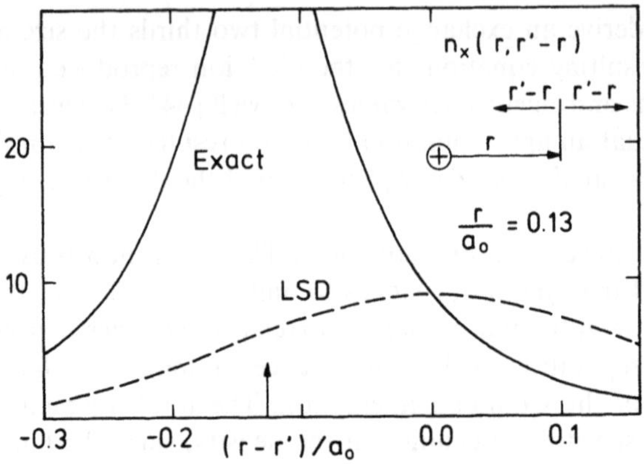

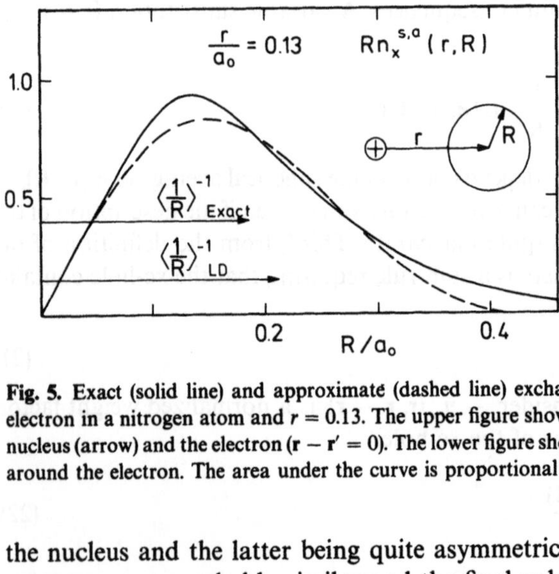

Fig. 5. Exact (solid line) and approximate (dashed line) exchange hole $n_x(\mathbf{r}, \mathbf{r}' - \mathbf{r})$ for a spin-up electron in a nitrogen atom and $r = 0.13$. The upper figure shows the hole along a line through the nucleus (arrow) and the electron ($\mathbf{r} - \mathbf{r}' = 0$). The lower figure shows the spherical average of the hole around the electron. The area under the curve is proportional to the exchange energy [Ref. 11]

the nucleus and the latter being quite asymmetric. Nevertheless, the spherical averages are remarkably similar and the final values of the exchange energies differ by only a few percent. The large differences in the exchange holes arise almost entirely from differences in the non-spherical components, and these contribute exactly zero to the exchange energy. This should also be true if we include correlation effects, since the sum rule also applies to the "exchange-correlation hole", and to other systems with inhomogeneous density distributions.

2.3.5b) Jellium Calculations for Large Clusters

The above arguments show that, under certain circumstances, a local density approximation can give a reasonable value of the exchange-correlation energy even in a system where the density is far from uniform. We now ask whether there are circumstances under which an even simpler model – a spherical cluster with a uniform positive charge – can describe features of the electronic structure. This "jellium" model has been widely studied as a model of bulk systems and we shall now see that it can describe some of the irregularities observed in cluster abundances for "simple" (*sp*-bonded) metals.

The abundance patterns of alkali metal clusters show discontinuities as a function of the number of atoms in the cluster, with clusters of special stability occurring in Na and K for $N = 8$, 20, 40, 58 and 92 [45]. *Knight* et al. [45] showed that these discontinuities could be understood in terms of the above model, by assuming a spherically symmetrical potential well of the form

$$U(r) = -\frac{U_0}{\exp[(r - r_0)/\varepsilon] + 1}, \tag{24}$$

where U_0 is the sum of the Fermi energy E_F and the work function ϕ of the bulk; r_0 is the effective radius of the cluster and assumed to be $r_s N^{1/3}$, where r_s is the radius of the sphere containing one electron in the bulk. The parameter ε determines the variation of the potential at the surface of the sphere.

The Schrödinger equation is solved numerically for this potential, leading to discrete electronic energy levels that shift down slowly as N increases. If the electronic energy is associated with the sum of the eigenvalues, clusters of relative stability should result when an energy level is just filled at a certain N and the next orbital starts to be occupied in the cluster with $N + 1$ atoms. The prominent peaks found in the mass spectra can then be correlated with the electronic properties, without invoking structural packing patterns. The observations of *Knight* et al. confirmed the prediction by *Ekardt* [46] that the shell structure in this model could be a real effect.

This is an interesting observation, and the correlation with the experimental spectra is clearly apparent. The model can also be refined to allow for deformations from spherical symmetry when the shells are only partially filled [47]. In the case of small Al clusters, *Upton* [48] found that the perturbative inclusion of the electron-ion interaction could account for trends found in ab initio calculations. This is encouraging, as the method can be applied to systems presently beyond the capability of more accurate approaches. Nevertheless, the arguments are most appropriate for simple metals and there is a wealth of experimental evidence, particularly for transition metal clusters, which points to more complicated structural variations with changing cluster size.

2.3.6 Concluding Remarks

The structure of clusters is the focus of much of the work described in this book. The preparation, identification and spectroscopy of clusters are difficult tasks, and cooperation between experimentalists and theoreticians is highly desirable.

In principle, theory can contribute much to the determination of cluster structures. The calculation of the energy surfaces should lead to the most stable isomers, but we have seen that there are severe problems: The determination of the energy E from the exact wavefunction of the cluster is impracticable in all but the smallest clusters, and there are usually many local minima in the energy surface.

In the present Chapter 1 have focused on an approach that addresses both aspects. The density functional formalism, with the local spin density approximation for the exchange-correlation energy, enables us to carry out energy calculations in an efficient manner. The use of finite temperature molecular dynamics prevents the cluster being trapped in unfavorable energy minima. The results to date have been very encouraging. The structures of small clusters of sulphur and selenium are described very well. In the case of S_7O we can simulate large structural changes involving energy barriers in excess of 5 eV, and the final structure (an O atom outside an S_7-ring) is in good agreement with the structure determined by X-ray diffraction. In the heterocyclic ring molecules Se_mS_n we have studied structures with very small energy differences and energy barriers that are small on a thermal scale. In the case of phosphorus clusters, we predict structures with plausible trends and with an interesting analogy to isoelectronic hydrocarbons.

The successes of this approach should not lead us to overlook its limitations. Even the most accurate calculation will reflect the approximations used, and the LSD approximation is an essential part of the calculations performed so far. Overestimates in the binding energy are evident in all the clusters discussed above. Improved approximations would be very valuable, as long as they can be implemented in a systematic way. Attempts in this direction have not been particularly successful so far and further work is necessary. Moreover, there are many local minima in the energy surfaces even for clusters containing less than 10 atoms. This remains true even after the implementation of molecular dynamics techniques, and one can not guarantee finding the global energy minimum, or even all the most important minima, with this scheme. The development of alternative optimization schemes will be important in the future.

Many of the calculations described here were supported by a grant by the German Supercomputer Centre (Höchstleistungsrechenzentrum, HLRZ) of computer time on the Cray Y-MP 832 in the Forschungszentrum Jülich.

References

1. C. Coulson: Rev. Mod. Phys. **32**, 170 (1960)
2. A. Cayley: Phil. Mag. (4) **47**, 444 (1874)
3. M.R. Hoare, J.A. McInnes: Adv. Phys. **32**, 791 (1983)
4. L.T. Wille, J. Vennik: J. Phys. A **18**, L419, L1113 (1985)
5. M.R. Garey, D.S. Johnson: *Computers and Intractability: A Guide to the Theory of NP-Completeness* (Freeman, San Francisco, 1979)
6. P. Hohenberg, W. Kohn: Phys. Rev. **136**, B864 (1964)

7. M. Levy: Proc. Nat. Acad. Sci. USA **76**, 6062 (1979)
8. W. Kohn, L.J. Sham: Phys. Rev. **140**, A1133 (1965)
9. O. Gunnarsson, B.I. Lundqvist: Phys. Rev. **13**, 4274 (1976); See also J. Harris, R.O. Jones: J. Phys. F **4**, 1170 (1974)
10. S. Vosko, L. Wilk, M. Nusair: Can. J. Phys. **58**, 1200 (1980)
11. R.O. Jones, O. Gunnarsson: Rev. Mod. Phys. **61**, 689 (1989)
12. R.O. Jones: J. Chem. Phys. **82**, 325 (1985)
13. R.O. Jones: J. Chem. Phys. **84**, 318 (1986)
14. J.E. Rice, R.D. Amos, N.C. Handy, T.J. Lee, H.F. Schaefer III: J. Chem. Phys. **85**, 963 (1986)
15. S. Kirkpatrick, C.D. Gelatt, M.P. Vecchi: Science **220**, 671 (1983)
16. R. Car, M. Parrinello: Phys. Rev. Lett. **55**, 2471 (1985)
17. See, for example, F. Stillinger, T.A. Weber, R.A. LaViolette: J. Chem. Phys. **85**, 6460 (1986)
18. J. Donohue: *The Structures of the Elements* (Wiley, New York, 1974), Chaps 8 [group Va] and 9 [group VIa]
19. R. Steudel: In *Studies in Inorganic Chemistry*, ed. by A. Müller, B. Krebs (Elsevier, Amsterdam, 1984) Vol. 5, p. 3
20. R. Steudel, E.M. Strauss: Adv. Inorg. Chem. Radiochem. **28**, 135 (1984)
21. R. Cooper, J.V. Culka: J. Inorg. Nucl. Chem. **29**, 1217 (1967)
22. M. Schmidt, E. Wilhelm: Z. Naturforsch. **25 B**, 1348 (1970)
23. R. Steudel, E.M. Strauss: In *The Chemistry of Inorganic Homo- and Heterocycles* (Academic, London, 1987) Vol. 2, p. 769; R. Steudel, R. Laitinen: Topics in Current Chemistry **102**, 177 (1982); *Selenium: Gmelin Handbuch der Anorganischen Chemie*, 8. Aufl., ed. by H. Bitterer (Springer, Berlin, Heidelberg, 1984), Ergänzungsband B2
24. R. Laitinen, L. Niinistö, R. Steudel: Acta Chem. Scand. A **33**, 737 (1979)
25. D. Hohl, R.O. Jones, R. Car, M. Parrinello: J. Chem. Phys. **89**, 6823 (1988)
26. R.O. Jones, D. Hohl: Int. J. Quantum Chem. Symposium [in press]
27. D. Hohl, R.O. Jones, R. Car, M. Parrinello: J. Am. Chem. Soc. **111**, 825 (1989)
28. R.O. Jones, D. Hohl: J. Am. Chem. Soc. **112**, 2590 (1990)
29. R. Laitinen, T. Pakkanen: J. Mol. Struct. (Theochem) **91**, 337 (1983); **124**, 293 (1985)
30. T. Maekawa, T. Yokokawa, K. Niwa: Bull. Chem. Soc. Japan **46**, 761 (1973).
31. J. Drowart, S. Smoes: J. Chem. Soc., Faraday Trans II **73**, 1755 (1977)
32. H.H. Eysel, S. Sunder: Inorg. Chem. **79**, 2626 (1979)
33. D.E.C. Corbridge: *The Structural Chemistry of Phosphorus* (Elsevier, Amsterdam, 1974); D.E.C. Corbridge: *Phosphorus. An Outline of its Chemistry, Biochemistry and Technology*, Third Edition (Elsevier, Amsterdam, 1985)
34. L. Kerwin: Can. J. Phys. **32**, 757 (1954); J.D. Carette, L. Kerwin: Can. J. Phys. **39**, 1300 (1961)
35. H. Bock, H. Müller: Inorg. Chem. **23**, 4365 (1984)
36. K. Raghavachari, R.C. Haddon, J.S. Binkley: Chem. Phys. Lett. **122**, 219 (1985); R. Ahlrichs, S. Brode, C. Ehrhardt: J. Am. Chem. Soc. **107**, 7260 (1985)
37. T.P. Martin: Z. Phys. D **3**, 221 (1986)
38. R.O. Jones, D. Hohl: J. Chem. Phys. **92**, 6710 (1990); R.O. Jones, G. Seifert: J. Chem. Phys. **96**, 7564 (1992)
39. H. Thurn, H. Krebs: Acta Cryst. B **25**, 125 (1969)
40. P.E. Eaton, T.W. Cole Jr: J. Am. Chem. Soc. **86**, 962, 3157 (1964)
41. L. Cassar, P.E. Eaton, J. Halpern: J. Am. Chem. Soc. **92**, 6366 (1970)
42. P.A.M. Dirac: Proc. Cambridge Philos. Soc. **26**, 376 (1930)
43. J.C. Slater: Phys. Rev. **81**, 385 (1951); **82**, 538 (1951)
44. R. Gáspár: Acta Phys. Hung. **3**, 263 (1954)
45. W.D. Knight, K. Clemenger, W.A. de Heer, W.A. Saunders, M.Y. Chou, M.L. Cohen: Phys. Rev. Lett. **52**, 2141 (1984); W.D. Knight, W.A. de Heer, W.A. Saunders, K. Clemenger, M.L. Cohen: Chem. Phys. Lett. **134**, 1 (1987)
46. W. Ekardt: Phys. Rev. B **29**, 1558 (1984)
47. W. Ekardt, Z. Penzar: Phys. Rev. B **38**, 4273 (1988)
48. T.H. Upton: J. Chem. Phys. **86**, 7054 (1987)

2.4 Transition from van der Waals to Metallic Bonding in Clusters

G.M. Pastor and *K.H. Bennemann*

2.4.1 Introduction

The physics of clusters studies the characteristic properties of aggregates which size is intermediate between the atom and the solid, and thus bridges the gap between atomic and bulk-like behavior. It is a cross disciplinary field which links, improves, and profits from our present understanding of atoms, molecules, surfaces and solids. In the past, clusters were mainly used as models for calculating surface or bulk electronic properties. However, in the past years, significant interest in the clusters themselves has developed due to their fundamental importance in both basic and applied science [1, 2].

One of the fundamental problems in cluster research is to understand how the physical properties vary as the electrons of a single atom become part of a group of atoms and how bulk-like behavior is achieved. Particularly interesting are metal clusters, for which a transition from localized, atomic-like to delocalized, band-like electronic states occurs with increasing cluster size. This transition will be exhibited in most of the electronic properties, for example: in the nature of the chemical bonding, as a transition from van der Waals to covalent or metallic bonding; in the optical properties, as a transition from atomic lines to collective excitations (e.g., plasmons); in the magnetic properties, as a transition from localized magnetic moments to magnetic moments due to itinerant electrons; or in the catalytic activity and related quantities like ionization potential, electron affinity, etc.

Interesting size dependent properties are expected for Hg_n and other divalent metal clusters (e.g., Be_n, Mg_n). These elements have a closed-shell atomic configuration causing the very small clusters to be insulating and bounded through weak van der Waals forces, in contrast to the corresponding bulk material which has metallic properties. Thus, one expects a transition from van der Waals to metallic bonding to occur with increasing cluster size, as illustrated in Fig. 1.1. Experimentally, metal-insulator transitions have been recently observed in expanded divalent-metal liquids. For example, in liquid-Hg a rather gradual metal-semiconductor transition occurs in the density range $\rho \simeq 8$–$9\ g/cm^3$ [3]. Comparing the average coordination number in the liquid at the critical density ($z \simeq 6$–7) with that of Hg_n, *Tománek* et al. [4] estimated that an insulator to metal transition should take place in Hg_n-clusters with $20 \lesssim n \lesssim 50$. This physical

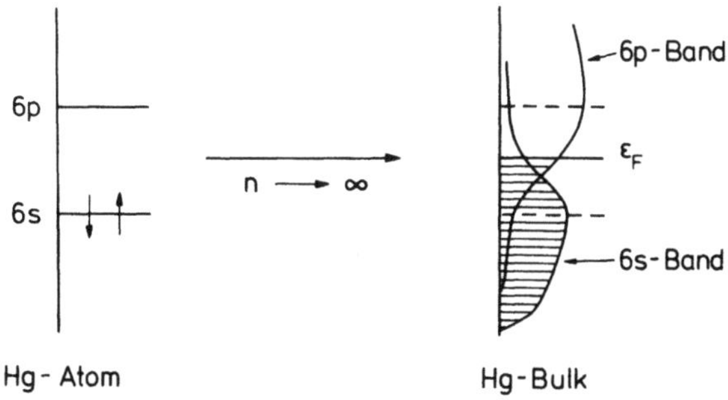

Fig. 1.1. Illustration of the electronic structure of Hg-atom and Hg-bulk. Note the atomic close shell configuration ([Xe]$4f^{14}5d^{10}6s^2$) and the metallic-like bulk configuration with overlapping $6s$- and $6p$-bands ([Xe]$4f^{14}5d^{10}6s^{1.4}6p^{0.6}$)

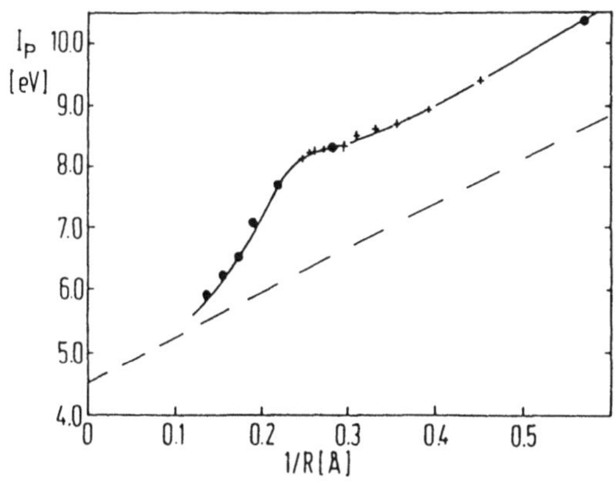

Fig. 1.2. Experimentally determined ionization energy of Hg_n-clusters as a function of $1/R$, where R refers to the approximate cluster radius ($R = (3v_b/4\pi)^{1/3}n^{1/3}$) (after *Rademann* et al. [5])

picture seems to be confirmed by recent experimental results for the ionization energy [5] and $5d \to 6p$ autoionization spectra [6] of Hg_n-clusters.

In Fig. 1.2 experimental results for the ionization energy of Hg_n-clusters are shown as a function of $1/R$, where R refers to the cluster radius [5]. For $n \lesssim 13$ a smooth decrease of I_n with increasing n is observed ($I_n - I_1 \simeq e^2/2R(n) - e^2/2R(1)$). However, I_n shows in this cluster size range ($n \lesssim 13$) a size dependence which seems not to converge towards bulk-like behavior ($I_n(n \to \infty) \simeq \phi + e^2/2R$, where ϕ refers to the bulk work function). Furthermore, by extrapolating I_n for $n \lesssim 13$ to the bulk limit one obtains $I_n(n \to \infty) \simeq 6.5$ eV, a value about 2 eV larger than the bulk work-function $\phi = 4.49$ eV. For $n \gtrsim 13$ a qualitative change in the size dependence of I_n is observed: I_n decreases much more rapidly and approximately converges to the predictions of the classical spherical-droplet model ($I_n(n \to \infty) = \phi + e^2/2R$) for $n \simeq 140$ [5, 7]. It is a challenge for theory to explain why I_n behaves atomic like for $n \lesssim 13$ and what causes the relative sudden transition for $n \gtrsim 13$.

Other very interesting experimental results regarding this transition are the inner-shell autoionization spectra of Hg_n-clusters obtained by *Bréchignac* et al.

[6]. In this experiment a $5d$ core electron is excited into an unoccupied $6p$-state after the absorption of synchrotron radiation:

$$Hg[5d^{10}6s^2(^1S_0)] + \hbar\omega \rightarrow Hg^*[5d^9 6s^2 (D_{5/2,\,3/2})6p] \ . \quad (1.1)$$

If the absorbed energy is larger than the ionization potential, the excitation decays mainly through an Auger-like process in which the $5d$ hole is filled back and a valence electron is emitted:

$$Hg^*[5d^9 6s^2 6p] \rightarrow Hg^+[5d^{10} 6s] + e^- \ . \quad (1.2)$$

This kind of transitions probes mainly the electronic structure of the valence states of neutral Hg_n and particularly its localized or delocalized character, since the Auger-like process occurs more likely when the $5d$-hole and excited $6p$-electron are on the same atom. The experimentally recorded photoionization efficiency curves are shown in Fig. 1.3. A transition from atomic-like to band-like behavior can be recognized in the size dependence of the line shapes. For

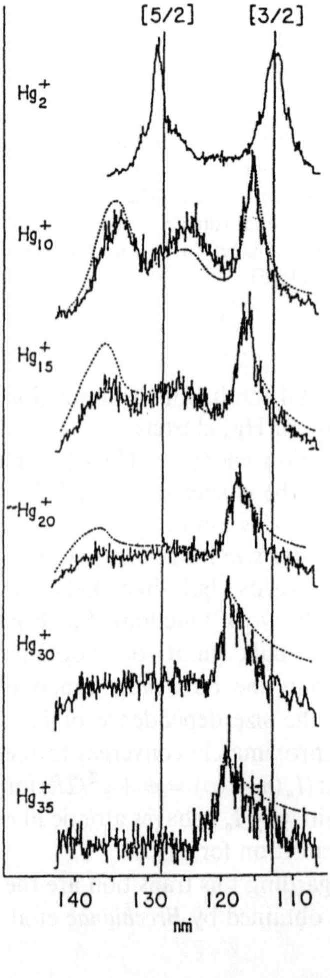

Fig. 1.3. Experimentally determined $5d \rightarrow 6p$ photoionization efficiency curves of Hg_n-clusters. The dashed lines indicate the normalized intensity and the vertical lines the position of the atomic transitions (after *Bréchignac* et al. [6])

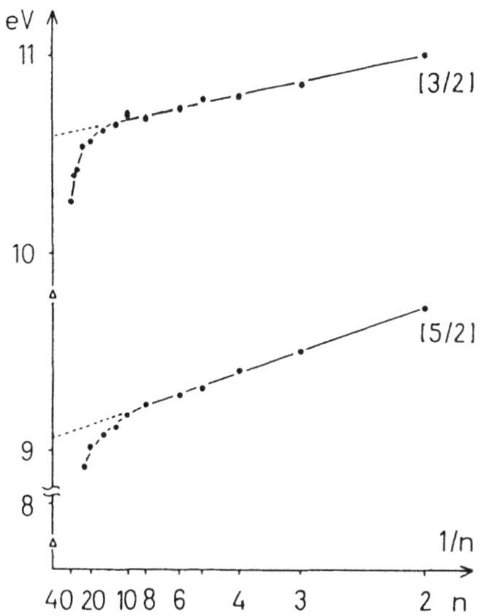

Fig. 1.4. Experimentally determined autoionization resonance energy δ_n correlated to the $5d \rightarrow 6p$ transitions given in Eq. (1.1) (after *Bréchignac* et al. [6])

$n \lesssim 15$ the lines are symmetric, like in an atomic transition, and the linewidth $\Delta\lambda_n$ is approximately independent of cluster size ($\Delta\lambda_1 \simeq \Delta\lambda_{10}$). For $n \gtrsim 15$ the [3/2]-line becomes asymmetrically broader indicating the formation of a $6p$-band. Notice that the [5/2]-line is poorly resolved for larger clusters and that the resonance disappears for $n \gtrsim 30$.

The transition from atomic-like to band-like behavior can be also clearly seen in the size dependence of the $5d \rightarrow 6p$ autoionization resonance frequency δ_n of Hg_n. Experimental results for δ_n are given in Fig. 1.4 as a function of $1/n$. For small clusters ($n \lesssim 15$) δ_n decreases rather slowly with increasing n, the red shift being approximately proportional to $1/n$ (i.e., $(\delta_1 - \delta_n) \propto (1 - 1/n)$). Notice that extrapolating this linear dependence to the bulk limit one does not obtain the experimental result for bulk-Hg but a value more than 1 eV larger. For $n \gtrsim 15$ the size dependence changes, and δ_n decreases much more rapidly yielding the expected convergence to the bulk limit. It is the task of theory to explain the observed changes in structure, position and shape of the photoionization efficiency curves of Hg_n as a function of cluster size.

Clearly, it is of considerable interest to understand the electronic origin of the qualitative changes in the physical properties of Hg_n-clusters (I_n, δ_n, $\Delta\lambda_n$, etc.) observed at the critical size $n_{cr} \simeq 13$–20. One of the goals of this work is to shed light on this point.

In the following we develop a tight binding type electronic theory applicable to *sp*-bonded clusters as well as to other clusters (e.g., transition-metal clusters) in order to explain most of the recent experimental results discussed above. We determine as a function of cluster size for Hg_n, as an example for divalent metal clusters:

- the transition from van der Waals to metallic bonding,
- the cohesive energy,

- the ionization energy,
- the autoionization spectra,
- the electronic structure (density of states), etc.

The aim of these calculations is to improve our understanding of the microscopic changes occurring along the transition from atomic- to bulk-like behavior. From comparison with experiment one may also learn about the accuracy of the tight-binding method and necessary improvements regarding future studies.

In Section 2.4.2 we describe our theoretical method. In Section 2.4.3 numerical results for various properties of Hg_n-clusters are presented and discussed in detail. The conclusions presented in Section 2.4.4 discuss limitations and possible extensions of the present approach. In the appendices details and extensions of the theory are also given, useful for a critical appraisal of this work and also of interest for future studies.

2.4.2 Theory for the Electronic Properties of Divalent-Metal Clusters

In order to study the characteristic physical behavior of clusters which size is intermediate between atom and solid, we develop a tight-binding type electronic theory. This approach seems particularly appropriate for studying size dependent cluster properties, since tight-binding theory describes correctly not only the atomic and bulk-like behavior, but also the changes in the electronic structure resulting from changes in the local environment, as is well known from studies on surfaces and alloys. The formalism allows to treat *s*, *p* and *d* electrons on equal footing and thus can be used to investigate a wide variety of elements and compounds throughout the periodic table. We apply this theory to study *divalent-metal clusters* (e.g., Hg_n), for which a qualitative change in the nature of the chemical bonding (from van der Waals to metallic like bonding) is expected to occur with increasing cluster size. Applications of this theory to other systems (e.g., transition-metal clusters) have been reported elsewhere [8, 9].

In order to determine for Hg_n the transition from van der Waals to metallic bonding, the *sp*-band formation, the ionization energy I_n, the $5d \rightarrow 6p$ autoionization resonance energy δ_n, and other cluster properties, we consider the tight-binding Hamiltonian for the 6*s*- and 6*p*-valence electrons:

$$\hat{H} = \sum_{i,\alpha,\sigma} \varepsilon\alpha \hat{n}_{i\alpha\sigma} + \sum_{\substack{i \neq j \\ \alpha,\beta,\sigma}} t_{ij}^{\alpha\beta} \hat{c}_{i\alpha\sigma}^{+} \hat{c}_{j\beta\sigma} + \frac{1}{2} \sum_{\substack{i\alpha\sigma \\ j\beta\sigma'}}{}' U_{ij} \hat{n}_{i\alpha\sigma} \hat{n}_{j\beta\sigma'} \,. \tag{2.1}$$

Here $\hat{c}_{i\alpha\sigma}^{+}$ ($\hat{c}_{i\alpha\sigma}$) refers to the usual creation (annihilation) operator of an electron with spin σ at atomic site i in the orbital α ($\alpha \equiv s, p_x, p_y, p_z$), and $\hat{n}_{i\alpha\sigma} = \hat{c}_{i\alpha\sigma}^{+} \hat{c}_{i\alpha\sigma}$ to the corresponding electron number operator. $\varepsilon_{i\alpha}^{0}$ stand for the bare orbital energies (i.e., excluding electron-electron interactions), and $t_{ij}^{\alpha\beta}$ for the hopping integrals between sites i and j. The first term in Eq. (2.1) describes electrons in the localized atomic-like orbitals, the second term the hopping of electrons between neighboring atoms leading to electron delocalization, and the third term the Coulomb interactions between electrons (approximated by the Hubbard-like form [10]), which in small clusters tend to suppress electron delocalization (see Appendix A).

2.4.2.1 Criterion for Covalent Bond Formation

In order to estimate the critical cluster size n_{cr} for the transition from van der Waals bonding with localized, atomic-like electronic states to covalent bonding with delocalized, band-like electronic states, we consider the microscopic process of electron delocalization in some detail. When the electrons delocalize and thereby jump from an atom to one of its neighbors, pairs of charged atoms ($Hg^+ \leftrightarrow Hg^-$) arise (see Fig. 2.1). This causes an increase in Coulomb energy approximately given by $\Delta E = U_{sp} - e^2/d$. Due to the uncertainty relation for energy and time, an electron could stay in this state only for a limited time $\tau_{fl} \simeq \hbar/\Delta E$ after which it would jump back into the localized atomic-like state. The delocalization can take place only if the other electrons can screen or reneutralize the charge fluctuation within the time τ_{fl}. This is the case if the hopping time τ_{hop} is shorter than τ_{fl}:

$$\hbar/W_s \simeq \tau_{hop} \lesssim \tau_{fl} \simeq \hbar/\Delta E \qquad (2.2)$$

or equivalently,

$$W_s(n_{cr}) \gtrsim U_{sp} - e^2/d \ , \qquad (2.3)$$

where W_s refers to the s-bandwidth. This criterion corresponds to Mott's criterion for metal-insulator transitions [11] and reflects the usual competition between kinetic and Coulomb energy. As in Mott's theory, delocalization occurs when a critical bandwidth and efficient screening are achieved. The main difference to Mott's famous "Gedankenexperiment" is that for clusters the critical bandwidth is related to a critical coordination number instead of being related to a critical interatomic distance (critical density). Equation (2.3) is estimating the transition to covalent bonding. For $n < n_{cr}$ covalent bonding is expected to be energetically not favorable and van der Waals bonding should dominate. From our calculations we obtain $n_{cr} \simeq 13$ for Hg_n. However, the

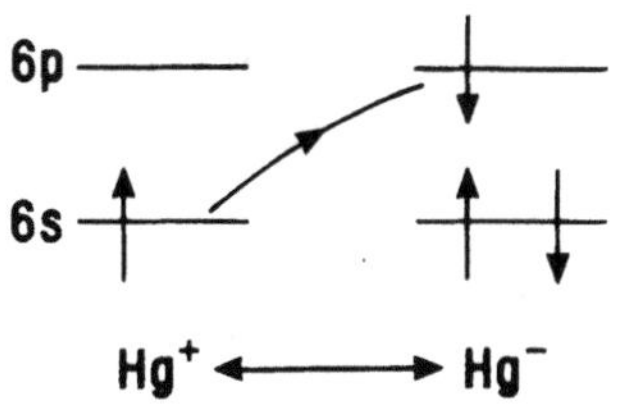

Fig. 2.1. Illustration of a charge fluctuation due to electronic sp-transitions (hopping) between neighboring atoms

quantitative aspect of Eq. (2.3) is not as important as the qualitative conclusions one derives from it: Eq. (2.3) shows that for $n < n_{cr}$ the properties of Hg_n-clusters result mainly from localized electrons occupying atomic-like orbitals whereas for $n > n_{cr}$ they result from delocalized electrons occupying molecular orbitals. This is an important fact which must be considered explicitly in our calculations. More strictly, the transition from localized, van der Waals bonding to delocalized, covalent bonding should result from a more accurate calculation including a many-body extension of Eq. (2.1). In Appendix A we outline a slave-boson method [12], which provides a framework for developing a more general theory describing as a function of cluster size n the interplay between kinetic energy terms t_{ij}, Coulomb interaction terms U_{ij}, and van der Waals terms, via a size dependent renormalization of the hopping integrals: $t \to t_{\text{eff}}(U, n)$ with $t_{\text{eff}} < t$ [13].

2.4.2.2 Cohesive Energy

The transition from van der Waals to covalent/metallic clusters at n_{cr} is most directly calculated from the cohesive energy $E_{\text{coh}}(n)$. Since Coulomb interactions tend to suppress covalent bond formation in small clusters, we expect covalent/metallic like bonding only when $E_{\text{coh}}^{\text{cov}} > E_{\text{coh}}^{\text{vdW}}$, where $E_{\text{coh}}^{\text{cov}}$ refers to the cohesive energy due to covalent or metallic bonding and $E_{\text{coh}}^{\text{vdW}}$ to the cohesive energy due to van der Waals like bonding.

The binding energy $E_{\text{coh}}^{\text{cov}}$ is given by [14, 15]

$$E_{\text{coh}}^{\text{cov}}(n) = \sum_{\alpha} \int_{-\infty}^{\varepsilon_F} d\varepsilon\, N_\alpha(\varepsilon)(\varepsilon_s - \varepsilon) - E_{\text{corr}} - E_R\,. \tag{2.4}$$

The first term in Eq. (2.4) represents the energy gained upon band formation (i.e., electron delocalization). $N_\alpha(\varepsilon)$ ($\alpha \equiv s, p_x, p_y, p_z$) denotes the electronic density of states per atom calculated from Eq. (2.1) in the Hartree–Fock approximation using the recursion method [16] or numerical diagonalization. E_{corr} results from charge fluctuations and correlations which in small clusters tend to suppress the delocalization of the s-electrons. Approximately $E_{\text{corr}} \simeq (U_{sp} - e^2/d)\,\nu_p/\kappa$, where the average number of p-electrons $\nu_p = \int_{-\infty}^{\varepsilon_F} N_p(\varepsilon)\,d\varepsilon$ gives approximately the probability for a charge fluctuation to occur, and the size dependent dielectric constant κ takes into account the screening of the charge fluctuations due to correlations. For the dielectric constant κ we use the model $\kappa = 1 + 4\pi\chi$, with $\chi = (z_n/z_b)(\gamma/v)(\varepsilon_p - \varepsilon_s)/\Delta_n$, where γ refers to the atomic polarizability, z_n to the average coordination number (z_b = bulk coordination number), v to the volume per atom and Δ_n to the gap between occupied and unoccupied states.[1] The Born–Mayer repulsive energy E_R is given

[1] The expression for χ, although oversimplified, yields proper limiting values: no screening ($\kappa = 1$) for an isolated bond and perfect screening ($\kappa = \infty$) in a metal. Furthermore, it gives good results for bulk semiconductors.

by $E_R = (1/2n)\sum_{i=1}^{n} z_i A \exp\{-p(d/d_b - 1)\}$. Here, z_i refers to the coordination number for an atom at site i, and d to the bond-length (d_b = bulk bond-length). The parameters A, p are fitted as usual to bulk compressibility and bulk atomic volume v_b. The distance dependence of the hopping integrals t_{ij} is given by $t_{ij} \propto d_{ij}^{-2}$ [17]. Notice that the approximate power law $t_{ij} \propto d^{-2}$ is valid only for rather small deviations ($\sim 10\%$) from bulk interatomic distance [17], since for larger distances t_{ij} decays exponentially like the overlap between neighboring orbitals. We then maximize $E_{\text{coh}}^{\text{cov}}$ with respect to d (uniform relaxation) and obtain the average equilibrium bond-length d_n and cohesive energy $E_{\text{coh}}^{\text{cov}}(n)$.

The interaction between Hg-atoms in Hg_n with $n < n_{\text{cr}}$ is probably more complicated than simple van der Waals attraction, due to the non-negligible overlap of the valence wavefunctions. However, we will still denote this bonding as van der Waals. Anyway, as long as electron delocalization is prevented we expect pairwise interactions between the atoms, and $E_{\text{coh}}^{\text{vdW}}$ to be approximately proportional to the average coordination number z_n. Thus, we calculate $E_{\text{coh}}^{\text{vdW}}$ from

$$E_{\text{coh}}^{\text{vdW}}(n) = (z_n/2)\, E_{\text{coh}}^{\text{exp}}(2)\,, \tag{2.5}$$

where $E_{\text{coh}}^{\text{exp}}(2)$ stands for the observed [18] binding energy of Hg_2 including in addition to van der Waals attraction all other bonding contributions. This simple bond model interpolates well between the cohesive energy of Xe_2 and bulk-Xe, for example (see also results in [13]).

2.4.2.3 Ionization Energy

The ionization energy I_n, which is sensitive to the 6s-band formation and thus to electron delocalization, is given by

$$I_n = E_n(v_t - 1) - E_n(v_t)\,, \tag{2.6}$$

where $E_n(v_t)$ refers to the total electronic energy of a n-atom cluster with v_t valence electrons. Approximately, (see Appendix B), we have

$$I_n = -\varepsilon_{\text{HOS}}(n) + \frac{e^2}{2R}\,, \tag{2.7}$$

where $\varepsilon_{\text{HOS}}(n)$ stands for the energy of the highest occupied state, and $R = (3v/4\pi)^{1/3} n^{1/3}$ for the radius of the cluster approximated by a sphere ($v = v_b(d_n/d_b)^3$) [8, 14]. The first term in Eq. (2.7) contains the changes in I_n resulting from size dependent changes in the electronic structure, and the second term is the electrostatic contribution due to the shift of the energy levels occurring for $v_t \to (v_t - 1)$. As illustrated in Fig. 2.2, for $n < n_{\text{cr}}$ the electron is emitted from an atomic-like state and thus $\varepsilon_{\text{HOS}}(n) \simeq \varepsilon_s(\text{atom})$, whereas for $n > n_{\text{cr}}$ the electrons are delocalized and $\varepsilon_{\text{HOS}}(n)$ refers to the energy of the highest occupied molecular orbital (HOMO) resulting from our Hartree–Fock calculations.

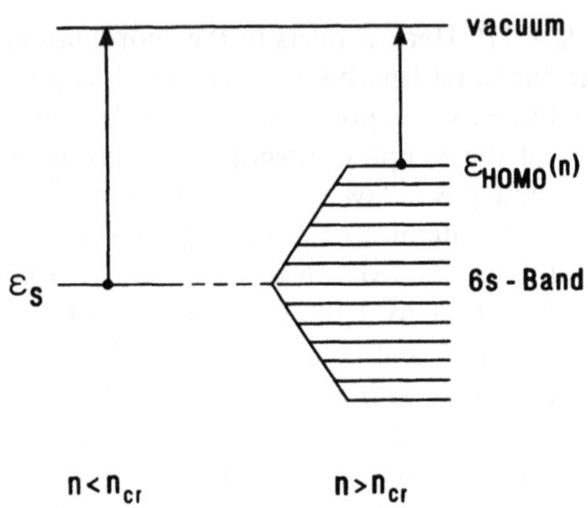

Fig. 2.2. Illustration of the ionization process for localized ($n < n_{cr}$) and delocalized ($n > n_{cr}$) electronic states in Hg_n

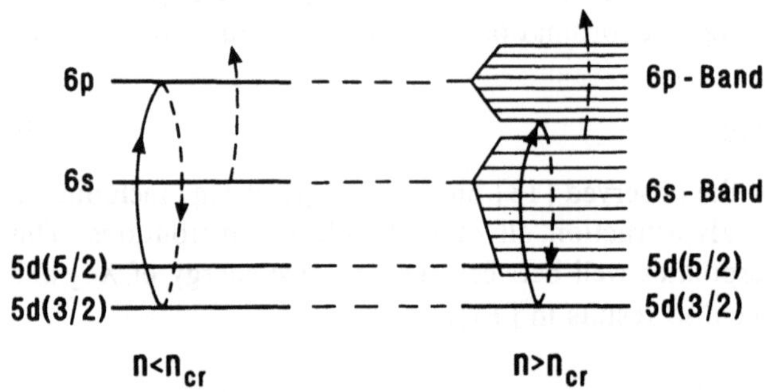

Fig. 2.3. Illustration of the autoionization process due to $5d \rightarrow 6p$ transitions in Hg_n-clusters. For $n < n_{cr}$ the excited $6p$ electron stays localized in an atomic like state (Frenkel exciton), whereas for $n > n_{cr}$ it occupies the lowest unoccupied molecular orbital (Mott–Wannier exciton)

2.4.2.4 Autoionization Resonance Energy Due to $5d \rightarrow 6p$ Transitions

The autoionization resonance resulting from the excitation of a $5d$ core electron into an unoccupied $6p$ state (see Fig. 2.3) is expected to reflect the $6p$-electron delocalization and $6p$-band formation since the involved Auger process occurs most likely if the $5d$-hole and excited $6p$-electron are on the same atom. Thus, the position δ_n and width of the autoionization resonance are expected to exhibit clearly the transition from van der Waals to metallic clusters [6]. The resonance position is given by

$$\delta_n = \varepsilon_{6p} - \varepsilon_{5d} + \varepsilon_{pd} \,, \tag{2.8}$$

where ε_{pd} denotes the interaction energy between the excited $6p$-electron and the $5d$-hole, and ε_{6p} refers to the energy of the lowest unoccupied $6p$-state. For smaller clusters with $n < n_{cr}$ the excited $6p$-electron stays localized in an atomic-like state (Frenkel exciton, see Fig. 2.3) whose energy is shifted due to attractive Coulomb interactions and polarization of neighboring atoms. Hence, we use the Ansatz $\varepsilon_{6p} \simeq \varepsilon_{6p}^0 - \alpha z_n$. Also then ε_{pd} is approximately constant ($\varepsilon_{pd} \simeq U_{pd}$). This

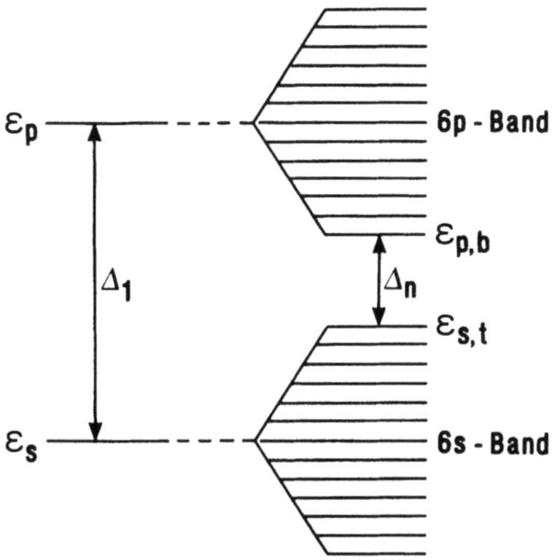

Fig. 2.4. Illustration of the size dependent gap Δ_n between occupied s-like and unoccupied p-like states in Hg_n

gives $\delta_n \simeq \delta_1 - \alpha z_n$. The constant α is fitted to the observed red shift in Hg_2 [6]. For larger clusters with $n > n_{cr}$ p-level broadening becomes increasingly important and ε_{6p} refers then to the energy of the lowest unoccupied $6p$-state given by our results for $N_p(\varepsilon)$ (Mott–Wannier exciton, see Fig. 2.3). Thus, $\varepsilon_{pd} \simeq -e^2/\kappa R$ for larger clusters with delocalized excited $6p$-electron and localized $5d$-hole.

2.4.2.5 Band Gap Δ_n

A measure for metallic like behavior is also the gap

$$\Delta_n = (\varepsilon_{p,b} - \varepsilon_{s,t}) \tag{2.9}$$

between the s- and p-states. Here, $\varepsilon_{p,b}$ refers to the bottom of the p-band and $\varepsilon_{s,t}$ to the top of the s-band (see Fig. 2.4). Δ_n is controlled mainly by the hopping integrals t_{ij} and thus depends to some extent on interatomic distances and cluster relaxation. Calculating $N_\alpha(\varepsilon)$ from Eq. (2.1) one obtains $\varepsilon_{p,b}$ and $\varepsilon_{s,t}$. The resulting Δ_n should clearly exhibit the transition from van der Waals to metallic clusters. It is physically relevant (e.g., for optical properties) to distinguish between covalent and metallic bonding. Metallic properties are expected for $\Delta_n \approx 0$.

2.4.3 Properties of Hg_n-Clusters as a Function of Cluster Size: The Transition from van der Waals to Covalent to Metallic Bonding

We present and discuss now results for various properties of divalent-metal clusters obtained by using the tight-binding theory developed in Section 2. The aim of these calculations is to determine for some representative examples the changes occurring in the electronic structure along the transition from atomic- to bulk-like behavior, and to understand how these changes result in the characteristic size dependence of different cluster properties. Particularly we discuss the transition from van der Waals to covalent to metallic Hg_n-clusters. This specific problem was chosen because of its relevance for fundamental topics in cluster research, like the transition from localized, atomic-like to delocalized, band-like electronic states in metal clusters, the development of band structure with increasing cluster size, or the relation between cluster structure and different observable properties like ionization energy, cohesive energy, etc. Furthermore, these physical systems (e.g., Hg_n) are particularly interesting, since for them detailed experimental results are available. Thus, it is a challenge for theory to explain these results microscopically. Comparison with experiment provides in addition unique information about the accuracy of the tight-binding theory and necessary improvements concerning future studies.

In order to determine the transition from van der Waals to metallic bonding expected to occur in Hg_n-clusters with increasing cluster size, and how this transition reflects in various cluster properties, we calculate:

- ionization energy I_n, which is sensitive to the 6*s*-band formation and thus to electron delocalization,
- autoionization resonance energy δ_n due to $5d \rightarrow 6p$ transitions, which measures the 6*p*-band formation as a function of cluster size *n*,
- cohesive energy $E_{\mathrm{coh}}(n)$, from which the dominant bonding mechanism (van der Waals, covalent or metallic) is determined as a function of *n*,
- band gap Δ_n between occupied *s*-like and unoccupied *p*-like states, which allows to distinguish between insulating, covalent and metallic behavior, and
- electronic densities of states, which reflect the 6*s*- and 6*p*-band structure and are relevant for photoemission experiments.

The parameters ε_s, ε_p, ε_{5d}, t_{ij}, and U_{sp} used for the calculations are determined as follows. The *sp*-promotion energy $\varepsilon_p - \varepsilon_s = 5.8$ eV is obtained from the average between the ^3P and ^1P atomic $6s^2 \rightarrow 6s6p$ transitions [19]. The hopping integrals t_{ij} are fitted to bulk-Hg. For simplicity and in order to keep the number of parameters small, we use the following relation between the hopping integrals [17]: $t_{ss\sigma} = -1.32t$, $t_{sp\sigma} = 1.42t$, $t_{pp\sigma} = 2.22t$, $t_{pp\pi} = -0.63t$, and $t = 0.44$ eV which was fitted to reproduce the average number of *p*-electrons per atom $\nu_p = \int_{-\infty}^{\varepsilon_F} d\varepsilon N_p(\varepsilon)$ obtained in APW calculations [3] for fcc bulk-Hg.[2] The position of the 6*s*-level (ε_s) and the 5*d*-level (ε_{5d}) with respect to vacuum are fitted to the bulk work function $\phi = 4.5$ eV and bulk $5d \rightarrow 6p$ transition energy [21], respectively. The Coulomb integral $U_{sp} = 7$ eV is taken from atomic Hartree-Fock calculations [22]. From other sources [23] we obtain $U_{sp} = 6.5$ eV. Using this value or determining the atomic-like parameters ($\varepsilon_s - \varepsilon_p$) and U_{sp} from bulk properties of Hg, would not change qualitatively our results shown in Figs. 3.1–7. Since both van der Waals clusters and Hg-bulk have close-packed

[2] The lattice structure of Hg-bulk is taken to be fcc since the actual rhombohedral structure can be seen as slightly distorted fcc structure [20].

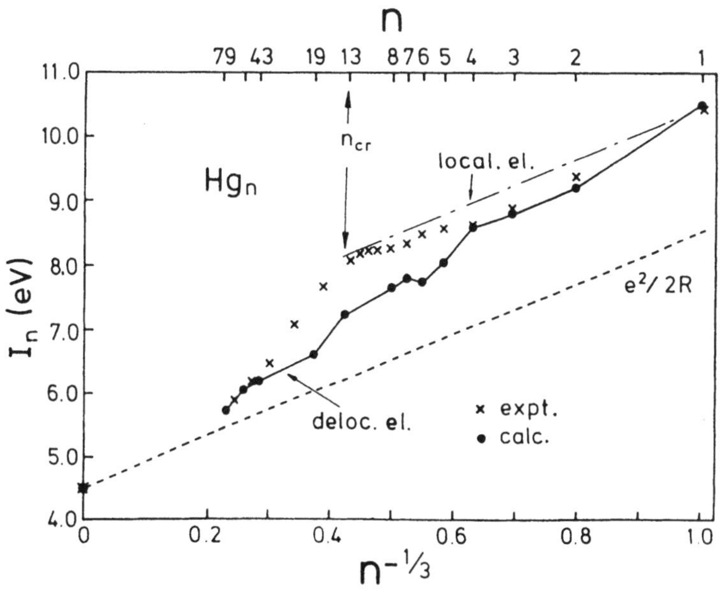

Fig. 3.1. Size dependence of the ionization energy I_n of Hg_n-clusters. Crosses indicate experimental results [5, 24], while dots the calculated ones. The dashed curve is obtained by setting $t = 0$ to simulate the "blocking" of the hopping integrals due to correlations. For $n \lesssim n_{cr}$ corrections to the mean-field results due to correlations are necessary [13]

structures we assume for Hg_n such cluster geometries: for $n \leq 13$ those shown in the inset of Fig. 3.2 and for $n \geq 13$ fcc-like structures. For small clusters several structures were examined. The most stable ones are shown in Fig. 3.2. Note that small van der Waals like clusters seem to have icosahedral structure. However, such differences in structure cause almost negligible changes in the properties we calculate.

In Fig. 3.1 results are given for the size dependence of the ionization energy I_n of Hg_n. For bulk-like clusters ($n \gtrsim 80$–135) the position of the energy of the highest occupied state ε_{HOS} with respect to the s-level energy ε_s is approximately the same as for the bulk, i.e., $\varepsilon_{HOS}(n) \simeq \varepsilon_{HOS}(b) = \varepsilon_F(b)$, where $\varepsilon_F(b) = -\phi$ refers to the bulk Fermi energy ($\phi =$ bulk work function). Therefore, we have $I_n \simeq \phi + e^2/2R$ ($n \gtrsim 80$–135). For $n < 80$ a gap Δ_n opens between occupied s-like and unoccupied p-like states, as a result of the reduction of the coordination number z_n and of the s- and p-band width occurring with decreasing cluster size ($W_s \sim \sqrt{z_n}$, $W_p \sim \sqrt{z_n}$, see Figs. 3.5–7). Then we have $\varepsilon_{HOS}(n) < \varepsilon_F(b)$, where now $\varepsilon_{HOS}(n)$ refers to the energy of the top of the s-band. Therefore, for $n < 80$, $I_n > \phi + e^2/2R$.[3]

[3] In contrast, alkali-metal clusters follow approximately the R^{-1}-law down to much smaller cluster sizes. This can be understood recalling that alkali-metal clusters with one valence electron per atom have a half-filled band and thus $\varepsilon_{HOS}(n) \simeq \varepsilon_s$ even though the band width W_s decreases with decreasing cluster size.

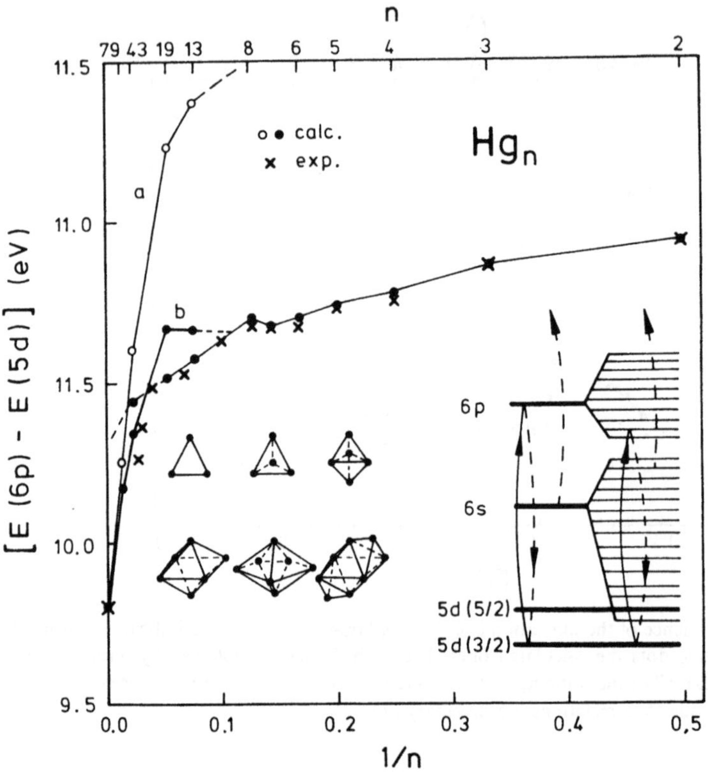

Fig. 3.2. Cluster size dependence of the autoionization resonance $(5d^{10}6s^2(^1S_0) \rightarrow 5d^9 6s^2(D_{3/2})6p)$ calculated from $\delta_n = \varepsilon_{6p} - \varepsilon_{5d} + \varepsilon_{pd}$. We use for $n \lesssim 13$ $\varepsilon_{6p} \simeq \varepsilon^0_{6p} - \alpha z_n$ and $\varepsilon_{pd} \simeq \text{const.} \simeq U_{pd}$, since electrons are localized (curve c). The constant α is fitted to the red shift observed for Hg_2. For $n \gtrsim 13$ we use for ε_{6p} the energy of the lowest unoccupied p-state (curve a: $\varepsilon_{pd} = 0$; curve b: $\varepsilon_{pd} = -e^2/\kappa R$). Crosses indicate experimental results [6]. The inset figures illustrate the assumed cluster geometries for $n \leq 13$, for larger clusters fcc structure is assumed

For Hg_n with $n < 80$ we have a full, mainly s-like, band and thus the reduction of W_s causes I_n to increase beyond the electrostatic contribution $e^2/2R$ (see Fig. 3.1). However note, we underestimate the reduction of W_s in small Hg_n clusters with $n \lesssim 20$, since we neglected electron correlations (beyond Hartree–Fock approximation), which tend to suppress electron hopping and thus band broadening in small Hg-clusters. Electron-electron interactions may result in a renormalization of the hopping integrals: $t \rightarrow t_{\text{eff}}(U, n)$, $t_{\text{eff}} < t$. As shown in Appendix A, t_{eff} should decrease with decreasing cluster size n, since the kinetic energy $E_{\text{band}}(n)$ gained upon delocalization and the screening constant $\kappa(n)$ decrease with decreasing n, whereas the unscreened Coulomb interaction energy due to an electronic charge fluctuation $\Delta E \simeq U_{sp} - e^2/d$, being a local property, remains essentially unchanged. In Hg_n with n smaller than the critical size n_{cr} the Coulomb interactions U_{ij} between the electrons favor the highly correlated van der Waals state where electrons are localized instead of the covalent bonded state where charge fluctuations are necessarily present. We expect van der Waals bonding to dominate (i.e., $t_{\text{eff}} \to 0$) when the hopping time $\tau_{\text{hop}} \simeq \hbar/W_s$ becomes larger than the lifetime of a charge fluctuation $\tau_{fl} \simeq \hbar/\Delta E$, i.e., for $W_s(n) < W_s(n_{\text{cr}}) \simeq U_{sp} - e^2/d$ (see Section 2). From our calculations for $W_s(n)$ we obtain that this should occur for $n < n_{\text{cr}} \simeq 13$. This situation can be simulated by setting $t_{ij} = 0$ for Hg_n with $n < n_{\text{cr}}$. Clearly, one obtains as expected $\varepsilon_{\text{HOS}}(n) = \varepsilon_s(\text{atom})$ and thus $I_n = -\varepsilon_s + e^2/2R$ improving the agreement with the experimental results for $n \leq 13$ (see Fig. 3.1). Comparison with experiment [5, 24] suggests therefore that the electrons in Hg_n with $n < n_{\text{cr}} \simeq 13$ occupy approximately localized states. Note that for $n \gtrsim 50$ good agreement between experiment and our Hartree–Fock calculations is obtained, indicating that $t_{\text{eff}} \simeq t$ for these larger clusters due to the increasing importance of kinetic energy and improved screening.

The transition from van der Waals to metallic bonding reflected in the experimentally recorded ionization energy I_n^{exp} seems to be quite sharp. This is most clearly seen by discussing how the energy of the highest occupied state $\varepsilon_{\text{HOS}}^{\text{exp}}(n) = e^2/2R - I_n^{\text{exp}}$ derived from experiment depends on the average coordination number z_n, which is a characteristic of the size dependent local environment.[4] For $z_n < z_{\text{cr}} \simeq 5.5$ ($z_{\text{cr}} = z_{n_{\text{cr}}} \simeq z_{13}$) we have $\varepsilon_{\text{HOS}}^{\text{exp}} \simeq \varepsilon_s(\text{atom}) = -6.5$ eV. For $z > z_{\text{cr}}$, ε_{HOS} increases rapidly reaching the bulk value $\varepsilon_{\text{HOS}}^{\text{exp}}(b) = -\phi = -4.5$ eV already for $z \simeq z_{Me} \simeq 8.5$, where z_{Me} refers to the coordination number required for metallic behavior ($z_{Me} \simeq z_{80} \simeq 8.5$). Remarkably, as we go from the atom ($z_1 = 0$) to the bulk ($z_b = 12$) $\varepsilon_{\text{HOS}}^{\text{exp}}$ seems to remain approximately constant for most values of z_n, varying from the atomic value -6.5 eV to the bulk value -4.5 eV only in the rather small intermediate range $5.5 \lesssim z_n \lesssim 8.5$.

In Fig. 3.2 results are shown for the size dependence of the $5d(D_{3/2}) \to 6p(P_{1/2})$ autoionization resonance energy of Hg_n-clusters. The transition from van der Waals to covalent/metallic clusters is again clearly seen. For small clusters $n \lesssim 20$–25 good agreement with experiment is obtained with our Ansatz $\delta_n \simeq \delta_1 - \alpha z_n$, indicating that the excited $6p$-electron occupies an atomic-like state (Frenkel exciton). For larger clusters the experimental observations are close to the results obtained with our tight-binding calculations. Thus, the transition is like a Mott–Wannier exciton with the excited $6p$-electron occupying one of the low lying delocalized states above ε_F. These two types of transitions are illustrated in the inset of Fig. 3.2. From these results we conclude that the delocalization of the excited $6p$-electron and band formation occurs in Hg_n-clusters consisting of approximately 20–25 atoms, in good qualitative agreement with our results for I_n. However, note that the localization $\to$ delocalization transition need not to manifest at the same cluster size for I_n and δ_n. In the autoionization resonance experiment the relevant $5d \to 6p$ absorption occurs in the neutral cluster, whereas in the ionization experiments the spectrum of

[4] For correlating cluster size and coordination number z_n we assume close packed structures (e.g., fcc-like). Since this need not to be the case for the experimentally studied clusters, the following discussion is somewhat speculative.

the positive ion is also involved. Since the kinetic energy gained upon delocalization is larger for Hg_n^+ than for Hg_n (due to the missing antibonding electron), it seems plausible that the critical size n_{cr} required for delocalization is smaller for the positive ion than for the neutral cluster. Therefore it is not surprising that we derive a smaller value for n_{cr} from our calculations for I_n.

The red shift for the two transitions $5d(D_{3/2}) \to 6p(P_{1/2})$ and $5d(D_{5/2}) \to 6p(P_{3/2})$ is expected to be different since the spin-orbit splitting of the $6p(P_{3/2})$ and $6p(P_{1/2})$ levels (~ 1 eV) becomes less important for increasing cluster size and delocalization of the excited $6p$-electron. The width ω' of the resonance is expected to be smaller than the width W_{6p} of the $6p$-band, due to ε_{pd} and the fact that the Auger process involved in autoionization occurs most likely when $5d$-hole and excited $6p$-electron are on the same atom. The width of the resonance results mainly from the lifetime τ_p during which the p-electron and $5d$-hole remain on the same atom. Furthermore, we expect the resonance width for the [5/2]-transition to become broader when the $5d(D_{5/2})$ state lies within the $6s$-band. We estimate this to occur for $n \gtrsim 40$. The autoionization resonance involving the final state $5d^9(D_{5/2})$ should disappear for $n > 40$, since the $5d$-hole is then rapidly delocalized. The probability for the Auger-like autoionization process to occur is strongly reduced if the excited $6p$-electron and the $5d$-hole are not on the same atom.

Results for E_{coh} are shown in Fig. 3.3. Obviously, for fully occupied s-states and empty p-states van der Waals like bonding dominates. Due to E_{corr} metallic cohesion is suppressed for $n < n_{cr}$. For $n > n_{cr} \simeq 15$ the transition to covalent and then metallic like cohesion with $E_{coh}^{cov}(n) > E_{coh}^{vdW}(n)$ occurs. The tight-binding calculations (using Hartree–Fock approximation, curve (a)) yield correctly for small clusters a relatively rapid increase in E_{coh}^{cov} due to covalent bonding for n up to 13–19 atoms, where almost 50% of the bulk cohesive energy is obtained, and then for larger clusters a much smaller increase of E_{coh}^{cov}. However, neglecting

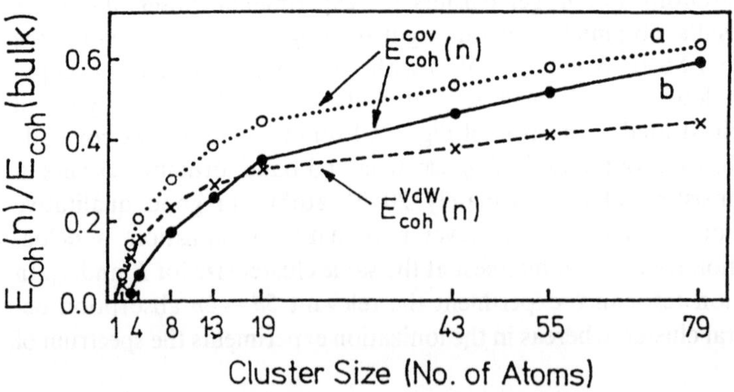

Fig. 3.3. Size dependence of the cohesive energy E_{coh}^{vdW} and E_{coh}^{cov} due to van der Waals- and covalent-bonding, respectively. In curve (a) for E_{coh}^{cov} charge fluctuations and electron correlations are neglected ($E_{corr} = 0$). Curve (b) is calculated including $E_{corr} = (U_{sp} - e^2/d)\, v_p/\kappa$

E_{corr} results in an overestimation of E_{coh}^{cov} suggesting wrongly covalent bonding for very small Hg_n-clusters with $n < n_{cr}$. Such overestimations of E_{coh} are also obtained when performing local density functional calculations for small Be-clusters [25]. The crossing of E_{coh}^{vdW} and E_{coh}^{cov} (curve (b)), illustrates the transition from van der Waals to covalent behavior. However note that the transition need not be extremely sharp, since covalent bonding, though weakened, might be also present to some extent for $n < n_{cr}$, and van der Waals bonding might be non negligible for $n > 15$. Although we calculated E_{corr} within a simple model, we expect these results to be of general validity, as supported by the results derived using slave-boson theory [13] (see Appendix A). Furthermore, we expect E_{coh}^{cov} to be properly estimated although we obtained the covalent bonding energy and E_{corr} using results for $N_\alpha(\varepsilon)$ calculated in the Hartree–Fock approximation, due to cancellation of errors in the first term in Eq. (2.4) and E_{corr}. However, notice that I_n depends more sensitively on $N_\alpha(\varepsilon)$ and thus correlations might play a more important role there.

In Fig. 3.4 results for the size dependence of the bond-length d_n are shown. We obtain that d_n increases with decreasing cluster size. The calculated nearest neighbor distance in the dimer is $d_2 = 1.06 d_b$ ($d_b = 3.03$ Å), which compares qualitatively well to the value inferred from experiment ($d_2^{exp} = 1.11 d_b$) [26]. This is in contrast to the behavior of alkali- and transition-metal clusters where, instead of an expansion, a contraction with respect to bulk is observed. This can be understood recalling that for simple- and transition-metals the repulsive energy E_R ($E_R \propto z_i$) decreases faster than the binding energy E_{band} ($E_{band} \propto \sqrt{z_i}$)

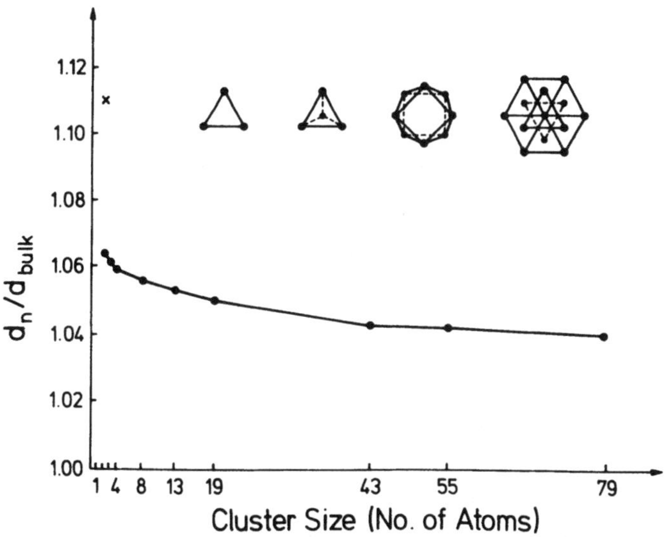

Fig. 3.4. The size dependence of the equilibrium bond-length d_n in Hg_n as obtained by maximizing the cohesive energy $E_{coh}^{cov}(n)$. The cross refers to the experimental result for Hg_2 [26]. The inset figures illustrate the assumed cluster geometries for $n \leq 13$, for larger clusters fcc structure is assumed

with decreasing cluster size, whereas for divalent-metal clusters E_{band} decreases more rapidly than E_R, since antibonding s-states are gradually occupied as we approach the closed shell atomic configuration (i.e., $E_{band} \sim [W_s v_s(2 - v_s) + W_p v_p(6 - v_p)] \to 0$, since $v_s \to 2$ and $v_p \to 0$ for $n \to 1$). This larger reduction of the binding contribution to the cohesive energy causes the expansion. For small clusters ($n \lesssim n_{cr}$) a further increase of d_n should result from Coulomb interactions which tend to weaken covalent bonding (for example, by renormalizing the hopping integrals: $t \to t_{eff} < t$). This would improve quantitative agreement with experiment for the dimer.

In Fig. 3.5 the size dependence of the gap Δ_n between occupied and unoccupied states is shown. Δ_n decreases with increasing n as the occupied s-like band and the unoccupied p-like band broaden and finally overlap. It is interesting that we obtain a more rapid decrease of Δ_n from $n \simeq 8$ on. For $n \leq 8$, Δ_n is large ($\Delta_8 \simeq 3.2$ eV). Therefore, these very small clusters behave like insulators. As discussed before, the electronic states in Hg_{19} and larger clusters are extended over the whole cluster. However, the gap is still large ($\Delta_{19} \simeq 1.8$ eV) so that no metallic, but rather semiconducting like optical properties are expected. The gap in the density of states (DOS) around ε_F, which is still clearly seen in Hg_{43}, for example, reduces for larger clusters. As shown in Figs. 3.6 and 3.7, the lowest unoccupied peak in the p-DOS gets closer and closer to the highest occupied peak in the s-DOS. For $n \gtrsim 80$ the gap ($\Delta_{79} \simeq 0.35$ eV) is close to its limiting value $[nN(\varepsilon_F)]^{-1}$ and low energy sp-band excitations start to become possible. Furthermore, for $n = 79$ the gap in the DOS at the Fermi energy is similar to the level spacing in the DOS within the s-band resulting from finite size effects (see Fig. 3.7). Therefore, we conclude that Hg_n-clusters with $n \gtrsim 80$ should already exhibit metallic properties, similar to those observed for alkali-metal clusters.

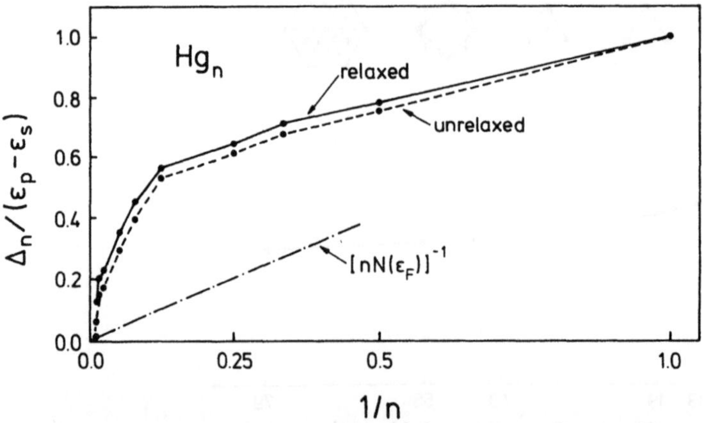

Fig. 3.5. Cluster size dependence of the gap Δ_n between occupied s-like and unoccupied p-like states determined from the density of states $N_s(\varepsilon)$ and $N_p(\varepsilon)$, which are calculated from Eq. (2.1) in the Hartree–Fock approximation. "Unrelaxed" and "relaxed", respectively, refer to calculations using the bulk bond-length d_b and the size dependent bond-length d_n resulting from maximizing $E_{coh}^{cov}(n)$

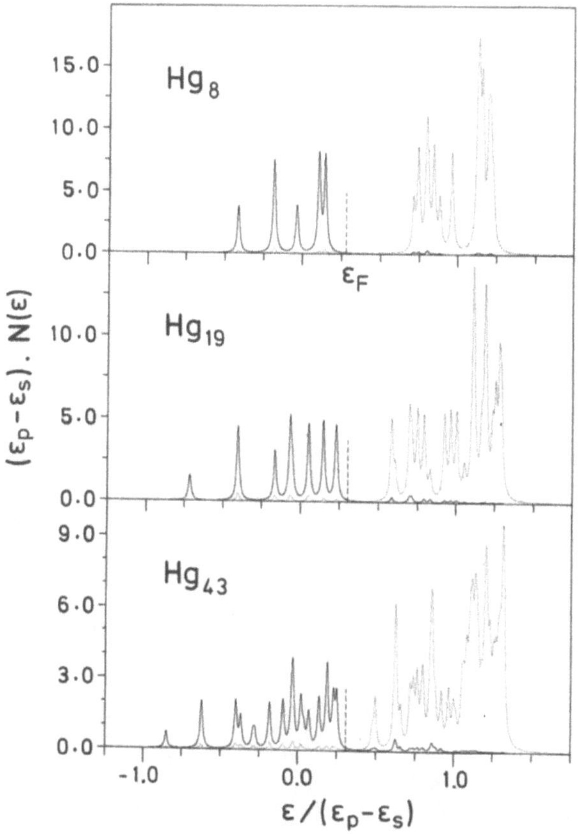

Fig. 3.6. The average density of states (DOS) of Hg_n is shown. $N_s(\varepsilon)$ (full line) and $N_p(\varepsilon)$ (dotted line) refer to the *s*- and *p*-DOS, respectively. A Lorentzian was used to broaden the discrete cluster levels (width $\gamma = 0.06$ eV). Note the reduction of the gap Δ_n between occupied and unoccupied states and the increasing *sp*-mixing with increasing cluster size

For larger clusters the *sp* gap closes completely ($\Delta_{135} \lesssim 0.1$ eV) and the Fermi energy ε_F lies within the *sp*-band as in Hg-bulk (see Fig. 3.7). For clusters with $n \gtrsim 40$ the results in Fig. 3.5 for the unrelaxed clusters seem more physical, since an uniform relaxation is not expected for large clusters.

The size-dependence of the electronic structure (DOS) of Hg_n-clusters is shown in Figs. 3.6 and 3.7. For very small clusters (e.g., Hg_8) we obtain small *s*- and *p*-band broadening and consequently a large band gap. Also the hybridization of *s*- and *p*-states is small, resulting in a small cohesive energy gain. For larger clusters (e.g., Hg_{19}, Hg_{43}) the formation of a *s*-like and a *p*-like band, although yet non-overlapping, can be recognized. Notice the larger *sp*-mixing which is responsible for the increase in $E_{coh}^{cov}(n)$ (Fig. 3.3), and for covalent bond formation and electron delocalization. For $n \simeq 80$ the main features of the bulk DOS are already present. Larger clusters (e.g., Hg_{135}) show no further

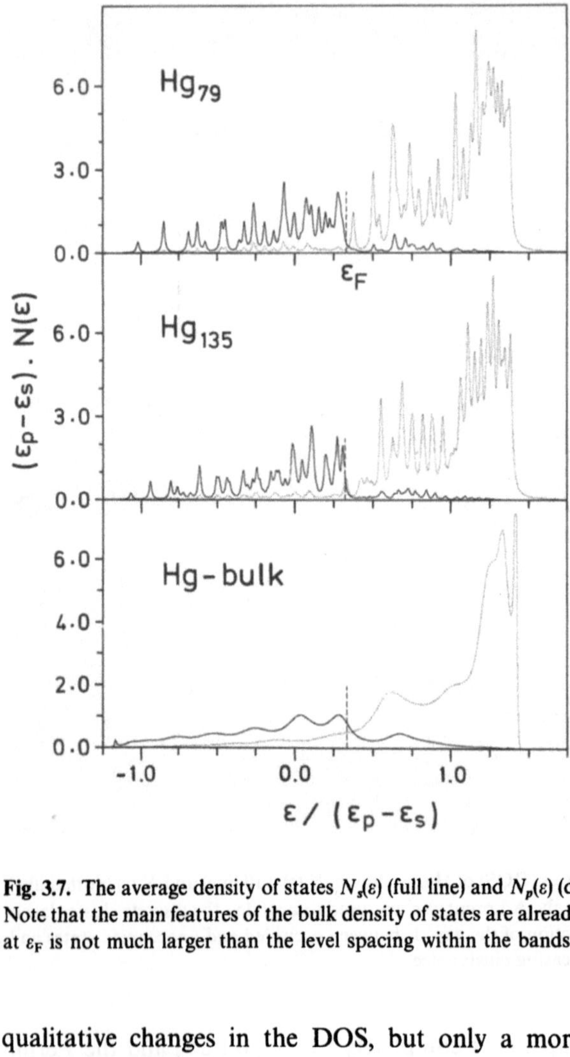

Fig. 3.7. The average density of states $N_s(\varepsilon)$ (full line) and $N_p(\varepsilon)$ (dotted line) for larger Hg_n-clusters. Note that the main features of the bulk density of states are already present in Hg_{79} and that the gap at ε_F is not much larger than the level spacing within the bands resulting from finite size effects

qualitative changes in the DOS, but only a more continuous energy level distribution [27].

2.4.4 Summary and Outlook

A tight-binding type electronic theory has been developed to study the characteristic physical behavior of clusters, which size is intermediate between the atom and the infinite solid. The formalism allows to treat *s*, *p* and *d* electrons on equal footing and thus can be applied to a wide variety of elements and compounds throughout the periodic table. *Divalent-metal clusters* (e.g., Hg_n) were investigated in detail, since for these clusters a dramatic change in the nature of the chemical bonding (from van der Waals to metallic) takes place with increasing cluster size. The changes occurring in the electronic structure as a function of cluster size have been determined in order to understand microscopically how they correlate with changes in the local environment, and how the characteristic size dependence of different cluster properties comes about. Particularly for Hg_n we calculated:

- cohesive energy $E_{coh}(n)$, which allows to determine the dominating bonding mechanism (van der Waals, covalent or metallic) as a function of cluster size n,
- ionization energy I_n, which is sensitive to the 6s-band formation and thus to electron delocalization,
- autoionization resonance energy δ_n due to $5d \to 6p$ transitions, which measures the 6p-band formation as a function of n,
- band gap Δ_n, which allows to distinguish between insulating, covalent and metallic behavior, and
- electronic densities of states, which reflect the 6s- and 6p-band structure and are relevant for photoemission experiments.

Our results for the cohesive energy $E_{coh}(n)$, ionization energy I_n, and $5d \to 6p$ autoionization resonance energy δ_n indicate that electron delocalization and covalent bond formation takes place in Hg_n-clusters with $n > n_{cr} \simeq 13$–20. For $n < n_{cr}$ van der Waals bonding dominates, whereas for $n > n_{cr}$ covalent and for larger clusters metallic bonding dominate. From our calculations for the band gap Δ_n and density of states $N(\varepsilon)$ we obtain that Hg_n should have for $20 \lesssim n \lesssim 80$ semiconducting-like optical properties, and for $n \gtrsim 80$ metallic properties. The discrepancy between experiment and our Hartree–Fock calculations for I_n, for example, may indicate the significance of electron correlations, which favor electron localization and "blocking" of covalent bonding (hopping integrals t_{ij}) for $n < n_{cr}$. Presently it remains experimentally and theoretically unclear whether the transition from van der Waals to covalent bonding occurs gradually or is relatively sharp (e.g., due to suppressing band formation as suggested by the ionization potential experiments [5]).

The physical origin of a possible band narrowing can be qualitatively understood as resulting from the competition between kinetic-energy and Coulomb interaction energy contributions. In order to reduce their kinetic energy the electrons tend to delocalize throughout the lattice and to form bands. However, as discussed in Section 2.4.2, delocalization necessarily involves local charge fluctuations which increase the Coulomb-interaction energy. Actually a compromise is made: The electrons move in a correlated fashion by trying at one time to delocalize in order to reduce the kinetic energy, and to avoid or screen local charge fluctuations in order to reduce the Coulomb interaction energy. Eventually, electron delocalization is prevented if the kinetic energy E_{band} gained upon delocalization is not enough to overcome the Coulomb energy increase E_{Coul}, or if other bonding mechanisms involving no local charge fluctuations (e.g., van der Waals attraction) are energetically more favorable. Due to size dependent screening one expects for small clusters that the role of

Coulomb interactions should be even more important than for the corresponding bulk material, since screening improves with increasing cluster size n and the energy involved in a local charge fluctuation is approximately given by $\Delta E_{\mathrm{Coul}} \simeq (U - e^2/d)/\kappa(n)$. Furthermore, the discreteness of the energy spectrum, and the reduction of coordination number and mobility of the electrons in small clusters should hinder the "backflow" of density excitations responsible for dynamic screening of charge fluctuations. Clearly, electron correlations are important for determining the transition from localized atomic-like to delocalized band-like electronic states occurring in clusters with increasing cluster size. Therefore, it is of considerable interest to show how size dependent changes in the local environment can cause changes in the electronic structure resulting from electron correlations. In particular it is important to explain how strong the "blocking" of the hopping integrals, which is expected to occur in small Hg_n-clusters with $n < n_{\mathrm{cr}} \simeq 13$–$19$, is. For this purpose we describe in Appendix A the slave-boson approach to the Hubbard Hamiltonian (as proposed by Kotliar and Ruckenstein for solids [12]) to study small clusters and surfaces. We show using a simple one band model that the effective hopping integrals t_{eff}, from which the quasiparticle density of states is derived, should decrease with decreasing n. Eventually $t_{\mathrm{eff}} \to 0$, if the energy gained upon band formation is not enough to overcome the Coulomb energy due to charge fluctuations present in the covalent/metallic bonded state. Several improvements and extensions are worthwhile. For example, one should include explicitly s- and p-bands and the competing van der Waals interactions in order to study systematically the transition from van der Waals to metallic divalent-metal clusters (e.g., Hg_n, Be_n, etc.). Furthermore, an attempt should be made to include size dependent screening effects due to interatomic Coulomb interactions. Research in these directions is currently in progress [13].

From our calculations discussed in Section 2.4.3, and summarized above we conclude that tight binding theory is a reliable method to study the properties of small clusters. Heretofore unexplained experimental results have been accounted for and brought into a consistent physical picture [14, 27]. The transparency of the method was exploited to correlate the observed size dependence of various cluster properties with the changes in the local environment of the atoms, which are ultimately responsible for all characteristic cluster properties. Necessary improvements and extensions of the theory (e.g., slave-boson theory for electron correlations in small clusters, etc.) have been discussed. Further studies on divalent-metal clusters (Hg_n, Be_n, Mg_n, etc.) should clarify the localized character of $s \to p$ transitions and hole excitations as a function of cluster size. For example, what is the lifetime of a s-state hole in small divalent-metal clusters as a function of cluster size? Does Mott's delocalization transition occur in small clusters? When do such clusters become metals? Studies of the electronic spectral density should help to answer these questions.

Acknowledgement. Helpful discussions with Dr. P. Stampfli are gratefully acknowledged.

Appendix A: Slave-Boson Approach to Electron Correlations in Small Clusters

For small clusters Coulomb interactions are generally expected to become more important. Thus, it is of general importance to develop a theory treating carefully the interplay of kinetic and Coulomb energy. For this purpose and for future work we extend here the slave-boson approach proposed by *Kotliar* and *Ruckenstein* [12] for solids to study small clusters and surfaces. We consider the single band Hubbard Hamiltonian

$$\hat{H} = \sum_{i,\sigma} \varepsilon_i^0 \hat{n}_{i\sigma} + \sum_{i \neq j} t_{ij} \hat{c}_{i\sigma}^+ \hat{c}_{j\sigma} + U \sum_i \hat{n}_{i\uparrow} \hat{n}_{i\downarrow} , \tag{A.1}$$

where $\hat{c}_{i\sigma}^+$ ($\hat{c}_{i\sigma}$) refers to the usual creation (annihilation) operator of an electron of spin σ at atomic site i ($\hat{n}_{i\sigma} = \hat{c}_{i\sigma}^+ \hat{c}_{i\sigma}$), ε_i^0 to the orbital energy, t_{ij} to the hopping integrals between atoms i and j, and U to the intra-atomic Coulomb repulsion integral. Let us remark that the simple form taken for $\hat{H}$ is not an intrinsic limitation of the theory, but was assumed here only to simplify the explanation of the method and approximations involved. Several improvements (e.g., inclusion of several bands, interatomic Coulomb interactions U_{ij}, van der Waals interactions, etc.), necessary for a quantitative description of real systems (e.g., Hg_n-clusters), are discussed below. Following *Kotliar* and *Ruckenstein* [12] we change the representation of the electronic states by including a set of four boson operators at each cluster site, which explicitly contain the information about the state of occupation of the site. The boson creation (annihilation) operators $\hat{e}_i^+$ ($\hat{e}_i$), $\hat{p}_{i\uparrow}^+$ ($\hat{p}_{i\uparrow}$), $\hat{p}_{i\downarrow}^+$ ($\hat{p}_{i\downarrow}$) and $\hat{d}_i^+$ ($\hat{d}_i$), create (annihilate) respectively an empty, singly occupied (with spin and down) or doubly occupied electronic state at site i. Notice that, among all possible states in the enlarged Hilbert space (with bosons and fermions), only those which have consistent boson and fermion occupation numbers have physical sense. Thus, we remove the unphysical states by imposing the conditions

$$\hat{e}_i^+ \hat{e}_i + \hat{p}_{i\uparrow}^+ \hat{p}_{i\uparrow} + \hat{p}_{i\downarrow}^+ \hat{p}_{i\downarrow} + \hat{d}_i^+ \hat{d}_i = 1 , \tag{A.2}$$

$$\hat{p}_{i\sigma}^+ \hat{p}_{i\sigma} + \hat{d}_i^+ \hat{d}_i = \hat{n}_{i\sigma} . \tag{A.3}$$

The completeness relation (A.2) indicates that one and only one of the four possible boson states must be (singly) occupied at each cluster site. Consequently, $\hat{e}_i^+ \hat{e}_i$, $\hat{p}_{i\sigma}^+ \hat{p}_{i\sigma}$, $\hat{d}_i^+ \hat{d}$ act in the physical subspace as projection operators onto the states with 0, 1, or 2 electrons at site i, respectively.

The form of any physical operator in the new representation is directly obtained once the transformation law for $\hat{c}_{i\sigma}$ is known. The general form of such a transformation is

$$\hat{c}_{i\sigma} \longrightarrow \hat{c}_{i\sigma} \hat{z}_{i\sigma} , \tag{A.4}$$

where $\hat{z}_{i\sigma}$ is an operator acting only on the boson variables, which takes care that the proper boson occupation state is obtained after annihilation. It is easy to see that

$$\hat{z}_{i\sigma} = (1 - \hat{d}_i^+ \hat{d}_i - \hat{p}_{i\sigma}^+ \hat{p}_{i\sigma})^{-1/2} (\hat{e}_i^+ \hat{p}_{i\sigma} + \hat{p}_{i\bar{\sigma}}^+ \hat{d}_i)(1 - \hat{e}_i^+ \hat{e}_i - \hat{p}_{i\bar{\sigma}}^+ \hat{p}_{i\bar{\sigma}})^{-1/2} \quad \text{(A.5)}$$

fulfills these conditions, since $\hat{c}_{i\sigma}\hat{z}_{i\sigma}$ has the same matrix elements in the physical subspace as $\hat{c}_{i\sigma}$ in the original fermion Hilbert space (see Eqs. (A.2) and (A.3)) [12].

Using Eqs. (A.2–5) the Hamiltonian can be written as:

$$\hat{H} = \sum_{i,\sigma} \varepsilon_i^0 \hat{n}_{i\sigma} + \sum_{i \neq j} t_{ij} \hat{c}_{i\sigma}^+ \hat{c}_{j\sigma} \hat{z}_{i\sigma}^+ \hat{z}_{j\sigma} + U \sum_i \hat{d}_i^+ \hat{d}_i . \quad \text{(A.6)}$$

The simplest nontrivial approach to Eq. (A.6) is the so called saddle-point approximation. Here all boson operators are set equal to constant numbers, which are to be determined by minimizing the resulting free energy $F = \mu N_t - k_B T \ln Z$, under the boundary conditions (A.2) and (A.3). Here N_t stands for the total number of electrons. In this approximation Z is given by

$$Z = \mathrm{Tr}[\exp\{-\beta(\hat{H}' - \mu\hat{N})\}]$$
$$\times \exp\left\{-\beta\left(U\sum_i d_i^2 + \sum_j \varepsilon_j^{(1)}(\alpha_j^{(1)} - 1) - \sum_{i\sigma} \varepsilon_{i\sigma}^{(2)} \alpha_{i\sigma}^{(2)}\right)\right\}, \quad \text{(A.7)}$$

where $\alpha_j^{(1)} = e_i^2 + p_{i\uparrow}^2 + p_{i\downarrow}^2 + d_i^2$ and $\alpha_{i\sigma}^{(2)} = p_{i\sigma}^2 + d_i^2$. The effective Hamiltonian $\hat{H}'$ is given by

$$\hat{H}' = \sum_{i,\sigma} \varepsilon_{i\sigma}' \hat{n}_{i\sigma} + \sum_{i \neq j} t_{ij}^{\sigma\prime} \hat{c}_{i\sigma}^+ \hat{c}_{j\sigma} , \quad \text{(A.8)}$$

$$\varepsilon_{i\sigma}' = \varepsilon_i^0 + \varepsilon_{i\sigma}^{(2)} , \quad \text{(A.9)}$$

$$t_{ij}^{\sigma\prime} = q_{ij}^\sigma t_{ij} . \quad \text{(A.10)}$$

Here $q_{ij}^\sigma = \langle \hat{z}_{i\sigma}^+ \hat{z}_{j\sigma} \rangle$ refers to the hopping renormalization factor, and $\varepsilon_j^{(1)}$, $\varepsilon_{i\sigma}^{(2)}$ to the Lagrange multipliers associated to the boundary conditions (A.2) and (A.3). $\hat{H}'$ describes the electrons as if they were independent particles (quasi-particles) having, as a result of electron-electron interactions, shifted energy levels ($\varepsilon_i^0 \to \varepsilon_i^0 + \varepsilon_{i\sigma}^{(2)}$) and renormalized hopping elements ($t_{ij} \to q_{ij}^\sigma t_{ij}$). The equilibrium values of $\varepsilon_{i\sigma}^{(2)}$ and q_{ij}^σ are obtained by minimizing $F = F(\varepsilon_{i\sigma}^{(2)}, \varepsilon_i^{(1)}, d_i, p_{i\sigma}, e_i)$.

In order to discuss the physics behind Eq. (A.7) and particularly how a size dependent renormalization of the hopping elements comes about we consider the total energy $E = F(T = 0)$ given by

$$E = E_{\mathrm{Coul}} - E_{\mathrm{band}} , \quad \text{(A.11)}$$

where

$$E_{\mathrm{Coul}} = U \sum_i d_i^2 , \quad \text{(A.12)}$$

and

$$E_{\text{band}} = \sum_{i,\sigma} \int_{-\infty}^{\varepsilon_F} (\varepsilon'_{i\sigma} - \varepsilon) N'_{i\sigma}(\varepsilon)\, d\varepsilon \,. \tag{A.13}$$

Here, $N'_{i\sigma}(\varepsilon)$ denotes the local density of states given by $N'_{i\sigma}(\varepsilon) = -(1/\pi)\text{Im}\{G'_{i\sigma,i\sigma}(\varepsilon)\}$, where $G'_{i\sigma,i\sigma}(\varepsilon)$ refers to the diagonal element of the Green's function operator $\hat{G}'(\varepsilon) = [\hat{H}' - \varepsilon]^{-1}$. E_{band} represents the energy gained due to band formation (i.e., electron delocalization). Clearly E_{band} increases ($E_{\text{band}} > 0$) with increasing q_{ij} ($0 \le q_{ij} \le 1$), since the effective band width $W(i)$ of the local density of states $N'_{i\sigma}(\varepsilon)$ increases ($W(i) \simeq \sqrt{\sum_j t'^2_{ij}}$, $t'_{ij} = q_{ij} t_{ij}$). In the weak coupling limit ($U = 0$) we have $E_{\text{Coul}} = 0$, and thus $E = -E_{\text{band}}$. Therefore, the minimum E is achieved by setting d_i such that $q_{ij} = 1$, i.e., the hopping elements are not renormalized, as physically expected. For $U > 0$ however, E_{Coul} is non negligible. This term is equal to U times the probability of having doubly occupied sites, and thus represents the energy due to charge fluctuations, which are necessarily present when electrons delocalize. The competition between kinetic energy (band energy) and Coulomb energy contributions to the total electronic energy E is clearly seen. On the one side an increase of double occupations d_i^2 (i.e., electron delocalization) causes E_{band} to increase, since q_{ij} increases with increasing d_i^2. On the other side a larger d_i^2 causes E_{Coul} to increase. Finally, at the saddle-point we have $E_{\text{band}} < E_{\text{band}}(q_{ij} = 1)$ and $E_{\text{Coul}} \ge 0$, with $q_{ij} < 1$, i.e., the hopping integrals are reduced by electron correlations ($t'_{ij} < t_{ij}$).

It is important to note that q_{ij}, and thus t'_{ij}, decrease with decreasing cluster size, since E_{band} decreases with decreasing coordination number z_n ($E_{\text{band}} \sim \sqrt{z_n}$). To illustrate this we show in Fig. A.1 results for the average renormalization factor $q = \langle q_{ij} \rangle$ as a function of the average coordination number z_n. These were obtained by calculating the local density of states in the paramagnetic case, using the second moment approximation and setting the number of electrons per atom equal to 1 (i.e., half filled band). Note how $t' = qt$ decreases with increasing U and decreasing coordination number. For sufficiently large U (e.g., for $U = U_c = 7.5t$) $d_i \to 0$ when $z \to z_{\text{cr}}$. As before, z_{cr} refers to the critical coordination number, below which the energy E_{band} gained upon band formation is not enough to overcome the Coulomb energy E_{Coul} due to charge fluctuations. For $z < z_{\text{cr}}$ it is energetically more convenient for the electrons not to delocalize, and other bonding mechanisms, e.g., van der Waals or ionic bonding, should dominate.[5]

Physically, E_{Coul} has the same meaning as the energy $E_{\text{corr}} \simeq (U_{sp} - e^2/d) v_p/\kappa$, which was found to be important for determining the transition from van der Waals to metallic bonding in Hg_n-clusters (see Sections 2 and 3). However note that no dielectric constant κ appears in Eq. (A.12), since we neglected several

[5] If magnetic solutions are allowed, the effective hopping t'_{ij} decreases with decreasing n, but remains finite for all values of U and n in the model Hamiltonian (A.1). The expected "blocking" of t'_{ij} is obtained when van der Waals interactions are included [13].

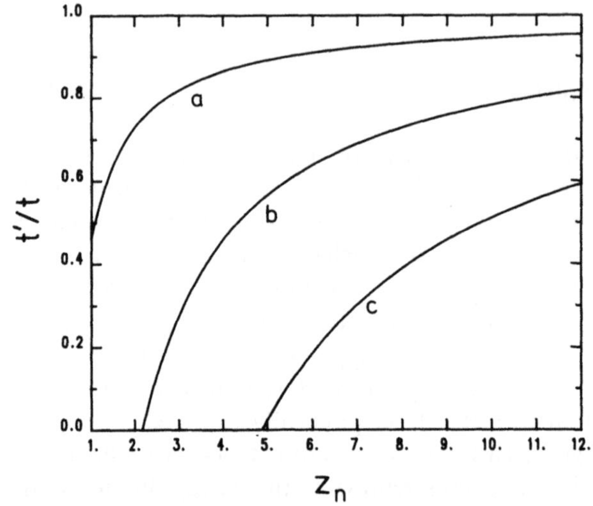

Fig. A.1. The hopping renormalization factor $q = t'/t$, representing the band narrowing due to correlations, as a function of the average cluster coordination number z_n. z_{cr} refers to the critical coordination number. For $z < z_{cr}$, $q = 0$ implies electron localization. The results were obtained by taking q_{ij} to be independent of cluster site and spin (i.e., $q^{\sigma}_{ij} = q$), and by calculating the density of states in the second moment approximation. a, b, and c refer to different values of the intra-atomic Coulomb repulsion integral U ($U_a/t = 2.5$, $U_b/t = 5$, $U_c/t = 7.5$). The number of electrons per atom is $v_t = 1$ (half filled band).

Coulomb interaction terms (U_{ij}, etc.) in $\hat{H}$ (see Eq. (A.1)). The present theory provides an alternative derivation of the equations used in Sections 2.4.2 and 2.4.3 for calculating $E^{\mathrm{cov}}_{\mathrm{coh}}(n)$ of Hg_n. The main advantage of the slave-boson approach is that it allows to calculate selfconsistently the corrections to E_{band} and E_{corr} as due to the renormalization of the hopping integrals. This is particularly important for a correct description of Hg_n in the intermediate size range $13 \lesssim n \lesssim 50$, where the electrons are delocalized (i.e., $d_i^2 > 0 \Rightarrow t'_{ij} > 0$), but effects due to the band narrowing may still be significant. The simple model Hamiltonian A.1 already contains the fundamental competition between kinetic and Coulomb energy. However, for a proper description of divalent-metal clusters one must improve on Eq. (A.1) by including *s*- and *p*-bands, interatomic Coulomb interactions, and van der Waals and excitonic interaction terms [13]. Also, the theory presented here is still of mean-field character which may neglect important fluctuations significant in small clusters. As a consequence one may obtain results for band formation (electron delocalization) which indicate a too sharp transition.

Appendix B: On the Size Dependence of the Ionization Energy of Small Clusters

In the following we derive Eq. (2.7) in the framework of density functional theory (DFT). An alternative derivation using the tight-binding formalism, particularly

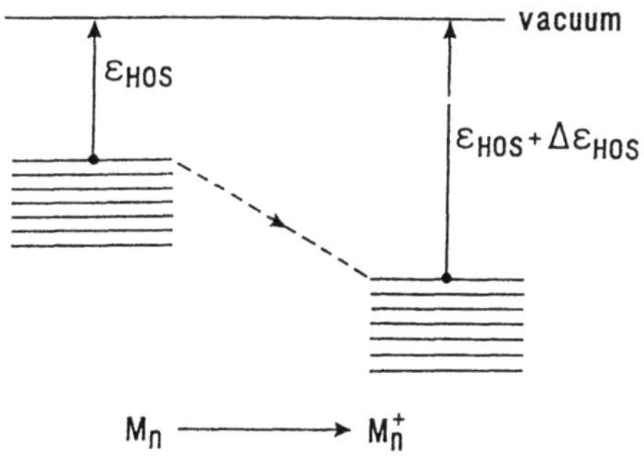

Fig. B.1. Illustration of the energy levels of the neutral and ionized cluster M_n. $\varepsilon_{\mathrm{HOS}}$ stands for the energy of the highest occupied state of the neutral cluster, and $\Delta\varepsilon_{\mathrm{HOS}}$ for the relaxation shift occurring upon ionization

including magnetic contributions (important e.g., for *3d* transition metal clusters), is given in Ref. [15].

Let ε_α be the αth-eigenvalue of the Kohn–Sham equations [28, 29] and let $E_n(\nu_e)$ be the total energy of a *n*-atom cluster with ν_e electrons, which can be viewed as a functional of the occupation numbers ν_α of the αth eigenstate ($\nu_e = \sum_\alpha \nu_\alpha$, $0 \leq \nu_\alpha \leq 1$). $E_n(\nu_e) = E_n(\{\nu_\alpha\})$, where $\{\nu_\alpha\}$ refers to the set of all occupation numbers ν_α. The eigenenergies ε_α can be obtained from the energy functional $E_n(\{\nu_\alpha\})$ as[6]

$$\frac{\partial E_n}{\partial \nu_\alpha} = \varepsilon_\alpha \, . \tag{B.1}$$

In particular for the highest occupied state α_h holds

$$\frac{\partial E_n}{\partial \nu_e} = \varepsilon_{\alpha_h} \, , \tag{B.2}$$

since $d\nu_e = d\nu_{\alpha_h}$ for $0 < \nu_{\alpha_h} < 1$. Notice that the energies ε_α, and particularly ε_{α_h}, depend on all occupation numbers ν_α and thus on ν_e. The dependence of ε_α on $\{\nu_\alpha\}$ is a consequence of the many-body character of the electron-electron interactions and gives rise to the relaxation shifts occurring upon ionization (see Fig. B.1).

The ionization energy I_n is given by

$$I_n = E_n(\nu_t - 1) - E_n(\nu_t) \, , \tag{B.3}$$

or equivalently

$$I_n = \int_{\nu_t}^{\nu_t - 1} \frac{\partial E_n}{\partial \nu_e} \, d\nu_e \, , \tag{B.4}$$

where ν_t refers to the total number of electrons in the neutral cluster. Using Eq. (B.2) in Eq. (B.4) and introducing the more convenient variable $\nu_h = \nu_t - \nu_e$,

[6] The following derivation rests on the assumption that the energy functional $E_n(\{\nu_\alpha\})$ is differentiable with respect to ν_α. This condition is satisfied by the usual approximations to DFT (e.g., LDA, WDA), but does not need to hold for the exact functional $E_n(\{\nu_\alpha\})$.

giving the number of holes in the system, we obtain

$$I_n = -\int_0^1 \varepsilon_{\alpha_h}(\nu_h)\, d\nu_h \,. \tag{B.5}$$

I_n is thus given by minus the average of the energy $\varepsilon_{\alpha_h}(\nu_h)$ of the highest occupied state of the cluster with ν_h holes, for $\nu_h = 0 \rightarrow \nu_h = 1$. Note that Eq. (B.5) is exact, and that none of the usual approximations to DFT (e.g., LDA) is involved.[7]

For an infinite system the changes in the self-consistent potential caused by the missing ν_h electrons are negligible ($0 < \nu_h < 1$). Thus ε_{α_h} is independent of ν_h, yielding

$$I_\infty = -\varepsilon_{\alpha_h}(0) = -\varepsilon_{\text{HOS}} \,, \tag{B.6}$$

where ε_{HOS} refers to the energy of the highest occupied state of the (neutral) solid. For a finite cluster, however, the missing ν_h electrons cause a non-negligible shift of the energy levels of the cluster. The exact dependence of $\varepsilon_{\alpha_h}(\nu_h)$ on ν_h is unknown, since no closed expression for the functional $E_n(\{\nu_\alpha\})$ is available. However, for $0 < \nu_h < 1$ one can expand $\varepsilon_{\alpha_h}(\nu_h)$ in Taylor series about $\nu_h = 0$,[7] whence

$$\varepsilon_{\alpha_h}(\nu_h) = \varepsilon_{\alpha_h}(0) - \frac{2e^2}{R} A \nu_h + \ldots \,, \tag{B.7}$$

where $\varepsilon_{\alpha_h}(0) = \varepsilon_{\text{HOS}}(n)$ refers to the energy of the highest occupied state of the neutral n-atom cluster. The prefactor $2e^2/R$ has been added so that $A = 1/2$ corresponds to the classical approximation for the potential shift inside the cluster. Here R refers to the radius of the cluster approximated by sphere ($R = (3v/4\pi)^{1/3} n^{1/3}$, v = volume per atom). This is a reasonable approximation at least for large clusters, for which the changes in the electronic density ρ, and thus the changes in the potential, caused by ionization are small and approximately localized at the cluster surface ($\Delta\rho \propto R^{-2}$). Substituting Eq. (B.7) in (B.5) and performing the integration one obtains

$$I_n \simeq -\varepsilon_{\text{HOS}}(n) + \frac{Ae^2}{R} \,. \tag{B.8}$$

Quantum mechanical effects are expected to cause deviations from the classical value $A = 1/2$. For example, the self-consistent potential $\phi(r)$ acting on the electrons raises beyond the cluster surface more rapidly for the cluster with ν_h holes than for the neutral cluster (e.g., $\phi(r) \simeq (1 + \nu_h)e^2/r$ for $r \rightarrow +\infty$). This could cause a stronger localization of the α_h wavefunction and thus a kinetic energy increase yielding $\varepsilon_{\alpha_h}(\nu_h) > \varepsilon_{\alpha_h}(0) - \nu_h e^2/R$. Furthermore, the positive hole density is not strictly localized at the cluster surface but extends somewhat below and above it. This causes the potential within the cluster to be higher than $-\nu_h e^2/R$, and thus one should also have $\varepsilon_{\alpha_h}(\nu_h) > \varepsilon_{\alpha_h}(0) - \nu_h e^2/R$. These two

[7] Numerical calculations using the local density approximation indicate that already the linear term of this expansion usually gives a very good approximation to $\varepsilon_{\alpha_h}(\nu_h)$ for $0 < \nu_h < 1$ [30]. However, this expansion is not valid for the eigenvalue $\varepsilon^{\text{ex}}_{\alpha_h}$ derived from the unknown exact functional E_n, for which it holds $I_n = -\varepsilon^{\text{ex}}_{\alpha_h}(0)$.

effects should be most important for metallic clusters, for which the electronic density extends appreciably beyond the cluster surface. Thus, at least for metal clusters $A < 1/2$ seems plausible. Indeed, experiments [31, 32] and local density functional calculations [33] on simple metal clusters (e.g., Na_n) suggest a smaller value of A ($A \simeq 3/8$) to be more appropriate than $A = 1/2$.

References

1. "Physics and Chemistry of Small Clusters", NATO ASI Series B: Physics Vol. 158, ed. by P. Jena, B.K. Rao, S.N. Khanna, Plenum Press, New York, 1987
2. "Small Particles and Inorganic Clusters", ed. by C. Chapon, M.F. Gillet, C.R. Henry: Z. Phys. D **12** (1989)
3. L.F. Mattheiss, W.W. Warren: Phys. Rev. B **16**, 624 (1977) and references therein
4. D. Tománek, S. Mukherjee, K.H. Bennemann: Phys. Rev. B **28**, 665 (1983)
5. K. Rademann, B. Kaiser, U. Even, F. Hensel: Phys. Rev. Lett. **59**, 2319 (1987)
6. C. Bréchignac, M. Broyer, Ph. Cahuzac, G. Delacretaz, P. Labastie, L. Wöste: Chem. Phys. Lett. **120**, 559 (1985), and Phys. Rev. Lett. **60**, 275 (1988)
7. H. Haberland: Oral contribution to the colloquium on the Physics of Anorganic Clusters, Hirschegg/Kleinwalsertal, Austria, Sept. 1988
8. G.M. Pastor, J. Dorantes-Dávila, K.H. Bennemann: Chem. Phys. Lett. **148**, 459 (1988)
9. G.M. Pastor, J. Dorantes-Dávila, K.H. Bennemann: Phys. Rev. B **40**, 7642 (1989)
10. J. Hubbard: Proc. Roy. Soc. London, Ser. A **276**, 238 (1963)
11. N.F. Mott: "Metal-Insulator Transitions", Taylor and Francis Ltd., London, 1974
12. G. Kotliar, A.E. Ruckenstein: Phys. Rev. Lett. **57**, 1362 (1986)
13. M.E. Garcia, G.M. Pastor, K.H. Bennemann: P.R.L. (1992)
14. G.M. Pastor, P. Stampfli, K.H. Bennemann: Europhys. Lett. **7**, 419 (1988)
15. G.M. Pastor: "Theory for the electronic properties of clusters", PhD. Thesis, Freie Universität Berlin, 1989, unpublished
16. R. Haydock: Solid State Physics **35**, 215 (1980)
17. W.A. Harrison: Phys Rev. B **27**, 3592 (1983) and references therein
18. K. Hilpert: J. Chem. Phys. **77**, 1425 (1982)
19. C.E. Moore: Atomic Energy Levels, Nat. Bur. Stand. Circ. No. 467 (1958)
20. R.W.G. Wickoff in "Crystal structures", 2nd ed., Interscience, 1963
21. S. Svensson, N. Martensson, E. Basilier, P.A. Malmqvist, U. Gelius, K. Siegbahn: J. Electron Spectry. **9**, 51 (1976)
22. J.B. Mann, Atomic Structure Calculations, Los Alamos Sci. Lab. Rept. LA-3690 (1967)
23. S. Fraga, J. Karwowski, K.M.S. Saxena: "Atomic Energy Levels", Phys. Sci. data **4**, Elsevier/North-Holland Inc. (New York, 1979)
24. B. Cabaud, A. Hoareua, P. Melinon: J. Phys. D **13**, 1831 (1980)
25. M.R. Press, B.K. Rao, S.N. Khanna, P. Jena in: "Physics and Chemistry of Small Clusters", NATO ASI Series B: Physics, Vol. 158, P. Jena, B.K. Rao, and S.N. Khanna, eds. (Plenum Press, New York, 1987), p. 431
26. S.H. Linn, C.L. Liao, C.X. Liao, J.M. Brom, C.Y. Ng: Chem. Phys. Lett. **105**, 645 (1984)
27. G.M. Pastor, P. Stampfli, K.H. Bennemann: Physica Scripta **38**, 623 (1988)
28. W. Kohn, L.J. Sham: Phys. Rev. **140**, A1133 (1965)
29. P. Hohenberg, W. Kohn: Phys. Rev. **136**, B864 (1964)
30. J.C. Slater: "The Self-Consistent Field for Molecules and Solids", Quantum Theory of Molecules and Solids, Vol. 4, Mc Graw Hill, 1974, chapter 2
31. W.D. Knight et al.: Phys. Rev. Lett. **52**, 2141 (1984)
32. W.D. Knight, W.A. de Heer, W. Saunders: Z. Phys. D **3**, 109 (1986), and references therein
33. D.R. Snider, R.S. Sorbello: Surface Sci. **143**, 204 (1984)

2.5 Analytic Cluster Models and Interpolation Formulae for Cluster Properties

H. Müller, H.-G. Fritsche, and *L. Skala*

2.5.1 Introduction

Small particles or clusters lie in an intermediate region between atoms or molecules and condensed matter (see Fig. 1). This mesoscopic region of a highly dispersed matter which is neither microscopic or macroscopic is very interesting not only from a purely scientific point of view [1–4] (the "fifth state of matter" [5]) but also from a technological one (catalysis, micro-electronics, high-tech ceramics etc.) [6]. Figure 1 shows that there are two possible theoretical approaches within the mesoscopic region. (i) Using quantum chemistry via the "giant molecules" approach or (ii) applying solid state physics methods via the "small solids" approach. The main problem of such approaches is that giant molecules are much smaller than small solids and there is a "no man's land" between them where no adequate methods have been developed till now. The usual computational methods of quantum chemistry and solid state physics cannot be used for finite systems with a very large number of atoms (mesoscopic particles). The importance of the analytically solvable models in this respect is obvious. Such models discussed in this article are the quantum chemical analytic cluster model (QACM, paragraph 3) and topological analytic cluster model (TACM, paragraph 4). Both approaches enable to introduce simple interpolation formulae for the size dependence of various stationary cluster properties. In paragraph 5, we make a few remarks on the generality of the interpolation formulae and the problem of non-stationary properties.

2.5.2 Special Role of the Analytic Cluster Model (ACM)

We investigate a cluster property $G(N)$ as a function of the cluster size N (for example, the number of atoms in a metallic cluster Me_N). For $N > N_{\mathrm{crit}}$, all conventional quantum chemical approaches fail (more sophisticated methods earlier, semiempirical ones later). The situation is similar also from the side of the solid state methods (see Fig. 2). It is obvious that the only way of going from a single atom or a small cluster through Me_N for a general N to the solid is the analytic cluster model (ACM) [7–16]. Because of the analytic results, there is no

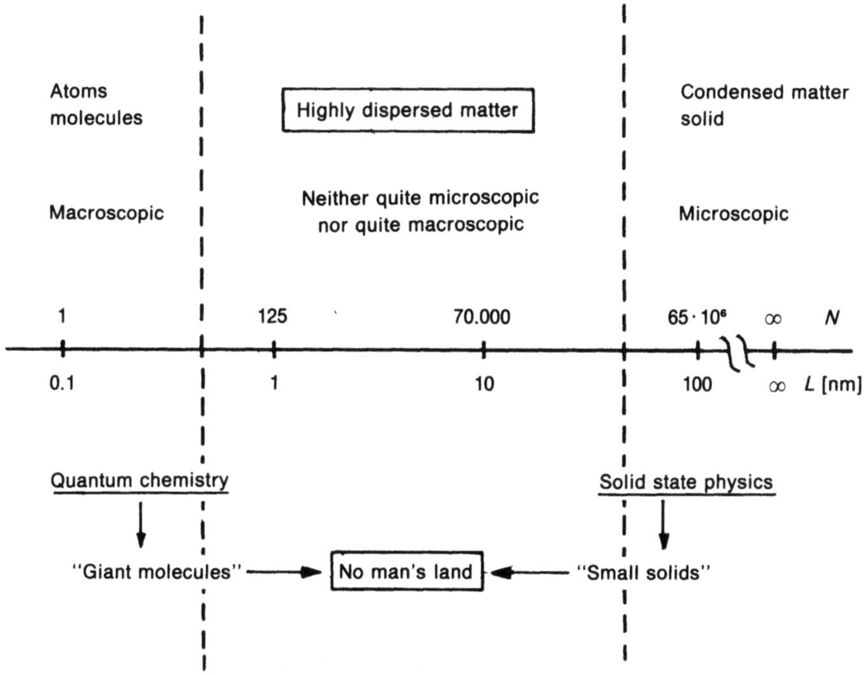

Fig. 1. The highly dispersed matter, its position between atoms or molecules, condensed matter or solid

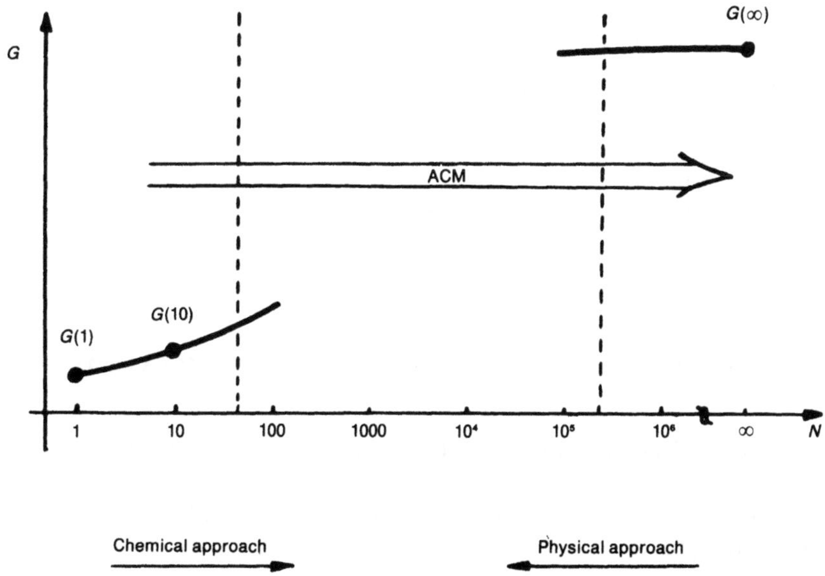

Fig. 2. The analytic cluster model (ACM) as a bridge between the quantum chemistry, solid state physics methods

restriction on the value of N in the ACM. This advantage is partly paid for by some simplifications made in the ACM.

2.5.2.1a) Quantum Chemical Analytic Cluster Model (QACM)

If the most simple quantum chemical methods are applied to the clusters of certain regular shapes the corresponding Schrödinger equation is in some cases analytically solvable i.e. closed analytic formulae for the eigenvalues and eigenfunctions including their N-dependence can be obtained. This can be done by solving (i) the differential equation analytically (the electron gas method, see e.g. [9]) or (ii) by solving the Schrödinger equation in the matrix representation (the tight binding model, see e.g. [11, 12]). Using the QACM it is possible to calculate the property G of a cluster of arbitrarily large size

$$G = G(N), \quad 1 \leq N \leq \infty \ . \tag{1}$$

2.5.2.1b) Topological Analytic Cluster Model (TACM)

In the TACM, the pair interaction potential of the cluster units (atoms, molecules) is assumed in a simple analytic form. For the short range potentials (e.g. for the van der Waals clusters) only the nearest neighbor contributions are important. Then, for highly symmetrical clusters, their topology may be used to calculate analytically their total energy and other properties (see e.g. [17, 18]).

2.5.2.1c) Classical Analytic Cluster Model (CACM)

In this category we can include classical phenomenological models like the classical droplet model.

We note at the end of this paragraph that these models overlap in some respects.

2.5.3 Quantum Chemical Analytic Cluster Model (QACM)

2.5.3.1 Analytic Calculation of a Cluster Property $G(N)$. An Example

As an example we calculate the condensation energy $\Delta E(N)$ ($= G(N)$) corresponding to the process

$$Me_N + Me \Rightarrow Me_{N+1} \tag{2}$$

for the one-dimensional solid in the free electron model (FEM). The condensa-

Table 1. Relative error F [%] of $\Delta E(N)$ given by Eq. (5)

N	1	3	5	7	9	11
F [%]	7.4	0.8	0.3	0.2	0.15	0.13

tion energy can be calculated analytically [9]

$$\Delta E(N) = 0.361 + 1/4\,(N^3 + 47/12\,N^2 + 29/6\,N + 93/48)/(N^2 + 2N + 3/4)^2 \ . \tag{3}$$

For large N, the asymptotical expansion of $\Delta E(N)$ equals

$$\Delta E(N) = 0.361 + 0.250/N - 0.0208/N^2 - 0.0834/N^3 + \cdots \ . \tag{4}$$

If N is sufficiently large we can neglect high order terms and write approximately

$$\Delta E(N) = \ \approx 0.361 + 0.250/N \ . \tag{5}$$

The relative error F [%] of Eq. (5) in comparison with the exact $\Delta E(N)$, Eq. (3), is shown in Table 1. It is obvious that Eq. (5) gives quite good results even for very small N. For two- and three-dimensional clusters we get a more general expression [19]

$$G(N) = c_0 + c_{-1}/N^{1/d} + c_{-2}/N^{2/d} + \cdots \tag{6}$$

$$G(N) \approx c_0 + c_{-1}/N^{1/d} \ . \tag{7}$$

Here, d denotes the number of dimension in which the cluster grows. Equations (4) and (5) are special case of Eqs. (6) and (7) for $d = 1$.

For large clusters it is convenient to convert N into the radius R of a spherical particle. For this conversion only the mass density ρ is needed. For three-dimensional clusters Me_N we use the following two asymptotical expansions

$$G(N) = c_0 + c_{-1}/N^{1/3} + \cdots \tag{8}$$

$$G(R) = c_0 + k_{-1}/R + \cdots; \quad k_{-1} = c_{-1}(3M/4\pi N_A \rho)^{1/3} \ . \tag{9}$$

Here, M is the molar mass of the cluster and N_A is the Avogadro number.

2.5.3.2 Fitting the Asymptotical Expansion of $G(N)$ to Experimental Data. Simple Interpolation Formulae

The concrete form of Eqs. (8) and (9) as they follow from the ACM is influenced by (i) the cluster topology reflected in the mathematical form of the expansion (useful information) and (ii) quantum chemical approximations reflected mainly in the numerical values of the coefficients (it may be questionable part of the result). In order to get better agreement with the experimental data we replace the expansion coefficients obtained from the ACM by the following "physical" parametrization [20]. First we assume an infinite cluster. It is obvious that c_0 is the bulk value of G.

$$c_0 = \lim_{N \to \infty} G(N) = \lim_{R \to \infty} G(R) = G(\infty) . \tag{10}$$

If we consider a cluster with n atoms we get the following equations for determining $c_{-1}(n)$ or $k_{-1}(n)$

$$\begin{aligned} G(n) &= G(\infty) + c_{-1}(n)/(n)^{1/3} \\ G(r_n) &= G(\infty) + k_{-1}(n)/r_n \end{aligned} \qquad n = 1, 2, 3, \ldots . \tag{11}$$

Fitting the first two expansion coefficients and neglecting all higher terms we get simple interpolation formulae

$$G(N) = c_0 + c_{-1}/N^{1/3} \tag{12a}$$

$$G(N) = c_0 + k_{-1}/R , \tag{12b}$$

where

$$\begin{aligned} c_0 &= G(\infty) , \\ c_{-1} &= c_{-1}(n) = [G(n) - G(\infty)]n^{1/3} \\ k_{-1} &= k_{-1}(n) = [G(r_n) - G(\infty)]\, r_n \end{aligned} \qquad n = 1, 2, 3, \ldots . \tag{13}$$

These interpolation formulae can be used as simple rules of thumb to describe and predict size effects. For practical application of these formulae we need only (i) the bulk value $G(\infty)$ tabulated for many properties in literature and (ii) value $G(n)$ or $G(r_n)$ for a cluster containing n atoms or having radius r_n. The values $G(n)$ or $G(r_n)$ must be measured or calculated for the cluster Me_n for some selected n. This means that only a single measurement is needed to get qualitative information on the convergence of $G(N)$ on the whole region of $N = 1, \ldots, \infty$. One can of course increase the accuracy of the interpolation formulae by going to the next terms in the asymptotic expansions (8) and (9). Going to the quadratic terms in $1/N^{1/3}$ or $1/R$ we get

$$G(N) = c_0 + c_{-1}/N^{1/3} + c_{-2}/N^{2/3} , \tag{12c}$$

$$G(R) = c_0 + k_{-1}/R + k_{-2}/R^2 . \tag{12d}$$

In most cases, however, $1/N^{1/3}$ and $1/R$ are the leading terms (see paragraph 5.2) so that this step is not necessary. The advantage of Eqs. (12c, d) is that it can describe cases when $G(N)$ or $G(R)$ is not only monotonic but has a more complex form. An important case may be that of the experimental data exhibiting even symmetry, $c_{-(2m+1)} = k_{-(2m+1)} = 0$. In such a case, $c_{-2}/N^{2/3}$ or k_{-2}/R^2 becomes the leading term and the convergence of $G(N)$ is faster.

2.5.3.3 Verification of the Interpolation Formulae

We have verified the validity of our interpolation formulae (12a, b) for many electronic, geometrical, thermodynamic, kinetic, optical, electrical and magnetic properties [20–23].

2.5.3.3a) Ionization Potential IP for Metal Clusters Me_N

The first example will be given in more detail. Figure 3 shows the ionization potential $IP^K(N)$ of potassium clusters K_N. Full circles are experimental data by Knight et al. [24]. The solid line is the interpolation formula (15a) for the parameters $G(\infty) = IP^K(\infty) = \phi^K_{bulk} = 2.3$ eV = work function of the potassium metal and $G(1) = IP^K_1 = IP^K = 4.44$ eV = ionization potential of the potassium atom. We see that the interpolation formula fits the experimental data very well. The work function and atomic ionization potential are known for almost all metals in the Periodic Table (see e.g. [29]) so that we can predict the size dependence of the ionization potential of many metal clusters

$$IP^{Me}(N) = \phi^{Me}_{bulk} + (IP^{Me} - \phi^{Me}_{bulk})/N^{1/3} \ , \tag{14a}$$

$$IP^{Me}(R) = \phi^{Me}_{bulk} + (4\pi N_A \rho^{Me}/3M)(IP^{Me} - \phi^{Me}_{bulk})/R \ . \tag{14b}$$

For potassium clusters, Eqs. (14) read

$$IP^K(N) = 2.3 + 2.04/N^{1/3} \quad [eV] \ , \tag{15a}$$

$$IP^K(R) = 2.3 + 5.35/R \quad [eV, Å] \ . \tag{15b}$$

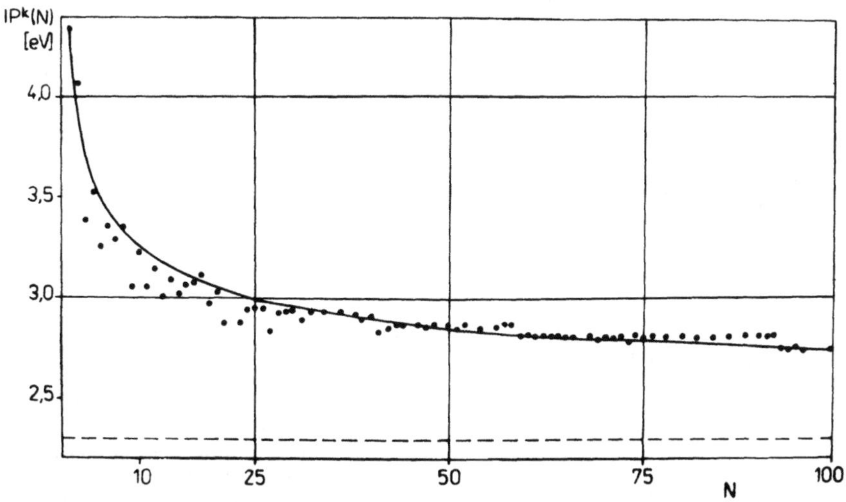

Fig. 3. The ionization potential $IP^K(N)$ of the potassium clusters as a function of the cluster size N. (●) experimental data [24], (——) interpolation formula (15a) (QACM), (---) bulk value

The size dependence of the work function or ionization potential of small metallic spheres following from the classical image potential theory (example of CACM) is given by [25, 26]

$$\mathrm{IP}^{Me}_{\mathrm{class}}(R) = \phi^{Me}_{\mathrm{bulk}} + (3e^2/8)/R \ . \tag{16a}$$

For potassium clusters, Eq. (16a) reads

$$\mathrm{IP}^{Me}_{\mathrm{class}}(R) = 2.3 + 5.40/R \quad [\mathrm{eV, Å}] \ . \tag{16b}$$

We see that Eqs. (15b) and (16b) give very similar result. Strictly speaking, Eqs. (16) can be applied in the classical region only ($R \gg R_0$, R_0 – average interatomic distance). However, Eqs. (16) are sometimes successfully used even for a few atomic clusters (see e.g. [25–28]). Interpolation formulae (15) show that such extension of the applicability of the classical result (16) to small R_0 is possible.

2.5.3.3b) Binding Energy of Metal Clusters

Similarly as above we can use the following values of the binding energy per atom $\mathrm{BE}(\infty) = \mathrm{BE}_{\mathrm{bulk}}$ = binding energy per atom of the bulk metal and $\mathrm{BE}(2) = 1/2\, D_e$, where D_e is the binding energy of the metallic dimer Me_2. Again, these values are well known for many metals for which the size dependence of the binding energy may be predicted

$$\mathrm{BE}^{Me}(N) = \mathrm{BE}^{Me}_{\mathrm{bulk}} + 2^{1/3}(1/2\, D_e - \mathrm{BE}^{Me}_{\mathrm{bulk}})/N^{1/3} \ , \tag{17a}$$

$$\mathrm{BE}^{Me}(R) = \mathrm{BE}^{Me}_{\mathrm{bulk}} + (2\pi N_A \rho^{Me}/3M)^{-1/3}\,(1/2\, D_e - \mathrm{BE}^{Me}_{\mathrm{bulk}})/N^{1/3} \ . \tag{17b}$$

Figure 4 shows the interpolation of the binding energy of Li clusters as an example ($D_e^{\mathrm{Li}} = 1.05$ eV, $\mathrm{BE}^{\mathrm{Li}}_{\mathrm{bulk}} = 2.01$ eV) [30]

$$\mathrm{BE}^{\mathrm{Li}}_{\mathrm{CNDO}}(N) = 2.01 - 1.87/N^{1/3} \quad [\mathrm{eV}] \ . \tag{18}$$

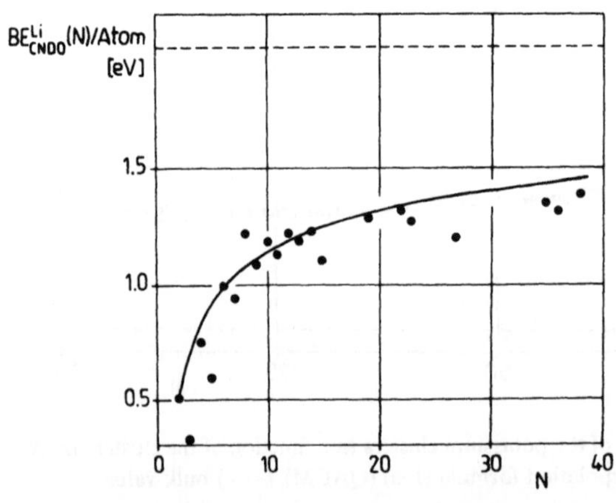

Fig. 4. The binding energy per atom $\mathrm{BE}^{\mathrm{Li}}_{\mathrm{CNDO}}(N)$ of the lithium clusters Li_N as a function of N. (●) CNDO/BW calculation [30], (——) interpolation formula (18) (QACM), (---) bulk value

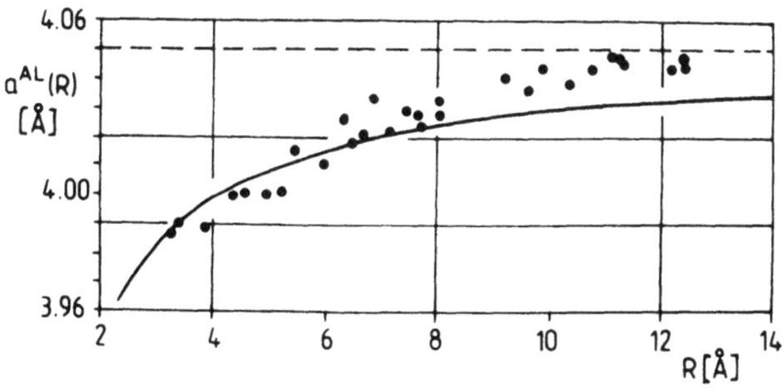

Fig. 5. The lattice constant $a^{Al}(R)$ of the aluminum particles as a function of the cluster size R. (●) experimental data [31], (——) interpolation formula (19) (QACM), (---) bulk value

2.5.3.3c) Lattice Parameter of Aluminum Particles

Nepijko et al. have measured the lattice parameter $a^{Al}(R)$ of aluminum particles Al(R) with radii between 25 and 150 Å [31]. Figure 5 shows the experimental data and their interpolation by means of the formula ($a^{Al}(\infty) = 4.049$ Å, $a^{Al}(32.4\ \text{Å}) = 3.987$ Å)

$$a^{Al}(R) = 4.049 - 2.01/R \quad [\text{Å}] , \tag{19}$$

which fits the experimental data reasonably well.

To describe thermodynamic properties, the quantum calculations must be combined with the statistical approaches. In this sense, the interpolation formulae for thermodynamic properties go beyond the ACM. It appears, however, that the interpolation formulae of the form (6) can be used for various thermodynamic properties, too.

2.5.3.3d) Free Energy, Entropy and Einstein Temperature of Metal Clusters

Hasegawa et al. developed a microscopic theory to explain the size dependence of the melting point of small metallic particles [32]. They calculated various thermodynamic properties for a large number of metal clusters. As an example, we see in Figs. 6a, b and c the free energy $\Delta F(R) = F(R) - F(\infty)$, entropy $\Delta S(R) = S(R) - S(\infty)$ and Einstein temperature $\Theta^2(R)/\Theta_b^2$ as a function of R (Θ_b is the bulk value of the theoretical melting point). Solid lines shown in Figs. 6 follow from the interpolation formulae

$$\begin{aligned} \Delta F^{Al}(R) &= 0.276/R \quad [\text{a.u., Å}] , \\ \Delta S^{Al}(R)/k &= 7.83/R \quad [\text{Å}] , \\ (\Theta^{Al}(R)/\Theta_b)^2 &= 1 - 3.75/R \quad [\text{Å}] \end{aligned} \tag{20}$$

corresponding to $\Delta F(15\ \text{Å}) = 18.4 \times 10^{-3}$ a.u., $\Delta S(15\ \text{Å})/k = 0.52$ and $(\Theta^{Al}(15\ \text{Å})/\Theta_b)^2 = 0.75$. The agreement with the data from [32] is excellent.

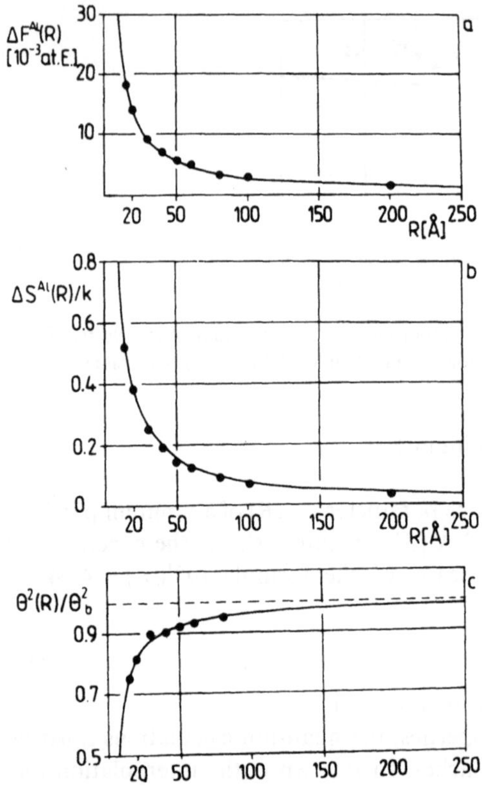

Fig. 6. The free energy $\Delta F^{Al}(R)$ (**a**), entropy $\Delta S^{Al}(R)$ (**b**), Einstein temperature $\Theta^{Al}(R)$ (**c**) of the aluminum particles as a function of their radius R. (●) calculated data [32], (——) interpolation formulae (20) (QACM), (– – –) bulk value

2.5.3.3e) Melting Temperature of Gold Particles

The size dependence of the melting temperature of small particles has been predicted as early as in 1909 by *Powlow* [33]. The experimental verification of this effect was first made in 1954 [34]. Figure 7 shows the experimental melting points of small gold particles [35] as a function of their radius together with our interpolation formula

$$T_m^{Au}(R) = 1336.15 - 5543.65/R \quad [\mathrm{K, Å}] \tag{21}$$

corresponding to the parameters $T_m(\infty) = 1336.5\,\mathrm{K}$, $T_m(9.7\,\mathrm{Å}) = 764.64\,\mathrm{K}$. Again, the agreement of the interpolation formula with the experimental data is satisfactory.

2.5.3.3f) Mie Resonance Absorption of Silver Particles

An explanation of this optical size effect was given at the beginning of this century by Mie in terms of classical electrodynamics [36]. Figure 8 shows

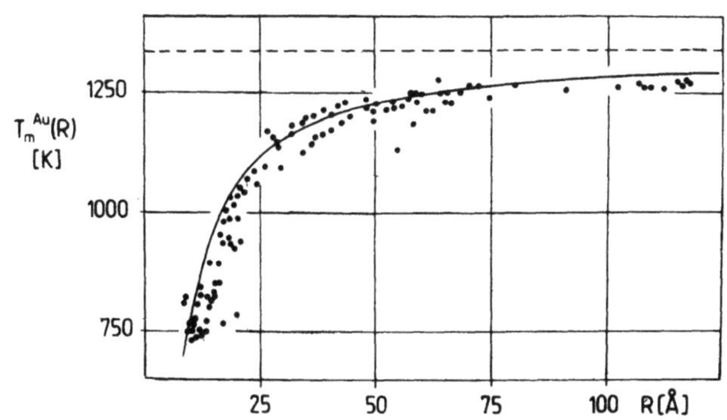

Fig. 7. The melting temperature $T_m^{\mathrm{Au}}(R)$ of the gold particles as a function of their radius R. (●) experimental data [35], (——) interpolation formula (21) (QACM), (---) bulk value

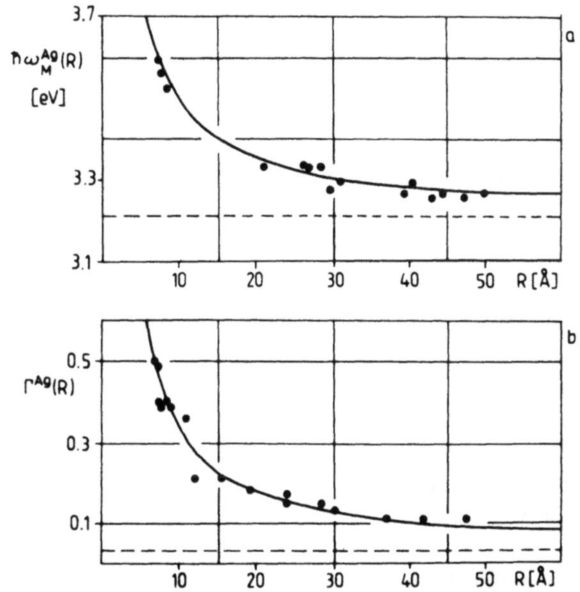

Fig. 8. The position $\hbar\omega_M^{\mathrm{Ag}}(R)$ (a), width $\Gamma^{\mathrm{Ag}}(R)$ (b) of the optical resonance absorption peak in the Mie absorption of the silver particles in the argon matrix as a function of the cluster radius R. (●) experimental data [37], (——) interpolation formulae (22), (---) bulk value

experimental results for the position $\hbar\omega_M^{\mathrm{Ag}}$ and width Γ^{Ag} of the resonance absorption peak as a function of the particle radius R for Ag clusters in Ar matrix [37]. The interpolation formulae

$$\begin{aligned} \hbar\omega_M^{\mathrm{Ag}}(R) &= 3.21 + 2.77/R \quad [\mathrm{eV, Å}] \,, \\ \Gamma^{\mathrm{Ag}}(R) &= 0.03 + 2.94/R \quad [\mathrm{eV, Å}] \end{aligned} \tag{22}$$

corresponding to the values $\hbar\omega_M(\infty) = 3.21$ eV, $\hbar\omega_M(7.3\ \mathrm{Å}) = 3.59$ Å, $\Gamma(\infty) = 0.03$ eV and $\Gamma(6.4\ \mathrm{Å}) = 0.49$ eV describe the size effect very well. In this case, the gap between the macroscopic and microscopic description has been

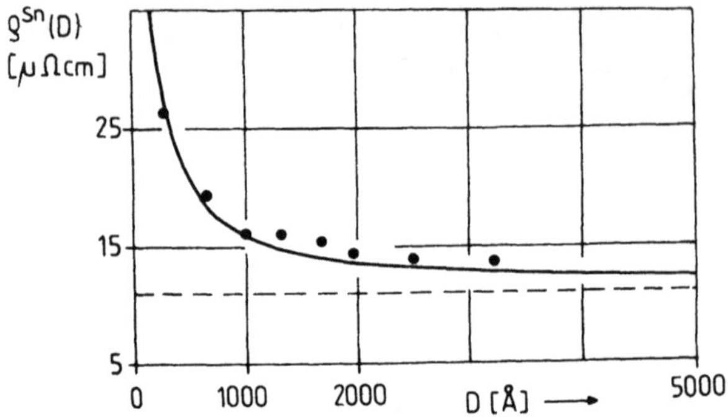

Fig. 9. The specific electrical resistance $\rho^{Sn}(D)$ of the tin films as a function of the film thickness D. (●) experimental data [41], (——) interpolation formula (23), (– – –) bulk value

successfully filled by the calculations of *Bonačič-Koutecký* et al. (see Chapter 2.1 of this volume).

2.5.3.3g) Specific Electric Resistance ρ of Tin Films

We show in this last example that the analytic thin film model (AFM) [38–40] is able to produce simple interpolation formulae in a full analogy with the ACM. *Abou-Saif* et al. [41] measured $\rho^{Sn}(D)$ of Sn films in dependence of their thickness D (see Fig. 9). In this case our interpolation formula corresponding to $\rho(\infty) = 11.15\ \mu\Omega\text{cm}$ and $\rho(315\ \text{Å}) = 26.5\ \mu\Omega\text{cm}$ reads

$$\rho^{Sn}(D) = 11.15 + 4900/D \quad [\mu\Omega\text{cm, Å}] \tag{23}$$

and describes the entire size dependence of ρ^{Sn} quite well.

2.5.3.4 Discussion of Particle Size Effects. Some Predictions and Applications

2.5.3.4a) Mass Density ρ of Metallic Particles

Knowing the size dependence of the lattice parameter $a = a(R)$ of the metallic particles (see Eq. (19)) we can also determine the size dependence of their density $\rho = \rho(R)$

$$\rho_{fcc}(R) = 4M/N_A a^3(R) \ , \tag{24a}$$

$$\rho_{bcc}(R) = 2M/N_A a^3(R) \ . \tag{24b}$$

For example, for aluminum particles Eq. (24a) reads [23]

$$\begin{aligned} \rho^{Al}(R) &= 6.642\ M^{Al}(4.049 - 2.01/R)^{-3} \quad [\text{g cm}^{-3}, \text{g, Å}] \\ &\approx 2.7 + 4.02/R \quad [\text{g cm}^{-3}, \text{Å}] \ . \end{aligned} \tag{25}$$

2.5.3.4b) Contact Potential ΔU Between Metal Clusters

QACM is able to modify the standard definition of the contact potential between two different metals I and II

$$\Delta U^{\mathrm{I,II}}_{\mathrm{bulk}} = 1/e(\phi^{\mathrm{I}}_{\mathrm{bulk}} - \phi^{\mathrm{II}}_{\mathrm{bulk}}) \tag{26}$$

and describe its size dependence for two metallic clusters with N_1 and N_2 atoms

$$\Delta U^{\mathrm{I,II}}_{\mathrm{bulk}}(N_1, N_2) = \Delta U^{\mathrm{I,II}}_{\mathrm{bulk}} + 1/e\,[(\mathrm{IP}^{\mathrm{I}} - \phi^{\mathrm{I}}_{\mathrm{bulk}})/N_1^{1/3} - (\mathrm{IP}^{\mathrm{II}} - \phi^{\mathrm{II}}_{\mathrm{bulk}})/N_2^{1/3}] \; . \tag{27}$$

We have shown in [22] that the contact potential between two differently large clusters ($N_1 \neq N_2$) of the same metal ($Me^{\mathrm{I}} = Me^{\mathrm{II}}$) does not vanish

$$\Delta U^{\mathrm{I=II}}(N_1, N_2) = 1/e(\mathrm{IP} - \phi_{\mathrm{bulk}})(N_1^{-1/3} - N_2^{-1/3}) \; . \tag{28}$$

To give a numerical example, we give the contact potential between the tungsten cluster with $N_1 = 10^6$ and tungsten metal ($N_2 = \infty$): $e\Delta U^{W=W}(10^6, \infty) = 35$ meV.

2.5.3.4c) Binding Energy Per Atom BE of Cluster Ions

As a consequence of the Born–Haber cycle applied to the ionization ($Me_N \rightarrow Me_N^+ + e$; IP($N$)) and fragmentation ($Me_N \rightarrow Me_{N-p} + Me_p$; $\Delta E_{\mathrm{diss}}(N, p)$) of metal clusters one gets the following general energy relation

$$\Delta E_{\mathrm{diss}}(N, p)^+ = \Delta E_{\mathrm{diss}}(N, p) + \mathrm{IP}(N-p) - \mathrm{IP}(N) \; . \tag{29}$$

As a special case we get the binding energy per atom $\mathrm{BE}(N)^+$ of the cluster ions Me_N^+

$$\mathrm{BE}(N)^+ = \mathrm{BE}(N) + 1/N\,[\mathrm{IP}(1) - \mathrm{IP}(N)] \; . \tag{30}$$

Equations (14a) and (17a) make possible to derive the expression [43, 44]

$$\mathrm{BE}^{Me}(N)^+ = \mathrm{BE}^{Me}_{\mathrm{bulk}} + 1.26(1/2\,D_e^{Me} - \mathrm{BE}^{Me}_{\mathrm{bulk}})/N^{1/3} + (\mathrm{IP}^{Me}(1) - \phi^{Me}_{\mathrm{bulk}})/N - 1.26(\mathrm{IP}^{Me}(2) - \phi^{Me}_{\mathrm{bulk}})/N^{4/3} \tag{31}$$

For the iron cluster ions Fe_N^+ Eq. (31) reads

$$\mathrm{BE}^{\mathrm{Fe}}(N)^+ = 4.29 - 4.74/N^{1/3} + 3.37/N - 2.27/N^{4/3} \quad [\mathrm{eV}] \; . \tag{32}$$

We note that this equation is the example of the expansion to higher orders of $1/N^{1/3}$. Figure 10 shows that Eq. (32) agrees with the experimental data [45] well. Compare also with a similar analysis of alkali metal clusters data by *Brechignac* et al. [46].

There are also other examples which cannot be discussed in detail here: Discussion of the competition between the fragmentation and ionization of metal clusters [42–44]; introduction of the size dependent parametrization in

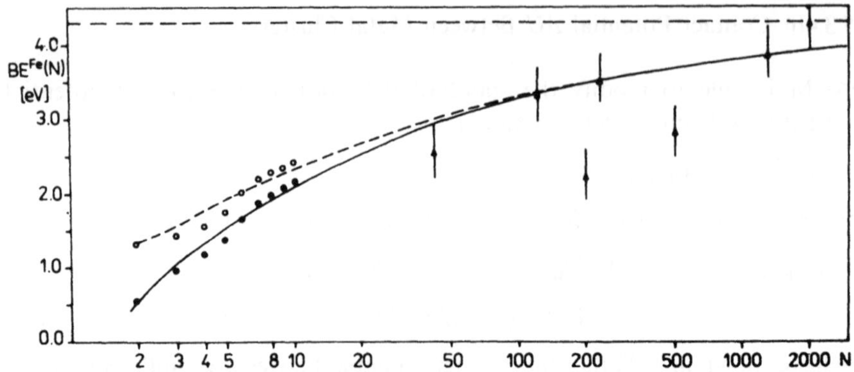

Fig. 10. The binding energies per atom $BE^{Fe}(N)^+$ and $BE^{Fe}(N)$ of the iron cluster ions Fe_N^+ and neutral clusters Fe_N as a function of the number of atoms N. (○), (●, ▲) are experimental data for $Be^{Fe}(N)^+$ and $BE^{Fe}(N)$, (---) and (——) correspond to the interpolation formulae (32) and (17a)

quantum chemical semiempirical methods [47]; extension of the range of applications from classic-phenomenological laws to the field of matter in a highly dispersed state [48]; the size induced metal-insulator transition with possible implications in microelectronics [49] and size dependence of the phase transition temperature important in producing advanced ceramics [50].

2.5.4 Topological Analytic Cluster Model (TACM)

2.5.4.1 Form of Particles

We consider clusters Y_N consisting of N atomic or molecular units Y which are assumed to be touching spheres. These units are regularly arranged in clusters of different geometrical shapes listed in Table 2.

In most cases, the clusters can be conceptually isolated from the corresponding crystals. The number of the units in the cluster is given by

$$N = a_3 n^3 + a_2 n^2 + a_1 n + a_0, \quad n = 1, 2, 3, \ldots, \tag{33}$$

Table 2. Considered shape of clusters and the corresponding parameters

Cluster shape	a_3	a_2	a_1	a_0	f_A	f_v	δ_e
Simple cubic	1	3	3	1	6	1	1
Face centered cube (fcc)	4	6	3	1	6	1	$\sqrt{2}/2$
Cuboctahedron (fcc)	10/3	5	11/3	1	$2(3+\sqrt{3})$	$10/(3\sqrt{2})$	$\sqrt{2}/2$
Anticuboctahedron (hcp)	10/3	5	11/3	1	$2(3+\sqrt{3})$	$10/(3\sqrt{2})$	$\sqrt{2}/2$
Rhombohedral							
Dodecahedron (bcc)	4	6	4	1	$8\sqrt{2}$	$16/(3\sqrt{3})$	0.61
Icosahedron	10/3	5	11/3	1	$5\sqrt{3}$	2.182	0.66

where n is the number of the cluster shells. The surface edge length equals

$$r = cR(n + \delta) , \tag{34}$$

where R is the shortest distance between the units and $c = \sqrt{2}$ for the face centered cube, $c = 1$ otherwise [51]. The parameter δ determines the location of the surface. Two limits are known in the literature: (i) $\delta = \delta_c = 0$, "molecule center dividing surface A_c" (the faces of the polyhedron go through the centeres of the surface units) and (ii) $\delta = \delta_e$, "equimolecular dividing surface A_e" (the cluster is limited by the tangential planes of the surface units; see Table 2) [52, 53]. The surface area is

$$A = f_A r^2 . \tag{35}$$

The volume of the cluster is given by

$$v = f_v r^3 . \tag{36}$$

We introduce also an intermediate surface A_x by the following definition

$$v_x - v_c = \sum_{i=1}^{N_s} \bar{v}_i , \tag{37}$$

where $\bar{v}_i$ is the part of the volume of the i-th surface unit outside of A_c and N_s is the number of units at the surface. This surface has the property $\delta_c < \delta_x < \delta_e$. In contrast to δ_c and δ_e, δ_x depends on the size of the particle (see Fig. 11).

2.5.4.2 Interaction Potential

The total energy E is calculated as a sum of pair potentials V_{ij} of the units Y_i and Y_j at the distance R_{ij}

$$E = \sum_{i<j} V_{ij}(R_{ij}) . \tag{38}$$

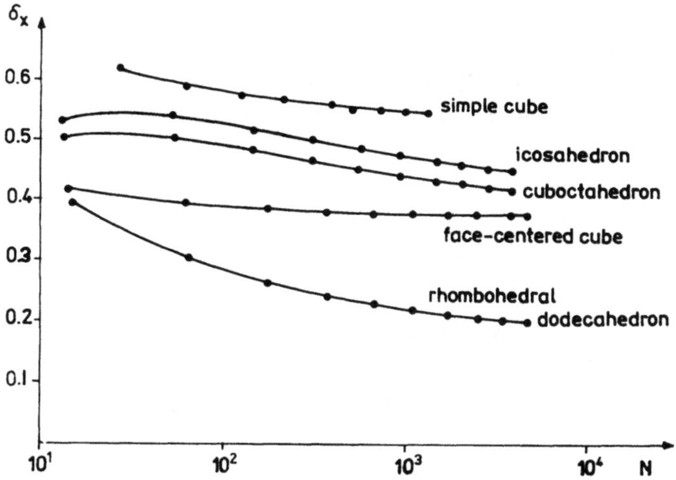

Fig. 11. The surface location parameter δ_x as a function of the number of units N

Table 3. The range of the potential in Ar_N clusters and the Ar crystal. $Z^{(k)}$ is the multiple of the distance $R_{ij}^{(k)}$, $(Z^{(k)} V(R^{(k)}))/(Z^{(1)} V(R^{(1)}))$ is the contribution of the k-th pair potential to that of the nearest neighbors

	$N = 13$	$N = 55$	$N = \infty$ (fcc)	coordination number
$Z^{(2)}$		12	90	6
$Z^{(3)}$		24	234	12
$Z^{(4)}$		6	93	
$Z^{(2)} V(R^{(2)}))/(Z^{(1)} V(R^{(1)}))$		7.8%	9.8%	11.7%
$Z^{(3)} V(R^{(3)}))/(Z^{(1)} V(R^{(1)}))$		4.8%	7.8%	7.2%
$Z^{(4)} V(R^{(4)}))/(Z^{(1)} V(R^{(1)}))$		0.5%	1.3%	

2.5.4.2a) $Y =$ Rare Gas Atoms

The attractive potential is assumed to be proportional to R^{-6} and the short range repulsion is taken to be proportional to R^{-12}. Table 3 shows the range of the attractive potential for the cuboctahedral clusters Ar_N and in the Ar crystal.

2.5.4.2b) $Y =$ Molecule

The molecular units are supposed to have a rigid structure. We assume therefore that the sum of the intramolecular energies of the units is constant. The main contributions to the intermolecular pair potentials are (see Table 4): (i) The orientation energy of permanent dipoles (multipoles)

$$V_1 \sim R^{-3} \quad \text{for } R \text{ and } T \text{ small}$$

$$\sim R^{-6} \quad \text{for } R \text{ and } T \text{ large (Keesom),}$$

(ii) the induction energy of permanent dipoles (multipoles) (Debye) $V_2 \sim R^{-6}$ and (iii) the quantum mechanical dispersion energy (London) $V_3 \sim R^{-6}$.

We see that the dispersion energy V_3 dominates for weakly polar molecules.

Table 4. The relative contributions of the dipole orientation energy V_1, dipole induction energy V_2 and dispersion energy V_3 at $T = 293$ K for molecules with the permanent dipole moment

Molecular unit Y	Ratio of dipole moments $\mu(CO)/\mu(Y)$	$V_1 : V_2 : V_3$
CO	1	1:17:19853
HJ	3.2	1:5:1091
HBr	6.5	1.5:1:43
HCl	8.6	3.4:1:19

2.5.4.2c) Topologicl Model

It is obvious that the pair interaction of the nearest neighboring units is the main contribution to the interaction energy of the van der Waals clusters. As an illustration we consider the simplified fcc like lattice of the carbon monoxide crystal (for the clusters of the rare gas atoms see Table 3). The distances between the CO molecules are as follows: $R_{ij}^{(1)} = 2^{-1/2}a$ (the first neighbors); $R_{ij}^{(2)} = a$ (the second neighbors) and $R_{ij}^{(3)} = 2^{1/2}a$ (the third neighbors). Here, a denotes the fcc lattice constant. We get for these distances

$$V(R_{ij}^{(1)}) : V(R_{ij}^{(2)}) : V(R_{ij}^{(3)}) = 64:8:1 \; . \tag{39}$$

For this reason, we consider only the nearest neighbor interaction

$$V_{ij}(R_{ij}) \approx \begin{cases} V(R), & \text{if } R_{ij} = R_{ij}^{(1)} = R \\ 0 & \text{otherwise} \; . \end{cases} \tag{40}$$

As a second approximation, the distance R of the nearest neighboring units is supposed to be a constant for all the considered clusters. We assume also that R is independent of the size of the clusters. These assumptions mean that we neglect small bond length differences (5% for icosahedra) as well as the size dependent changes of the bond length (< 2% for the rare gases). The total interaction energy of the units depends then on their coordination numbers CN [17, 52]

$$E = 0.5 \; V(R) \sum_{i=1}^{N} \mathrm{CN}(i) \; . \tag{41}$$

Taking into consideration the topological equivalence of the units Y the following dependence of the energy E on n can be derived

$$E = V(R)[b_3 n^3 + b_2 n^2 + b_1 n + b_0] \; . \tag{42}$$

Using Eq. (33) this result can be transformed [18, 55] into the form of the interpolation formula

$$E(N) = E(\text{bulk})(1 + c_1/N^{1/3} + c_2/N^{2/3} + c_3/N + \cdots) \; . \tag{43}$$

Table 5. Coefficients b_i and c_i

Particle shape	b_3	b_2	b_1	b_0	c_1	c_2	c_3
Simple cube	3	6	3	2	− 1.00	0.00	0.00
Face centered cube	24	12	0	0	− 1.59	0.63	− 0.50
Cuboctahedron/	20	12	4	0	− 1.34	0.00	0.45
Anticuboctahedron	20	12	4	0	− 1.34	0.00	0.45
Rhombohedral							
Dodecahedron	16	12	4	0	− 1.19	0.00	0.25
Icosahedron	20	15	7	0	− 1.12	0.00	0.38

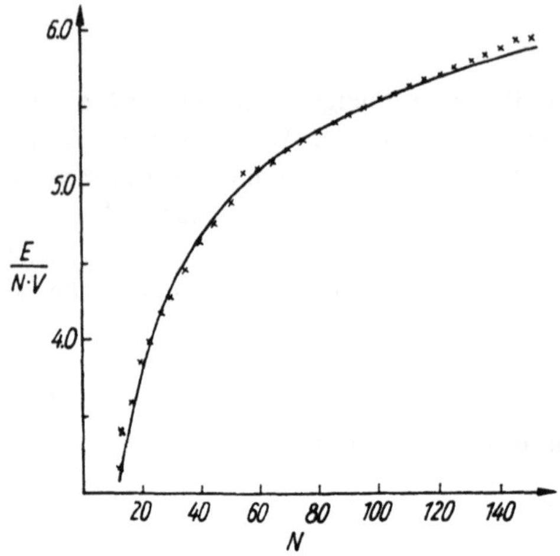

Fig. 12. The reduced pair potential energies $E/(NV)$ of the argon particles Ar_N [56]. (×) sum of the all pair contributions, (——) least square fit by the function $E/(NV) = 8.02–11.34/N^{1/3}$

The coefficients appearing in Eqs. (42), (43) are given in Table 5. For clusters in which more than two shells are filled ($n > 2$) only the terms proportional to $1/N^{1/3}$ are significant [18]

$$E(N) \approx E(\text{bulk})(1 + c_1/N^{1/3}) \ . \tag{44}$$

Figure 12 shows the interpolation of the binding energies of Ar_N clusters calculated by *Northby* [56] by Eq. (44).

2.5.4.3 Applications

2.5.4.3a) Magic Numbers

It has been shown by many authors (see eg. [17, 57–59]) that the structure of the van der Waals clusters is often determined by the optimum coordination of their units Y. In this sense, adding of a new unit to the cluster takes place on the highest coordinated site of the cluster. This concept explains most of the significant peaks in the mass spectra of many van der Waals clusters (Ne_N^+ [60], Ar_N^+ [57], Xe_N^+ [61], Kr_N^+ [62], $(N_2)_N^+$ [63], $(CO)_N^+$ [64], $(CH_4)_N^+$ [65]). Such clusters have usually the "magic number" N^* units and their structure is often the closed shell icosahedron. Now we illustrate the application of the TACM to the problem of the magic numbers. We consider the energy of adding the N-th unit to the cluster Y_{N-1}

$$\Delta E(N) = 0.5V(R)\left[\sum_{i=1}^{N} \mathrm{CN}(i) - \sum_{i=1}^{N-1} \mathrm{CN}(i)\right] = 0.5V(R)\mathrm{CN}(N) \ . \tag{45}$$

The last step of the formation of the icosahedron consists in closing the pentagonal cap. Here, the added unit has the 6-fold coordination to the remaining part of the cluster. The topologically preferred site for the next growth of the cluster is the site over the center of the triangular facet of the icosahedron surface (3-fold coordination). Equation (45) enables to compare the energies of the both steps

$$\Delta E(N^*):\Delta E(N^*+1) = 2:1 \ . \tag{46}$$

This equation explains why the particles $Y_{N^*}^+$ are more stable than $Y_{N^*+1}^+$ in agreement with the experimental intensities of the corresponding mass spectrum $I(N^*) \gg I(N^*+1)$. The magic numbers have therefore topological origin [54, 66]. We note also that the van der Waals clusters prefer usually the minimum surface/volume ratio [67].

2.5.4.3b) Compressibility

The compressibility at zero temperature is defined as

$$\kappa = 1/v\, \partial^2 v/\partial E^2 \ . \tag{47}$$

We introduce into Eq. (41) the Lennard-Jones potential for the nearest neighbors

$$V(R) = 4\varepsilon[(\sigma/R)^{12} - (\sigma/R)^6] \ , \tag{48}$$

determine the equilibrium volume v_0 from the condition $\partial E/\partial R = 0$ and get the energy $E = E(v_0)$ [68]. The corresponding compressibility is

$$\kappa(n) \sim [\sigma(n+\delta)]^3/(An^3 + Bn^2 + Cn) \ , \tag{49}$$

where $A = 20$, $B = 15$, $C = 7$ for icosahedra and $A = 20$, $B = 12$, $C = 4$ for cuboctahedra. The compressibility for very large clusters equals

$$\kappa(\text{bulk}) = \lim_{n\to\infty} \kappa(n) \sim \sigma^3 \ . \tag{50}$$

In order to estimate the effect of the topological approximation we compare $\kappa(\text{bulk})^{\text{topol}}$, Eq. (50), with the compressibility of the fcc lattice of the rare gas atoms $\kappa(\text{bulk})^{\text{exact}}$ where all the pair potentials are considered [69]

$$(\kappa^{\text{topol}}/\kappa^{\text{exact}})_{\text{bulk}} = 90.3\% \ . \tag{51}$$

Finally, the size dependence of the compressibility is given by

$$\kappa(n) = \kappa(\infty)\, A(n+\delta)^3/(An^3 + Bn^2 + Cn) \ . \tag{52}$$

2.5.4.3c) Surface Energy and Surface Stress

In contrast to the bulk units the forces among the surface units and their neighbors do not compensate. As a result, a surface force directed to the bulk appears. The surface energy E_s is

$$E_s = 0.5\, V(R) \sum_{i=1}^{N_s} [\text{CN(bulk)} - \text{CN}(i)] \ , \tag{53}$$

where CN(bulk) is the coordination number of the bulk units. The surface energy per unit surface area (the surface tension σ) and the isotropic surface stress γ are defined as

$$\sigma = E_s/A \ , \tag{54}$$

$$\gamma = \sigma + A\, d\sigma/dA \ . \tag{55}$$

Using Eqs. (35) and (38) we get for the cuboctahedra, icosahedra and rhombohedral dodecahedra [51, 52, 55]

$$E_s(n) = V(R)(k_2 n^2 + k_1 n + k_0) \ , \tag{56}$$

$$\gamma(n) = 1 - (\delta - 0.5)/(n + \delta) \ . \tag{57}$$

The last equation can be converted into the N-dependent form

$$\gamma(N) \approx \gamma(\text{bulk})(1 + \alpha_1/N^{1/3} + \alpha_2/N^{2/3} + \alpha_3/N + \cdots) \tag{58}$$

(the coefficients are given in Table 6). The size dependence of γ is strongly

Table 6. The coefficients to Eqs. (56)–(58)

	k_2	k_1	k_0	α_1	α_2	α_3
Cuboctahedron/anticuboctahedron	18	18	6	– 0.31	0.10	– 0.11
Rhombohedral dodecahedron	12	12	4	– 0.18	0.03	– 0.04
Icosahedron	15	15	6	– 0.24	0.06	– 0.08

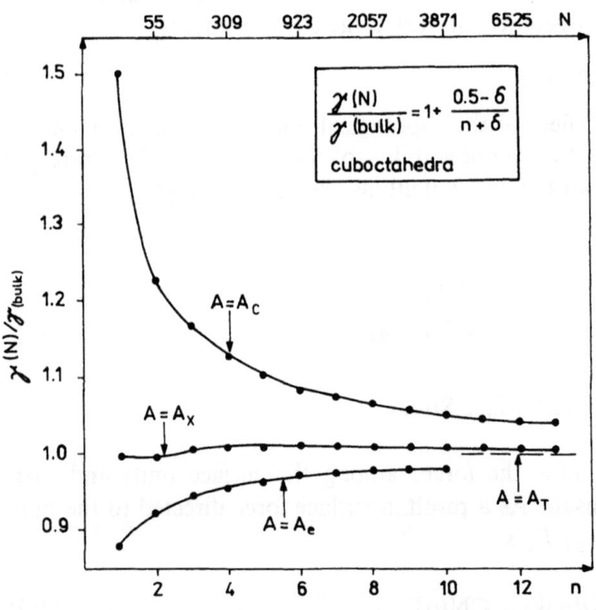

Fig. 13. The surface stress $\gamma = \gamma(n)$ for different surfaces.

influenced by the cluster surface [51–53]. For $\delta = 0.5$, γ does not depend on the size of the cluster at all. This condition defines the "surface of tension" A_T postulated by Gibbs long time ago. The surface A_T is very close to the intermediate surface A_x for which δ_x is also close to 0.5 (compare Figs. 11 and 13).

2.5.4.3d) Pressure and Chemical Potential

Now we consider a system of a cluster and its vapor phase [51, 53]. The solid phase is denoted by s and the vapor phase by g. At constant temperature, the free energy of the system is given by

$$df = -p_s dv_s - p_g dv_g + \bar{\mu}_s dN_s + \bar{\mu}_g dN_g + \gamma dA \ , \tag{59}$$

where p is the pressure and $\bar{\mu}$ is the chemical potential. At the constant volume we have $dv_s = -dv_g = dv$. If the numbers of atoms in the two phases N_s and N_g remain constant the equilibrium condition $df = 0$ (for T, v_t, N_s, N_g is constant) leads to the Young–Laplace equation of the mechanical equilibrium

$$\begin{aligned} p_s - p_g &= \Delta p = \gamma \, dA/dv \\ &= \text{const}\ \gamma_\infty/[R(n+\delta)]\,[1-(\delta-0.5)/(n+\delta)] \ , \end{aligned} \tag{60}$$

where const = 2.449 for the cuboctahedron and 2.646 for icosahedron. For the diffusion equilibrium ($dN_s = -dN_g = dN$) we get the Kelvin equation

$$\Delta\bar{\mu} - v_s\Delta p = \gamma \, dA/dN = \text{const}\ \gamma_\infty[R(n+\delta)]^2[1-(\delta-0.5/(n+\delta)] \tag{61}$$

with const = 5.774.

We note at the end of this paragraph that the numerical calculations on highly symmetrical clusters $(NaCl)_N$ and $(CsCl)_N$ ($4 \leq N \leq 500$) by means of the Born–Mayer pair potential [70] show that the size effects for the ionic bonded clusters can be also described by the interpolation formulae analogous to those discussed above.

2.5.5 Theoretical Background of Interpolation Formulae

2.5.5.1 Interpolation Formulae in Other Fields of Physics and Chemistry

A few interesting results were achieved by *Weyl* at the beginning of this century [71–78]. Weyl showed that asymptotically leading terms in the integrated density of states of the scalar wave equation depend only on the surface of the considered region but not on its shape. The answer to the question if one can hear the shape of a drum must be therefore answered in a negative sense. For a sufficiently large drum its spectrum does not depend on its shape. It has been also shown (see [79]) that it is impossible to hear the style of a cathedral in which it is built (i.e. the curvature of its surface, singularities at the surface etc.) since the spectrum of the concert does not depend, at least for a sufficiently large cathedral, on the surface details. Weyl considered also the spectrum of the electromagnetic waves in a cavity and elastic vibrational problem. All three problems lead to the same conclusion: The integrated density of states $I(E)$ i.e. the number of eigenvalues of the problem depends asymptotically (for very large volume V or surface S for two-dimensional systems) only on V or S and is independent of the form of the boundary region. This result is closely related to the existence of the *Planck* formula [80] for the black body radiation or *Debye* T^3 law for the specific heat [81] which are valid only for very large volume. The first correction to these formulae for a finite volume is proportional to the ratio surface/volume similarly to the results mentioned above [79]. Analogous results were obtained also for the *Stefan–Boltzmann* law [82, 79]. We note that deviations from these laws for finite volumes can be observed experimentally (see [79]).

In other fields of physics similar results were obtained, too. We should mention the acoustics of rooms, thermodynamic properties of ideal gas in a finite volume, semiempirical formula for the nuclear binding energy, autocorrelation functions in statistical physics, Bose–Einstein condensation in a thin film and vibrational contribution to the specific heat of small crystallic particles (see review [79]). Further examples from different fields of physics can be found in [83, 79, 85–90].

2.5.5.2 Dependence of Interpolation Formulae on $1/N^{1/3}$

As we see from the examples given above, the formulae analogous to (6) appear both in classical and quantum physics and can be found both in non-relativistic and relativistic physics (for the relativistic photon gas see [79]). It is obvious, that their existence must be related to some general property which is common to all these fields. Regardless of the classical or quantum formulation, we search for the size dependence of physical properties of a system in a finite volume. The corresponding equations of motion may be different depending on the problem solved. It may be the diffusion equation, wave equation, Schrödinger equation or another equation of motion. The unifying point of view can be found transforming this equation of motion into the matrix form in a suitably chosen localized basis. In case of clusters and Schrödinger equation it may be for example the LCAO basis. When increasing the cluster size we can suppose that the matrix elements corresponding to the basis functions at large distance from the surface change negligibly in contrast to those where at least one basis function lies near the surface. As a result, we get in all cases a hermitian matrix eigenvalue problem of the form

$$Ax = \lambda Bx \ , \tag{62}$$

where B is a positive definite "overlap" matrix ($B = 1$ for the orthogonal basis).

λ can have for example the meaning of energy (stationary Schrödinger equation) or the square of the frequency (wave equation). Some matrix elements of the matrices A and B change negligibly with the increasing size of the system (we denote this set as a) while the others are surface depended (set b). The problem (62) can be then partitioned in the form

$$A = \begin{bmatrix} A_{aa} & A_{ab} \\ A_{ba} & A_{bb} \end{bmatrix}, \quad B = \begin{bmatrix} B_{aa} & B_{ab} \\ B_{ba} & B_{bb} \end{bmatrix}, \quad x = \begin{bmatrix} x_a \\ x_b \end{bmatrix} . \tag{63}$$

Denoting the number of eigenvalues of the problem (62) less than a real number μ (the integrated density of states) as $I\{A, B; \mu\}$ and using the so-called negative eigenvalue theorem [91–94, 10, 84] closely to the old result of Weierstrass [95] we get

$$I\{A, B; \mu\} = I\{A_{aa}, B_{aa}; \mu\} + I\{\tilde{A}_{bb} - \mu B_{bb}, 1; 0\} . \tag{64}$$

Here, the μ-dependent matrix $\tilde{A}_{bb}$, sometimes called the effective Hamiltonian equals

$$\tilde{A}_{bb} = A_{bb} - (A_{ba} - \mu B_{ba})(A_{aa} - \mu B_{aa})^{-1}(A_{ab} - \mu B_{ab}) . \tag{65}$$

Without going into detail (see for example [10, 84]) we see that the contribution of the second term in (64) to $I\{A, B; \mu\}$ cannot be larger than the number of the basis functions n_b present in the surface region b

$$0 \leq I\{\tilde{A}_{bb} - \mu B_{bb}, 1; 0\} \leq n_b . \tag{66}$$

Increasing the size of the system, the number of the basis functions in the region a increases as the volume V

$$n_a \sim V \tag{67}$$

while n_b depends on the surface S

$$n_b \sim S . \tag{68}$$

It means therefore that the relative contribution of the second surface term in Eq. (64) to the integrated density of states cannot be larger than a constant times S/V. The form of the surface is not important here. This result is very general as it does not depend on the values of the matrix elements of the matrices A and B and is valid for any problem of the form (62). Equation (66) gives the lower and upper bound to $I\{\tilde{A}_{bb} - \mu B_{bb}, 1; 0\}$. The actual contribution of the surface term to $I\{A, B; \mu\}$ can be considerably smaller. More detailed discussion can be found in [19, 10]. We note that the *Ledermann* theorem [96] follows from Eq. (64) as a special case.

According to this result, the upper bound to the relative contribution of the surface term to the eigenvalue spectrum is proportional to S/V, in agreement with closely related results of Weyl. The real dependence of the quantities derived from the eigenvalue spectrum on S/V depends of course on their character. It is the reason that some cluster properties depend on $1/N^{1/3}$

(binding energy) while the others have $1/N^{2/3}$ dependence (band width, the lowest occupied MO). We note that for small clusters better agreement with experiment may be achieved if instead of $S/V \sim 1/N^{1/3}$ the parameter proportional to $S/(V-S)$ is taken [88]. We note also that for fractal surfaces the number of atoms at the surface is proportional to its fractal dimension df, $S \sim N^{df}$, so that $S/V \sim 1/N^{3-df}$. In such a case, the exponent in the interpolation formula (6) can be different from 1/3 or 2/3.

2.5.5.3 Non-Stationary Properties

In case of non-stationary properties, the problem is complicated by the existence of further "degree of freedom", time. This complicates both the theoretical analysis and experimental investigation (necessity to perform time resolved experiments). The amount of information in this respect is therefore limited.

First we consider the solution of the Schrödinger equation

$$H\psi_n = E_n\psi_n \tag{69}$$

with the time independent Hamiltonian H. A general solution of the time Schrödinger equation equals

$$\psi(t) = \sum_n c_n \exp[E_n t/(i\hbar)]\psi_n \tag{70}$$

and the average value of a time independent operator L is

$$\bar{L} = \sum_{n,m} c_n^* c_m \exp[(E_m - E_n)t/(i\hbar)]L_{nm} , \tag{71}$$

where

$$L_{nm} = \int \psi_n^* L \psi_m d\tau . \tag{72}$$

Size dependence of $\bar{L}$ appears in the time dependent part of (71) through $E_m - E_n$ and the matrix elements L_{mn}. Also the coefficients c_n given by initial conditions are different for a different size of the system. The combination of these factors complicates the analysis. In some cases, the problem can be simplified.

For example, the average energy equals

$$E = \sum_m |c_m|^2 E_m . \tag{73}$$

This result is similar to the expression for the binding energy in the analytic cluster model [10, 13, 14]. Therefore, except for the dependence on initial conditions, a similar size dependence of E as for the binding energy can be expected.

A special importance have propagators i.e. the response of the system to a localized excitation at $t = 0$. For example, the propagator of the Schrödinger equation for the infinite linear chain equals [97]

$$\psi_{np}(t) = 1/2\pi \int_{-\pi}^{\pi} e^{in\vartheta} e^{-ip\vartheta} e^{-i/\hbar 2V\cos\vartheta\, t} d\vartheta$$
$$= (-i)^{n-p} J_{n-p}(2V\cos\vartheta t) , \quad (74)$$

where J is the Bessel function. $|\psi_{np}(t)|^2$ gives the probability of finding the excitation at time t at site n if the excitation at $t = 0$ was localized at site p ($|\psi_{np}(0)|^2 = \delta_{np}$). For general initial conditions, the solution has the form

$$\psi_n(t) = \sum_p \psi_p(0)\, \psi_{np}(t) . \quad (75)$$

For a finite linear chain with N atoms we get [97]

$$\psi_{np}(t) = \frac{2}{(N+1)} \sum_{k=1}^{N} \sin[nk\pi/(N+1)]\sin[pk\pi/(N+1)]$$
$$\times e^{1/(i\hbar)2V\cos[k\pi/(N+1)]t} . \quad (76)$$

The comparison of Eqs. (76) and (74) shows that the size dependence of the propagator (76) is mathematically equivalent to the problem of replacing the integration by a discrete sum.

Similar conclusions can be drawn also for the incoherent and partially coherent motion of the excitation. For example, for the Pauli kinetic equation

$$dP_n/dt = \sum_{n'} F_{nn'} P_{n'} , \quad (77)$$

where F is a matrix of rate constants, the propagator for the infinite linear chain equals [97]

$$\psi_{np}(t) = (2\pi)^{-1} \int_{-\pi}^{\pi} e^{in\vartheta} e^{-ip\vartheta} e^{2F(-1+\cos\vartheta)t} d\vartheta = e^{-2Ft} I_{n-p}(2Ft) , \quad (78)$$

where I is the modified Bessel function. For a finite linear chain we get [99]

$$P_{np}(t) = \sum_{k=0}^{N-1} h_n^{(k)} h_p^{(k)} \exp(\lambda_k t) , \quad (79)$$

where λ_k and $h_n^{(k)}$ denote the eigenvalue and the n-th component of the corresponding eigenvector of the matrix F. We see that the result (79) is similar to (76) except for the missing imaginary unit in the exponential. This is characteristic for the transition from the coherent to incoherent motion. Analogous results can be obtained also for the Liouville equation [98].

Theory of the time resolved spectroscopy of the photosynthetic systems shows that, similarly to the stationary properties discussed above, an important parameter is $1/N$, where N denotes the total number of active molecules in photosynthetic unit [100]. More results for the fluorescence intensity, the corresponding quantum yield and other quantities can be found in [100–101].

We have seen a few concrete examples of the size dependence of time dependent properties of finite systems. In more general cases, the results are analogous to those given above.

Although the actual dependence of the properties of clusters, thin films, linear macromolecules and other finite systems on their size may be complicated, the discussions given above shows that the negative eigenvalue theorem provides a general argument that the interpolation formulae depending on the parameter $1/N^{1/d}$, where d is usually 1, 2, 3 but can be also small non-integer, are reasonable approximate approaches for describing the size dependence of properties of finite systems.

2.5.6 Concluding Remarks

The investigation of the physical and chemical properties of clusters belonging to the mesoscopic region is difficult not only theoretically but also experimentally. We have overcome these difficulties by means of the analytic cluster models which enable to calculate analytically numerous properties of arbitrary large clusters. Using the results following from the analytic cluster models we have introduced simple interpolation formulae enabling to describe and predict size dependence of physical and chemical cluster properties in agreement with experimental data. We have shown that the interpolation formulae are simple and theoretically justified tool for describing size dependence of clusters and other mesoscopic systems.

Acknowledgements. Financial support of this work by Deutsche Forschungsgemeinschaft and the Fonds der Chemischen Industrie is gratefully acknowledged.

References

1. J.A.A.J. Perenboom, P. Wyder, F. Meier: Phys. Rep. (Rev. Sect. Phys. Lett.) **78**, 173 (1981)
2. H. Müller: Wiss. Z. Humboldt-Univ. Berlin, Math.-Nat. R. **34**, 78 (1985)
3. H. Müller: in: Metal Clusters, Ed. F. Traeger, G. zu Pulitz, Springer Verlag, 1986 (p. 133); Z. Phys. **D3**, 233 (1986)
4. L. Skala, H. Müller: in: Lecture Notes in Physics, Vol. 269, PDMS and Clusters, Eds. E.R. Hilf, F. Kammer, K. Wien, Spinger Verlag, 1987 (p. 144)
5. G.D. Stein: The Physics Teacher, Nov. 1979, p. 503
6. H.-G. Fritsche, P. Kadura, L. Künne, H. Müller, E. Bauwe, S. Engels, W. Mörke, G. Rasch, P. Birke, H. Spindler, M. Wilde, H. Lieske, J. Völter: Z. Chem. **24**, 169 (1984)
7. T.A. Hoffmann: Acta Phys. Hungar. **1**, 1 (1951); **2**, 97 (1952)
8. R.P. Messmer: Phys. Rev. **B15**, 1811 (1977)
9. H. Müller: Z. Chem. **12**, 475 (1972)
10. L. Skala: Czech. J. Phys. **B27**, 171 (1977)
11. O. Bilek, P. Kadura: phys. stat. sol. (b) **85**, 225 (1978)
12. P. Kadura, L. Künne: phys. stat. sol. (b) **88**, 537 (1978)
13. O. Bilek, L. Skala: Czech. J. Phys. **B28**, 1003 (1978)

14. L. Künne, O. Bilek, L. Skala: Czech. J. Phys. **B29**, 1030 (1979)
15. O. Bilek, L. Skala, L. Künne: phys. stat. sol. (b) **117**, 675 (1983); L. Künne, L. Skala, O. Bilek: phys. stat. sol. (b) **118**, 173 (1983)
16. L. Salem: J. Phys. Chem. **89**, 5576 (1985)
17. H.-G. Fritsche, D. Bonchev, O. Mekenyan: Z. phys. Chem. Leipzig **270**, 469 (1989); J. Less-Common Metals **141**, 137 (1988)
18. H.-G. Fritsche: Cryst. Res. & Technol. **24**, 629 (1989)
19. L. Skala: phys. stat. sol. (b) **127**, 567 (1985)
20. H. Müller: Ch. Opitz, K. Strickert, L. Skala: Z. phys. Chem. Leipzig **268**, 625 (1987)
21. H. Müller, L. Skala, Ch. Opitz, K. Strickert: Wiss. Z. Univ. Jena, Math.-Nat. R. **36**, 531 (1987); Ch. Opitz, H. Müller: ibid. **36**, 535 (1987); S. Romanowski, P. Mlynarski, Ch. Opitz, H. Müller: ibid. **36**, 539 (1987)
22. H. Müller, Ch. Opitz, S. Romanowski, L. Skala: phys. stat. sol. (b) **148**, K11 (1988)
23. H. Müller, Ch. Opitz, L. Skala: J. Mol. Catal. **54**, 389 (1989)
24. W.D. Knight, W.A. De Heer, W.A. Saunders: Z. Phys. **D3**, 109 (1986)
25. I. Smith: J. Amer. Inst. Aeronaut. Astronaut. **3**, 648 (1965)
26. F. Wood: Phys. Rev. Lett. **46**, 749 (1981)
27. E. Schumacher, M. Kappes, K. Marti, P. Radi, M. Schär, B. Schmidhalter: Ber. Bunsenges., Phys. Chem. **88**, 220 (1984)
28. M.M. Kappes: Chem. Rev. **88**, 369 (1988)
29. Handbook of Chemistry, Physics, Ed. R.C. Weast, CRC Press, Inc. Boca Raton, Florida 33431, 1979/80
30. L. Skala: phys. stat. sol. (b) **107**, 351 (1981); **109**, 733 (1982); **110**, 299 (1982)
31. S.A. Napijko, E. Pippel, I. Woltersdorf: phys. stat. sol. (a) **61**, 469 (1980)
32. M. Hasegawa, K. Hoshimo, M. Watabe: J. Phys. **F10**, 619 (1980)
33. P. Powlow: Z. phys. Chem. Leipzig **65**, 545 (1909)
34. M. Takagi: J. Phys. Soc. Jap. **9**, 359 (1954)
35. Ph. Buffat, J.-P. Borel: Phys. Rev. **A13**; 2287 (1976)
36. G. Mie: Ann. Phys. (Leipzig) **25**, 377 (1908)
37. K.-H. Bennemann, S. Reindl: Ber. Bunsenges., Phys. Chem. **88**, 276 (1984)
38. L. Künne, R. Berndt: Wiss Z. Friedrich-Schiller-Univ. Jena, Math.-Nat. R. **25**, 757 (1976)
39. L. Künne: Z. phys. Chem. Leipzig **265**, 745 (1984)
40. L. Künne, H. Müller: Z. Chem. **26**, 345 (1986)
41. E.A. Abou-Saif, A.A. Mohamed, M.G. El-Khodary: Thin Sol. Films **94**, 133 (1982)
42. H. Müller, Ch. Opitz, S. Romanowski: Wiss. Z. Univ. Jena, Math.-Nat. R. **39**, 173 (1990)
43. H. Müller, Ch. Opitz: unpublished results
44. H. Müller, Ch. Opitz: Z. phys. Chem. NF **169**, 51 (1990)
45. S.K. Loh, D.A. Hales, L. Lian, P.A. Armentrout: J. Chem. Phys. **90**, 5466 (1989)
46. C. Brechignac, Ph. Cahuzac, F. Carlier, M. de Frutos, J. Leygnier: J. Chem. Soc. Faraday Trans. **86**, 2525 (1990)
47. H. Müller, Ch. Opitz, S. Romanowski: Z. phys. Chem. Leipzig **270**, 33 (1989)
48. H. Müller, Ch. Opitz: J. Mol. Catal. **82**, 361 (1993)
49. P. Marquardt, G. Nimtz, B. Mühlschlegel: Solid State Comun. **65**, 539 (1988)
50. G. Petzow: Jahrbuch der Max-Planck-Ges., Vandenhoeck & Ruprecht, Göttingen, p. 62 (1986)
51. E. Müller, W. Vogelsberger, H.-G. Fritsche: Cryst. Res & Technol. **23**, 1153 (1988); in Lecture Notes in Physics 309, P.E. Wegner, G. Vali (Eds.), Atmospheric Aerosol, Nucleation, Proceedings, Wien (1988), Springer Verlag, Berlin, Heidelberg, 1988
52. W. Vogelsberger, H.-G. Fritsche, E. Müller: phys. stat. sol. (b) **148**, 155 (1988)
53. W. Vogelsberger, E. Müller: phys. stat. sol. (b) **155**, K17 (1989)
54. H.-G. Fritsche, T. Mittelbach, E. Müller, W. Vogelsberger: Z. Phys. **D20**, 357 (1991)
55. T. Mittelbach: Diplomarbeit, Univ. Jena, Faculty of Chemistry (1990)
56. J.A. Northby: J. Chem. Phys. **87**, 6166 (1987)
57. J.A. Harris, R.S. Kidwell, J.A. Northby: Phys. Rev. Lett. **53**, 2390 (1984)
58. J. Farges, M.F. de Feraudy, B. Raoult, G. Torchet: Surf. Sci. **156**, 370 (1985)

59. B.W. van de Waal: Z. Phys. **D20**, 349 (1991)
60. T.D. Maerh, P. Scheier: J. Chem. Phys. **87**, 1456 (1987)
61. O. Echt, K. Sattler, E. Recknagel: Phys. Rev. Lett. **47**, 1121 (1981)
62. W. Miehle, O. Echt, P. Kandler, T. Leisner, E. Recknagel: Z. Phys. **D12**, 273 (1989)
63. L. Friedman, R.J. Beuhler: J. Chem. Phys. **78**, 4669 (1983)
64. O. Kandler, T. Leisner, O. Echt, E. Recknagel: Z. Phys. **D10**, 295 (1988)
65. O. Echt, O. Kandler, T. Leisner, W. Miehle, E. Recknagel: J. Chem. Soc. Faraday Trans. **86**, 2411 (1990)
66. H.-G. Fritsche, D. Bonchev, O. Mekenyan: Z. Chem. **27**, 234 (1987)
67. H.-G. Fritsche, D. Bonchev, O. Mekenyan: phys. stat. sol. (b) **148**, K101 (1988)
68. H.-G. Fritsche: Cryst. Res. & Technol. **24**, K47 (1989)
69. Ch. Kittel: Einführung in die Festköperphysik, Leipzig, Akademie-Verlag 1973, p. 143
70. H.-G. Fritsche: phys. stat. sol. (b) **154**, 603 (1989)
71. H. Weyl: Gottinger Nachr., 110, 1911
72. H. Weyl: Math. Ann. **71**, 441 (1912)
73. H. Weyl: J. f. reine u. angew. Math. **141**, 1 (1912)
74. H. Weyl: J. f. reine u. angew. Math. **141**, 163 (1912)
75. H. Weyl: J. f. reine u. angew. Math. **143**, 177 (1913)
76. H. Weyl: Rend. Circ. Mat. Palermo **39**, 1 (1915)
77. H. Weyl: Bull. Amer. Math. Soc. **56**, 115 (1950)
78. H. Weyl: Gesammelte Abhandlungen, Vol. 1, Abhdlg. Nr. 13, 16–19, 22 ed. K. Chandrasekharan, Springer, Berlin, 1968
79. H.P. Baltes, E.R. Hilf: Spectra of Finite Systems, Bibliographisches Institut AG, Zurich 1976
80. M. Planck: Verhdl. Deut. Phys. Ges. **2**, 202 (1900); M. Planck, Physikalische Abhandlungen und Vortrage, Vol. 3, Braunschweig Viewig, 125, 1900
81. P. Debye: Ann. Phys. (Leipzig) **39**, 789 (1912)
82. L. Boltzmann: Ann. Phys. (Leipzig) **258**, 291 (1884)
83. R. Kubo, A. Kawabata, S. Kobauaski: Ann. Rev. Mat. Sci. **14**, 49 (1984)
84. L. Skala, P. Pancoska: Chem. Phys. **125**, 21 (1988)
85. V.R. Pandharipande et al.: Phys. Rev. Lett. **50**, 1676 (1983)
86. J. Myers, W.J. Swiatecki: Nuclear Physics **81**, 1 (1966)
87. N.M. Hugenholtz, D. Pines: Phys. Rev. **116**, 489 (1959)
88. W.C. Stwalley, M. de Llano: Z. Phys. **D2**, 153 (1986)
89. H.R. Le et al.: Phys. Rev. **B39**, 2822 (1989)
90. H. Neuberger, T. Ziman: Phys. Rev. **B39**, 2608 (1989)
91. P. Dean: Rev. Mod. Phys. **44**, 127 (1972)
92. J. Hori: Spectral Properties of Disordered Chains, Lattices, Pergamon, Oxford, 1968
93. J. Ladik et al.: Int. J. Quant. Chem. **29**, 597 (1986)
94. K. Litzmann, P. Rosza: Proc. Roy. Soc. **85**, 285 (1965)
95. J. Jeffreys, B.S. Jeffreys: Methods of Mathematical Physics, Cambridge University Press, Cambridge, 1946, p. 127
96. W. Ledermann: Proc. Roy. Soc. **A182**, 362 (1944)
97. V.M. Kenkre, P. Reineker: Exciton Dynamics in Molecular Crystals, Aggregates, Springer, Berlin, 1983
98. L. Skala, V.M. Kenkre: Z. Phys. **B63**, 259 (1986)
99. L. Skala: unpublished result
100. L. Skala, V. Kapsa: Chem. Phys. **137**, 77 (1989)
101. L. Skala, P. Jungwirth: Chem. Phys. **137**, 93 (1989)

2.6 Shell Structure in Atoms, Nuclei and in Metal Clusters

S. Bjørnholm

2.6.1 Quantum Shells in Spherical Fermion Systems

A prominent and much celebrated feature of simple metal clusters is the occurrence of magic numbers and their interpretation in terms of quantal shells; see Section 6.1 and [1–4]. These notions were originally introduced to describe the electron system in atoms and to explain the Periodic System of the Elements some seventy years ago. When one thinks of the far reaching importance that the understanding of atomic constitution has had and still has, it is indeed very interesting to realize that small drops of a simple metal are forming a periodic system of their own. One can speak of such clusters as quasiatoms, or giant atoms. By way of example, a sodium droplet with 92 atoms and hence 92 conduction electrons is closely analogous to an atom of uranium. The main difference is that in uranium all the 92 positive charges are concentrated in the nucleus, while in the cluster the positive charges are evenly distributed among the electrons. That is of course an important difference. The Na_{92}-cluster is not indivisible the way uranium is. This in turn is related to another difference, namely that sodium quasiatoms presumably grow in size in proportion to the number of constituents, while the real atoms, from hydrogen to uranium, are about equal in size. In other words, the (valence) electron density in metal clusters tends to be constant, while the density varies strongly in atoms as one proceeds through the Periodic System. Nevertheless it is quite striking that in both cases one has to do with a fully quantized system of electrons in a spherical mean field.

Historically, the idea of quantum shells made its second appearance some forty years ago in connection with the efforts to understand the structure of atomic nuclei. In this case the idea had been much slower in gaining acceptance [5, 6]. The constituent particles, protons and neutrons, are spin-one-half particles and thus subject to the Pauli exclusion principle like the electron. But apart from that nuclei are very different from atoms. The forces are attractive and the size grows in proportion to the number of constituent nucleons. Here it is due to the fact that the force between two nucleons has its maximum value at a characteristic range, like a van der Waals force, and unlike the long range Coulomb forces between the electric charges in an atom. It is therefore less obvious that a particular nucleon will feel the attraction from all the other nucleons in terms of a time-independent average mean field, as one imagines with the atom. From the similarity to a van der Waals cluster one should rather expect that nuclei would be *crystalline* in their ground state.

With regard to this question there is an interesting analogy to noble gas clusters. The nucleus is such a compact object that the zero-point energy of even the lowest frequency excitation mode one could possibly imagine is much larger than the thermal energies for temperatures occurring in our province of the universe. One should therefore compare the nucleus to aggregates of noble gas atoms at ultra low temperatures. Macroscopic amounts of the heavier noble gases, xenon, krypton, argon, and neon, indeed crystallize upon cooling. As will be shown in Section 6.7, also smaller quantities form ordered close-packed clusters, where the atoms are well-localized. With helium, on the other hand, the situation is different. The specific range of the attractive van der Waals forces between atoms of helium compels it to form a medium with about the same number density as the other noble gases, if it is to form a condensed phase at all. But the helium atoms are so light and the forces so weak that the quantal zero-order kinetic energy of an atom held at a fixed, crystal-like position exceeds the potential energy of mutual attraction. Hence there is no net binding in a crystal. A quantal wave-like-ordering, on the other hand, where each atom is delocalized within a large volume, lowers the zero-point energy while still allowing the attractive forces to act efficiently. (They are harder to visualize, but they are there.) In this form the system can be bound. Helium actually forms a quantum liquid at ultra low temperatures with a density given by the range of the force between two atoms. It does not crystallize. The degree of delocalization is related to the mean free path, which in liquid helium three increases proportional to the inverse temperature. For clusters of helium three, the mean free path can easily exceed the linear dimension of the droplets. Each atom can then be described by a wavefunction that extends throughout the entire cluster volume. And ^{3}He-atoms are fermions, because the nucleus in the ^{3}He-atom has spin one-half, while the two electron spins are paired to zero in the ground state. The atom, regarded as a whole, has therefore a total spin of one half. It will obey the Pauli exclusion principle, like the individual nucleons in the nucleus. Only two particles – one with spin up, the other with spin down – are allowed to have the same spatial wave pattern.

Like ^{3}He, nuclei fail to show any sign of crystallization. They quite convincingly behave like liquid drops, as illustrated for example when they undergo fission. There is thus a close analogy between ^{3}He-droplets and the nuclear droplets. The recognition of shell structure in nuclei by the physicists of that time hinged initially on the question of the length of the mean free path compared to the nuclear dimensions [5]. It is indeed a touch and go, but the systematics of low-lying excitations, together with periodicities in the ground state binding energies gradually made it clear that to some approximation one could think of the nucleons as individually bound particles, moving quasi-independently in a common mean field [6]. Whether ^{3}He-atoms or nucleons, one can obtain an approximate picture of the spatial form of each possible wave pattern and the associated energy by finding the stationary solutions to Schrödinger's wave equation for *one particle* in a potential. This potential is likely to be flat inside the droplet with walls rising up to a certain value where the cluster surface is. Such a potential is shown in Fig. 1 for the case of a spherical cluster of ^{3}He with seventy atoms. Each energy eigenstate of the wave equation is characterized by an angular momentum quantum number l, and the total quantum state is then pictured by filling in ^{3}He atoms from below with $2(2l + 1)$ atoms in each energy eigenstate – or single-particle level as it is called. Picturing the cluster structure in this way is equivalent to assuming that the attraction, which all other atoms exert on a particular atom, is constant in time. It is assumed to be the same whichever atom one chooses, even though their wave patterns are quite different. The static potential represents the effect of this averaged attraction. Filling atoms into the eigenstates of this potential without further ado means assuming that they do not interact with each other at all. The interactions are *completely* described by the potential – or mean field.

As one sees from Fig. 1, the various l-states are bunched together into bands with large gaps in between. The "magic" number of particles required to fill the bands up to a particular gap are indicated in the figure. They represent closed shells. Such a picture may describe the ordering in the cluster pretty well, especially of the least bound atoms. Differences between the binding energies of two adjacent clusters will also be given with reasonable accuracy, while the total binding energy of a cluster cannot be realistically described in this way.

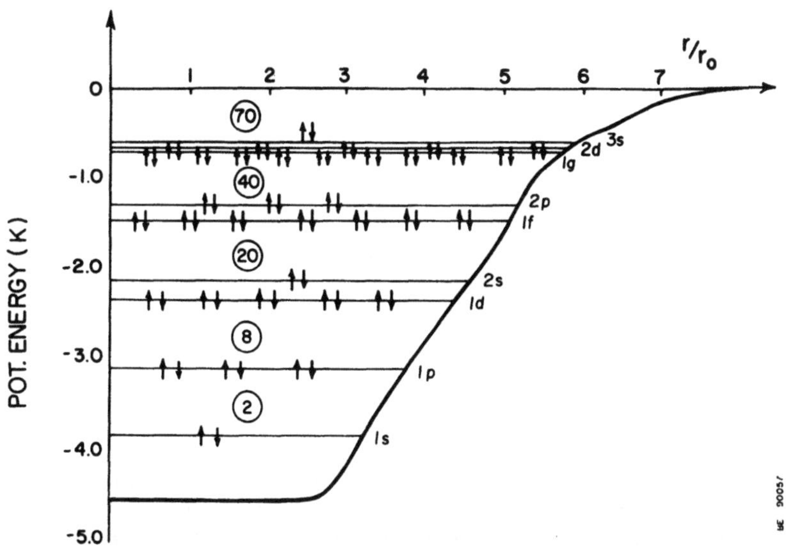

Fig. 1. ^{3}He eigenvalue spectrum in a single-particle mean field of a spherical drop with 70 ^{3}He atoms. The parameters of the radial field can be estimated from the values of density, binding energy, surface tension, compressibility, and heat capacity known for bulk ^{3}He-liquid at low temperatures. From Ref. [19]

For nuclei, one has similar theoretical pictures of the shell structure, although the shell gaps occur at somewhat different magic numbers. Unlike the case of ^{3}He just shown, the theory of nuclear shell structure is supported by very large body of experimental data [9, 10].

Returning to the metallic quasiatoms and their shell structure it is now possible to characterize them a little closer. Figure 2 attempts to illustrate this.

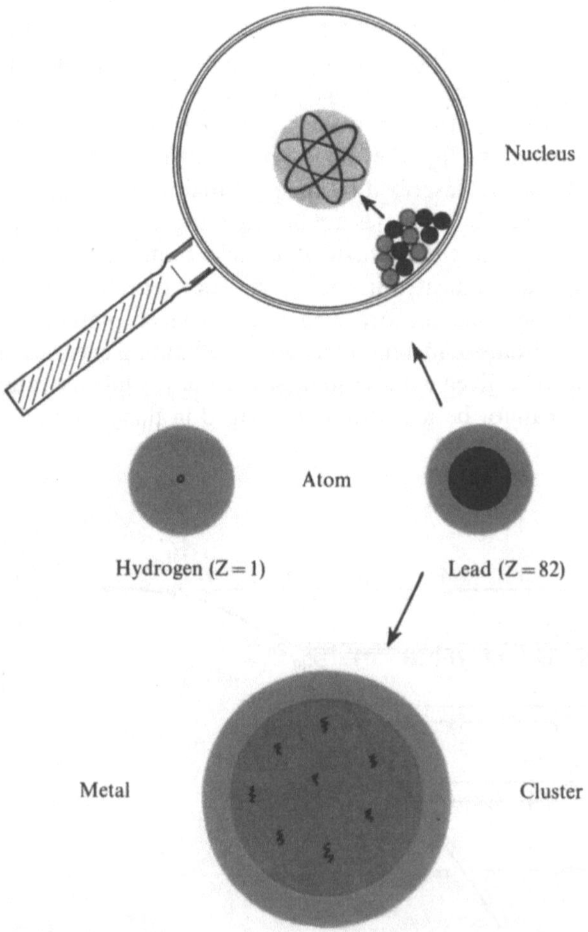

Fig. 2. The atom, the nucleus, and a simple metal cluster. The atom is an example of a quantum gas, whereas the nucleus is a quantum liquid. Both are described in terms of delocalized wave-like Fermi-particles. This is illustrated by ellipsoidal "orbits" in the nucleus above (as opposed to the naive particle picture, indicated at the rim of the looking glass); and by the shading of the atoms, centre. Their first excited quantum states lie quite high compared to normal temperatures. In the metal cluster below, the conduction electrons also have a definite quantum many-body ground state with an electron density that extends beyond the positive charge, as in the atom, but the excited states lie closer together. The positive background can be idealized as an inert medium, jellium. In reality it is a classical many-body medium (black "worms") that readily absorbs heat

The atoms can be thought of as a quantized electron gas at zero temperature. A gas needs an external agent to be confined, otherwise it disperses completely. The Coulomb field of the nucleus plays that role, just as the earth's field is the agent holding our atmosphere in place. The gas can have any density, depending on the strength of the external holding field. Since the electrons repel each other, the electron gas will not condense upon cooling. It will quantize directly in the gaseous state, becoming perfectly ordered with zero entropy and strong shell structure. Our usual ambient temperatures are low enough to make that happen. The nucleus on the other hand, and helium three likewise, needs no external agent to produce condensation. They will form ordinary liquids with a characteristic density, but the binding forces are too weak to allow crystallization as the next step. They quantize in the liquid state as the temperature is reduced below the lowest quantal energy fluctuation possible in the system. For spherical droplets there will be pronounced shell structure.

The metal clusters are like hybrids between atomic and nuclear quantum systems. At the same time they are slightly more complicated, because they are made from two different kinds of particles with very different masses. In the molten state they can be looked at as plasmas, held together by the long range forces between the oppositely charged particles, and saturating at a density that is a characteristic constant for each element. Because of the enormous difference in mass, the positive ions form a classical liquid medium, while the electrons at the same temperature form a quantized medium in, or very near to, its ground state. It is the electrons that can give rise to shell structure and magic numbers, provided they can be assumed to form a *spherical* many-body state. The actual observation of shell structure in experiments [1, 2] where the clusters in all likelihood are above melting temperatures, shows directly that the magic clusters are spherical and that the energy of the electronic subsystem dominates the variations in total energy of the cluster, despite the large amount of heat energy present in the ionic subsystem and the multitude of "isomeric" configurations represented in any experimental ensemble of cluster droplets of a given size. The ions thus appear to form a uniform or slowly varying background medium of relatively constant density and shape, in which the electrons quantize. To a first approximation one can describe this in terms of a static mean field for the electrons, brought about in some selfconsistent way by the attractive and repulsive Coulomb forces between the constituents, and including the kinetic pressure from the electrons. Figure 3 shows as an example a concrete calculation [3] of the mean field appropriate to a spherical Na_{20}-cluster. It is based on the jellium approximation to the positive background.

It follows from this description that the metallic quasiatoms have as much, if not more, in common with the nuclear quantum systems (or ^3He) as with ordinary atoms. In the metal clusters the density is constant, and the mean field results from the cohesive forces as a collective effect. There is no external agent as in the atom. Also, the mean field is flat inside, rising more or less smoothly at the surface. This is precisely how the nuclear shell model describes nuclei. In the following section we shall therefore discuss a few well-known facts about nuclei, having their relevance to metal clusters in mind.

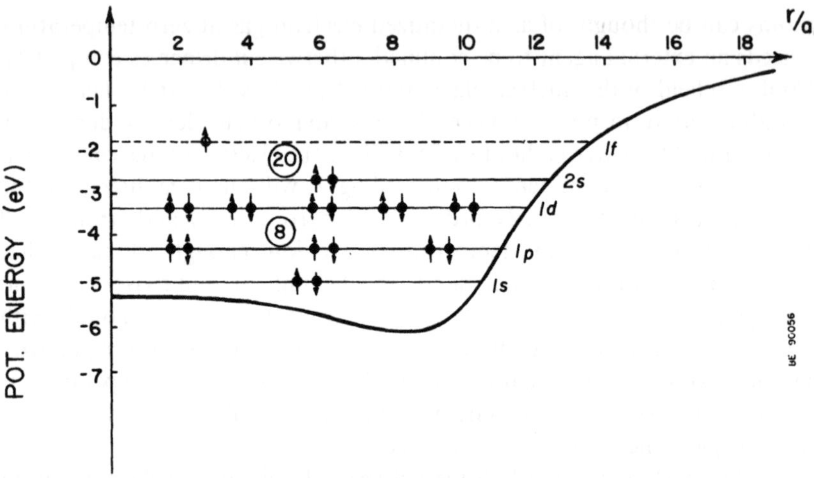

Fig. 3. Ekardt's self-consistent electron potential, calculated for a spherical sodium cluster with twenty atoms. The calculated single-electron energy levels are also shown. Filled circles indicate electrons occupying the lowest levels; the open circle shows where a 21st electron would have to go. In this calculation, the positive sodium ions are assumed to form a uniform, positive, spherical charge distribution with a sharp surface and a density equal to the bulk density of sodium metal (jellium model). From Ref. [3]

2.6.2 Nuclear Shell Structure and Deformations

The mean fields that the nuclear physicists use to describe shell structure in nuclei are of the same type as the potential shown in Fig. 1. The degree of sophistication entering in the derivation of a selfconsistent potential from basic knowledge of the nuclear forces and other nuclear properties varies considerably. For the present purpose it is enough to quote some simple potentials, especially the Woods–Saxon potential [7] and the modified harmonic oscillator Nilsson potential [8]. They are both purely phenomenological constructs, with parameters fitted by requiring agreement with experimental data, such as neutron or proton separation energies, nuclear radii, and also spectroscopic data, i.e. spins and parities of ground- or low lying excited states in complex nuclei. The spherical Woods–Saxon potential has the following form:

$$V(r) = \frac{V_0}{1 + \exp((r - R)/a)} \ . \tag{1}$$

Typical parameters are: $V_0 = -50 \cdot 10^6$ eV, $R = r_0 A^{1/3}$, with the equivalent of the Wiegner–Seitz radius $r_0 = 1.2 \cdot 10^{-15}$ m, A being the total number of nucleons, i.e. the mass number and $a = 0.65 \cdot 10^{-15}$ m. Figure 4 shows the potential for the special case of $A = 100$. Also shown in this figure is a harmonic oscillator

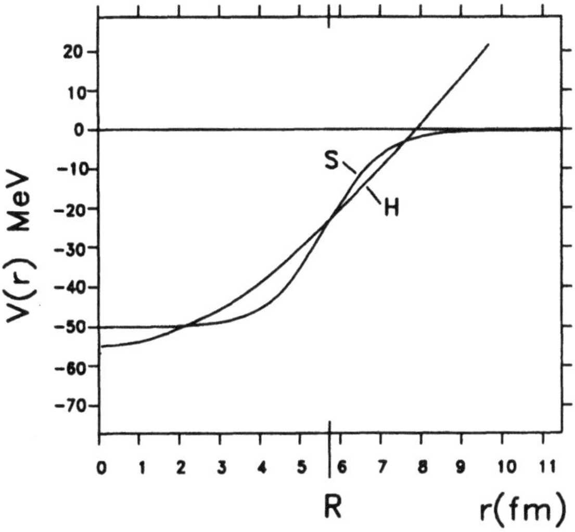

Fig. 4. The radial Woods–Saxon potential (S) Eq. (1) and the simpler harmonic oscillator potential (H) with parameters that give the best fit to the Woods–Saxon potential. The two potentials have the radial extent corresponding to particle number A equal to 100. From Ref. [9]

potential

$$V(r) = \frac{1}{2} M\omega^2 r^2 - 55 \cdot 10^6 \,\text{eV} \ . \tag{2}$$

With M the nucleon mass, and ω given by $\hbar\omega = 8.6 \cdot 10^6$ eV. The parameters are chosen to give the best possible approximation to the Woods-Saxon potential [9].

The potential Eq. (1) resembles the more sophisticated potential shown on Fig. 1. They describe particles moving more or less freely in the interior while being gently reflected by the spherical surface, which encloses a volume proportional to the number of particles in question. In both cases the least bound particle is only one seventh of V_0 from being unbound – as opposed to the cluster potential, Fig. 3, where the least bound particle is just half way up from the bottom of the potential. While the volume of the potential scales with the number of particles it is supposed to accommodate, the diffuseness parameter, a, does not. This reflects a physical assumption about the quantum liquid, namely that its surface has the same "thickness" or diffuseness whatever size the droplets.

It is not difficult to imagine that in this situation the Woods–Saxon potential will come closer to the harmonic oscillator potential with diminishing size; whereas it will approach a square well-potential, or a spherical cavity, when the radius R becomes large compared to the diffuseness, a. Neglecting the finiteness of the surface repulsion, the square well potential reads,

$$V(r) = 0 \quad \text{for } r < R \tag{3a}$$

$$V(r) = \infty \quad \text{for } r > R \tag{3b}$$

with $R = r_0 A^{1/3}$ as before.

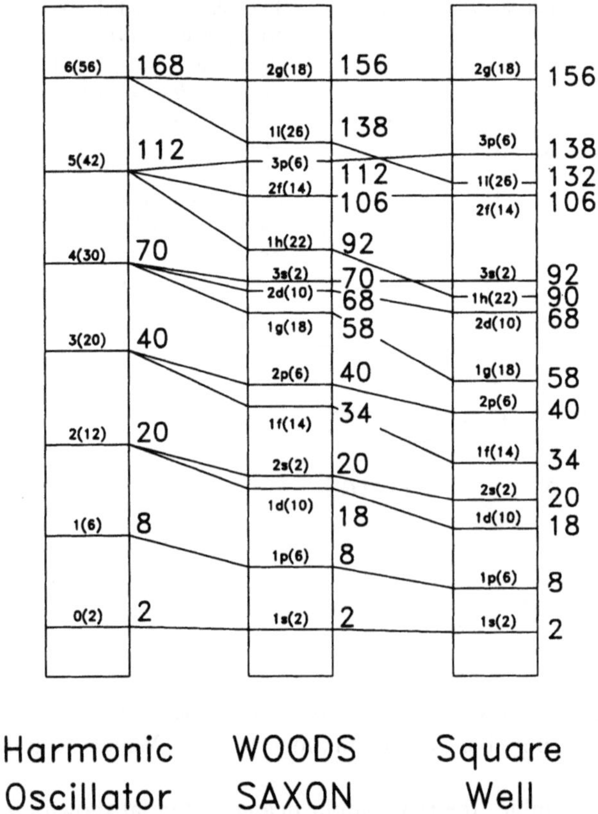

Fig. 5. Energy eigenvalue spectra for the three different spherically symmetric potentials, Eqs. (1, 2, and 3). The spectroscopic notation and the degeneracy is indicated for each level. The numbers to the right of each column describe the shell or subshell closures. From Ref. [6]

The result [6] of solving the Schrödinger equation for the three potentials and plotting the eigenvalues is shown in Fig. 5 for the case of a system with = 160 (like) particles. In all three cases the $2(2l + 1)$ degenerate levels are bunched into more or less narrow bands with empty gaps in between. This is the shell structure. It is most pronounced with the harmonic oscillator, where the various l-states are completely degenerate within a shell. The intermediate Woods–Saxon potential allows the high angular momentum eigenstates from the harmonic oscillator shell (circular orbits) to come lower in energy. That is readily understood by looking at Fig. 4. A particle in a circular orbit is clearly more strongly attracted by the Woods–Saxon potential than by the harmonic oscillator potential. A similar effect can also be seen from the ordering of the l-states within each shell on Fig. 1 (and Fig. 3). Computationally, the harmonic oscillator potential is by far the most convenient to handle. This has inspired the introduction of a modified harmonic oscillator potential, i.e. the Nilsson potential [8]. While still simple to work with, it corrects for the increased binding of high-l orbitals by adding an l^2-dependent term.

$$V(r) = \frac{1}{2} M\omega_0^2 r^2 - U\hbar\omega_0(l^2 - \langle l^2 \rangle_n) , \tag{4}$$

where U is a (positive) parameter, and $\langle l^2 \rangle_n$ is the mean squared l-value of oscillator shell no. n. The frequency is size dependent, typically [9]

$$\hbar\omega_0 = 40\, A^{1/3} \cdot 10^6 \text{ eV} . \tag{5}$$

In the following section, we shall return to the idealized harmonic oscillator and square well potentials. The more realistic potentials Eqs. (1) and (4) form the basis for the description of shell structure and spectroscopy in actual nuclei. In principle it should be possible with their help to predict nuclear ground state properties and excitation spectra. The trouble is that the observed properties, the magic numbers for example, do not agree with such simple model predictions. It was a decisive step in the development of the nuclear shell model when it was realized [6] that the eigenenergies of the nuclear states depend strongly on the relative orientation of the orbital and spin angular-momenta, l and s. States with $l \cdot s$ positive are less bound than states with $l \cdot s$ negative. A correction to the eigenvalue spectrum based on Eq. (1), with splittings $-\Delta\xi_{ls} = 24(l \cdot s)A^{2/3} \cdot 10^6$ eV added, leads to predictions in much better agreement with observations.

First of all, the nuclear magic numbers (2, 8, 20, 28, 50, 82, and 126) come out right. In addition, the j-values and parities of spherical nuclei with an odd number of protons or neutrons are correctly predicted. The ground state of the entire A-particle system, with A odd, turns out to have angular- and parity quantum numbers equal to those expected for the single, least bound, odd nucleon. This is a considerable simplification compared to the atomic case. Taking Fig. 1 as an example, the odd-numbered "nuclei" with A between 59 and 67 all have at total l-value of $L = 2$ (and $S = 1/2$), while the even-numbered "nuclei" have $L = S = 0$. One should contrast this to the atomic case, where the spins add up to the maximum possible value according to Hund's first rule. For truly independent particles there is no rule. The relative orientation of spins and l-vectors has no influence on the total energy of a system of independent particles in degenerate eigenstates. Hund's rule reflects the mutual Coulomb repulsion between the atomic electrons. It is less for electrons with parallel spin. The nuclear rule, conversely, reflects the action of residual net attraction between nucleons – beyond the general attraction from the mean field. Two nucleons in time-reversed states (and opposite spins) have maximum overlap and exploit the attraction maximally. That is why all even-numbered nuclei have total spin zero and why there is such a simple rule for odd-numbered nuclei. (In molecules there is also a tendency for the electrons to pair off the spins – for a very different reason, however. The symmetry in typical molecules is so low that orbits with degeneracy greater than two do not occur, or only in exceptional cases. The electron spins actually pair off for the same reason that the two electrons in a helium atom pair off.)

The attraction between the nucleons, reflecting its character of a self-bound fluid, has another interesting consequence. Although small classical droplets always assume a spherical shape at equilibrium, small droplets of a quantum liquid are not necessarily spherical in their ground state. A large number of nuclei are found to be non-spherically deformed, looking more like an american football i.e. like a prolate spheroid. The recognition of nuclear deformation effects was another milestone in the development of nuclear structure physics [10].

There are various ways of visualizing how a drop of the nuclear quantum liquid can take on a non-spherical shape. Starting from the experimental fact that closed-shell nuclei are spherical, one may use the $^3He_{70}$-cluster in Fig. 1 as a convenient example of a spherical, magic droplet. Imagine that the spherical mean field is deformed to form an ellipsoid. (It is very nearly like taking a three dimensional harmonic oscillator potential and allowing the three stiffness constants, or the three different oscillator frequencies to become unequal.) This will lead to a splitting, or fanning out, of the otherwise degenerate l-subshells, see also Fig. 6. For a completely filled shell, gain and loss in energy of the individual particles is likely to balance to zero in first order. (The second order contribution is bound to be a restoring term, otherwise the equilibrium shape could not be spherical.) With a partly filled l-shell the situation is different. Here the particles will tend to fill the orbitals corresponding to a gain in energy and avoid the loss orbitals, resulting in a net gain upon distortion to first order. (The second order restoring force will ensure that the distortion remains finite). As a result,

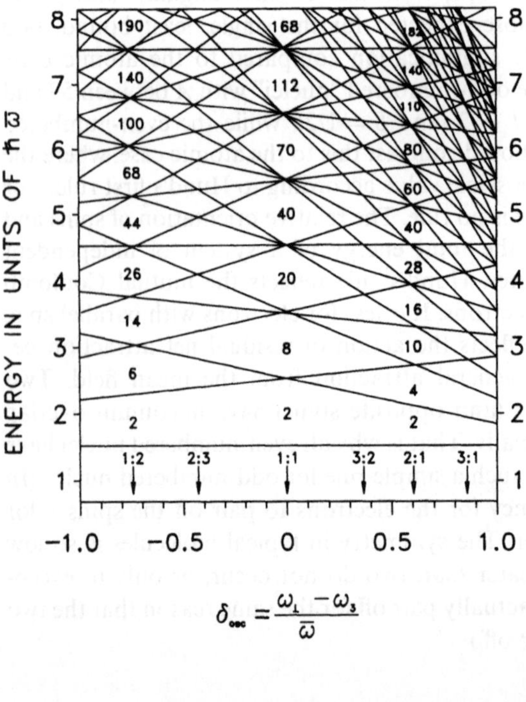

Fig. 6. Eigenvalue spectrum for a harmonic oscillator potential with an axially symmetric shape. The abscissa describes the elongation or flattening compared to a sphere ($\delta_{osc} = 0$) in terms of the mean frequency $\bar{\omega}$ and the difference between the frequency ω_3 along the distortion axis and perpendicular to it, $\omega_\perp$. Degenerate shells occur for simple ratios of the two frequencies and hence axes. Numbers corresponding to shell closures are indicated. From Ref. [10]

a droplet with a partly filled shell will assume a non-spherical equilibrium shape. This description seems to be nothing else than an explanation in terms of the Jahn–Teller effect, which is familiar from molecular physics. In a way this is what it is, but it raises the question of the generality of the Jahn-Teller mechanism for symmetry breaking.

This mechanism is not effective in the case of an atom with a partially filled l-shell. Why then in a nucleus and (presumably) in a ^3He droplet? The answer is once again, that the nature of the residual forces between the particles in the field is important. Only for particles that attract each other does it make sense to deform the selfconsistent mean field and expect the matter (i.e. particle-) distribution to follow the distortion self-consistently such that the change in eigenvalues can be used to estimate gain or loss upon deformation.

A still simpler picture of the deformation driving effect of nucleons outside a closed shell is also based on the attractive residual force. If one nucleon is added to a spherical closed shell configuration it will occupy one of the m-levels in the lowest vacant l-shell. The next nucleon is not entirely free and independent in its choice of m-level. It is attracted by the first nucleon and will occupy its time reversed orbit, because this allows the largest overlap. The third nucleon will also find an orbit with a large overlap; and thus a non-spherical "bulge" outside the closed shell core will begin to form. This bulge will in turn distort the spherical core, thereby amplifying the deformation.

Described in this way, nuclear deformations are so to say the reverse of the stabilizing effect of closed shells. A partly filled shell gains energy by distortion. In this picture, there is one definite equilibrium shape – spherical or deformed – for each individual nuclear size. In fact the true picture of nuclear structure is more rich than that. A single nucleus may have two (even three) distinct and well defined equilibrium shapes, with quite different values of the distortion parameter [11]. In uranium 238 for example there is an isomeric state with prolate deformation and a ratio of axes equal to 1 : 1 : 2, while the ground state has an axis ratio of 1 : 1 : 1.2. The ground state decays by alpha particle emission with a half life of 10^9 years, while the isomer undergoes spontaneous fission within a few microseconds.

The discovery of shape isomers in the nuclear quantum droplet has lead to the realization that shell structure and the special stability associated with closed shells has to be considered more broadly. It is not restricted to spherically symmetric systems alone.

Taking the three-dimensional harmonic oscillator potential as an example makes it particularly simple to demonstrate this but, as the nuclear shape isomers reveal, the phenomenon is more general. The eigenvalues of a generalized, three dimensional harmonic oscillator, expressed in (x, y, z)-coordinates, are

$$E_n = \hbar\omega_x\left(n_x + \frac{1}{2}\right) + \hbar\omega_y\left(n_y + \frac{1}{2}\right) + \hbar\omega_z\left(n_z + \frac{1}{2}\right) . \tag{6}$$

In the spherical case, $\omega_x = \omega_y = \omega_z$, there will be $1/2\,(n + 1)(n + 2)$ degenerate levels (disregarding spin), corresponding to the number of ways it is possible to sum integers n_x, n_y and n_z to a given total value n. The magic numbers including the spin degeneracy will be $1/3\,(n+1)(n+2)(n+3)$, viz. 2, 8, 20, 40, 70, etc. Clearly, if the frequencies ω_x, ω_y and ω_z in Eq. (6) are in simple integer ratios to each other, then there will also be several ways of obtaining the same energy E_n. The eigenstates will be bunched together and there will be shell structure. In the simplest example of setting $\omega_x{:}\omega_y{:}\omega_z = 1{:}1{:}2$, the shell closures will occur for sizes $N = 2$, 4, 10, 16, 28, 40 etc. The situation is illustrated in Fig. 6, where several other cases of non-spherical shell structures appear. This figure is based on the simple non-isotropic harmonic oscillator. As said before, the evidence from nuclear shape isomerism makes it clear that the picture of non-spherical shell structure is more general. It can occur with many different types of potential. For shell structure to arise it is not necessary to have an exact degeneracy of eigenstates, as in Fig. 6. All that is required is that the density of single particle states around the Fermi level varies in a systematic way between bunching and debunching as the distortion parameter or alternatively, the particle number changes.

The evidence for nuclear shell structure discussed here is based to a very large extent on the binding energies, i.e. stabilities, of the nuclear ground or isomeric states, and on the spectroscopy of low lying excited states. The description in terms of a mean field, with its characteristic spectrum of single particle eigenstates, actually goes much further. In principle, a representation as the one shown in Fig. 1 "predicts" the exact removal energies of particles from any of the filled orbitals – not just the least bound ones. The mean field description of the atom is similar in this respect, and it is well known that one observes a sharp threshold in photoionization experiments (with X-rays) resulting in the creation of a deep lying hole in the K-shell of a heavy atom. Afterwards it is possible to observe how the hole is filled again with an electron from a higher shell under the emission of relatively sharp K-X-ray lines. From the line width it is possible to infer that the hole has a lifetime that is some hundred times longer than the period of an electron in a K-state.

The situation with regard to nuclear "K-X-ray" spectroscopy is very different. It is possible to remove deeply bound nucleons, but there are no clear thresholds to be observed and no sharp lines that can be associated with the filling of a deep lying hole. On the contrary, the nuclear spectroscopist is lead to conclude that a K-hole in the nucleus enjoys a fleeting existence indeed. It is filled by a less bound nucleon within a fraction of a period. The mean free path, in other words, is much shorter than the diameter of the nucleus. This is another reflection of the nature of a densely packed quantum liquid as opposed to the quantum gas of electrons surrounding the nucleus. One of the interesting questions that remain to be answered about metal clusters is where they are placed on this scale. Are they like atoms with well defined single particle excitations at all energies, or like nuclei where such excitations can only be defined for quite low excitation energies, involving states in the immediate vicinity of the Fermi level.

This section on nuclear shell structure and spectroscopy will be grossly incomplete without mentioning rotational and vibrational excitations, in addition to the particle excitations discussed so far.

The closed shell nuclei have truly spherical, rotation invariant shapes and there are no excitations corresponding to rotation of the nucleus as a whole. The nonspherically deformed nuclei, on the other hand, have rotational bands built on the ground state and each of the intrinsic single-particle excitations. In both cases there are also excitations that correspond to dynamical shape vibrations, as one should indeed expect from a system like a liquid drop.

A classification like this into rotations, vibrations and particle excitations has great similarity to molecular spectroscopy, and is indeed inspired in many ways by molecular spectroscopy. The big difference, on the other hand, between nuclei and molecules is the absence in the nucleus of the contrast between the heavy positive charges and light electrons, as one knows it from molecules. In the nucleus there are only particles of the same mass. The result is that the typical energies for exciting vibrational and single particle motion in nuclei are very nearly equal, with rotations being just a little less energy craving. For the discussion of shell structure and shell effects, this aspect of nuclear structure is actually of minor importance. It is mentioned here for the sake of completeness.

2.6.3 Shells and Supershells in Large Fermion Systems

Atoms and nuclei are limited in size to a few hundred constituent fermions. This is not so with metallic clusters. The shell structure encountered in these clusters is therefore potentially more than just a repeat of previously known physical effects. Metal clusters indeed add new dimensions to the phenomenon of shell structure. In this section we will discuss a particularly striking aspect, the supershells. In essence, this aspect amounts to following a pure quantum phenomenon towards the limit of large quantum numbers where – according to *Niels Bohr* – a correspondence between classical and quantal motion will become apparent.

To begin with, we may look at a pure quantum mechanical calculation – a very simple one. Approximating real clusters as systems of independent electrons in a common mean field of the Woods–Saxon type one can generate the energy eigenvalues ε_i by solving the one particle Schrödinger equation with the radial potential of Eq. (1). The sum of eigenvalues will represent a measure of the total binding energy $E(N)$ of a spherical cluster of size N.

$$E(N) = \sum_{i=1}^{N} \varepsilon_i \ . \tag{7}$$

Plotting this quantity as a function of N one discovers that it can be approximated rather accurately by

$$\tilde{E}(N) = -aN + bN^{2/3} \ , \tag{8}$$

with a and b being positive numbers. This expression describes the binding of a classical liquid drop, a is the binding energy per particle inside the drop (volume energy), while $bN^{2/3}$ is a correction term, the surface energy, accounting for the reduced binding of the particles on the surface of the drop. The approximation is not perfect. The difference between $E(N)$ and $\tilde{E}(N)$ is finite as it must be, since $E(N)$ oscillates as a result of the non-uniform distribution of eigenvalues associated with shell structure while $\tilde{E}(N)$ is perfectly smooth. This difference is a measure of the modulation of the total binding energy due to shell structure, viz.

$$E_{\text{shell}} = E(N) - \tilde{E}(N) \ . \qquad (9)$$

Figure 7 is a plot of this quantity, taken from Ref. [12]. The downward cusps occur at shell closings and represent the magic numbers. As one sees, the shells appear to be regularly spaced. They lie almost equidistantly in the plot where the abscissa is a measure of the cluster radius, proportional to the cube root of the size N.

In addition to the ordinary periodic shell structure, the calculation in Fig. 7 also shows a higher-order periodicity. The shell oscillations reach a maximum in amplitude around the tenth period; they then tend to die out at period fifteen,

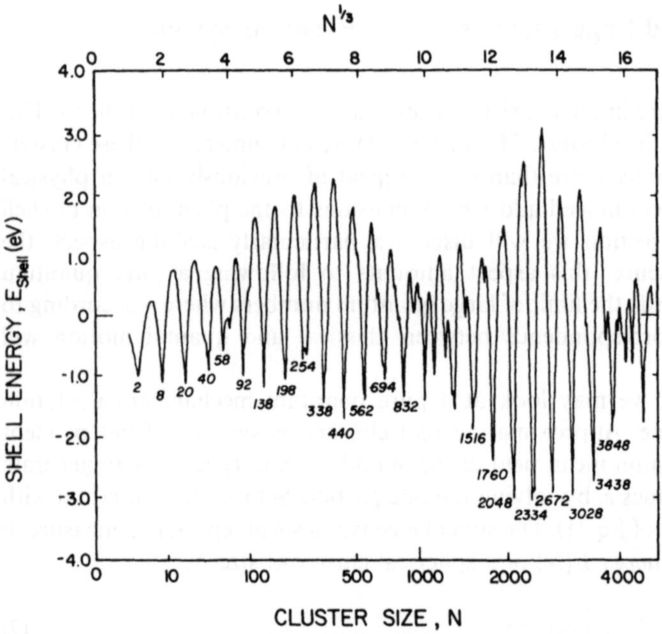

Fig. 7. The periodically varying contribution to the calculated binding energy of a spherical sodium cluster. The *total* binding energy is about one electron volt (eV) per atom, i.e. $E_{\text{tot}} \approx N$ eV. The *periodic* variations are due to the quantized motion of the conduction electrons in a field of the type shown in Figs. 1, 3 and 4. They are small compared to the total binding energy, but they are responsible for the intensity variations observed in experiment [1, 2]. From Ref. [12]

only to rise again to a new maximum at the twentieth period, around size 2300. This higher-order beat mode actually keeps repeating as long as the calculation goes, i.e. to a size of about ten thousand. The calculation thus predicts a sequence, where the shell structure appears to terminate at about size one thousand. Here the model clusters exhibit bulk-like, structureless electronic properties. In fact, however, this is just a passing stage in the development. The shell structure reappears.

These two kinds of periodicity, the primary shell structure and the beat mode, or supershell, appear here as the result of the purely numerical exercise of solving the one-particle Schrödinger equation with the Woods–Saxon potential. How can one understand the two periods? As always when one has to do with a beat mode, one can either think of it in terms of two close-lying frequences or wavenumbers, or in terms of their mean value and their difference. The ratio of the latter two is 10–15 in the present case, i.e. the two base frequencies or wavenumbers differ by some 7–10 pct. How does that come about?

To answer this it is helpful to express the periodicities in terms of the relation between magic numbers N_0 and the shell number n and then compare with some more familiar cases. As one sees from the plot in Fig. 7, the shell closings, i.e. the magic numbers N_0 are equidistantly spaced. The abscissa is linear in $N^{1/3}$, so $N_0^{1/3}$ is linear in n, or N_0 is proportional to n^3 to leading order. The numerical coefficient can be estimated by inserting the calculated magic numbers between 100 and 1000 to obtain, for the Woods–Saxon potential:

$$N_0 \approx 0.21\, n^3 + \cdots , \tag{10}$$

where contributions from lower powers of n are suppressed.

Let us compare this with the eigenvalue spectrum of the hydrogen atom, or more generally the Kepler problem of a particle in a $-r^{-1}$-potential. Each shell is perfectly degenerate with all odd and even numbered l-values represented in a given shell n up to $l_{\max} = (n-1)$. The number of (Fermi) particles in a shell is equal to

$$\sum_{l=0}^{l_{\max}} 2(2l+1) = 2(l_{\max}+1)^2 = 2n^2 \quad . \tag{11}$$

The magic numbers for the $-r^{-1}$ potential are:

$$N_0 = \sum_{\nu=1}^{n} 2\nu^2 = \frac{2}{3}n\left(n+\frac{1}{2}\right)(n+1) \approx \frac{2}{3}n^3 + \cdots , \tag{12}$$

with $n = 1, 2, 3, \ldots$. Thus there are more levels per shell than with the Woods–Saxon potential.

In the spherical harmonic oscillator potential each (perfectly degenerate) shell holds alternating even (0, 2, 4, . . .) or odd (1, 3, 5, . . .) l-values. As mentioned in Section 2.6.2 the magic numbers (harmonic oscillator) are:

$$N_0 = \frac{1}{3}(n+1)(n+2)(n+3) \approx \frac{1}{3}n^3 + \cdots , \tag{13}$$

with $n = 0, 1, 2, 3, \ldots$. Here the number of particles per shell is intermediate between the Kepler potential and the Woods–Saxon potential. On the other hand, the third power dependence is common to all three spherical potentials; likewise the subshell l-degeneracies, which increase as l^2 or, effectively, as n^2.

As a next step we may look at the three dimensional harmonic oscillator Eq. (6). It is equivalent to three independent one-dimensional harmonic oscillators. The motion is separable in the three coordinates, and the total energy is just the sum of three independent contributions. Expressing the change in total energy Eq. (6) in terms of the changes in individual quanta, expressed in units of action $\hbar$, leads to:

$$\begin{aligned}\delta E_n &= \frac{\partial E_n}{\partial(\hbar n_x)}\delta(\hbar n_x) + \frac{\partial E_n}{\partial(\hbar n_y)}\delta(\hbar n_y) + \frac{\partial E_n}{\partial(\hbar n_z)}\delta(\hbar n_z) \\ &= \omega_x \delta(\hbar n_x) + \omega_y \delta(\hbar n_y) + \omega_z \delta(\hbar n_z) \ . \end{aligned} \tag{14}$$

The particular way of writing this example shows how the derivative of the quantum energy E_n with respect to action quantum numbers $\hbar n_i$ is equal to the frequency of the motion of the *classical* harmonic oscillator. This observation is actually not as particular as one may suspect. It is valid for any quantum system, where the corresponding classical motion is periodic with characteristic frequencies ω_i. Another simple example is circular motion in a radial potential. The quantum energy is $E_l = l(l+1)\hbar^2/2mR^2$. Thus, with the angular momentum $mvR = \hbar(l + 1/2)$,

$$\frac{\partial E_l}{\partial(\hbar l)} = \frac{\hbar(l+1/2)}{mR^2} = \frac{mvR}{mR^2} = \frac{v}{R} = \omega_l \ . \tag{15}$$

Returning to the question of shell structure, a perfectly degenerate shell in a three-dimensional harmonic oscillator requires that δE_n in Eq. (14) is equal to zero for several values of δn_x, δn_y, δn_z, i.e.

$$\omega_x \delta n_x + \omega_y \delta n_y + \omega_z \delta n_z = 0 \ , \tag{16}$$

with δn_x, etc. being positive or negative integers. Clearly, this is equivalent to requiring $\omega_x : \omega_y : \omega_z = a^{-1} : b^{-1} : c^{-1}$ where a, b, and c are integers, cf. Fig. 6.

Integer ratios for the three independent periods implies that the classical motion is described by *closed orbits*. After a period divisible with the three individual periods, the motion will be back at the initial point. Thus, in this example, there is a direct connection between closed classical orbits and quantal shell structure. In addition, we have seen that the partial derivative of the total eigenenergy with respect to a quantized action variable is equal to the frequency of the corresponding classical motion.

The situation with the three dimensional harmonic oscillator is particularly transparent because not only is the motion separable in x, y, and z; the total energy is also a linear sum of three independent contributions, each governed by its specific quantum number. (A particle in a rectangular box is another slightly more complicated example.)

The motion in the $-r^{-1}$ potential is also separable in the polar coordinates (r, θ, φ), but the total eigenvalue is not a sum of three independent contributions. (The radial wave equation depends on l.) Still, one can write the eigenenergy E_n for hydrogen:

$$E_n = Ry\, n^{-2} = Ry(n_r + l)^{-2} \,, \tag{17}$$

where $(n_r - 1)$ denotes the number of nodes in the radial wavefunction. (For a given l, the m-states are always degenerate, and the radial wavefunction is independent of m.) Using the generalized result that the derivative of the total eigenenergy with respect to an action quantum number equals the classical frequency and that $(n_r + l) = \text{const.}$ for any shell leads to

$$\omega_r = \omega_l \tag{18}$$

or that the angular period and the radial period of the classical motion are equal. The Kepler ellipses describing orbiting in the r^{-1}-potential with its center at one of the focal points have exactly this property, whatever their size or eccentricity. Thus the occurrence of highly and fully degenerate eigenstates reflects the existence of closed elliptical orbits as the only kind of motion for a bound state in a r^{-1}-potential.

The motion in the three-dimensional, spherical harmonic oscillator potential is also elliptic. Here, the center of the orbits coincide with the center of the potential. This means that the radial motion goes through two complete periods (in-out-in-out) for each angular turn. Thus, classically:

$$\omega_r = 2\omega_l \tag{19}$$

for the spherical, harmonic oscillator. This is again reflected in the eigenvalue equation. In polar coordinates it can be written

$$E_n = \hbar\omega_0\left(n + \frac{3}{2}\right) = \frac{1}{2}\hbar\omega_0\left(2n_r + l + \frac{3}{2}\right) , \tag{20}$$

where n_r describes the number of nodes in the radial wavefunction. With $\partial E_n/\partial n_r \equiv \hbar\omega_r$ and $\partial E_n/\partial l \equiv \hbar\omega_l$ we again obtain Eq. (19); this time from the quantal description.

Figure 8 summarizes and illustrates the correspondence between closed classical orbits and quantal shell structure. To illustrate further one can think of plots of E_{shell} versus $N^{1/3}$ for the hydrogen atom and the spherical harmonic oscillator in analogy to Fig. 7. Compared to this figure such plots (not shown) differ in three respects. *First*, the shell spacings are different, Eqs. (12, 13 and 10)

$$\text{Hydrogen:} \qquad N_0 \approx \frac{2}{3}n^3 + \cdots; \quad \omega_l = \omega_r \tag{21a}$$

$$\text{Harm. osc.:} \qquad N_0 \approx \frac{1}{3}n^3 + \cdots; \quad \omega_l = \frac{1}{2}\omega_r \tag{21b}$$

$$\text{Woods–Saxon:} \qquad N_0 \approx 0.21 \cdot n^3 + \cdots; \quad \omega_l \approx \frac{1}{3}\omega_r? \tag{21c}$$

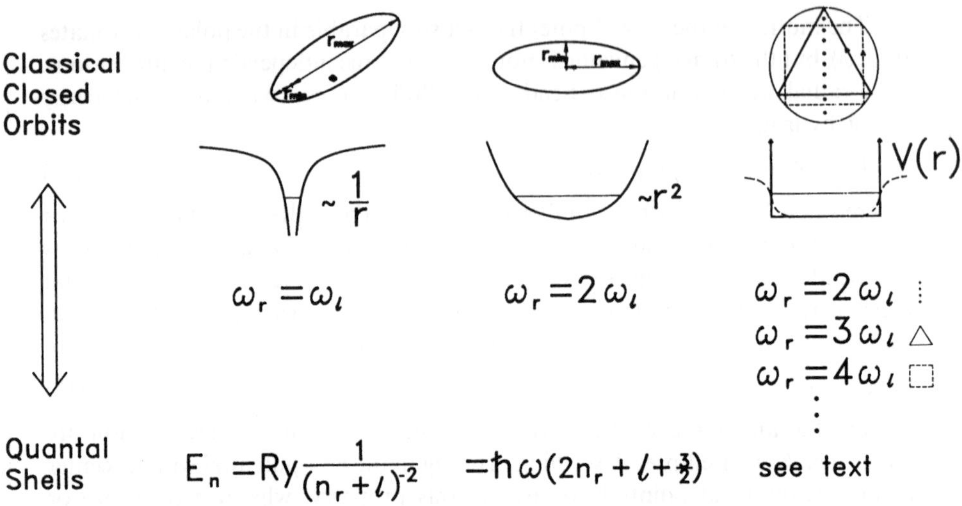

Fig. 8. For spherically symmetric potentials the motion is separable in the radial and angular coordinates. Classically one can therefore speak about separate frequencies in the radial and angular motion, respectively. If the frequencies stand in a (simple) integer ratio to each other, the orbits will be closed. Quantum mechanically, the energy eigenvalues can be labelled with both radial and an angular quantum numbers. Closed periodic orbits in the classical case, and degeneracies of levels with different radial and angular quantum numbers, (shell structure) in the quantal case are closely related to each other. In the two cases, left and center, there is only one type of closed orbit. In the third case, right, several different orbits are possible

Comparing these spacings with the classical frequency relations Eqs. (18 and 19) seems to suggest that classical triangular or pear-shaped orbits are playing a role in the case of the Woods–Saxon potential, or equivalently that one out of three l-values contribute to each shell. *Secondly*, the hydrogen spectrum and the harmonic oscillator exhibits no beat pattern, as opposed to Fig. 7. This reflects the fact that there is only one type of orbit in each case and only one frequency relation. A beat pattern requires two different orbits with two different frequency relations, i.e. periods. In other words, this is a hint that triangular orbits cannot alone account for the shell and supershell periodicities obtained with the Woods–Saxon potential. *Thirdly*, the amplitudes in plots analogous to Fig. 7 are larger for the hydrogen and harmonic oscillator cases and also more regular. This reflects the perfect degeneracies within a shell in these two models. The eigenenergy spectrum of the Woods–Saxon model looks more like the examples shown in Figs. 1 and 3. There is bunching, but no perfect degeneracies.

These qualitative considerations lead naturally to ask what can be said about classical orbits in a Woods–Saxon potential, Eq. (1). Without much loss of generality one may instead think simply of a spherical cavity, Eq. (3).

There is a large number of orbits possible. Some are closed, others never return to the starting point. The closed orbits may be simple, closing after a single angular turn, or they may form more complicated, starlike figures. The situation is not at all as simple as in the hydrogen or harmonic oscillator cases.

Nevertheless one can order the orbits according to increasing length. The simplest are the pendulating orbits going through the origin. There is a two-dimensional multitude of those, corresponding to the two polar angles. Next comes the triangular and then the square orbits, representing a three-dimensional multitude each (corresponding to the three Euler angles), and so forth. What is needed is a theory that connects this system of orbits with the eigenvalue spectrum of stationary waves in the cavity. Such a theory actually exists. It is due to *Balian* and *Bloch* [13] and to *Gutzwiller* [14]. In Ref. [13] the theory is applied to the cavity, while in Ref. [12] it is extended to the spherical Woods-Saxon potential. In essence, the complete set of delta functions describing eigenvalues as a function of energy, viz. Fig. 5, can be expressed as a sum over contributions from an infinite series of closed classical orbits of increasing length. The major periodicities, the ones seen in Fig. 7, on the other hand, emerge already from the sum truncated after the third i.e. square, orbit. The pendulating orbits contribute little, and thus the shell and supershell periodicities can be understood as the quantum signatures of the classical triangular and square orbits contributing with about equal strength to the density of levels in the eigenvalue spectrum.

The beat pattern can be qualitatively visualized by looking at a particle of definite speed, v_F, in a cavity of increasing radius $R \sim N^{1/3}$, where N is the N'^s eigenstate, degeneracies included. If $R = r_{WS} \cdot N^{1/3}$, where r_{WS} is the Wiegner–Seitz radius, it corresponds to looking at an electron in a metallic drop where the electron moves with Fermi velocity. The action J of such a particle in a closed orbit of length L is (2 · energy · time) = $mv^2\tau$, τ being the period of the orbit equal to L/v_F. This action is quantized in units of h. For a triangular orbit, $L_3 = 3\sqrt{3} \cdot R$. For a square orbit $L_4 = 4\sqrt{2} \cdot R$. Hence

$$J_3 = n_3 h = mv_F^2 \frac{L_3}{v_F} = 3\sqrt{3} m v_F R_3 \ , \tag{22a}$$

$$J_4 = n_4 h = mv_F^2 \frac{L_4}{v_F} = 4\sqrt{2} m v_F R_4 \ , \tag{22b}$$

or,

$$R_3 = \left(\frac{h}{mv_F}\right) \frac{n_3}{3\sqrt{3}}; \quad n_3 = 1, 2, 3, \ldots \tag{23a}$$

$$R_4 = \left(\frac{h}{mv_F}\right) \frac{n_4}{4\sqrt{2}}; \quad n_4 = 1, 2, 3, \ldots . \tag{23b}$$

If only triangular orbits were contributing to shell structure, closed shells would occur for drop sizes shown in the top right of Fig. 9. Analogously, if square orbits were dominating, the sizes at the top left would represent closed shells. If both orbits contribute equally it results in the size pattern shown below in Fig. 9. The triangular and square orbits are alternatingly being in phase and out of phase, resulting in the beat pattern shown. Simple quantization of the

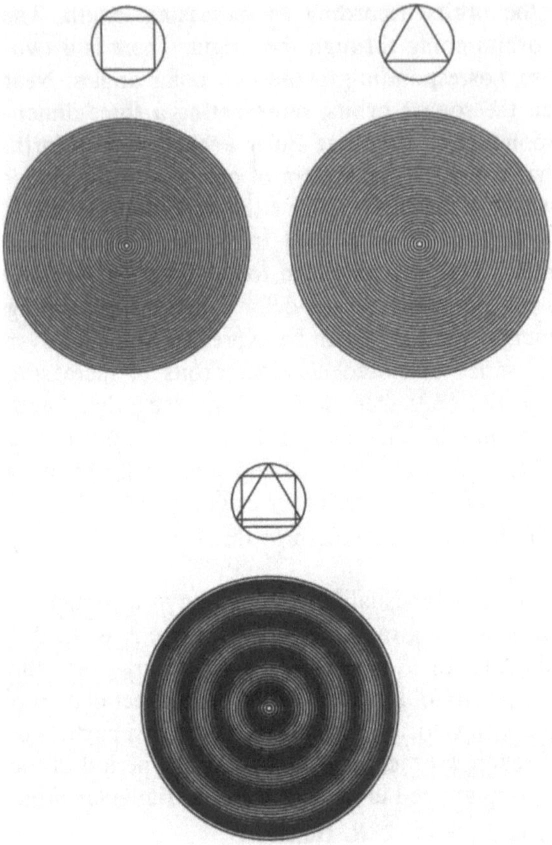

Fig. 9. If two classically closed orbits e.g. triangles and squares contribute equally to the quantal shell structure, it will give rise to a higher periodicity, supershells

action of classical triangular and square orbits having the same energy thus leads to the same results regarding the bunching of levels as solving the Schrödinger wave equation for a particle in the same potential. This is the correspondence principle.

The correspondence becomes more accurate as the action quantum number increases; and metal clusters offer special opportunities for observing systems with large quantum numbers.

The ultimate correspondence with classical orbital motion of a well-localized particle with continuously variable total energy requires that one goes beyond a description in terms of stationary energy eigenvalues. By combining wavefunctions from neighbouring shells one can build wave packets moving like a classical particle, as was originally demonstrated by *Niels Bohr* in connection with the hydrogen problem. With increasing quantum numbers, the relative uncertainty in momentum and position of the wave packet decreases, converging in this way towards the classical limit.

2.6.4 Further Reading

W. Pauli's article [15] in Handbuch der Physik from 1926 is a good starting point for studying the old quantum theory and its relation to the classical mechanics of multiply periodic systems in the Hamiltonian or Jacobian forms. The article is translated to English in Ref. [16]. A more recent introduction to semiclassical quantum theory can be found in Landau and Lifschitz's textbook [17]. Application to the subject of this chapter is found in *Balian* and *Bloch's* seminal article, Ref. [13]. It is not easy reading. Nor is Gutzwiller's semiclassical formulations of the level density in general forms of potentials, Refs. [14]. More readily comprehensible are the articles by *Berry* and *Tabor* [18]. In the same way, *Bohr* and *Mottelson*'s generalized discussion of shell structure in spherical potentials, Ref. [10] is relatively straightforward. It focuses on the physics of real systems, nuclei in particular. The new article by *Nishioka* et al. [12] builds on *Gutzwiller* [14] and *Balian* and *Bloch* [13], but is nicely self-contained and it is directly concerned with metal clusters.

2.6.5 Recent Developments

The existence of shells and their beat pattern in terms of supershells was demonstrated experimentally in 1991 by an extension of the mass spectroscopic measurements of sodium clusters [1] to particles with up to 3000 atoms [20, 21]. Later, similar effects have been found with clusters of lithium [22] and also with the trivalent metal, gallium [23].

The theoretical treatment [12] has been extended to include the influence of temperature on the magnitude of the shell effects, Fig. 7, using the local density, Cohen–Sham, approximation (LDA) [24]. On this basis experimental results have been compared to theoretical predictions in a semiquantitative fashion [25]. Establishing a truly quantitative connection between theory and experiment remains, however, a major challenge.

References

1. W.D. Knight, K. Clemenger, W.A. de Heer, W.A. Saunders, M.Y. Chou, M.L. Cohen: Phys. Rev. Lett. **52**, 2141 (1984); W.A. de Heer, W.D. Knight, M.Y. Chou, M.L. Cohen: Solid State Physics **40**, 93 (1987)
2. I. Katakuse, T. Ichihara, Y. Fujita, T. Matsuo, H. Matsuda: Int. J. Mass Spectro. Ion Proc. **69**, 109 (1986)
3. W. Ekardt: Phys. Rev. **B29**, 1558 (1984)
4. D.E. Beck: Solid State Commun. **49**, 381 (1984).
5. N. Bohr: Collected Works Vol. **9**, ed. Rudolf Peirls, North Holland (Elsevier), Amsterdam, 1986

6. M.G. Mayer, J.H.D. Jensen: *Elementary Theory of Nuclear Shell Structure*, Wiley, New York, 1955
7. R.D. Woods, D.S. Saxon: Phys. Rev. **95**, 577 (1954)
8. S.G. Nilsson: Mat. Fys. Medd. K. Dan. Vidensk. Selsk. **29**, No. 16 (1955)
9. A. Bohr, B.R. Mottelson: Nuclear Structure, Vol. **I**, Benjamin, New York, pp. 208–241, 1969
10. A. Bohr, B.R. Mottelson: Nuclear Structure, Vol. **II**, Benjamin, London, pp. 578–607, 1975
11. S. Bjørnholm, E. Lynn: Rev. Mod. Phys. **52**, 725 (1980)
12. H. Nishioka, K. Hansen, B.R. Mottelson: Phys. Rev. **B42**, 9377 (1990)
13. R. Balian, C. Bloch: Ann. Phys. **69**, 76 (1971)
14. M.C. Gutzwiller: J. Math. Phys. **8**, 1979 (1967); ibid **11**, 1791 (1969); ibid **12**, 343 (1971)
15. W. Pauli, in Handbuch der Physik **23**, 1–278 (1926)
16. W. Pauli: *General Principles of Quantum Mechanics*, Springer, Berlin, 1980
17. L.D. Landau, E.M. Lifschitz: Course of Theoretical Physics, Vol. **3**, *Quantum Mechanics*, Pergamon Press, London, pp. 157–171 (1958)
18. M.V. Berry, M. Tabor: Proc. Roy. Soc. London **A349**, 101 (1976); **A 356**, 375 (1977)
19. D.S. Lewardt, V.R. Pandharipande, S.C. Pieper: Phys. Rev. **B37**, 4950 (1988)
20. J. Pedersen, S. Bjørnholm, J. Borggreen, K. Hansen, T.P. Martin, H.D. Rasmussen: Nature **353**, 733 (1991)
21. T.P. Martin, S. Bjørnholm, J. Borggreen, C. Bréchignac, Ph. Cahuzac, K. Hansen, J. Pedersen: Phys. Lett. **186**, 53 (1991)
22. C. Bréchignac, Ph. Cahuzac, M. de Frutos, J.Ph. Roux, K. Bowen: In *Physics and Chemistry of Finite Systems; from Clusters to Crystals*, eds. P. Jena, S.N. Khanna, B.K. Rao (Klüwer Academic Publications, 1992)
23. J. Lermé, M. Pellarin, J.L. Vialle, B. Baguenard, M. Broyer: Phys. Rev. Lett. **68**, 2818 (1992)
24. M. Brack, O. Genzken, K. Hansen: Z. Phys. **D21**, 65 (1991)
25. O. Genzken: Mod. Phys. Lett. **7**, 197 (1993)

2.7 Introduction to Statistical Reaction Rate Theories

M.F. Jarrold

2.7.1 Introduction

If a diatomic molecule is vibrationally excited to an energy above its dissociation threshold it will normally dissociate within a vibrational period. On the otherhand, a macroscopic piece of material (such as a few cm^3 of aluminum) at room temperature, contains thermal energy $\sim 10^{19}$ times the binding energy of an atom to the bulk, but the rate of evaporation is vanishingly small because the energy is distributed statistically among the $3n$ vibrational degrees of freedom or phonons of the bulk (n is the number of atoms). It is not surprising then to find that a cluster, with for example $n \sim 100$, will not instantaneously dissociate when excited to just above its dissociation threshold. Dissociation is important as the basis for a number of experimental techniques in cluster research. So an understanding of the factors which control the dissociation rates of energized clusters is clearly desirable. Most methods of determining the dissociation energies of clusters depend on preparing the clusters with a known amount of energy and noting whether dissociation occurs. In interpreting these experiments an accurate method for calculating cluster dissociation rates is essential.

The idea that there is a delay between activation of a molecule and its subsequent dissociation was first suggested in the 1920s in connection with thermally activated unimolecular reactions. The type of problem outlined above is also encountered in connection with nuclear fission. Statistical models appropriate for nuclear fission were developed in the late 1930s, and we will consider these models in more detail below. We will start by considering the simplest model for evaluating the dissociation rate of a molecule or cluster containing a known amount of energy, which is RRK theory.

2.7.2 RRK Theory

In 1927 *Rice* and *Ramsperger*, and *Kassel* [1–3] proposed a simple theory for the rate of dissociation of energized molecules. There were small differences in their approaches, but basically they started by considering a molecule to be a collection of s independent harmonic oscillators, which were strongly coupled so that energy flows between the oscillators. Reaction is assumed to occur when, by chance, an energy greater than some critical energy (E_0) accumulates in a particular oscillator. This oscillator can be related to the bond being broken in the reaction. The reaction probability is then given by the number of ways of arranging the energy among the oscillators with an energy greater than E_0 in the particular oscillator, divided by the number of ways of arranging the energy with any energy in the chosen oscillator. From classical high temperature statistics the number of ways of arranging an energy E among s oscillators is $E^{s-1}/(s-1)!$ so the reaction probability is given by

$$\frac{(E-E_0)^{s-1}/(s-1)!}{E^{s-1}/(s-1)!} = \left(\frac{E-E_0}{E}\right)^{s-1} \tag{1}$$

and the reaction rate constant is

$$k(E) = A\left(\frac{E-E_0}{E}\right)^{s-1} \tag{2}$$

where A is a constant. Kassel showed that A is the high pressure Arrhenius pre-exponential factor for a thermally activated unimolecular reaction. This constant is now usually related to a vibrational frequency (assuming that the energy is reshuffled ν times per second).

Though appealing for its simplicity, classical RRK theory under-estimates the dissociation rates by many orders of magnitude. *Kassel* [3] noted that good agreement with experimental data could only be achieved if the number of vibrational degrees of freedom, s, was reduced by around a factor of two. For some time it was thought that only a portion of the vibrational degrees of freedom were active in promoting dissociation. This was later shown to be incorrect. One obvious deficiency of RRK theory, as outlined above, is that the vibrational degrees of freedom are treated classically. Kassel has developed a quantum version of RRK theory [4]. Assuming that there are s oscillators of the same frequency, the number of ways of distributing j quanta among the oscillators is $(j+s-1)!/j!(s-1)!$, so if we define $p = E/h\nu$ and $q = E_0/h\nu$, the reaction probability is given by

$$\frac{(p-q+s-1)!/(p-q)!(s-1)!}{(p+s-1)!/p!(s-1)!} = \frac{p!(p-q+s-1)!}{(p+s-1)!(p-q)!} \tag{3}$$

and the reaction rate constant becomes

$$k(E) = A\frac{p!(p-q+s-1)!}{(p+s-1)!(p-q)!} \tag{4}$$

It is possible to extend the approach outlined above to include oscillators of different frequencies but the expressions soon get extremely complex. The quantum version of RRK is a considerable improvement over the classical version, but it is still inadequate.

2.7.3 RRKM Theory and the Transition State

RRKM theory was developed by *Marcus* [1, 5]. It is known as RRKM theory because it builds on the concepts of RRK theory. RRKM theory is also sometimes called quasi-equilibrium theory or QET. An important feature of RRKM theory is the use of transition state concepts [6] in the calculation of the rate constant. Since it is possible to separate electronic and nuclear motion, a chemical reaction can be viewed as occurring on an electronic potential energy surface, and the reaction can be represented by a trajectory on this surface. Since large molecules have many internal degrees of freedom the potential surface will be a multi-dimensional surface and the trajectories extremely complex. The idea behind transition state theory is that there exists a dividing surface, which separates reactants from products, that the trajectory crosses only once. This dividing surface is called the transition state, and it is usually located at the maximum potential energy along the reaction coordinate (which is the lowest energy path separating reactants from products). Since trajectories only pass through this transition state once, the reaction rate can be obtained by determining the rate at which trajectories cross the dividing surface. The reaction rate is evaluated in terms of the properties of the transition state which is treated thermodynamically even though it is unstable.

The two basic assumptions of RRKM theory are 1) all the states at a particular energy are accessible and equally likely to be populated; and 2) energy redistribution among the states occurs more rapidly than dissociation. These assumptions allow us to use statistical mechanics to evaluate the rate at which trajectories pass through the transition state. For simplicity, in the derivation of the RRKM expression described below we only consider vibrational energy. Rotational energy will be included later. Figure 1 shows a schematic reaction coordinate diagram for the dissociation of a molecule or cluster. Using statistical mechanics the dissociation rate can be written as a ratio of the density of

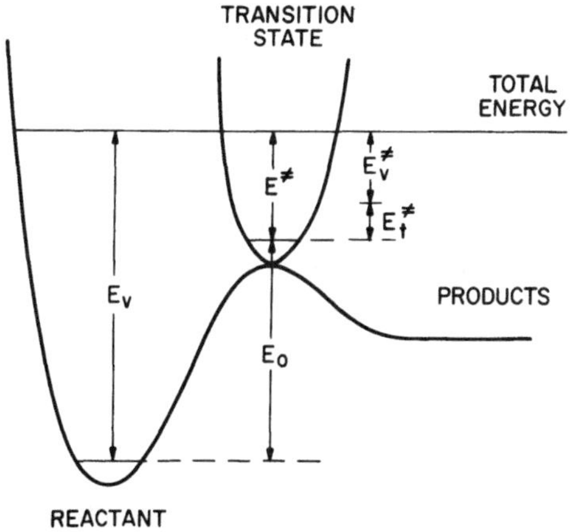

Fig. 1. Schematic reaction coordinate diagram showing the location of the transition state at the top of an activation barrier

quantum states at the transition state, $\rho(E^\ddagger)$, and the stable configuration, $\rho(E_v)$:

$$k(E) = \frac{\nu}{2}\frac{\rho(E^\ddagger)}{\rho(E_v)} . \tag{5}$$

ν in this expression is the frequency of passage through the transition state. The factor of two arises because under true equilibrium conditions trajectories pass through the transition state in both directions, and we only wish to count trajectories from reactants to products. For motion through the transition state one of the vibrational degrees of freedom is treated as an internal translational motion. So $\rho(E^\ddagger)$ can be expressed as

$$\rho(E^\ddagger) = \sum N(E_v^\ddagger)\,\rho(E_t^\ddagger) \tag{6}$$

where $N(E_v^\ddagger)$ is the sum of vibrational states at vibrational energy $E_v^\ddagger$ and $\rho(E_t^\ddagger)$ is the density of translational states at energy $E_t^\ddagger$ with $E_t^\ddagger = E^\ddagger - E_v^\ddagger$. We now need to evaluate $\rho(E_t^\ddagger)$. The energy levels for translation of a particle of mass μ in a one-dimensional box of length δ are given by

$$E_t^\ddagger = \frac{k^2 h^2}{8\mu\delta^2} . \tag{7}$$

The significance of μ and δ need not be considered in detail because they both cancel out of the final expression. The density of states $\rho(E_t^\ddagger)$ is obtained from

$$\rho(E_t^\ddagger) = \frac{dk}{dE_t^\ddagger} = \left(\frac{2\mu\delta^2}{h^2 E_t^\ddagger}\right)^{1/2} . \tag{8}$$

The frequency of passage through the transition state is given by the velocity of the particle of mass μ, in a one-dimensional box, divided by the length of the box,

$$\nu = \frac{(2E_t^\ddagger/\mu)^{1/2}}{\delta} . \tag{9}$$

Substituting Eqs. (6), (8), and (9) into Eq. (5) leads to the remarkably simple expression

$$k(E) = \frac{N(E^\ddagger)}{h\rho(E_v)} , \tag{10}$$

where $N(E^\ddagger)$ is the sum of vibrational states at the transition state, given by $\sum N(E_v^\ddagger)$. Equation (10) is the RRKM expression for the rate of dissociation of a molecule or cluster with total internal energy E. It is common to call $N(E^\ddagger)/h$ the flux, $F(E)$, through the transition state. Including rotational degrees of freedom the RRKM expression becomes

$$k(E, J) = \frac{F(E, J)}{\rho(E, J)} = \frac{\sigma(2J+1)N(E^\ddagger - B^\ddagger J(J+1))}{h(2J+1)\rho(E - BJ(J+1))} , \tag{11}$$

if the transition state and stable configuration are approximated as spherical tops and so B and $B^{\ddagger}$ are the geometric mean rotational constants. The spherical top approximation does not usually introduce significant errors. J in Eq. (11) is the total angular momentum which is conserved during the reaction, and σ is the reaction path degeneracy which is included to account for the fact that there are often several different ways a reaction can occur. For example, for the reaction $CH_4 \rightarrow CH_3 + H$, σ would be 4. The $2J + 1$ factors in Eq. (11) arise because of the degeneracy of the rotational levels of a spherical top. There is an additional $2J + 1$ degeneracy arising from different values of the z-component quantum number M. In the absence of perturbing fields M is conserved throughout the reaction and does not enter into the expression for the rate.

Several different methods are available to evaluate $N(E)$ and $\rho(E)$, the sum and the density of vibrational states [1]. The simplest approach is to treat the molecule as a collection of classical harmonic oscillators. However, the classical approximation is extremely poor, particularly at low energies. The most accurate method for determining $N(E)$ and $\rho(E)$ is to use a direct count, for which remarkably simple and efficient algorithms have been described [7]. For very large molecules or clusters, a direct count may not feasible at high energies. Several other methods are available for determining the sums and densities of states. One of the most widely used is the Whitten-Rabinovitch approximation [1] which is based on empirical corrections to the classical sums and densities of states. The Whitten–Rabinovitch approximation is almost exact for non-fixed vibrational energies $E_v > \sum h\nu_i/2$. For large molecules or clusters the best approach is to use the direct count method for $N(E_v)$ for $E_v < \sum h\nu_i/2$, but to employ the Whitten–Rabinovitch approximation for $\rho(E_v)$ and $N(E_v)$ for $E_v > \sum h\nu_i/2$.

In order to determine the accuracy and reliability of RRKM theory it is necessary to compare the predictions with experimental results, and there have been many such comparisons [1]. To perform RRKM calculations vibrational frequencies and rotational constants are required for both the stable species and the transition state, along with values for the reaction path degeneracy, σ, and the zero point energy difference between the transition state and the stable species. Since most of this information is usually not available for clusters (in fact RRKM theory or other less sophisticated methods are often used to model experimental data in order to determine dissociation energies) it is not possible to perform a meaningful comparison between experiment and theory for cluster dissociation. So instead, as an example, we will consider the dissociation of $C_6H_6^+$ ions, which have been extensively studied [8]. Figure 2(a) shows a comparison of RRKM calculations with rate constants measured for the dissociation of $C_6H_6^+$ ions by photo-ion photo-electron coincidence (PIPECO). The RRKM rate constants shown in Fig. 2(a) by the line have been averaged over the angular momentum distribution and vibrational energy distribution of the dissociating ions. They are in good agreement with the measurements. Figure 2(b) shows a comparison of the predictions of RRKM theory and the relative abundances of the four main products from the dissociation of $C_6H_6^+$ ions ($C_6H_5^+$, $C_6H_4^+$, $C_4H_4^+$, and $C_3H_3^+$) also measured by PIPECO. The relative

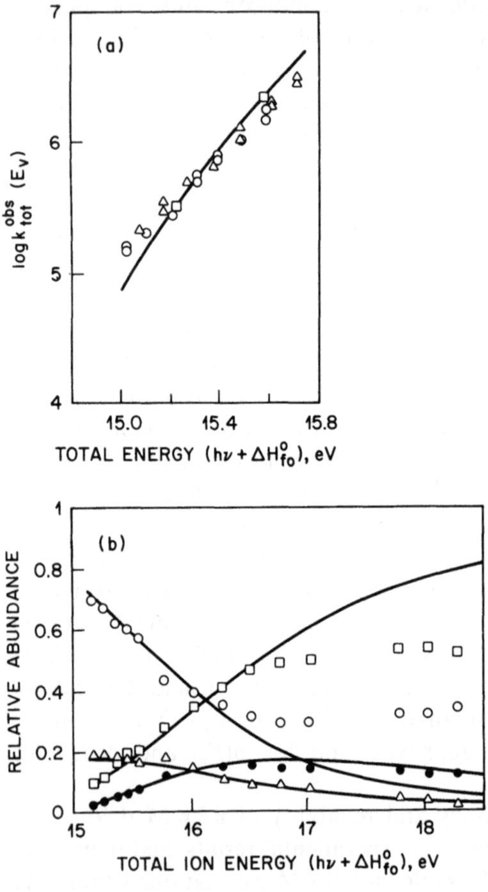

Fig. 2. Comparison of the predictions of RRKM theory with experimental data for the dissociation of $C_6H_6^+$ ions. Reactions rate constants (**a**) and product branching ratios (**b**) are shown. (Adapted from [8])

abundance of product i is given by

$$f_i(E_v) = \int_0^\infty dJ\, P(J) \frac{F_i(E_v, J)}{F_{tot}(E_v, J)} \tag{12}$$

where $F_i(E_v, J)$ is the flux through the transition state for vibrational energy E_v and angular momentum J to give product i and $F_{tot}(E_v, J) = \sum F_i(E_v, J)$. The disagreement between experiment and theory for total energies above 16.5 eV is believed to arise because with increasing energy, dissociation occurs more rapidly than isomerization to non-cyclic forms of $C_6H_6^+$ which are precursors to formation of $C_4H_4^+$. Note how dramatically the product distribution in Fig. 2(b) changes with energy, this is something to bear in mind when attempting to draw energetic conclusions from product distributions. For the $C_6H_6^+$ system most of the information required to perform RRKM calculations is available, except for values for the vibrational frequencies and rotational constants of the transition states. The transition states are transient species so it is not generally possible to determine this information spectroscopically. In some cases information about the transition state is available from ab initio calculations. But in most cases the parameters must be estimated. However, there are often very tight constraints on the transition state parameters even though their precise values are not known. For example, for the $C_6H_6^+$ system discussed above the main product at low energies is $C_6H_5^+$. The most stable form of $C_6H_6^+$ is the benzene ion and the $C_6H_5^+$ product is the phenyl ion. So the reaction coordinate corresponds to a C–H stretch of the benzene ion and all the other frequencies except for two C–H bends remain essentially the same (the C–H stretch vanishes at the transition state as it becomes a translational degree of freedom for motion through the transition state). The C–H bends vanish on going from reactants to products (the two bends and the C–H stretch become the three translational degrees of freedom of the departing atom). Modes which vanish like the C–H bends are called transitional modes and are usually assumed to lie between 25% and 75% of their values in the stable molecule.

2.7.4 Phase Space Theory

While the origins of RRKM theory can be traced to studies of thermally activated unimolecular reactions, the basic ideas behind phase space theory can be traced to the application of statistical methods to nuclear reactions and the development of the compound nucleus model [9, 10]. *Keck*, and *Light* and coworkers [11] pioneered the application of phase space theory to bimolecular reactions. *Klots* [12], and *Chesnavich* and *Bowers* [13] extended this approach to unimolecular reactions. RRKM theory is cast from the perspective of the dissociating molecule, phase space theory (as applied to unimolecular reactions) focuses on the products and hinges on the application of the principle of microscopic reversibility. Consider the reaction

$$A \rightleftarrows B + C \qquad (13)$$

represented by the reaction coordinate diagram shown in Fig. 3. At equilibrium we can write

$$k_1[A] = k_2[B][C] \qquad (14)$$

and by the principle of microscopic reversibility

$$R_{A \to B+C}(E, J) = R_{B+C \to A}(E - E_0, J) \;, \qquad (15)$$

where $R_{A \to B+C}(E, J)$ is the total rate of passage of reactants A with energy E and angular momentum J to products $B + C$. The forward rate can be written as the unimolecular rate constant for $A \to B + C$ at total energy E and angular momentum J, $k_A(E, J)$, times the density of states at energy E and angular momentum J

$$R_{A \to B+C}(E, J) = k_A(E, J)\,\rho_v^A(E - E_r^A)\,S_r^A \qquad (16)$$

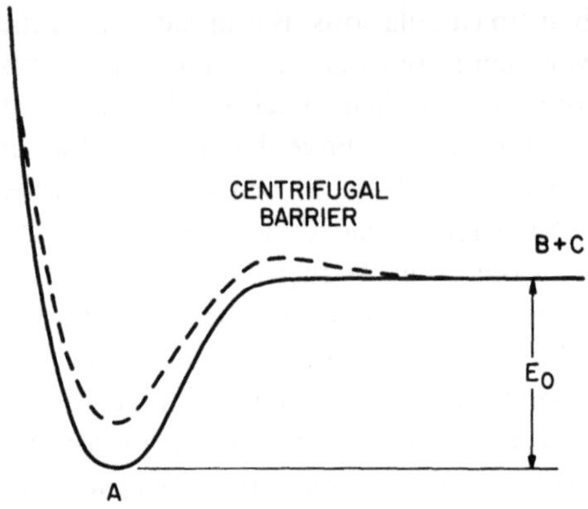

Fig. 3. Schematic reaction coordinate diagram for a reaction with no activation barrier except the centrifugal barrier

where $\rho_v^A(E - E_r^A)$ is the density of vibrational states of A with total energy E and rotational energy E_r^A, and S_r^A is the rotational degeneracy factor (for a spherical top S_r^A is $(2J + 1)^2$, consisting of a $2J + 1$ spatial degeneracy and a $2J + 1$ internal degeneracy). For the reverse reaction the products $B + C$ not only have rotational and vibrational degrees of freedom but also have relative translational degrees of freedom since a collision must occur. Thus [13]

$$R_{B+C\to A}(E - E_0, J) = \frac{\sigma_A}{\sigma_B\sigma_C} S_r' \int\int k_{BC}(E_r, E_t, J)\,\rho_t(E_t)\,dE_t \times \rho_v(E - E_0 - E_{tr})\,dE_{tr}\ , \tag{17}$$

where $k_{BC}(E_r, E_t, J)$ is the partial rate constant at relative translational energy E_t and rotational energy E_r for formation of a collision complex with angular momentum J, $\rho_t(E_t)$ is the density of translational states per unit volume at energy E_t, $\rho_v(E - E_0 - E_{tr})$ is the vibrational density of states of $B + C$ at energy $E - E_0 - E_{tr}$ where $E_{tr} = E_t + E_r$ and E_0 is the zero point energy difference between reactants and products, S_r' is the spatial rotational degeneracy factor $(2J + 1)$, and $\sigma_A/\sigma_B\sigma_C$ is the ratio of rotational symmetry numbers which accounts for the reaction path degeneracy. The integrals in Eq. (17) are over the available energy consistent with angular momentum conservation. Combining Eqs. (15)–(17) $k_A(E, J)$ is given by

$$k_A(E, J) = \frac{\sigma_A}{\sigma_B\sigma_C}\frac{S_r'}{S_r^A}\frac{\int\int k_{BC}(E_r, E_t, J)\,\rho_t(E_t)\,dE_t\,\rho_v(E - E_0 - E_{tr})\,dE_{tr}}{\rho_v^A(E - E_r^A)}\ . \tag{18}$$

In the collision between B and C the total angular momentum J is composed of contributions from the orbital angular momentum L and the rotational angular momenta of B and C, J_r^B and J_r^C. The total rotational angular momentum J_r is obtained from the vector addition of J_r^B and J_r^C, so the allowed values of J_r are

$$|J_r^B - J_r^C| \le J_r \le J_r^B + J_r^{C}\ . \tag{19}$$

The total angular momentum J is then obtained from the vector addition of J_r and L, and the allowed values of J are

$$|J_r - L| \leq J \leq J_r + L \tag{20}$$

subject to energetic constraints which will be described below. For reactions dominated by spherically symmetric long range intermolecular potentials the partial capture rate constant for forming a collision complex with angular momentum J is

$$k_{BC}(E_r, E_t, J) = \frac{v\pi\hbar^2}{2\mu E_t} P_{ro}(E_t, E_r, J) \ , \tag{21}$$

where $P_{ro}(E_t, E_r, J)$ is the density of rotational and orbital angular momentum states, and v is the velocity. The density of translational states per unit volume is

$$\rho_t(E_t)dE_t = \frac{\mu^2 v}{2\pi^2\hbar^3} dE_t \tag{22}$$

and the sum of the rotational and orbital angular momentum states at a given E_{tr} and J is defined as

$$\Gamma_{ro}(E_{tr}, J) = \int P_{ro}(E_t, E_r, J)dE_t \tag{23}$$

so $k_A(E, J)$ can be written as

$$k_A(E, J) = \frac{\sigma_A}{\sigma_B\sigma_C} \frac{S_r'}{S_r^A} \frac{\int_{E_{tr}^+}^{E-E_0} \rho_v(E - E_0 - E_{tr})\Gamma_{ro}(E_{tr}, J)dE_{tr}}{h\rho_v^A(E - E_r^A)} \ , \tag{24}$$

where E_{tr}^+ is the minimum value for E_{tr} for which $\Gamma_{ro}(E_{tr}, J) > 0$. This is the generalized phase space theory expression for unimolecular dissociation.

The vibrational densities of states, $\rho_v(E)$, can be calculated as described above. However, the evaluation of $\Gamma_{ro}(E_{tr}, J)$ the sum of rotational and orbital angular momentum states, is not so simple. For an interaction potential with a long range attractive term of the form $-C/r^m$ (where $m = 4$ for an ion-molecule reaction and $m = 6$ for a neutral-neutral reaction) the minimum translational energy that can overcome the centrifugal barrier is

$$E_t^+ = L^{2m/(m-2)}/\Lambda \ , \tag{25}$$

where Λ is a constant depending on the value of m. $\Gamma_{ro}(E_{tr}, J)$ is evaluated by integration of the rotational sum of states $\Gamma(E_r^*, J_r)$ over the L–J_r plane [13]

$$\Gamma_{ro}(E_{tr}, J) = \iint \Gamma(E_r^*, J_r)dJ_r dL \tag{26}$$

subject to conservation of angular momentum and energetic constraints. E_r^* in Eq. (26) is given by $E_{tr} - E_t^+$. The value of $\Gamma(E_r^*, J_r)$ depends on the number of rotational degrees of freedom possessed by B and C. As a simple example, if B is treated as a sphere and C is an atom $\Gamma(E_r^*, J_r)$ is given by $2J_r$. More complicated expressions result if both B and C have rotational degrees of freedom. Figure 4 shows the range of integration of $\Gamma(E_r^*, J_r)$ in the L–J_r plane. The

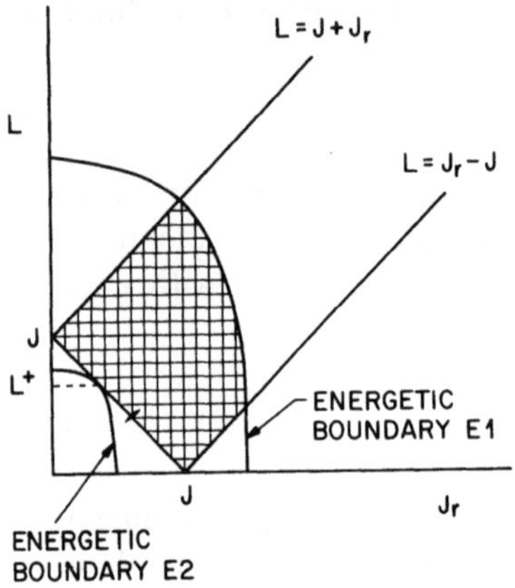

Fig. 4. Diagram showing the region of integration of $\Gamma(E_r^*, J_r)$ in the L–J_r plane

boundary labelled $E1$ is defined by the energetic constant

$$L^{2m/(m-2)}/\Lambda + B_r J_r^2 \leq E_{tr} , \tag{27}$$

where B_r is the reduced rotational constant of B and C. The other energetic boundary, $E2$, defines E_{tr}^+, which is given by

$$E_{tr}^+ = (L^+)^{2m/(m-2)}/\Lambda + B_r(J - L^+)^2 . \tag{28}$$

It is instructive to compare the phase space theory expression for unimolecular dissociation

$$k_A(E, J) = \frac{\sigma_A}{\sigma_B \sigma_C} \frac{S_r'}{S_r^A} \frac{\int_{E_{tr}^+}^{E-E_0} \rho_v(E - E_0 - E_{tr}) \Gamma_{ro}(E_{tr}, J) dE_{tr}}{h\rho_v^A(E - E_r^A)} \tag{29}$$

to the analogous RRKM expression

$$k_A(E, J) = \sigma \frac{\int_0^{E-E_0} \rho_v^{\ddagger}(E - E_0 - E_r - E_{rc}) dE_{rc}}{h\rho_v^A(E - E_r)} . \tag{30}$$

The expressions are quite similar and become identical if the transition state in RRKM approach is moved out to the centrifugal barrier and rigorous conservation of angular momentum is included. However, it is worth emphasising that these two expressions result from quite different perspectives. RRKM theory focusses on the local conditions at the transition state, whereas phase space theory is a statistical-dynamical approach which focusses on the initial conditions of the reactants for the reverse reaction.

Engelking [14] has recently proposed a simplified version of phase space theory for application to the unimolecular reactions of loosely bound ionic clusters. (Note that although Engelking describes his model as a reformulation of RRK theory it is, in fact, a phase space theory model.) According to Engelking's model the dissociation rate is given by

$$k(E) = 8\pi\sigma\mu v^3 S(s-1) \frac{(E - E_0)^{s-2}}{E^{s-1}} , \tag{31}$$

where S is the geometric cross section for forming the complex. Equation (30) can be obtained by inserting the classical vibrational density of states into Weisskopf's model for nuclear decay [9] and integrating over kinetic energy. There is no attempt to conserve angular momentum. Like RRK theory this model is appealing for its simplicity, however, like RRK theory it also underestimates the dissociation rate by many orders of magnitude (when compared to the more rigorous model).

A particularly appealing feature of phase space theory is that the rate is formulated in terms of the properties of the reactants and products which can all be directly measured. This is in contrast to RRKM theory where the rate is formulated in terms of a transition state, which by definition is a transient species and direct measurement of its properties are generally not possible. However, phase space theory and RRKM theory only give the same value for the dissociation rate if the transition state in the RRKM approach is located at the centrifugal barrier. In RRKM theory this would correspond to an extremely loose transition state (more like the products than the reactant). If a tighter transition state (one more like the reactant) is employed the dissociation rate predicted by RRKM theory will be much lower. So the obvious question is which approach provides the correct answer and agrees with the experimental data. This depends on the nature of the potential energy surface. If there is an activation barriers as shown in Fig. 1 then the answer is unambiguous, RRKM theory is appropriate. On the otherhand many potential energy surfaces do not have activation barriers. Activation barriers are usually absent for radical-radical recombination reactions and ion-molecule reactions, and in particular activation barriers are usually absent in cluster dissociation because the reverse association reactions occur efficiently (or cluster growth would not occur). Before discussing potential energy surfaces without an activation barrier it is instructive to consider what happens on a potential energy surface with an activation barrier. While it is clear that a transition state exists near the top of the activation barrier, it is also possible that a second orbiting transition state (located at the centrifugal barrier) occurs on the product's side of the activation barrier. Thus there could be two transition states (or more correctly minima in the local equilibrium flux along the reaction coordinate). If the activation barrier is sufficiently large that the flux through the tight transition state, $F^{\ddagger}$, is always less than the flux through the orbiting transition state, F^{orb}, then the overall rate of reaction will be controlled by the tight transition state. Since F^{orb} always increases more rapidly with energy than $F^{\ddagger}$ the above criteria is met if the activation barrier is larger than a few tenths of an electron-volt.

Experimentally it is known that bimolecular ion-molecule reactions and radical-radical recombination reactions (reactions which usually proceed without an activation barrier) often occur at close to the collision rate. Thus the rates of these bimolecular reactions are generally controlled by the orbiting transition state located at the centrifugal barrier. It follows that the reverse dissociation must also be controlled by the orbiting transition state. It is worth noting at this time that these reactions occur with total energies very close to the zero point energy difference between the reactants and products. In contrast to the experimental data on bimolecular reactions, studies of unimolecular reactions for systems which do not have significant activation barriers show that the measured rates cannot be accounted for by the orbiting transition state, and a tight transition state is required to fit the experimental data. These studies are often performed with total energies considerably larger than the zero point energy difference between reactants and products. For example, dissociation of $C_6H_6^+$ to $C_6H_5^+ + H$ is believed to occur without an activation barrier but the measured rate constants can only be fit assuming an RRKM model with a tight transition state. The dissociation threshold for $C_6H_6^+$ on the total energy scale shown in Fig. 2 is at 13.97 eV so the rate constants in Fig. 2(a) were measured for energies between 1.0 and 1.8 eV above the dissociation threshold.

From the above discussion of experimental data it appears that for potential energy surfaces in which there is no activation barrier the nature of the transition state controlling the rate of reaction changes with energy. At low energies an orbiting transition state is most appropriate and at energies considerably above threshold an RRKM model assuming a tight transition state fits the experimental data. Several different approaches are available to handle this problem. The simplest approach is to use phase space theory to describe the dissociation of complexes formed in collisions, but to employ RRKM theory to describe the dissociation of molecules with internal energies significantly above their dissociation threshold. The most rigorous approach is to use the methods of variational transition state theory [15] and calculate the local equilibrium flux along the reaction coordinate and locate the transition state at the point of minimum flux. Such procedures are feasible for small molecules but require detailed knowledge of the potential energy surface and so are not practical for more complicated systems. A third approach, the transition state switching model, has been proposed by *Chesnavich* and *Bowers* [16]. This model, which was formulated for ionic systems, uses *Miller*'s unified statistical theory [17] as a starting point.

It is useful to consider how a tight transition state can arise on a potential surface with no activation barrier. On moving from reactants to products the local equilibrium flux along the reaction coordinate decreases as vibrational energy is converted to potential energy. However, as the products begin to separate vibrational modes are converted into rotational and translational degrees of freedom and the flux increases again. A transition state thus arises at the minimum in the flux that results from the competition between these two factors. This transition state will occur at energies $E \leq E_0$. In other words the transition state could lie at energies slightly below the dissociation threshold. In fact, since $F^{\ddagger}(E, J)$ is nearly always less than $F^{\text{orb}}(E, J)$, if the orbiting transition state is to control the rate close to threshold the tight transition state must be located slightly below the dissociation threshold. This has important implications for determining dissociation energies by modeling experimental data using RRKM theory. This approach will give the energy of the transition state which may be slightly smaller (by up to 0.2 eV) than the true dissociation energy.

In the previous section the predictions of RRKM theory were compared with experimental data to illustrate the application of the theory and demonstrate its reliability. Since it seems useful to continue this practice we will now consider an example of the application of phase space theory. As noted above phase space theory is appropriate to describe the dissociation of complexes formed in collisions. Thus this approach is ideal for describing association reactions or clustering reactions resulting in the growth of clusters [18]:

$$A^+ + B \rightarrow AB^+ \ . \tag{32}$$

These reactions are believed to occur by a two step process. First a metastable adduct is formed. This metastable adduct is then either stabilized by collisions or by radiative emission, or dissociates back to reagents. Thus we can write

$$A^+ + B \rightleftarrows AB^{+*} \tag{33}$$

$$AB^{+*} + X \rightarrow AB^+ + X \tag{34}$$

$$AB^{+*} \rightarrow AB^+ + h\nu \tag{35}$$

where X is either A, B or any other molecule. We will limit our discussion to ion-molecule reactions through the approach can be generalized to other systems. We will also only consider conditions in which the rate of collisional stabilization is much larger than the rate of radiative emission. Starting with the mechanism outlined above in Eqs. (33) and (34) we will make the reasonable assumption that forming the AB^{+*} metastable adduct from A^+ and B, and collisional stabilization both occur at the collision rate. Then the rate of formation of AB^+ only depends on the lifetime of the metastable adduct. If it survives long enough it will be collisionally stabilized. A steady state analysis for the mechanism discussed above yields a third order rate coefficient given by

$$k_3 = \frac{k_f k_s}{k_b + k_s[X]} \ , \tag{36}$$

where k_f is the rate constant for formation of AB^{+*}, k_s is the rate constant for stabilization, and k_b is the rate constant for unimolecular dissociation of AB^{+*}. The dissociation rate, k_b, is given by phase space theory as

$$k_b(E, J) = \frac{F^{\mathrm{orb}}(E, J)}{\rho(E, J)} \ , \tag{37}$$

where $F^{\mathrm{orb}}(E, J)$ is the flux through the orbiting transition state at energy E and angular momentum J, and $\rho(E, J)$ is the density of states in the AB^{+*} metastable adduct. In order to evaluate k_3 Eq. (37) is inserted into Eq. (36) and averaged over the E and J distributions of the metastable adduct. The result is

$$k_3 = \frac{\int e^{-E/k_BT} \int 2J\, F^{\mathrm{orb}}(E, J) \dfrac{k_f k_s}{k_b(E, J) + k_s[X]}\, dJ\, dE}{\int e^{-E/k_BT} \int 2J\, F^{\mathrm{orb}}(E, J)\, dJ\, dE} \ . \tag{38}$$

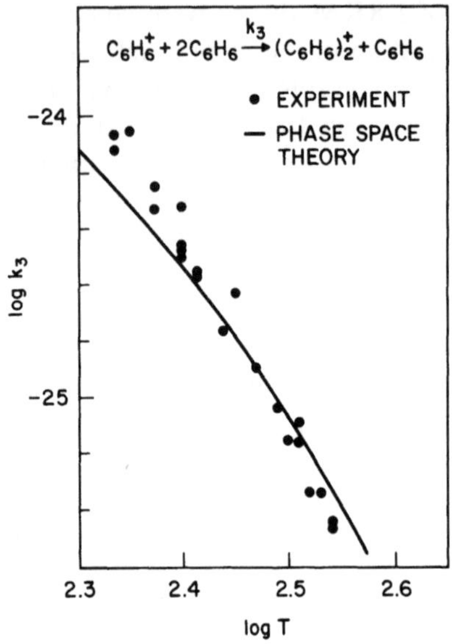

Fig. 5. Comparison of the predictions of phase space theory with experimental data for the association reaction of $C_6H_6^+$ with C_6H_6 (Adapted from [19])

Figure 5 shows a plot of log k_3 against log T for the clustering reaction of $C_6H_6^+$ with C_6H_6 to give $(C_6H_6)_2^+$ [19]. The points are experimental data and the line the result of phase space theory calculations. The experimental data was recorded by ion cyclotron resonance (ICR). There is clearly excellent agreement between experiment and phase space theory, and theory successfully accounts for the remarkably large temperature dependence of this reaction (k_3 is proportional to $T^{-6.3}$). As noted above phase space theory is formulated in terms of the reactants and products so in principal all the required parameters can be measured. In the case of the $C_6H_6^+/C_6H_6$ system discussed above, all the required parameters are well characterized except the frequencies of the dimer vibrations that arise from monomer transitions and rotations. However, estimates of these frequencies can be obtained from the ΔS for the reaction obtained from equilibrium studies.

2.7.5 Product Kinetic Energy Distributions

Product kinetic energy distributions are of interest for several reasons. They can provide information on the nature of the potential energy surface on which dissociation occurs, and can be used to estimate the excess energy (the energy above the dissociation threshold) possessed by the dissociating species. Furthermore, in a sequential evaporation process it is necessary to know how much energy is carried away as kinetic energy in order to determine whether further evaporation is possible. Product kinetic energy distributions are most easily measured for ionic systems were the presence of the charge not only permits

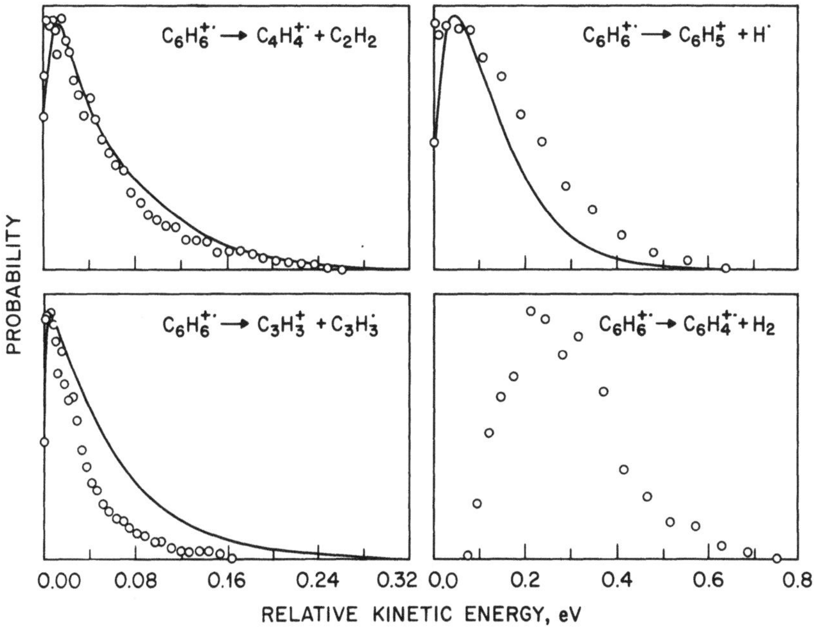

Fig. 6. Product kinetic energy distributions measured for the products from the dissociation of $C_6H_6^+$ ions. The points are the experimental data and the lines the result of phase space theory calculations. (Adapted from [8])

selection of a particular species but also allows energy analysis of the product ion.

Figure 6 shows product kinetic energy distributions measured for the dissociation of $C_6H_6^+$ into $C_6H_5^+$, $C_6H_4^+$, $C_4H_4^+$, and $C_3H_3^+$ [8]. Again, these systems are used as examples instead of clusters because they are well characterized and can provide a test of the theoretical procedures described here. The product kinetic energy distributions shown in Fig. 6 were recorded using mass analyzed ion kinetic energy spectrometry where a mass selected ion undergoes metastable fragmentation and the products are then energy analyzed. Product kinetic energy distributions can be divided into two groups. For the first group, represented by $C_6H_5^+$, $C_4H_4^+$, and $C_3H_3^+$ in Fig. 6, the product kinetic energy distributions peak at close to zero kinetic energy and then fall-off roughly exponentially with increasing kinetic energy. Kinetic energy distributions of this form are characteristic of dissociation on a potential energy surface without a significant reverse activation barrier. The second group, of which only $C_6H_4^+$ in Fig. 6 is a member, peak at kinetic energies considerably above zero kinetic energy. Kinetic energy distributions of this form suggest that there is a significant reverse activation barrier along the reaction coordinate (as shown in Fig. 1).

Several methods are available for calculating kinetic energy release distributions for reactions which occur on potential surfaces without a significant activation barrier. Simple classical equipartition arguments lead to an average kinetic energy of $k_B T^{\ddagger}$ where $T^{\ddagger}$ is defined by $(E - E_0)/[(3n - 3)k_B]$. However, measured average kinetic energies are generally larger than this value. It is straightforward to write down the relative kinetic energy distribution between the separating products at a tight transition state [20]

$$P(E; E_t^{\ddagger})dE_t = \frac{\rho(E - E_0 - E_t^{\ddagger})}{N(E - E_0)} dE_t \ . \tag{39}$$

But to relate this to the kinetic energy of the separated products, we must assume that energy redistribution abruptly stops at the transition state. Equation (39) also generally underestimates product kinetic energies.

On the otherhand, product state distributions can be evaluated from phase space theory without further assumptions. According to phase space theory the probability of products separating with kinetic energy E_t is given by [20, 21]

$$P(E, J; E_t)dE_t = \frac{\int \rho_v(E - E_0 - E_t - E_r) P_{ro}(E_t, E_r, J) dE_r}{\int \rho_v(E - E_o - E_{tr}) \Gamma_{ro}(E_{tr}, J) dE_{tr}} dE_t \ , \tag{40}$$

where the numerator is the flux through the orbiting transition state at kinetic energy E_t and the denominator is the total flux. $P_{ro}(E_t, E_r, J)$ is the density of rotational and orbital angular momentum states. This is obtained by integrating the density of rotational states, $P(E_r, J_r)$, over the L–J_r plane as described above for $\Gamma_{ro}(E_{tr}, J)$ except that the energetic boundaries are different. The maximum values of L and J_r for integration of $P(E_r, J_r)$ are given by [21]

$$L^* = (AE_t)^{(m-2)/2m} \tag{41}$$

$$J^* = (E_r/B_r)^{1/2} \ . \tag{42}$$

These angular momentum restrictions cause $P(E, J; E_t)$ to approach zero as E_t approaches zero, as observed experimentally (see Fig. 6).

As described above product kinetic energy distributions can be evaluated rigorously from phase space theory. However, this is only true if the rate of reaction is controlled by the orbiting transition state. As we have seen, accounting for dissociation rates for energies considerably above threshold usually requires a tight transition state. Thus in order to use phase space theory to calculate product kinetic energy distributions for dissociations controlled by a tight transition state it is necessary to assume that energy redistribution continues after passing through the tight transition state up to the orbiting transition state. Clearly one needs to be careful about using product kinetic energy distributions as a probe for determining whether a dissociation process occurs statistically.

Most measurements of product kinetic energy distributions sample dissociation within a particular time window. Thus in order to compare the predictions of phase space theory with experiment it is necessary to average $P(E, J; E_t)$ over the E and J distribution of the ions which dissociate on the experimental timescale. Thus

$$P_i(E_t) = \frac{\int dE_v P_v(E_v) \int dJ\, P_J(J)\, P_\tau(E_v, J) \dfrac{k_i(E_v, J)}{k_{\mathrm{tot}}(E_v, J)} P_i(E, J; E_t)}{\int dE_v P_v(E_v) \int dJ\, P_J(J) P_\tau(E_v, J) \dfrac{k_i(E_v, J)}{k_{\mathrm{tot}}(E_v, J)}}, \tag{43}$$

where $P_v(E_v)$ and $P_v(J)$ are the vibrational energy distribution and angular momentum distribution at $t = 0$, and $P_\tau(E_v, J)$ is the fraction which dissociates in the time window. The solid lines in Fig. 6 show the results of phase space theory calculations using the methods discussed above. The measured product kinetic energies for $C_4H_4^+$ are in good agreement with the phase space theory calculations, but for $C_6H_5^+$ phase space theory significantly underestimates the product kinetic energies, and for $C_3H_3^+$ the product kinetic energies are considerably smaller than the theoretical predictions. All of these three products are believed to arise from $C_6H_6^+$ without significant activation barriers, but there is not good quantitative agreement between experiment and theory. The discrepancies must arise from the influence of the details of the potential energy surface on which dissociation occurs, which is equivalent to saying that our assumption of energy redistribution up to the orbiting transition state does not appear to be valid in some cases.

No attempt have been made to compare the predictions of phase space theory with the measured kinetic energy distribution for $C_6H_4^+$. This is because it is clear from the shape of the experimental distribution that dissociation involves a significant activation barrier. Calculation of the product kinetic energy distributions under these circumstances is not straightforward, since it is now necessary to have information about the details of the potential energy surface and then investigate the dynamics on the surface. Calculations along these lines have been performed in some cases [22], but the theoretical approach is difficult to generalize and so will not be discussed further here.

As mentioned above Engelking [14] has proposed an approximate version of phase space theory using classical densities of states and ignoring angular momentum conservation. According to this model the probability kinetic energy release E_t is given by

$$P(E; E_t) = (s-2)(s-3)\, E_t \frac{(E - E_0 - E_t)^{s-4}}{(E - E_0)^{s-2}} \tag{44}$$

and the average kinetic energy release is given by

$$\langle E_t \rangle = \frac{2(E - E_0)}{s - 1}\,. \tag{45}$$

Equation (45) is quite similar to the empirical model of *Franklin* [23]. Franklin found that the average kinetic energy release for a number of reactions was approximately given by $(E - E_0)/0.44s$.

2.7.6 Evaporative Cooling

In the preceding sections expressions were obtained for the dissociation rates of energized clusters according to the two accepted formulations for unimolecular rate constants: RRKM and phase space theory. The derived expressions are microcanonical rate constants (they are a function of E and J), and in order to compare with experiment it is necessary to average over the E and J distributions. However, in many experiments the internal energy distribution is not well defined. Clusters are often generated with a large amount of internal energy and cool by evaporation with a rate roughly proportional to $1/t$, where t is the time since formation. *Klots* has coined the term "evaporative ensemble" to describe this situation [24]. Assuming that *all* clusters are generated at $t = 0$ with enough internal energy to evaporate several monomers, at time t the clusters will contain a distribution of internal energies that is bracketed by the lifetimes for evaporation. The upper limit on the distribution of internal energies for a cluster with n atoms, E'', is approximately defined by

$$k_n(E'') = 1/t \tag{46}$$

as clusters with $E > E''$ will have dissociated before time t. Since cluster n arises from cluster $n+1$ by evaporation the lower limit on the energy range is approximately defined by

$$k_{n+1}(E' + \Delta E) = 1/t \tag{47}$$

where $k_{n+1}(E)$ is the dissociation rate of cluster $n+1$ and ΔE is the energy lost from the cluster by evaporation. ΔE is given by $D_{n+1} + E_{KE}$ where D_{n+1} is the dissociation energy of cluster $n+1$ and E_{KE} is the relative kinetic energy of the products. The width of the energy distributions of cluster n, $E'' - E'$, is approximately ΔE. Thus it is possible to roughly characterize the internal energy distribution of the evaporating clusters. However, it is not possible to rigorously define the internal energy distribution without knowing both the cluster size distribution at $t = 0$ and the internal energy distributions of the clusters at $t = 0$. Assuming that the clusters undergo only one evaporation the probability that cluster n contains internal energy E at time t can be written as [24]

$$P_n(E, t) \approx P^0_{n+1}(E + \Delta E) \int_0^t \exp[-k_{n+1}(E + \Delta E)t'] \times k_{n+1}(E + \Delta E) \exp[-k_n(E)(t - t')]\,dt' \tag{48}$$

$$= P^0_{n+1} \frac{k_{n+1}(E + \Delta E)}{k_{n+1}(E + \Delta E) + k_n(E)} (\exp[-k_n(E)t] - \exp[-k_{n+1}(E + \Delta E)t]) \tag{49}$$

where $P^0_{n+1}(E + \Delta E)$ is the probability that the $n+1$ cluster has internal energy $E + \Delta E$ at $t = 0$. If this factor is ignored Eq. (49) defines an energy distribution which is roughly trapezoid in shape with upper and lower limits approximately as defined above by Eqs. (46) and (47).

If cluster n is now isolated after time t and allowed to undergo further metastable evaporation the relative abundance at time $t + t_x$ is

$$P_n(t_x) \approx \int (\exp[-k_n(E)t] - \exp[-k_{n+1}(E + \Delta E)t]) \times \exp[-k_n(E)t_x]dt_x \ . \tag{50}$$

Klots [24] has shown that this expression can be approximated as

$$P_n(t_x) \approx 1 - (C/\gamma^2) \ln\left(\frac{t}{t + (t_x - t)\exp[-\gamma^2/C]}\right) , \tag{51}$$

where C is the heat capacity of the cluster (in units of k_B) and γ is given by

$$\gamma = \frac{\Delta E}{k_B(T'T'')^{1/2}} \ . \tag{52}$$

T' and T'' are the temperatures before and after evaporation. This model is only appropriate for large clusters where $\Delta E \gg E$.

Figure 7 shows a comparison of the predictions of this model with experimental results for the metastable dissociation of Xe cluster ions. The heat capacity has been adjusted to fit the experimental data. The fraction of clusters which dissociate within the time window increases with cluster size. According to Eq. (51) this increase is due to the increase in the heat capacity of the cluster with cluster size. A more meaningful explanation of this behavior is that with increasing cluster size $k(E)$ increases less rapidly with increasing energy, so a larger fraction of the range of internal energies populated in the clusters (which is $\sim \Delta E$ wide) can dissociate within the available time.

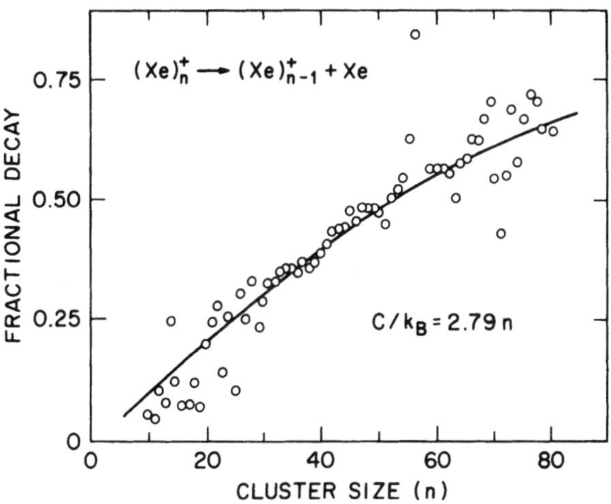

Fig. 7. Comparison of the measured fractional decay of Xe cluster ions with the predictions of the model of Kolts [24]. (Adapted from [24])

2.7.7 Determining Cluster Dissociation Energies

By far the most important application of the statistical theories outlined in this chapter is in the determination of cluster dissociation energies. As noted in the Introduction, there is a problem with the concept of determining dissociation energies by exciting a cluster with a known amount of internal energy and noting whether it dissociates. If the cluster is excited to just above its dissociation threshold, dissociation will only occur promptly for clusters with a few atoms. Figure 8 shows lifetimes predicted by RRKM theory for the dissociation of aluminum clusters with a range of sizes [25]. If the excited clusters do not dissociate relatively quickly infrared emission will occur and cool the clusters before they dissociate. Lifetimes for infrared emission from clusters are not known, but based on what is known for other molecules infrared emission could become a problem if the lifetime towards dissociation is $> 10^{-3}$ s. As can be seen from Fig. 8, in order to achieve dissociation lifetimes $< 10^{-4}$ s it is necessary to excite the larger clusters to considerably above their dissociation thresholds. The phenomena of requiring extra energy above the dissociation threshold in order for a dissociation process to occur within a limited timescale has been known for many years in connection with appearance energy measurements in mass spectrometry. In order to obtain dissociation energies from the energy required to cause dissociation on a particular timescale it is necessary to account for the excess energy using statistical theories. However, while there is little concern about the accuracy and suitability of the statistical approach, as

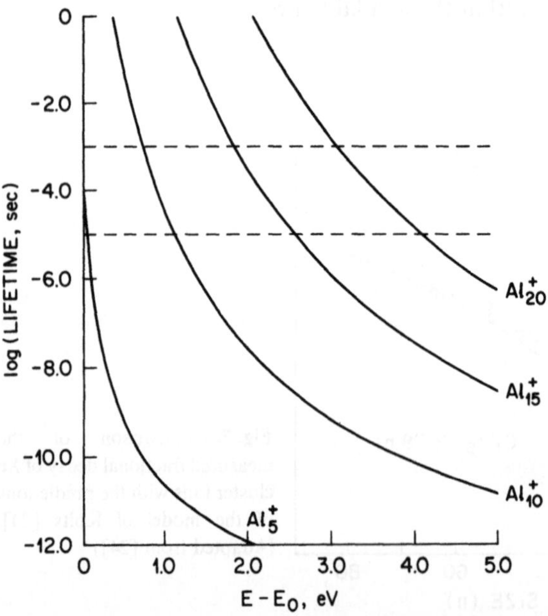

Fig. 8. Lifetimes predicted by RRKM theory for the dissociation of aluminum cluster ions of various sizes. (Adapted from [25])

will be discussed in the following section, most of the input data required to perform statistical calculations on clusters is not readily available and there remain some unresolved issues (for example, the role of excited electronic states). Thus statistical models are required to determine cluster dissociation energies, but the models cannot be rigorously tested without prior knowledge of the dissociation energy.

With these problems in mind, probably the most reliable way to determine cluster dissociation energies is to measure dissociation lifetimes over a broad energy range and then fit the lifetimes using an appropriate statistical model. The experimental measurements can be performed by photoexcitation of cold clusters [25]. The problems with measurements of this type are complications arising from multiphoton processes and the availability of light sources to scan a sufficiently high energy range. An alternative approach is to use collision induced dissociation (CID) [26]. It is relatively easy to control the collision energies of ions, however, in a collision a broad range of energies up to the collision energy are transferred into internal energy, and so measurements of the lifetimes are not possible. In CID experiments the cross sections for dissociation must be recorded as a function of collision energy and then fit with a threshold model which takes into account the finite lifetimes of the clusters.

It is possible to obtain some crude information on the dissociation energies of clusters from experiments where measurements are not performed as a function of energy by exploiting the properties of the "evaporative ensemble" outlined above. Assuming that the clusters undergoing metastable evaporation have a range of internal energies given approximately by ΔE, the energy above the dissociation threshold can be estimated from the kinetic energy of the fragments [14, 27]. Knowing the excess energy and the timescale for dissociation it is then possible to estimate the dissociation energy. The accuracy of this approach hinges on the ability to accurately predict the fraction of the available energy which is partitioned into kinetic energy. As we have seen our ability to do this is not guaranteed.

Even without measurements of the product kinetic energies it is possible to obtain crude information on the cluster dissociation energies from the fraction of clusters which undergo metastable evaporation in a particular time frame [28].

2.7.8 Problems Associated with the Application of Statistical Theories to Clusters

Often most of the information required to apply RRKM theory or phase space theory to the dissociation of clusters is not available, and must be estimated. Rotational constants can be estimated by treating the clusters as spheres and using the bulk density. In some cases vibrational frequencies for the stable cluster are available from ab initio calculations. But even when this information is available it is limited to small clusters. In cases where no information is available vibrational frequencies can be estimated from a simple Debye heat capacity model:

$$\nu_i = \left(\frac{i - 1/2}{3n - 6} \nu_{Dn}^3\right)^{1/3}, \tag{53}$$

where ν_{Dn} is the Debye frequency of cluster n which is obtained from an extrapolation between bulk Debye frequency and the dimer frequency.

The simplest reaction that can occur for a cluster is loss of an atom. We can relate the reaction coordinate for this reaction to motion of an atom away from the cluster surface. As a vibrational degree of freedom this coordinate would be viewed as vibration of the atom with respect to the cluster surface. This vibrational degree of freedom vanishes at the transition state (it is converted into a translational degree of freedom). In addition, two other vibrational degrees of freedom (transitional modes) vanish on going from reactants to products. These transitional modes can be related to wagging vibrations of the departing atom perpendicular to the stretch. The reaction coordinate and two transitional modes ultimately become the three translational degrees of freedom of the departing atom. Since we have no experimental information on the vibrational frequencies of the transition state the best approach is to use the same frequencies as for the stable cluster, and to remove the middle frequency as the reaction coordinate and to take its two neighbors to be the transitional modes and reduce them by a factor of two. For a medium-sized cluster the choice of the reaction coordinate and the transitional modes is not very important because they comprise only a small fraction of the total number of modes in the cluster. Another type of dissociation which frequently occurs with clusters is ejection of a dimer. For this reaction six modes change significantly on going from reactants to products, since the dimer has one vibrational and two rotational degrees of freedom in addition to three translational degrees of freedom. Thus the frequencies for the transition state for dimer ejection are expected to be slightly lower, on average, than the frequencies for atom ejection. This has important consequences. If the dissociation thresholds for atom ejection and dimer ejection are the same, then for energies significantly above threshold the sum of states for dimer ejection will be larger than the sum of states for atom ejection. Thus the main product observed will be the dimer, and one might be led to the incorrect conclusion that loss of a dimer is energetically favored. It has been argued that for the reaction path degeneracy, σ, either n or the number of surface atoms should be employed. However, not all surface atoms are equivalent and it is appropriate to use a reaction path degeneracy less than the number of surface atoms (for example, $n^{0.66}$). For a large cluster one can imagine that the atoms will evaporate preferentially from the edges rather than the facets. One parameter, the zero point energy difference between the stable cluster and the transition state, remains to be defined. As noted above cluster dissociation energies are generally not known, and RRKM theory is often used to model experimental data in order to deduce the dissociation energies.

We have not yet considered the effects of isomers and excited electronic states. Theoretical calculations suggest that structural isomers are likely to be quite common for clusters. Clusters are also expected to have a significant number of low lying excited electronic states. For example, it has been estimated that within the first 10000 cm^{-1} of the ground state separated atom energy, over 2500 molecular states arise for V_2, and nearly 1000 for Ni_2 [29]. The enormous electronic state densities of these dimers arise because they have open *d*-shells. Analogous calculations for C_2 yield only 50 states. The presence of excited electronic states poses no fundamental problems to RRKM theory if the assumptions that all states are equally likely to be populated and rapid energy redistribution among the states are not violated. Then it is only necessary to sum over the different electronic states to obtain the total sums and densities of states. Since vibrational sums and densities of states increase so rapidly with energy it is only necessary to consider the low lying excited electronic states. If the excited electronic states do not interconvert rapidly the problem becomes intractable as high quality theoretical calculations are required to determine the interconversion rates. Little is known about the electronic structure of metal clusters. In the absence of detailed information, however, it is a good approximation to assume that the contributions from excited electronic states to the stable species and the transition state cancel. Structural isomers can be treated in an analogous manner. If isomerization is more rapid than dissociation and the energies of the isomers known, their contribution to the sums and densities of states can be included. Again only the low lying isomers need to be considered. If isomerization is slow compared to dissociation it is necessary to include a transition state for isomerization since this is now the rate determining step. For clusters, in the absence of detailed information, the best approach is to assume that the contributions from structural isomers to the stable species and the transition state cancel.

2.7.9 Summary

In this chapter the two accepted methods for calculating the dissociation rates of energized molecules or clusters have been described, and the appropriate conditions for the use of the different methods outlined. Statistical methods for evaluating product kinetic energy distributions have also been described and the limitations of the methods pointed out. In many cases the internal energy distributions of dissociating clusters is not well known and the concept of the "evaporative ensemble" was described and shown to provide a rough bracketing of cluster internal energies, assuming that *all* clusters are generated with sufficient internal energy to rapidly evaporate at least on monomer.

By far the most important application (in cluster research) of the statistical theories outlined in this chapter is in the determination of cluster dissociation energies. The relative merits of the various approaches to determining cluster dissociation energies were briefly reviewed. Finally problems with obtaining and estimating the input data required for RRKM theory or phase space theory calculations on clusters were discussed and the potential uncertainties introduced by the high electronic state densities and structural isomers were mentioned.

References

1. P.J. Robinson, K.A. Holbrook: *Unimolecular Reactions* (Wiley, London, 1972); W. Forst, *Theory of Unimolecular Reactions* (Academic, New York, 1973); W.L. Hase, in *Dynamics of Molecular Collisions* (Part B), ed. by W.H. Miller (Plenum, New York, 1976); J.H. Beynon, J.R. Gilbert: *Application of Transition State Theory to Unimolecular Reactions: An Introduction* (Wiley, Chichester, 1984)
2. O.K. Rice, H.C. Ramsperger: J. Amer. Chem. Soc. **49**, 1617 (1927)
3. L.S. Kassel: J. Phys. Chem. **32**, 225 (1928)
4. L.S. Kassel: J. Phys. Chem. **32**, 1065 (1928)
5. R.A. Marcus, O.K. Rice: J. Phys. Colloid Chem. **55**, 894 (1951); R.A. Marcus: J. Chem. Phys. **20**, 359 (1952)
6. B.H. Mahan: J. Chem. Educ. **51**, 709 (1974); P. Pechukas: Ann. Rev. Phys. Chem. **32**, 159 (1981)
7. S.E. Stein, B.S. Rabinovitch: J. Chem. Phys. **58**, 2438 (1973); M.J.H. Kemper, J.M.F. van Dijk, H.M. Buck: Chem. Phys. Lett. **53**, 121 (1978)
8. M.F. Jarrold, W. Wagner-Redeker, A.J. Illies, N.J. Kirchner, M.T. Bowers: Int. J. Mass Spectrom. Ion Proc. **58**, 63 (1984)
9. V. Weisskopf: Phys. Rev. **52**, 295 (1937)
10. E. Femi: Prog. Theor. Phys. **5**, 570 (1950)
11. J.C. Light: Disc. Faraday Soc. **44**, 14 (1967)
12. C.E. Klots: J. Phys. Chem. **75**, 1526 (1971)
13. W.J. Chesnavich, M.T. Bowers: J. Chem. Phys. **66**, 2306 (1977); W.J. Chesnavich, M.T. Bowers: J. Amer. Chem. Soc. **99**, 1705 (1977)
14. P.C. Engelking: J. Chem. Phys. **85**, 3103 (1986); P.C. Engelking: J. Chem. Phys. **87**, 936 (1987)
15. D.G. Truhler, A.D. Isaacson, B.C. Garrett: in *Theory of Chemical Reaction Dynamics*, ed. by M. Baer (Chemical Rubber, Boca Raton, 1985)
16. W.J. Chesnavich, L. Bass, T. Su, M.T. Bowers: J. Chem. Phys. **74**, 2228 (1981); W.J. Chesnavich, M.T. Bowers: Prog. Reac. Kinet. **11**, 139 (1982)
17. W.H. Miller: J. Chem. Phys. **65**, 2216 (1976)
18. L.M. Bass, W.J. Chesnavich, M.T. Bowers: J. Amer. Chem. Soc. **101**, 5493 (1979); L.M. Bass, P.R. Kemper, V.G. Anicich, M.T. Bowers: J. Amer Chem. Soc. **103**, 5283 (1981)
19. S. Liu, M.F. Jarrold, M.T. Bowers: J. Phys. Chem. **89**, 3127 (1985)
20. C.E. Klots: J. Chem. Phys. **41**, 117 (1964); C.E. Klots: J. Chem. Phys. **64**, 4269 (1976)
21. W.J. Chesnavich, M.T. Bowers: J. Amer. Chem. Soc. **98**, 8301 (1976)
22. J.D. McDonald, R.A. Marcus: J. Chem. Phys. **65**, 2180 (1976); G. Worry, R.A. Marcus: J. Chem. Phys. **67**, 1636 (1977)
23. M.A. Haney, J.L. Franklin: J. Chem. Phys. **48**, 4093 (1969)
24. C.E. Klots: J. Chem. Phys. **83**, 5854 (1985); C.E. Klots: Nature **327**, 222 (1987); C.E. Klots: J. Phys. Chem. **92**, 5864 (1988); C.E. Kolts: Z. Phys. D **5**, 83 (1987)
25. U. Ray, M.F. Jarrold, J.E. Bower, J.S. Kraus: J. Chem. Phys. **91**, 2912 (1989)
26. S.K. Loh, D.A. Hales, L. Lian, P.B. Armentrout: J. Chem. Phys. **90**, 5466 (1989)
27. P.P. Radi, M.E. Rincon, M.T. Hsu, J. Brodbelt-Lustig, M.T. Bowers: J. Phys. Chem. **93**, 6187 (1989)
28. C. Brechignac, Ph. Cahuzac, J. Leygnier, J. Weiner: J. Chem. Phys. **90**, 1493 (1989)
29. E.M. Spain, M.D. Morse: Int. J. Mass Spectrom. Ion Proc. **102**, 183 (1990)

2.8 Melting and Freezing of Clusters: How They Happen and What They Mean

R.S. Berry

2.8.1 Introduction: The "Phases" of Clusters

Clusters, of even as few as seven atoms, may exhibit characteristics we associate with distinct, solid-like and liquid-like forms of matter [1]. Understanding how these forms behave at a microscopic level, how we identify them and characterize them, and what we can learn from them about phase changes more generally are the topics of this chapter.

The first issue to address is what "liquid" and "solid" should mean in the context of clusters. What is a liquid-like cluster or a solid-like cluster? What observable or calculable properties characterize liquid-like and solid-like forms? We take a first clue from the *compliance* of bulk liquids and the contrasting stiffness of solids. Bulk liquids, inelastic as they are, respond to even very small forces, deforming and conforming as the forces demand. Solids are, of course, stiff and elastic up to fairly high limits, beyond which they deform permanently or fracture. The compliance of liquids implies that they have some "soft modes" of motion, that is, some vibration-like modes whose natural frequencies are very low compared with those of typical lattice vibrations, e.g. lower than $10^{12}\ s^{-1}$ [2, 3]. This is not to say that liquids have no high-frequency modes; indeed, all the spectral evidence indicates that liquids have many modes with frequencies comparable to those of solids, but in addition have some of much lower frequencies. The characteristic deformability of liquids also implies that liquids can change their geometric forms easily. Translating both of these to apply to the world of small clusters, we can expect the vibrational spectrum of a liquid cluster to have a significant density of states at frequencies well below typical frequencies of solid-like clusters, and to be capable of passing readily – on the time scale of the periods of the soft modes – from the well around one minimum on the cluster's potential surface to the wells around other minima. In other words, the liquid cluster should have some slow, large-amplitude modes of motion so that it can relatively readily rearrange, isomerize and explore its potential surface.

Another kind of structural characteristic that should distinguish solid clusters from liquids are their radial and angular correlation functions. Solids, specifically crystals, show peaks in radial (pair) correlation functions corresponding to the mean nearest-neighbor distance and to successive second-nearest and further-removed neighbors. Liquids show broader nearest-neighbor peaks and rarely any well-defined peaks corresponding to more distant atoms or molecules. The same criterion should distinguish solid from liquid clusters, with the one ramification that clusters may have polyhedral as well as lattice-based geometries and still be solid-like [4–6]. The angular correlation function, being a three-body property rather than a two-body, is less frequently used than the radial or pair function [7, 8]. However it too is a useful diagnostic. Not only are maxima at particular angles sometimes helpful in identifying a geometry or ruling one out; minima, corresponding to nonappearance of certain angles, are strong indicators of solid-like behavior and in many cases of one structure rather than another. The appearance of a low minimum at 90° in the angular distribution functions of cold Ar_{13} and Ar_{55} indicate solid-like icosahedral structures; the face-centered cubic (fcc) close-packed geometry is ruled out because it has successive nearest neighbors forming 90° angles. More energetic Ar_{13} and Ar_{55} clusters have angular distribution functions with significant probabilities at 90° (but still this is a minimum) indicating some kind of breakdown of structural rigidity.

The adjacency matrix, showing which atoms are next to which others, has been used to distinguish stiff solids, soft solids and liquids [9]. Another related characteristic which naive intuition suggests might be an important property to distinguish liquids from solids is the ease with which identical atoms or molecules can permute positions with one another [10]. Indeed, the establishment of permutational equivalence of identical particles is a *sufficient* condition for liquid-like behavior. However the time scale required to achieve permutational equivalence may be far longer than is relevant for experimental or computational tests of "phase", so we should not require attainment of permutational equivalence as a necessary condition for liquid-like behavior. However we may wish at least to demonstrate that feasible paths are open to establish that equivalence. Alternatively, if only the atoms of certain sites of a cluster seem able to permute with one another, we may wish to denote such a situation as "surface melting" or some other appropriate, restricted behavior.

Clusters may but need not exhibit clear solid-like or liquid-like behavior [1]. They may, instead, in a range of temperature or energy that is in some sense "intermediate", be slush-like, with properties that fall between those of solids and liquids. Clusters of almost all substances are solid-like at low energies or temperatures and clusters of many substances are liquid-like at higher energies or temperatures. Between, a cluster may be slush-like or may exhibit clearly defined, observable, coexisting solid and liquid forms, like chemical isomers [11]. Both kinds of behavior seem to occur; at issue is what factors determine which kind of behavior a given cluster follows. Another issue is whether the solid or especially the more energized liquid can be relatively stable toward evaporation, at least enough so to be observed in experiments or simulations.

At present, our ideas concerning the "phases" of clusters come from theory and simulation. Important experimental results consistent with the theory and simulations have been obtained but as yet they cannot be called definitive or unambiguous demonstrations of the theory-based ideas. Historically, the first indications of distinguishable solid-like and especially liquid-like forms were obtained from simulations [3, 12–17]. The general theory of the thermodynamics of small systems has been well established some time ago [18] but the specifics that explain how real systems behave, and how their thermodynamic and dynamic behavior are related to each other and to the potential surface and energy levels of the cluster, have emerged more recently, stimulated in part by the simulations. Then, in part because of the new challenges stirred by theoretical advances, many more simulations have now been done. The picture is still far from complete but many of the main aspects of the subject now seem clear. Here, we describe the theory and the simulations, and what we learn from them of the freezing and melting of clusters, and of the implications of the behavior of clusters for phase stability and metastability of bulk matter.

In the next section we develop the theory from the viewpoint of its logical structure insofar as that can now be done, rather than from its historical evolution. We begin with a few comments regarding that evolution in order to illuminate some of the conceptual and technical difficulties. The third section addresses the use of simulations for studying clusters and particularly for studying their phase behavior. The final section treats the implications for phase equilibrium of bulk matter.

2.8.2 Theoretical Basis

The theory of phase equilibrium of clusters is really a balancing act between thermodynamics and dynamics, in which time scales become an overarching concern [19]. The very meaning of "equilibrium" comes under scrutiny in a way one very, very rarely encounters with conventional phase or chemical equilibrium. The reason is that the time scales for clusters to attain dynamic equilibrium are sometimes just the time scales characteristic of some of the experiments best suited for studying clusters, so that "what you get is what you look for." But before we explore the ramifications of time scales, let us examine the problem in what seem like static terms.

We begin our discourse on the thermodynamics of clusters with a description in terms of the Helmholtz free energy of a cluster in a canonical (isothermal) ensemble of clusters, all of a specific size, $F_N(T)$. This description is appropriate for the clusters of N particles, which we shall call N-clusters, under conditions of some but certainly not all experiments. That is, in some experiments, the size distribution becomes constant – "frozen in" – but the clusters are still in thermal equilibrium with a heat bath of surrounding atoms. This may occur with jets of carrier gas seeded with the species of interest, producing molecular beams of the clusters. By contrast, in jet experiments with little or no carrier, the clusters may, as a result of evaporative cooling without collisions, attain a distribution of energies far from a thermal distribution [20]. A canonical distribution is hardly appropriate for such a system; instead, one must describe the system in terms of an "evaporative ensemble". At an opposite extreme, in a static cell clusters of each size may not only approach thermal equilibrium distributions of energy; the entire sample may approach an equilibrium distribution of cluster sizes, so that the description of clusters of any given size must be described by a grand canonical ensemble, in which both energy and mass may be exchanged with the surroundings [21]. But with any of these descriptions, we can assume that the internal vibrational modes of the individual clusters are coupled strongly enough that we can define a mean internal temperature based on the mean kinetic energy of the atoms within the cluster. Moreover the reasoning we present can be extended readily, at least to evaporative ensembles and, with a little modification to allow for evaporation and condensation, to grand canonical ensembles.

Now we return to the Helmholtz free energy and to its connection to the quantum statistics of a system. The canonical partition function, $Q(T) = \sum_j g_j e^{-E_j/kT}$, is related to $F_N(T)$: $Q(T) = e^{-F_N(T)/kT}$. Furthermore all the other thermodynamic functions can be derived from $Q(T)$ by taking suitable derivatives. In other words, knowledge of the canonical partition function is sufficient to tell us all the equilibrium properties of the system of fixed mass number N at temperature T. For example if we know the vibrational and rotational energy levels E_j of a solid-like N-cluster and the degeneracies g_j of those levels, we can compute the free energy of that cluster.

In fact, if we have a reasonable approximation for those vibrational and rotational energies and degeneracies, we can compute rather reliable free energies and other thermodynamic properties simply because the thermodynamic properties are essentially weighted *average* values of negative exponentials of the energies, in units of the average kinetic energy. That is, we do better in estimating thermodynamic quantities than in estimating the individual energy levels and degeneracies, provided we represent their general pattern correctly.

But we are interested not in the solid alone or the liquid alone but in the comparison of the stabilities of the two. To make this comparison we must evaluate free energies for both forms. This means we need models for both, from which their free energies can be estimated. It looks at this point like two simple models are all we need. This would be correct if we could assume that both the solid and liquid forms are stable at whatever temperature interests us. This is a terribly strong assumption, too strong to be acceptable, although making it, one can go on to rationalize the results of the early simulations from rather simple quantum statistics [22]. Instead of making that assumption, we can take a much sounder step by asking, "Within a context consistent with our (presumed) models for solid and liquid clusters, what are necessary and sufficient conditions for the solid and liquid to both be stable?" This is a very productive question because it leads us to new physical insight [23].

The stability of a form of matter can always be expressed in terms of the existence of a minimum in a thermodynamic potential, such as a free energy, suitable for the constraints on the system of interest. This is an easy statement to make, but it avoids a vital but subtle problem: with respect to what quantity is the thermodynamic potential a minimum? In the present case, we suppose that the system may be characterized by a parameter indicating the degree of nonrigidity of the system, a parameter that acts much like an order parameter in the Landau theory of second-order phase transitions. At least two ways have been used to define this nonrigidity parameter, one a phenomenological quantity, the ratio of two spectroscopic frequencies and the other, a quantity based in the microscopic structure of the substance, the density of defects in the material when it is quenched to the geometry of lowest energy in its current potential well. The former, the ratio of the energy of the lowest rotational transition to the lowest transition with no rotational excitation, is a slight generalization of a parameter that had been used previously to characterize the degree of nonrigidity in triatomic molecules [24]. The latter is an adaptation of a quantity central to the Stillinger–Weber version of the defect model for liquid structure and melting [25]. Used in our context, both give the same result when both yield results, and the microscopic parameter leads to additional results [26].

Both ways of defining the extent of nonrigidity can be associated with ideal, extreme models at the rigid, solid-like and nonrigid, liquid-like ends of a scale. Calling γ the nonrigidity parameter and setting $\gamma = 0$ at the rigid end of the scale and $\gamma = 1$ at the nonrigid end (so the density of defects is defined on a relative scale), we construct the patterns of energy levels and degeneracies, i.e. the densities of states, for the two limiting cases. By introducing this parameter, we make the energies E_j, the partition functions and the free energies into functions of γ. Explicitly, we now write $F(T, \gamma)$ for the free energy; T is a physical variable, and γ is a parameter that allows us to tune the extent of nonrigidity of the system, in principle by varying the Hamiltonian in a suitable, continuous manner. We shall use this dependence of F on γ shortly.

Using the phenomenological parameterization, it is natural to choose for the solid either a very general, phenomenological model such as the Einstein or Debye crystal whose density of states is well known and derivable from the model Hamiltonian, or a reasonably realistic Hamiltonian such as a harmonic model based on diagonalizing the harmonic Hamiltonian representing small-amplitude oscillations around an assumed equilibrium structure. The Einstein model, with all the vibrational frequencies the same and independent of cluster size, is too crude to represent the solid-like clusters adequately, but almost any more refined model, even an Einstein model with a size-dependent single frequency for the lattice vibrations, seems adequate for describing the qualitative aspects of the phase equilibrium of clusters [22].

A suitable corresponding choice of a phenomenological model for the liquid is the *Gartenhaus–Schwartz* model [27], in which the interactions are identical, harmonic attractive forces between every pair of particles. This model, developed for nuclei, leads to a spectrum of equally spaced levels with the degeneracies of the totally symmetric representations of the unitary group $SU(3N - 3)$ for a cluster of N identical particles. This model supposes that there is so much empty space in the cluster that all the significant encounters between particles occur outside the radius of any hard-core repulsion, yet close enough that the attractions are strong. It is not particularly realistic, but that is irrelevant because we need not require the free energy to have its liquid-like minimum at the extreme limit.

The microscopic model based on the density of defects yields results for the solid essentially identical to those of the accurate derivation described above because they really describe the same situation [26]. The defect model for the liquid introduces, in addition to the solid-like modes, a configurational entropy and a set of energy terms in the Hamiltonian. The precise form of these terms depends on one's choice of detailed model. For example one may assume that the defects are independent of one another and of the vibrational modes of the host, or that the defects interact with each other or with the host modes, presumably to lower their frequencies.

For both kinds of model, the outcome is similar: the density of states of the solid-like cluster is the lower at the low end of the energy scale, but at higher energies, the density of states of the liquid-like cluster becomes the larger. Consequently if we connect the limiting cases, rigid and nonrigid, in a correlation diagram, all the energy levels at the low end of the energy scale slope upward from the rigid limit to the nonrigid, but the energy levels high on the energy scale must slope downward from the rigid limit to the nonrigid [28]. This is because every state appearing at one limit must also appear at the other, and, apart from avoided crossings forced by Ehrenfest's adiabatic theorem, the connections can be made in order from bottom up. An example of such a correlation diagram is shown in Fig. 1, for a cluster of five argon atoms [22]. The vertical scale is exaggerated by a factor of about 5 for the rotational level spacings of the rigid limit in order to make them visible.

Now it is time to follow the implications of the energy level patterns for the free energy [23]. At low temperatures, with only the low-energy levels populated, the free energy of the solid cluster is lower than that of the corresponding liquid; in fact, at low enough temperatures, because of the upward slope of the low-lying energy levels, the free energy must be a monotonic, increasing function of γ. However as the temperature increases, usually lowering the free energy of solid, liquid and everything between, the free energy near the nonrigid end of the scale decreases slower with T than that near the rigid end, because of the larger contribution to $T\Delta S$ of the many available levels of the nonrigid form. This makes the curve of $F(T, \gamma)$, for T above the lowest range, droop at the nonrigid end of the scale of γ. At some temperature, $F(T, \gamma)$ develops a flat spot at or near the nonrigid limit, a point of

$$\frac{\partial F(T, \gamma)}{\partial \gamma} = 0 \; ;$$

we call this temperature T_f, the "freezing temperature", because the solid is the only thermodynamically stable form at temperatures below T_f. Above this temperature, $F(T, \gamma)$ has two minima, one near the rigid end of the scale of γ and another near the nonrigid end. Each minimum corresponds to a locally stable

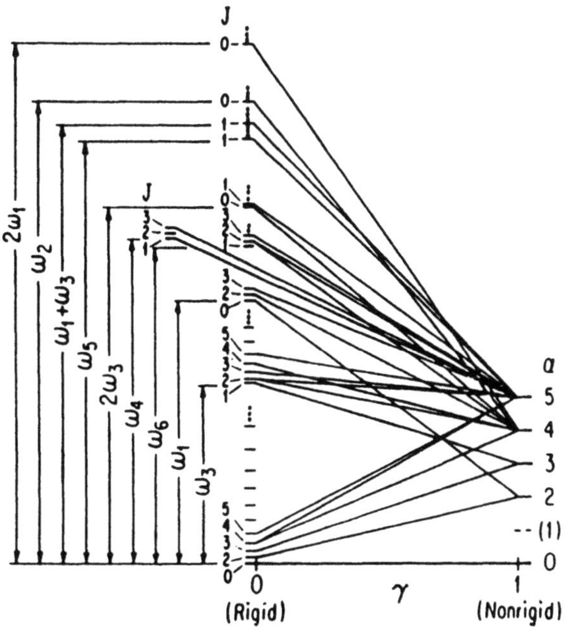

Fig. 1. A correlation diagram for the rotation-vibration levels of Ar_5 in which the horizontal scale is the degree of nonrigidity, γ. No promotional energy is assumed necessary to achieve the liquid-like form, which is a far stronger assumption than necessary. Nevertheless the qualitative form of the diagram, with upward-sloping levels at the bottom of the scale and downward-sloping levels at higher energies, carries the essentials for describing the phase equilibrium of solid and liquid clusters

form, analogous to chemical isomers. If a collection of N-clusters is in thermal equilibrium, the solid and liquid forms are both present at any temperature for which $F(T, \gamma)$ has two minima on the γ scale, and the ratio

$$\frac{(\text{liquid})}{(\text{solid})} \equiv K_{\text{eq}} = e^{(-\Delta F/kT)} ,$$

where $\Delta F = F_{\text{liq}} - F_{\text{sol}}$. The two forms coexist in a dynamic equilibrium; each cluster is either solid-like or liquid-like at any instant, but over time, passes back and forth between the two forms. At a temperature we can call T_{eq}, the free energies of the two forms are equal and the equilibrium ratio is unity. At temperatures below this but above T_f, the solid predominates. At temperatures above T_{eq} the liquid predominates. In fact, as the temperature increases, the curve of $F(T, \gamma)$ tips more and more toward the nonrigid end, until a temperature T_m is reached which has

$$\frac{\partial F(T, \gamma)}{\partial \gamma} = 0$$

near or at the rigid limit. Above T_m, only the liquid-like form is stable, and the

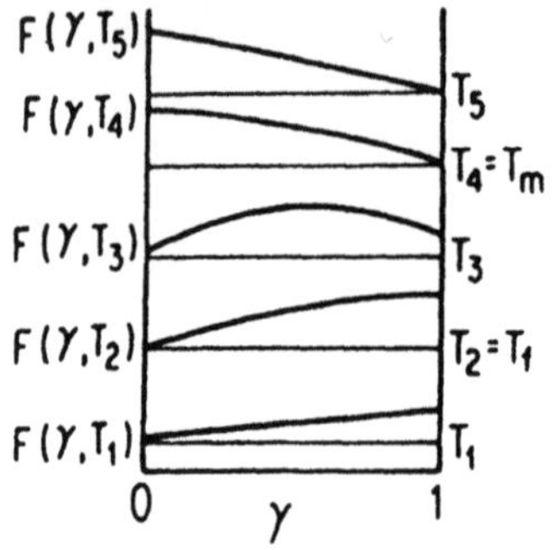

Fig. 2. A schematic representation of the free energy $F(T, \gamma)$ as a function of the nonrigidity parameter γ for several temperatures, from very low to high enough that the only stable form is liquid

curve of $F(T, \gamma)$ is monotonic and decreasing with γ. Figure 2 shows the free energy as a function of γ for several different temperatures including T_f and T_m.

The picture that has emerged here is one in which clusters may have sharp freezing temperatures, below which only their solid forms are stable, and sharp melting temperatures, above which only their liquid forms are stable, but the *freezing temperature and melting temperature are not the same*! Moreover this description has separated the freezing and melting temperatures logically; they need not be the same, when expressed in these terms. The equilibrium constant has two discontinuities, at the limits of thermodynamic stability of each "phase".

This description would be adequate if a) all clusters had distinct liquid-like and solid-like forms and b) all clusters spent long enough intervals in their solid-like and liquid-like forms to establish well-defined, observable equilibrium properties characteristic of each phase. In reality, only some clusters meet both these conditions [11, 29]. Some clusters, such as Ar_7, Ar_{13}, Ar_{15} and Ar_{19}, do meet both conditions. Others, such as Ar_8 and Ar_{14}, have potential surfaces with wells forming heavily-worn staircases, successions of wells with not very high barriers between one and the next, so these clusters do not have well-defined solid-like and liquid-like regions of their potential surfaces. Others, possibly Ar_{17}, pass back and forth between "phases" too rapidly to exhibit well-defined solid and liquid forms and, instead, display average properties that make such clusters seem slush-like. Figure 3 illustrates three kinds of potential surface, of which only the first kind gives rise to the phase equilibrium described above.

The matter of time scale is crucial here [19]: does the cluster pass from one "phase" to another at a rate fast or slow relative to the time required to establish well-defined properties of a single phase, and is this time long or short compared with the time we require to observe the "phase"? We cannot answer this question from thermodynamic arguments; we must turn to dynamics to address it. But notice that we have made a subtle twist to what we are calling "equilibrium". If the observation time is long relative to the time required to establish the dynamic equilibrium of the two "phases", then "equilibrium" means we observe a single kind of species with properties that are the average over both "phases"; if the observation time is short relative to the mean interval spent in a single phase and to the time required to establish stable, phase-like properties,

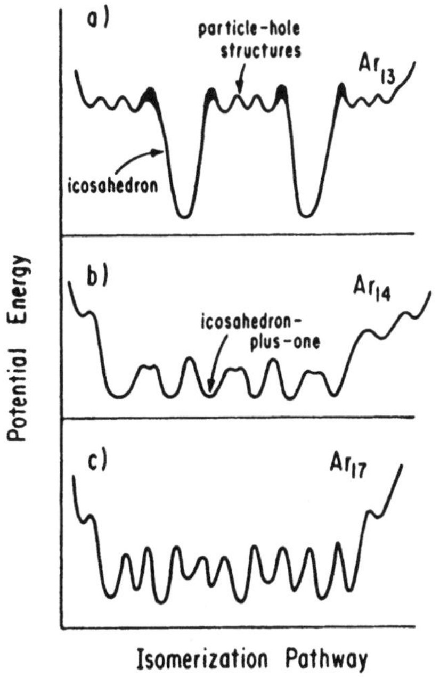

Fig. 3. Schematic cross sections of the potential energy surfaces for three kinds of argon cluster. Only the first kind, here illustrated by Ar_{13}, gives rise to separately observable solid-like and liquid-like phases and to sharp but unequal freezing and melting temperatures

then "equilibrium" means a mixture of two forms whose concentrations are in a characteristic, fixed ratio at each temperature. In other words, what we mean by "equilibrium" depends on how we observe the system.

2.8.3 Simulations and Experiments

Experiments to distinguish liquid and solid clusters have relied on two kinds of information: spectral line widths and electron diffraction data. The spectral analyses have been carried out thus far with argon clusters containing a single foreign molecule that acts like a probe. In one set of experiments, *Bösiger* and *Leutwyler* [30] used carbazole, a large, flat, disc-like molecule as the probe. They found a range of low-energy conditions in which the visible spectrum of carbazole with several Ar atoms attached appears sharp but shifted from the spectrum of pure carbazole. At higher energies, the carbazole bands become broader, which was interpreted as due to the availability of low-frequency modes of motion for liquid-like argons attached to the carbazole by weak van der Waals forces.

Other spectroscopic experiments were done by Gough, Knight and Scoles [31] with sulfur hexafluoride in argon and by Hahn and Whetten [32], with benzene in argon. The former were based on interpretation of the shift and width of vibrational bands of the SF_6 and the latter, on an electronic transition of benzene, again on the frequency shifts and widths of the transition in clusters of various sizes. At a fixed temperature in both of these experiments, the small clusters show broad bands which were interpreted as due to liquid-like behavior and the large clusters showed sharp bands, attributed to solid clusters. Clusters of intermediate size showed complex bands consisting of sharp, solid-like bands and broader, liquid-like bands. This was interpreted to support the idea of coexisting solid-like and liquid-like clusters of a given size. However both sets of experiments were reinterpreted – the infrared by Eichenauer and LeRoy [33] and the ultraviolet by Adams and Stratt [34] and by Fried and Mukamel [35] – to indicate not two phases but two sets of sites for the impurity molecules, one kind in the interior of the cluster and the other, on the surface. Other experiments, using fragmentation patterns of ionized clusters [36], were interpreted to imply that clusters may exhibit both solid-like and liquid-like forms. This interpretation has not been challenged or replaced.

Electron diffraction experiments have concentrated primarily on structural studies. However *Bartell* and coworkers have seen sharply peaked and more diffuse, broadly peaked patterns for carbon tetrachloride and then for other species, which they have interpreted as due to solid and liquid clusters [6]. Which appears in a given experiment depends on the conditions in the source from which the clusters come. Coexistence of these two forms has yet to be proven.

Until more experiments can be done which test the coexistence and sharp limits of existence of solid and liquid phases of clusters, simulations can be carried out to explore and test some aspects of the issue. Two kinds of simulations have dominated the research on phase changes in clusters. Some have been carried out by *Monte Carlo* (MC) methods [7, 15–18] and others, by molecular dynamics (MD) [3, 8–14, 19, 37], which is simply successive solution of the equations of motion from an assumed potential of interaction among the particles. Both MC and MD calculations may be carried out at constant energy or at constant temperature. However MC has been used far more for isothermal (constant-T) systems and MD, for constant-E systems because these correspond to time-invariant Hamiltonians and conservative systems. However it is possible to carry out MD calculations that give the same values for all equilibrium properties as a canonical ensemble gives. This method, introduced by *Nosé* [38], involves adding to the Hamiltonian one degree of freedom more than is found for the physical system. The "phantom" degree of freedom, if suitably chosen, acts like a heat bath, exchanging energy with all the physical variables. The entire system maintains a constant energy but the energy of the set of physical variables fluctuates just like an isothermal system. This procedure can be carried out with many different choices of the extra degree of freedom, but until now, only the choice made by *Nosé* has been exploited [39].

Fig. 4. Time histories of short-term average kinetic energies or vibrational temperatures for clusters of Ar_{13}: **a)** a cold solid; **b)** a warm liquid; **c)** within the coexistence range of energies, with both solid and liquid behavior for the same cluster, as it passes between the two forms as time proceeds

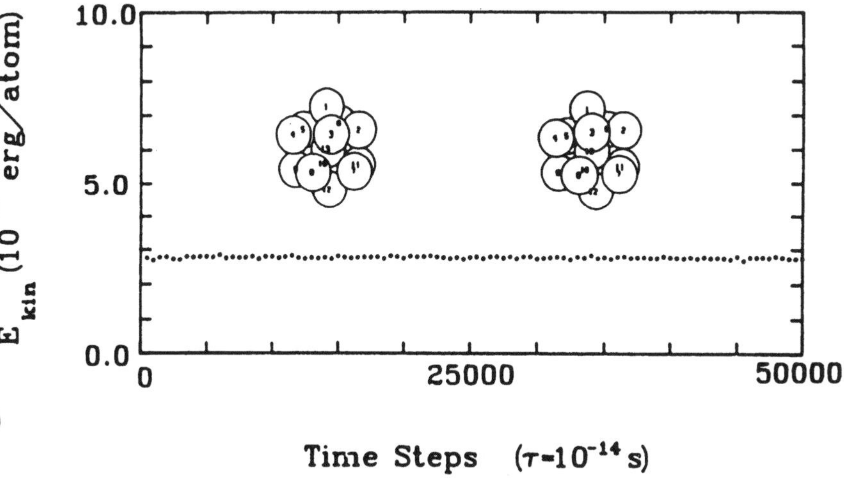
10.0
5.0
0.0
0
25000
50000
E_{kin} (10^{-15} erg/atom)
Time Steps (τ=10^{-14} s)

a)

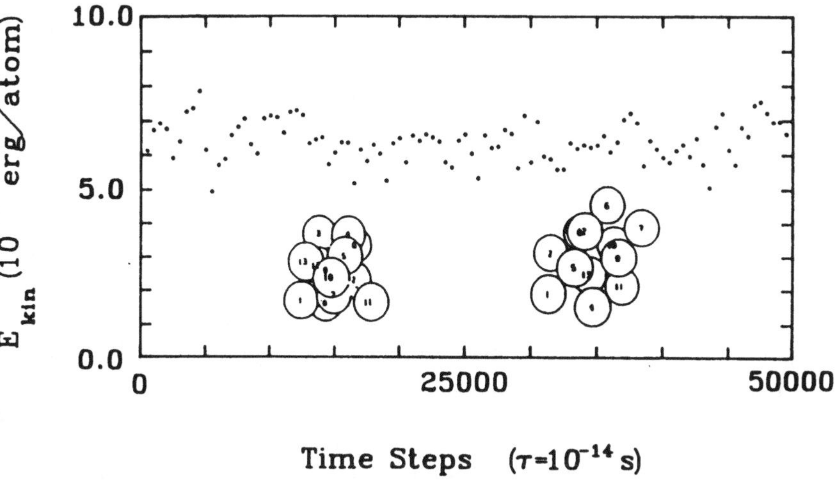
10.0
5.0
0.0
0
25000
50000
E_{kin} (10^{-15} erg/atom)
Time Steps (τ=10^{-14} s)

b)

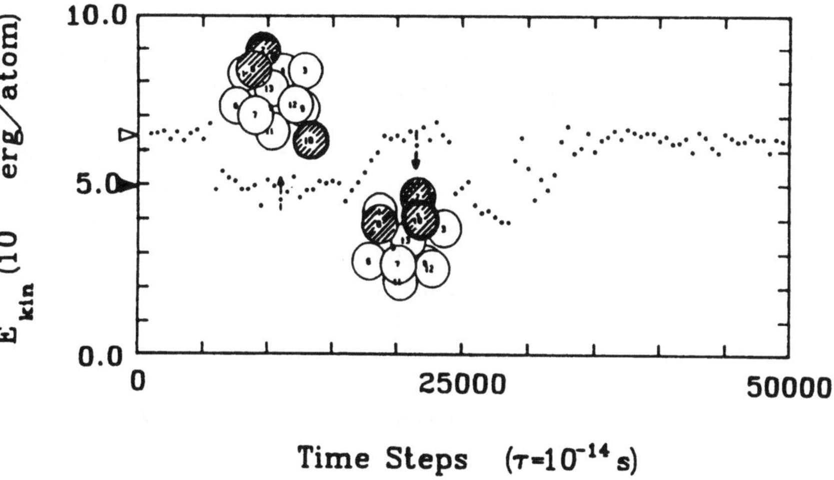
10.0
5.0
0.0
0
25000
50000
E_{kin} (10^{15} erg/atom)
Time Steps (τ=10^{-14} s)

c)

The simplest way to use simulations is to follow the time sequence of the mean kinetic energy of the atoms in a cluster held at constant energy. This is, in effect, a vibrational temperature. One carries out the averaging over enough steps of the integration in time to cover a few periods of vibration but not so long that many different regions of the potential surface are explored during the averaging interval. For argon clusters, a time step of $10^{-14}\,s^{-1}$ is suitable for most work and an averaging interval of 500 steps works out well. For systems with stronger forces, such as clusters of KCl, one must use a shorter time step.

A time history of a cluster of Ar_{13} at low energy is shown in Fig. 4a [37]; the temperature is nearly constant and low. In fact the mean temperatures for this system form a sharp Gaussian distribution. At a slightly higher temperature, one sees a slightly broadened distribution of short-term mean kinetic energies, and an occasional dip corresponding to passage from one potential well to another, that is, from one permutational isomer to another. At considerably higher energies, one sees a broad distribution. Furthermore the solid clusters all have approximately the geometry of the icosahedron, while the broad distribution of Fig. 4b is associated with what seems an amorphous structure, i.e. that of a liquid. Between the energy at which the cluster looks liquid and the lower energy at which it seems solid, one sees for many clusters a bimodal pattern of mean temperatures, a sharper distribution corresponding to a hot solid and a broader distribution corresponding to a cold liquid. Recall that these refer to simulations at constant energy, so the two parts of the distribution correspond to a high-potential energy region of liquid behavior and a deep-potential region in which the cluster has a high mean kinetic energy and temperature. Figure 4c shows an example of this.

The distribution of mean temperatures may be constructed in either of two ways. The mean temperatures may be collected into "bins" and the number in each bin plotted [11, 37], or the mean temperatures may simply be sequenced and a curve constructed by augmenting the ordinate one unit with each new incidence of a temperature as one goes up the temperature scale of the abscissa [40]. The former is, in principle, the derivative of the cumulative distribution constructed in the second method. The former is susceptible to errors of making the bins too large so that small peaks and shoulders in the distribution get overlooked. The latter may give false impressions of shoulders or small maxima because of noise in the data.

The passage back and forth between liquid and solid occurs only within a limited energy range, consistent with the predicted sharp limits on the stability of liquid and solid. However simulations cannot of course tell whether these limits are truly *sharp*. They can only give us an indication of bounds on the region within which the two phases may coexist.

To establish well-defined short-term temperatures requires only about $5\times10^{-12}\,s^{-1}$ or 500 time steps. To establish a stable distribution and mean of the short-term mean temperatures requires about 10,000 time steps. In the energy range within which both phases may coexist, one may be able to see

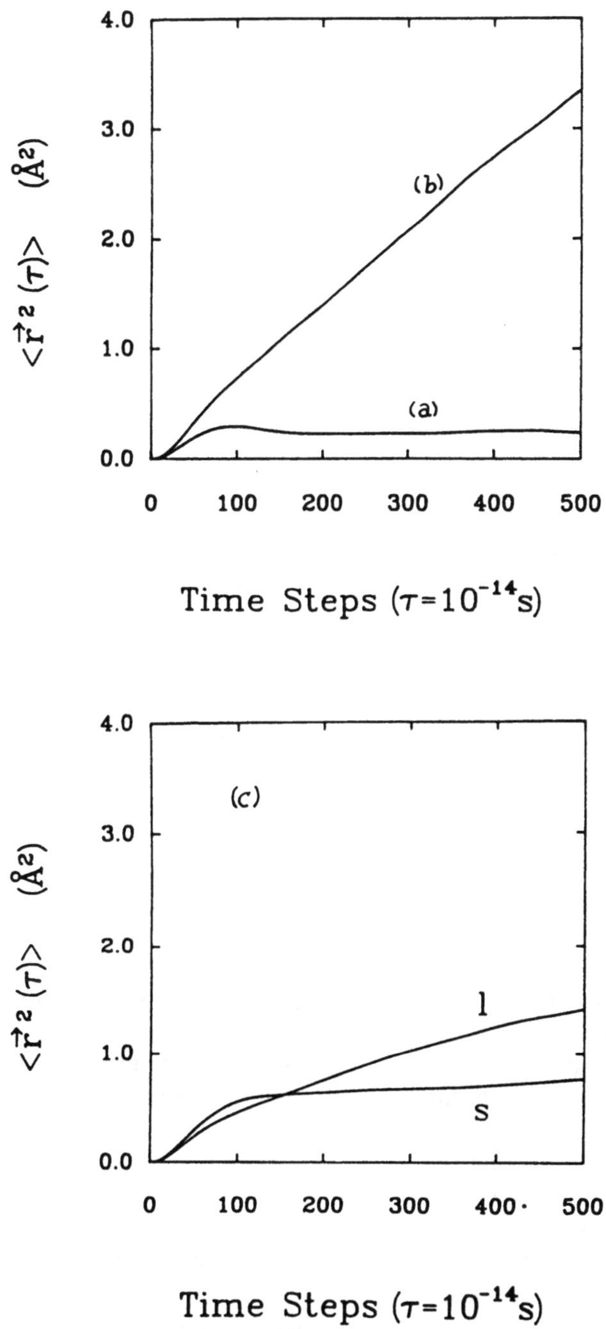

Fig. 5. Mean square displacements for Ar_{13} as functions of time: **a**) a cold solid as in Fig. 4a; **b**) a warm liquid, as in Fig. 4b; **c**) the mean square displacements for the coexistence region with the data for the two peaks in the distribution plotted separately

passage from one to another "phase" within perhaps 50,000 time steps or less. However to determine the equilibrium ratio of liquid to solid as the ratio of times spent by one cluster as liquid divided by the corresponding time spent as solid, i.e. to use the ergodic hypothesis, we require as many as 1–3 million time steps. Fortunately such long runs are now entirely feasible on modern computers.

Other diagnostics are also very useful and sometimes very enlightening. The mean square displacement as a function of time, $\langle r^2(t) \rangle$ is one of these. Its slope is six times the diffusion coefficient. Clusters at low energies show $\langle r^2(t) \rangle$ that are essentially horizontal, meaning that there is no diffusion. At higher energies, the cluster takes on the slope and diffusion coefficient of a liquid. In the region in which the temperature distribution is bimodal, the mean square displacements for each mode in the distribution can be computed. What one finds is a very warm solid with only very slow diffusion and a significantly cooler liquid with significant diffusion and a diffusion constant compatible with that of the bulk liquid. Of course all the mean square displacements reach a constant upper limit at the distance corresponding to the diameter of the cluster.

The radial and angular distributions have been used as diagnostics of simulations. The angular distributions in particular, as mentioned previously, give helpful indications about what structures the solid form may have [7, 8].

A common and helpful characteristic for distinguishing liquid-like behavior from solid-like is the root-mean-square deviation of the nearest-neighbor distance from its mean. It had been suggested by Lindemann that this quantity should show a sharp increase when the liquid forms, and that this should follow a slow increase with temperature up to a value of about 10 should jump. This criterion, shown in Fig. 6, is well supported.

The caloric curve of $\langle T(E) \rangle$ vs. energy for a constant-energy simulation or $\langle E(T) \rangle$ vs. temperature for a constant-temperature simulation is another useful way to extract information about phase equilibria from simulations and the data they generate [8, 11, 14, 37]. A smooth curve without inflection is a good indicator for a continuous transition from solid-like to liquid-like; a caloric curve with a flat spot such as that in Fig. 7, or even an S-like wiggle indicates equilibrium between two phases. However the presence of an inflection or wiggle only indicates phase equilibrium without disclosing what phases are involved.

That not all clusters show coexistence of well-marked phases emerges also from the simulations. Some, such as Ar_{14}, simply have no bimodal distribution of temperatures. Others, such as Ar_{17}, appear to have ready passage between solid and liquid. In either case, the clusters must act like slush, rather than like distinguishable liquid and solid phases. And still others, notably metal clusters and clusters of salt molecules, show "phases" much like soft solids, in which the cluster may pass among several structures without large displacements that permute atoms among all the equivalent sites. For example Cu_6 has a regular octahedral geometry in its configuration of lowest energy [9] (if the Jahn–Teller distortion, due to orbital degeneracy, is neglected); there are thirty of these

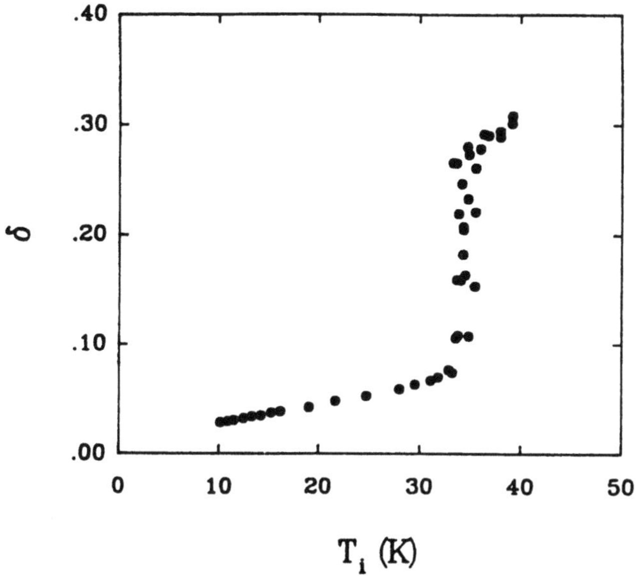

Fig. 6. The behavior of the root-mean-square variation in the nearest-neighbor distance as a function of temperature; the example is Ar_{13} but the sharp jump is a very generally found phenomenon

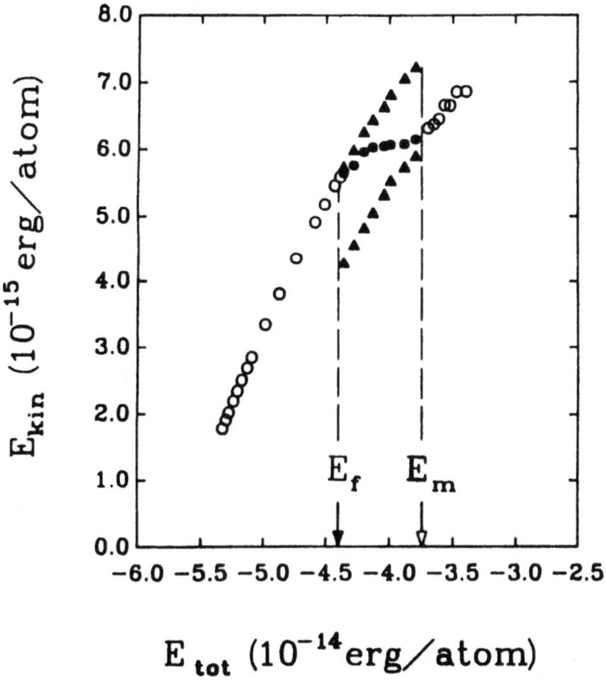

Fig. 7. A typical caloric curve of $\langle T(E) \rangle$ vs. E, in this case for Ar_{13}: circles represent averaging over the entire distribution of temperatures, while triangles represent averaging only within a single peak in the bimodal distribution of temperatures, in the range where bimodality occurs

octahedra that are permutationally distinct. And each octahedron can be distorted to another structure that corresponds to a local minimum on the potential surface well above the octahedron in energy; around each of the thirty octahedra, there are twelve of these distorted structures. Within a restricted range of energies a Cu_6 cluster may pass among the twelve distorted octahedra and the regular octahedron from which they are derived, without passing to any other octahedron or related distorted structure. This situation is a kind of soft solid behavior, which *Sawada* and *Sugano* [9] called the "fluctuating state." At higher energies, the cluster can pass among all the stable structures and permute all the atoms among all the sites, but in the energy range of the "fluctuating state" no such permutational equivalence can be established.

The connection between these various kinds of behavior and the potential surface or the energy levels of the cluster is a subject still very much in development. It is clear that if the cluster has one kind of deep, narrow potential well and a high, rolling region separated from the deep well by a moderate or high barrier, then the cluster must exhibit the two-phase coexistence described in the previous section. If the potential has a series of wells at successively higher energies, all separated by low barriers, like a staircase, then the behavior is slush-like and no clearly defined coexistence region will be found. A set of quantum-statistical models, based on constructed densities of states, were analyzed by *Bixon* and *Jortner* [41] to probe the coexistence question in terms of the two diagnostics of a) the number of maxima in the distribution of mean temperatures and b) the form of the caloric curve of mean temperature vs. energy.

2.8.4 Implications for Bulk Matter

The phase equilibrium behavior of clusters seems to produce a paradox for the phase equilibrium of bulk matter, but in fact if we follow carefully how this equilibrium depends on N, we can gain several insights into the behavior of bulk matter [26]. There are two things to question: first, why are the freezing and melting transitions the same sharp transition for bulk matter? Second, what becomes of the discontinuities at T_f and T_m as N becomes arbitrarily large? The first question was answered many years ago but we get a little new insight by answering it in the present context. The second is a new challenge whose answer tells us something about metastable, supercooled bulk matter.

The reason that freezing and melting occur at the same temperature for bulk matter is apparent when we write the free energy and the equilibrium constant with N explicit [18]. The free energy difference, $\Delta F_N = N \times \Delta\mu_N$ where $\Delta\mu_N$ is the difference in the mean chemical potentials of the liquid and solid clusters of N particles. The equilibrium constant

$$K = e^{-\Delta F_N/kT} = e^{-N\Delta\mu_N/kT}$$

which is dominated by the behavior of $N\Delta\mu_N$. The chemical potential difference $\Delta\mu_N$ changes from positive to negative as the temperature increases through T_{eq} where $\Delta\mu_N = 0$. This makes K swing from a small number to a large number. If N is relatively small, then the change of K from less than 1 to greater than 1 is gradual. If, on the other hand, N is large, for example 10^{10}, then K swings from very near zero to very, very large within a tiny interval of T. To see this, we can best use not K but $\mathscr{K} = (K-1)/(K+1)$, equivalent to [(liquid) – (solid)]/(total amount), which varies from -1 at low temperatures where the clusters are all solid to $+1$ at high temperatures where the clusters are all liquid. The quantity $\mathscr{K}$ has discontinuities at T_f and T_m just as K does, but it is easier to display. Figure 8 shows how $\mathscr{K}$ behaves for two values of N.

The result is that for large N, $\mathscr{K}$ switches continuously from a value very near -1 through 0 to a value very near $+1$ within an immeasurably small interval of temperature, so $\mathscr{K}$ and K look discontinuous at T_{eq}. The discontinuity associated with the first-order melting/freezing transition is thus a consequence of large N, and for any finite system $\mathscr{K}$ is, strictly speaking, continuous. Furthermore the *local* stability of each phase persists on both sides of T_{eq}. This means that a theory of the first-order melting and freezing transition must not be one in which one phase changes from unstable to stable while the other does the reverse at T_{eq}. The local stability prevents such behavior.

Now how about the other question? What happens to $\Delta T_c \equiv T_m - T_f$ as N becomes very large? To answer this, we return to the defect model of *Stillinger* and *Weber* [25], which gives us a way to write and differentiate $\Delta F(T, \gamma)$ for any

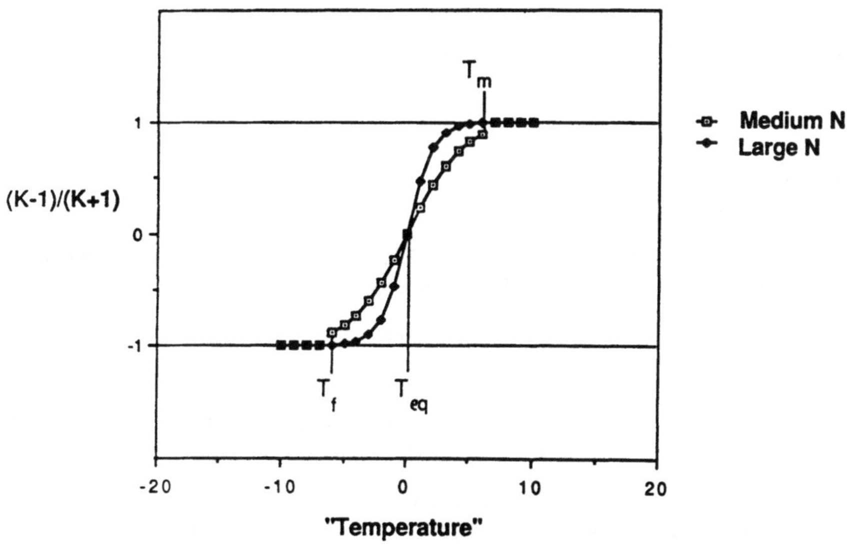

Fig. 8. Two schematic curves of $\mathscr{K}$, the more gradually sloping curve for moderate N and the steeper curve for larger N. Note that the discontinuities at T_f and T_m are visible for moderate N but not for larger N

γ. The details of this calculation are given elsewhere [40]; here we merely cite the conclusions.

If the defects do not interact with each other or with the lattice vibrations, then ΔT_c goes to zero as N becomes large. However if the defects attract each other or if, as is most usual, the defects lower the energy of the lattice modes, then ΔT_c remains finite as N becomes arbitrarily large. The *local* stability of the thermodynamically unfavored phase persists beyond T_{eq}. However if N is large, this local stability cannot be seen in a system *in equilibrium* because $\mathscr{K}$ is too near -1 or $+1$. Moreover the discontinuities at T_f and T_m cannot be seen because they are too small.

There is a way to observe the persistence of local stability, namely by preventing a system being cooled from coming to chemical and phase equilibrium. If a liquid is cooled below T_{eq} and no solid has been formed, one can cool that liquid further still. In fact the liquid can be cooled as low as T_f. Below that temperature, there is no locally stable liquid, whatever the size of the system. In other words, this analysis has shown that there is a sharp limit to supercooling of pure liquids, and that the locus of lowest temperatures for which the liquid branch of the p–V curve has negative slope, the so-called spinodal curve, has a sharp lower bound. This is an answer to a question that has been raised since the spinodal was recognized from the equation of state of van der Waals.

Acknowledgements. The author gratefully thanks his research associates and graduate students who have made our contribution to this subject possible: Franois Amar, Thomas Beck, Paul Braier, Hai-Ping Cheng, Heidi Davis, R.J. Hinde, Julius Jellinek, Grigory Natanson, John Rose and David Wales. Research described here done at The University of Chicago was supported by a Grant from the National Science Foundation.

References

1. For a recent review of this topic, see R.S. Berry, T.L. Beck, H.L. Davis, J. Jellinek.: In: *Evolution of Size Effects in Chemical Dynamics*, Part 2, Advances in Chemical Physics, Vol. LXX 1988, p. 75 (Wiley, New York)
2. J.M. Dickey, A. Paskin: Phys. Rev. **188**, 1407 (1969); ibid. Phys. Rev. B **1**, 851 (1970)
3. W. Damgaard Kristensen, E.J. Jensen, Martin, R.M. Cotterill: J. Chem. Phys. **60**, 4161 (1973)
4. J. Farges, M.F. de Faraudy, B. Raoult, G. Torchet: J. Chem. Phys. **59**, 3454 (1973); ibid. **78**, 5067 (1983); ibid. **84**, 3491 (1986); ibid. In: *Physics and Chemistry of Small Clusters*, P. Jena, B.K. Rao, S.N. Khanna (eds.), p. 15. (Plenum, New York); B. Raoult, J. Farges, M.F. de Feraudy, G. Torchet: Z. Phys. D **12**, 85, 93 (1989)
5. B.G. De Boer, G.D. Stein: Surf. Sci. **106**, 84 (1981)
6. E.J. Valente, L.S. Bartell: J. Chem. Phys. **80**, 1451, 1458 (1984); L.S. Bartell: Chem. Rev. **86**, 491 (1986); Y.Z. Barshad, L.S. Bartell: In: *Physics and Chemistry of Small Clusters*, P. Jena, B.K. Rao, S.N. Khanna (eds.), p. 31. (New York, Plenum, 1987); L.S. Bartell, L. Harsanyi, E.J. Valente: In: *Physics and Chemistry of Small Clusters*, P. Jena, B.K. Rao, S.N. Khanna (eds.), p. 37. (Plenum, New York, 1987)

7. N. Quirke, P. Sheng: Chem. Phys. Lett. **110**, 63 (1984)
8. H.L. Davis, J. Jellinek, R.S. Berry: J. Chem. Phys. **86**, 6456 (1987)
9. S. Sawada, S. Sugano: Z. Phys. D **12**, 189 (1989); S. Sawada: In: *Microclusters*, S. Sugano, Y. Nishina, S. Ohnishi (eds.) (Springer, New York, 1987)
10. F. Amar, R.S. Berry: J. Chem. Phys. **85**, 5943 (1986)
11. T.L. Beck, J. Jellinek, R.S. Berry: J. Chem. Phys. **87**, 545 (1987)
12. D.J. McGinty: J. Chem. Phys. **58**, 4733 (1973)
13. R.M. Cotterill, W. Damgaard Kristensen, J.W. Martin, L.B. Peterson, E.J. Jensen: Comput. Phys. Comm. **5**, 28 (1973)
14. C.L. Briant, J.J. Burton: Nature **243**, 100 (1973); ibid. J. Chem. Phys. **63**, 2045 (1975)
15. J.K. Lee, J.A. Barker, F.F. Abraham: J. Chem. Phys. **58**, 3166 (1973)
16. R.D. Etters, J.B. Kaelberer: Phys. Rev. A **11**, 1068 (1975); ibid. J. Chem. Phys. **66**, 5112 (1977); J.B. Kaelberer, R.D. Etters: J. Chem. Phys. **66**, 3233 (1977); R.D. Etters, R. Danilowicz, J. Dugan: J. Chem. Phys. **67**, 1570 (1977); R.D. Etters, R. Danilowicz, J.B. Kaelberer: J. Chem. Phys. **67**, 4145 (1977); R.D. Etters, R. Danilowicz: J. Chem. Phys. **71**, 4767 (1979)
17. V.V. Nauchitel, A.J. Pertsin: Mol. Phys. **40**, 1341 (1980)
18. T.L. Hill: *Thermodynamics of Small Systems*. New York: Benjamin Part 1, 1963; Part 2, 1964
19. R.S. Berry: In: *Large Finite Systems*, J. Jortner, A. Pullman, B. Pullman (eds.) (Riedel, Amsterdam, 1987); T.L. Beck, R.S. Berry: J. Chem. Phys. **88**, 3910 (1988); H.L. Davis, T.L. Beck, P.A. Braier, R.S. Berry: In: *The Time Domain in Surface and Structural Dynamics*, G.J. Long, F. Grandjean (eds.) (Kluwer, New York, 1988); R.S. Berry: Z. Phys. D **12**, 161 (1989)
20. C.E. Klots: J. Phys. Chem. **92**, 5864 (1988)
21. H. Reiss, P. Mirabel, R.L. Whetten: J. Phys. Chem. **92**, 7241 (1988)
22. G. Natanson, F. Amar, R.S. Berry: J. Chem. Phys. **78**, 399 (1983)
23. R.S. Berry, J. Jellinek, G. Natanson: Chem. Phys. Lett. **107**, 227 (1984); ibid. Phys. Rev. A **30**, 919 (1984)
24. K. Yamada, M. Winnewisser: Z. Naturforsch. A **31**, 134 (1976)
25. F.H. Stillinger, T.A. Weber: J. Chem. Phys. **81**, 5095 (1984)
26. R.S. Berry, D.J. Wales: Phys. Rev. Lett. **63**, 1156 (1989); D.J. Wales, R.S. Berry: J. Chem. Phys. **92**, 4473 (1990)
27. S. Gartenhaus, C. Schwartz: Phys. Rev. **108**, 482 (1957)
28. M.E. Kellman, R.S. Berry: Chem. Phys. Lett. **42**, 327 (1976); F. Amar, M.E. Kellman, R.S. Berry: J. Chem. Phys. **70**, 1973 (1979); M.E. Kellman, F. Amar, R.S. Berry: J. Chem. Phys. **73**, 2387 (1980); G.S. Ezra, R.S. Berry: J. Chem. Phys. **76**, 3679 (1980)
29. J.D. Honeycutt, H.C. Andersen: J. Phys. Chem. **91**, 4950 (1987)
30. J. Bösiger, S. Leutwyler: Phys. Rev. Lett. **59**, 1895 (1987)
31. T.E. Gough, D.G. Knight, G. Scoles: Chem. Phys. Lett. **97**, 155 (1983)
32. M.Y. Hahn, R.L. Whetten: Phys. Rev. Lett. **61**, 1190 (1988)
33. D. Eichenauer, R.J. LeRoy: Phys. Rev. Lett. **57**, 2920 (1986); ibid. J. Chem. Phys. **88**, 2898 (1988)
34. J.E. Adams, R.M. Stratt: J. Chem. Phys. **93**, 1358 (1990)
35. L.E. Fried, S. Mukamel: Phys. Rev. Lett. **66**, 2340 (1991)
36. A.J. Stace: Chem. Phys. Lett. **99**, 470 (1983)
37. J. Jellinek, T.L. Beck, R.S. Berry: J. Chem. Phys. **84**, 2783 (1986)
38. S. Nosé: J. Chem. Phys. **91**, 511 (1984); ibid. Mol. Phys. **52**, 244 (1984)
39. J. Jellinek: J. Phys. Chem. **92**, 3163 (1988); J. Jellinek, R.S. Berry: Phys. Rev. **A 38**, 3069 (1988)
40. D.J. Wales, R.S. Berry: J. Chem. Phys. **92**, 4283 (1990); ibid. **92**, 4473 (1990)
41. M. Bixon, J. Jortner: J. Chem. Phys. **91**, 1631 (1989)

3 Experimental Methods

H. Haberland

In this chapter some experimental methods of cluster science will be treated. Stress will be laid on representative examples, not on an exhaustive discussion. Most of the chapter will deal with the various sources used to produce clusters. Despite intense developments in the last years, this is still a difficult enterprise. New types of cluster sources are constantly developed. Some examples only will be discussed here. Problems with detectors for cluster ions are treated in Section 3.2. Electron diffraction from a cluster beam is discussed in Section 3.3. Methods to produce neutral mass selected clusters are reviewed in Section 3.4. Section 3.5 contains a brief overview of different types of mass spectrometers. Methods for optical and electron spectroscopy are discussed in 3.6 to 3.8.

A very simple experiment is shown in Fig. 1. The clusters are generated using a supersonic jet. They traverse a skimmer, are ionized by electron or photon impact, are separated according to their mass by a mass spectrometer, and are registered after having hit a detector. The different components of this apparatus will be discussed in detail below. Some more sophisticated set-ups for experiments on mass selected clusters are discussed in 3.5.

3.1 Sources

A simple but very inefficient way to produce clusters is using a *Knudsen cell* [1]. A solid or liquid is heated to so low a vapor pressure that the mean free path of the evaporated particles in the cell is larger than a hole through which the particles (atoms, molecules or clusters) can effuse. The hole should be so small that it does not perturb the thermodynamic equilibrium between the gaseous and condensed phase in the cell. Analyzing the beam one observes that besides the monomers, dimers, trimers, etc. can be present. The intensity decreases roughly exponentially with cluster size and can be calculated by standard thermodynamic expressions. This has been used to determine binding energies of small clusters. Deviations from the exponential behavior occurs when the binding energy does not vary monotonously with cluster size. A well known case is antimony where a larger signal on Sb_4 than on Sb_3 or Sb_5 is always observed. The intensity of the Knudsen source is not sufficient for most cluster experi-

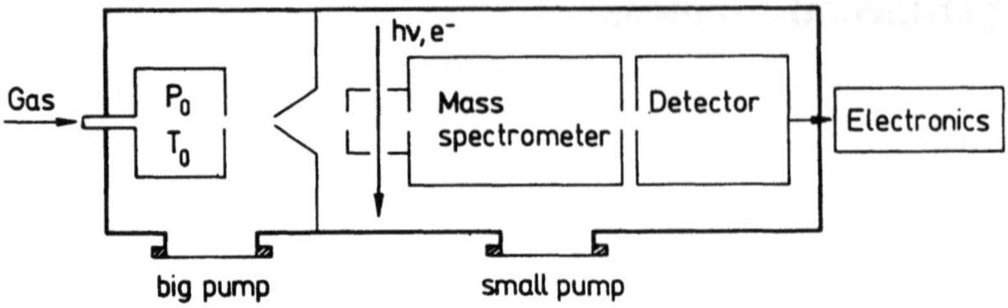

Fig. 1. Schematic of a simple apparatus. Clusters are produced in a supersonic expansion, they traverse a skimmer, are ionized by electron or photon impact, are mass selected and then hit a detector

ments. The sources used today for an efficient production of free clusters belong to three main categories:

1. Supersonic jets: A gas is expanded from a high pressure through a small hole. This adiabatic expansion reduces the relative velocity of the gas atoms or molecules dramatically and clusters are formed.
2. Gas aggregation sources: Atoms or molecules are injected into a stationary or streaming gas. Their velocity is slowed down and clusters are formed. These sources have also been colloquially called smoke sources. Smoke and cloud formation in nature is an example of this process.
3. Surface sources: Atoms or clusters are removed from a solid surface by photon or heavy particle impact or from a liquid surface by a high electric field.

A large variety and combinations of these three types have been constructed and used successfully. Several variants will be discussed below. The above classification is somewhat arbitrary and others have been given [2].

3.1.1 Supersonic Jets

The supersonic jet is the best understood cluster source. Its working is shown schematically in Fig. 2. Gas is expanded from a high pressure P_0 through a small hole with diameter D into vacuum. The random velocities (indicated by arrows in Fig. 2) inside the reservoir (sometimes called stagnation vessel) are nearly equalized by this expansion process. This makes for a very low temperature environment for a observer who travels along the beam with the same mean velocity. At a typical pressure of $P_0 = 10$ bar the mean free path is many order of magnitude smaller than a typical nozzle diameter D. The gas atoms make many collisions during the expansion and an aerodynamic treatment is necessary.

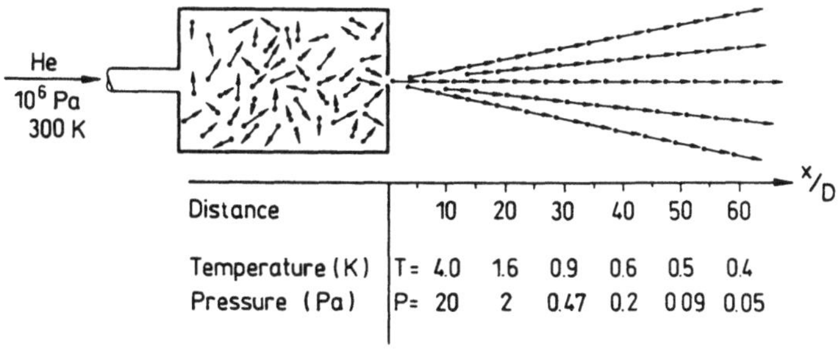

Fig. 2. A gas is expanded from a high vacuum through a small hole of diameter D into vacuum. The velocities of the atoms and molecules are indicated by arrows. They are random in direction and size prior to the expansion. Afterwards they equalize rapidly. Temperature and pressure on the center line of the beam are given. Note the extreme cooling very near to the nozzle. This low temperature surrounding leads to clustering

3.1.1.1 Ideal Gas Analysis of the Expansion

The theoretical treatment of the supersonic expansion can be considerably simplified if an ideal gas is assumed. We will further assume that no clusters are formed, and that turbulence and effects of heat conduction are unimportant. As discussed in [3, 4] the stagnation enthalpy $H_0 = c_p T_0$ of the gas at rest is converted into kinetic energy $\frac{1}{2}mu^2$ of the directed mass flow and a rest enthalpy H. Conservation of energy gives:

$$H_0 = H + \frac{1}{2}mu^2 = c_p T_0 \ . \tag{3.1}$$

Here c_p is the specific heat at constant pressure which is $c_p = \frac{5}{2}k$ for an atom. The subscript "0" refers always to the conditions before the expansion; the stagnation conditions as they are called in the jargon of this field. One can define a local temperature T along a stream line of the expanding gas. In this case Eq. (3.1) gives:

$$c_p T_0 = c_p T + \frac{1}{2}mu^2 \ . \tag{3.2}$$

Observing that $k = c_p - c_v$, one obtains after some rearrangements:

$$T = T_0 \left[1 + \frac{1}{2}(\gamma - 1)M^2\right]^{-1} \ . \tag{3.3}$$

The local Mach number M is defined as the ratio of the stream velocity u to the local speed of sound $c = (\gamma k T/m)^{1/2}$. The ratio of specific heats $\gamma = c_p/c_v$ at constant temperature and volume is assumed to be independent of the temperature. This is strictly valid only for atoms. The square of the Mach number, M^2, is a measure of the ratio of the directed mass flow to the thermal energy remaining in the beam. M increases dramatically along the expansion mainly due to the decrease of the velocity of sound c which decreases as $\sqrt{T}$. The velocities in a supersonic beam are not much higher than in a normal effusive beam (compare Fig. 5) but are large with respect to the local sound velocity. It is because of this property that these beams have been termed supersonic.

The temperature T cannot drop below zero. One obtains from Eq. (3.2) for $T = 0$ the highest possible kinetic energy:

$$E_{\max}^{\mathrm{kin}} = \frac{1}{2} m u_{\max}^2 = c_p T_0 \ , \tag{3.4}$$

which gives for an atomic beam:

$$\frac{1}{2} m u_{\max}^2 = \frac{5}{2} k T_0 \ . \tag{3.5}$$

So far only energy conservation has been used. One can show [4] that under the conditions stated above entropy is also conserved along each streamline, in other words the flow is isentropic. Under these conditions the Poisson equations for an adiabatic expansion hold as explained in every textbook on thermodynamics:

$$\left(\frac{P}{P_0}\right) = \left(\frac{\varrho}{\varrho_0}\right)^{\gamma} = \left(\frac{T}{T_0}\right)^{\gamma/(\gamma-1)} \ . \tag{3.6}$$

This equation couples the local thermodynamic variables (pressure P, density ϱ, and temperature T) to the corresponding stagnation values. Inserting Eq. (3.3) into Eq. (3.6) one obtains immediately

$$\begin{aligned} \frac{T}{T_0} &= \left(1 - \frac{\gamma-1}{2} M^2\right)^{-1} , \\ \frac{\varrho}{\varrho_0} &= \left(1 - \frac{\gamma-1}{2} M^2\right)^{1/(1-\gamma)} , \\ \frac{P}{P_0} &= \left(1 - \frac{\gamma-1}{2} M^2\right)^{\gamma/(1-\gamma)} . \end{aligned} \tag{3.7}$$

With these equations the local values for T, ϱ and P can be calculated at each point of the expanding beam once the local Mach number M is known. The dependence of M on the reduced distance x/D from the nozzle is shown in Fig. 3 for two different values for γ. To obtain this result the partial differential equations for the isentropic, compressible flow had to be solved. This result can be used to plot the values for P, T, and ϱ as a function of the reduced distance from the nozzle, as shown in Fig. 4. Temperature, density and pressure all decrease dramatically along the expansion, typically one order of magnitude in 10^{-6} to 10^{-7} s. The velocity distribution corresponding to the flux of particles on the axis is normally taken as a Maxwellian superimposed on the mean

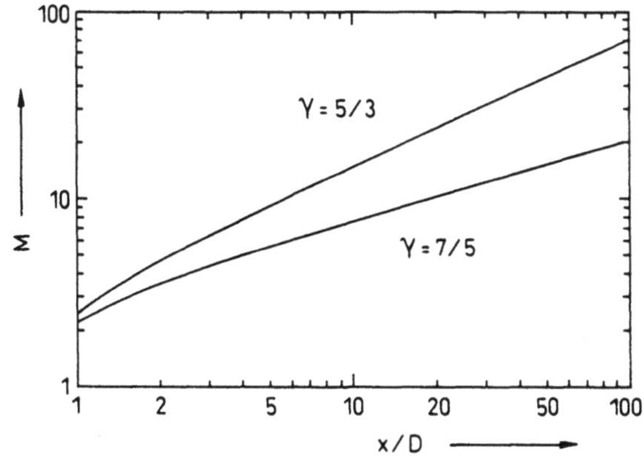

Fig. 3. Mach number M as a function of the reduced distance from the nozzle for two different values of γ

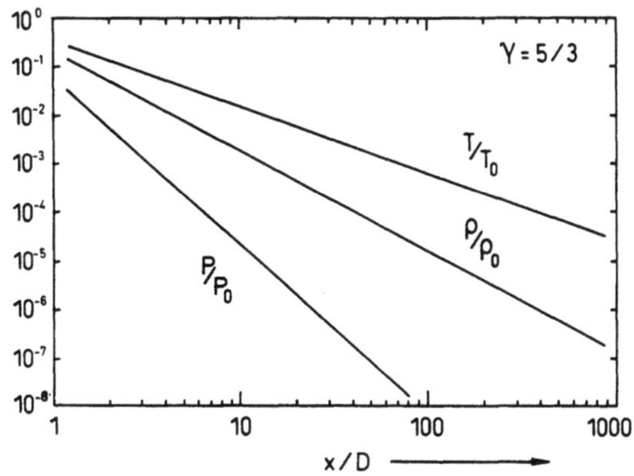

Fig. 4. Temperature T, density ϱ, and pressure P as a function of the reduced distance from the nozzle

velocity u:

$$f(v) = \left(\frac{v}{\alpha}\right)^3 \exp\left[-\left(\frac{v-u}{\alpha}\right)^2\right], \quad \text{with} \quad \alpha = \sqrt{\frac{2kT}{m}} \,. \tag{3.8}$$

Velocity distributions for different Mach numbers are shown in Fig. 5. They have been reasonably well confirmed by experiment [4, 5]. Note that Eq. (3.8) corresponds to a flux density (dimension: particles/s · cm^2); the velocity distribution corresponding to the density (dimension: particles/cm^3) has a square instead of a third power in the prefactor of the exponential [6]. According to Fig. 4 the density and pressure decrease rapidly. At some time the expansion runs out of collisions and the continuum assumption is no longer valid. It is this point which determines the Mach number. It has been termed the "freezing surface" as the continuum properties are frozen in at this point.

From Figs. 2 to 4, and Eqs. (3.6) to (3.7), it is obvious that a supersonic beam presents a very cold environment for a particle moving with the same mean velocity as the beam. This has been used to cool molecules to very low

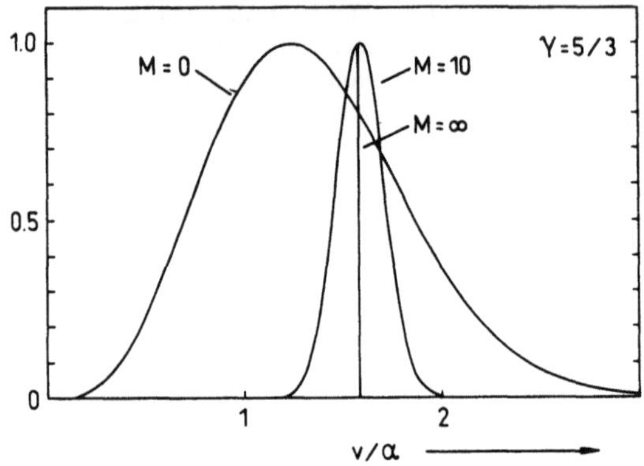

Fig. 5. Velocity distributions for different Mach numbers. One has $M = 0$ for an effusive beam from a Knudsen cell. For higher M the width of the velocity distribution becomes narrower, but the mean velocity does not increase much

vibrational and rotational temperatures. This can be done by diluting a small fraction of the molecule into the expansion. It will become apparent below that the beam temperature is increased if clustering occurs. Clusters from a neat (i.e. not seeded) supersonic expansion are always hot, as the heat of condensation liberated during cluster formation heats the beam.

Pulsed Supersonic Jets. Pulsing a supersonic beam has become quite popular as it enjoys several advantages [7]. Less gas and smaller pumps can be used: the experiment becomes cheaper and smaller. It takes about 10–20 μs for the gas going through the nozzle to reach steady state. This is the opening time for the fastest pulsed nozzles constructed. Commercially available nozzles have opening times of about 100 μs or longer. For less demanding applications or for a small budget, a diesel fuel injector will do a good job. Pulsed beams have one peculiarity, which can be an advantage sometimes. The amount of gas flowing out of the nozzle is smaller at the beginning and end of the pulse than in its middle. The distribution of cluster sizes and also the temperature of the clusters will therefore vary over the gas pulse.

Slit Nozzles. For absorption experiments a long optical path length is needed, and it might be advantageous in this case to use a slit instead of a hole for the expansion. A simple design having an adjustable slit width is given in Ref. [8]. Another design is reported in Ref. [9].

Skimmers. The skimmer serves to separate the regions of different pressure, as can be seen in Fig. 1. The shape and sharpness of the skimmer is of minor importance for simple nozzle holes. But for conical nozzles, or pulsed jets, the skimmer's edge should be very sharp and polished in order to avoid skimmer interferences which can destroy the beam.

The skimmer serves a second function if the residual gas pressure P_{res} is high in the first chamber. If the decreasing pressure in the expanding beam becomes similar to P_{res}, a shock wave is formed which heats the beam. A decrease in beam intensity and cluster size results. Numerical studies [4] show that the distance z_{shock} between nozzle and skimmer is given by $z_{shock}/D \simeq 0.67 \cdot (P_0/P_{res})^{0.5}$. For most cluster beams z_{shock} is larger than the nozzle-skimmer distance and no shock wave is formed in the beam.

3.1.1.2 Cluster Formation

Cluster formation in a supersonic beam is a complex process and no complete theory exists. But regarding Eqs. (3.6) to (3.7) as well as Figs. 2 to 4 it is plausible that clusters will be formed due to the very low temperature environment of the beam. If the local temperature in the beam becomes less than the binding energy of the dimer, it can be stabilized by a 3-body collision. This process can be written, taking Ar as an example:

$$\mathrm{Ar} + \mathrm{Ar} + \mathrm{Ar} \rightarrow \mathrm{Ar} + \mathrm{Ar}_2 \tag{3.9}$$

This dimer formation starts the condensation process if dimers or larger aggregates are not already present in the gas prior to the expansion. There are three atoms necessary for this process, otherwise energy and momentum conservation can not be fulfilled at the same time. The kinetic energy of the argon atom on the right hand side of Eq. (3.9) has to be so high that the other two atoms find themselves bound. Once dimers are formed they act as condensation nuclei for further growth. If the ratio of atoms to clusters is small the latter grow by monomer addition. For heavily clustered beams cluster-cluster aggregation becomes important. The supersonic expansion and the clustering within it has been treated by Molecular Dynamics and Monte Carlo Techniques [10, 11, and many references therein]. It is extremely difficult for these calculations to cover a time scale longer than a few nanoseconds. This is very short compared to the time between nozzle and the sudden freeze surface, so that a complete treatment of cluster formation by these methods has not been possible to date.

Alternatively *classical nucleation theory* can be used [12] which treats cluster formation in an isothermal, infinite gas. A supersonic beam is evidently a very different surrounding, and the applicability of the theory therefore limited. It can much better describe the condensation source discussed in Section 3.1.2 below. But the theory gives a good qualitative picture of the thermodynamics of cluster formation and is therefore treated here.

Cluster formation is treated as a gas-fluid phase transition. In a phase diagram (pressure P plotted against temperature T, see Fig. 6) the two phases are separated by the saturation vapor pressure line, which is approximately given by:

$$\ln P_\infty = A - \frac{B}{T}\,. \tag{3.10}$$

Here P_∞ is the vapor pressure of a plane fluid surface, and A and B are two material constants. Condensation takes a long time for a pure gas at P_∞. Some

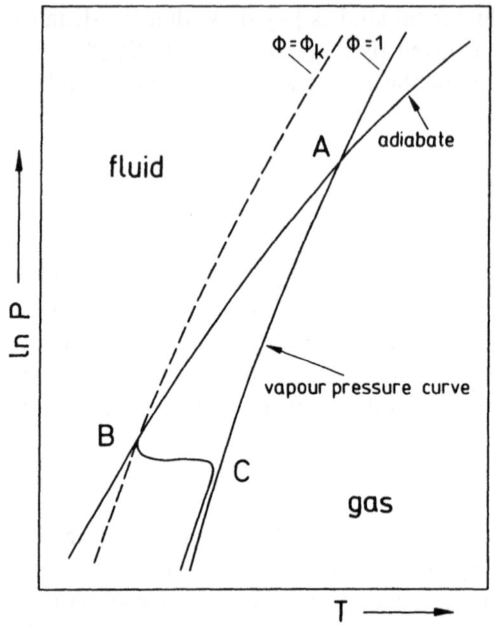

Fig. 6. The vapor pressure curve separates the gaseous and fluid phases. Pressure and temperature of a supersonic expansion follow the adiabate until point B, where cluster formation sets in. This heats the beam so that the p–T curve leaves the adiabate and goes to point C

supersaturation Φ_k with

$$\Phi_k = \frac{P_k}{P_\infty} \tag{3.11}$$

is necessary, so that the condensation starts on the timescale available in a jet. Here P_k is the vapor pressure determining Φ_k. A small particle or cluster of radius r has a larger vapor pressure P_r compared to that of a plane surface:

$$\ln \frac{P_r}{P_\infty} = \frac{2\sigma m}{kT\varrho} \cdot \frac{1}{r} , \tag{3.12}$$

where σ is the surface tension of the small drop or cluster, ϱ its density, and m its molecular or atomic weight. For $P_r = P_k$ and a given supersaturation Φ_k, one obtains a critical cluster size r^* with

$$r^* = \frac{2\sigma m}{kT\varrho} \cdot \frac{1}{\ln \Phi_k} . \tag{3.13}$$

Clusters having a smaller radius than r^* evaporate, larger ones grow. Figure 6 shows the vapor pressure curve for two values of the supersaturation Φ_k. The curve $\Phi_k = 1$ separates the thermodynamic stable regions of a fluid and a gas. According to Eq. (3.6), pressure and temperature in a beam follow an adiabate, which has a less steep slope than the vapor pressure curve. The two lines intersect at point A. Cluster formation is delayed till point B. The p–T curve of the beam leaves the adiabate, as the heat of condensation heats the beam.

Equation (3.9) describes cluster formation by monomer addition. The monomer must have by far the highest intensity and cluster-cluster collisions are rare. This gives a mass spectrum which decays roughly exponentially with cluster size. This is shown in the upper panel of Fig. 7. If the stagnation pressure is increased a large percentage of the beam can be aggregated. This results in cluster-cluster collisions and gives the large peak at higher masses. The solid line is a simulation which is in good agreement with the data [13].

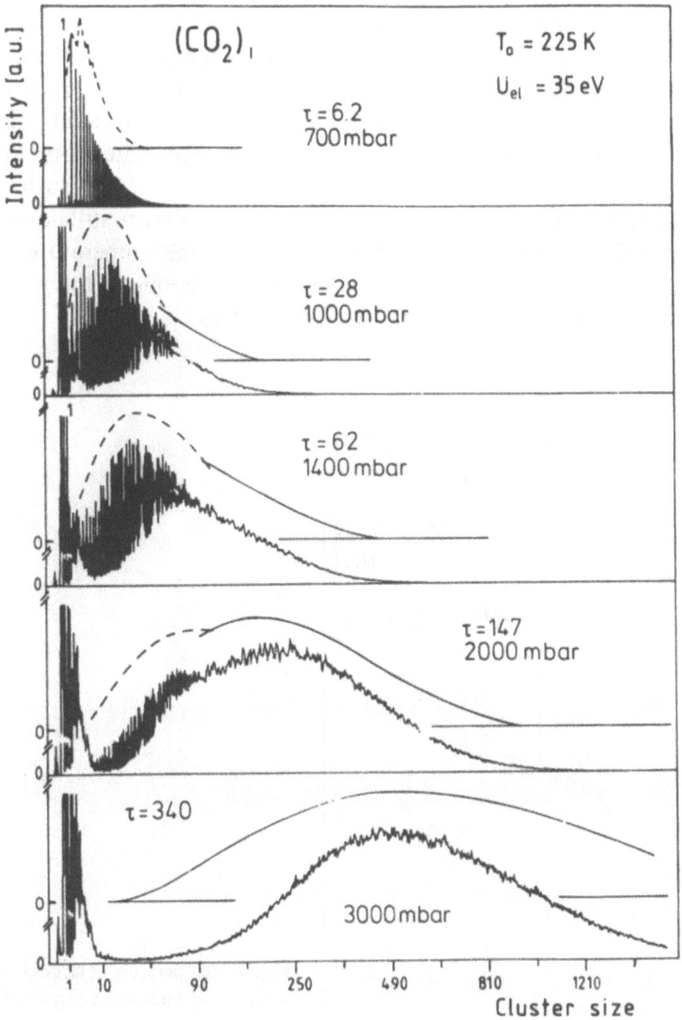

Fig. 7. Mass spectrum of CO_2 clusters. Under weak clustering conditions a roughly exponential mass spectrum is always observed, compare the upper panel. Under strong clustering conditions the clusters can grow by cluster-cluster collisions. This leads to the big hump. A theoretical analysis gives a good agreement with the data

Temperature of Clusters from Supersonic Beams. A supersonic beam with negligible clustering provides a very low temperature surrounding as discussed above. But a heavily clustered beam with a large condensate fraction produces clusters which are hot, probably "boiling" hot. The clusters are heated by their very formation process. Each time an atom or molecule condenses onto the cluster, the heat of formation is added to the cluster's internal energy. The cluster can cool by collisions with atoms in the beam and by evaporation. Once past the freezing zone or the skimmer, evaporation of atoms is the only cooling mechanism available on the time scale of most experiments. For a very long lifetime, longer than compatible with most experiments, emission of an infrared photon can contribute to the last cooling steps. Figure 8 shows the temperature of large clusters as a function of their dimer binding energy. The temperature was measured by electron diffraction (see Section 3.3) about 6.5 cm downstream from the nozzle [14]. The clusters are indeed "hot", i.e. they evaporate atoms on the timescale of most experiments. *Gspann* [15] has argued that rare gas clusters given by the temperature of Fig. 8 are solid but that metal clusters are still liquid. Note that for most of the experiments discussed in this book, the temperature or internal energy is not known. The temperature of a cluster is one of the most difficult experimental parameters to control for free clusters.

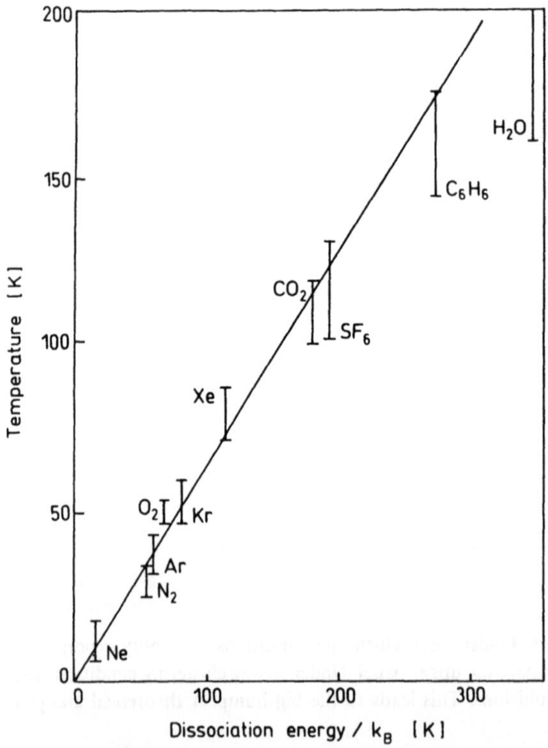

Fig. 8. Temperature of large clusters plotted against the dimer well depth. The temperature was measured by electron diffraction about 5 cm after the nozzle. Note that the clusters are not cold at all

For very small clusters, $n \leq 5$, a low temperature can be obtained by seeding the gas in a large surplus of rare gas. e.g. 10^4 Pa of Na in 10^6 to 10^7 Pa argon gives very cold Na_3 [16]. Also mixed clusters like Na_3Ar_m [16], $m = 0, 1, 2 \ldots$ are observed under these conditions. The week Na–Ar bond is a guarantee, that the internal energy is indeed low. But no one has succeeded so far to produce clusters like Na_nAr_m, with $n \geq 10$. This proves that these clusters can indeed not be produced cold in a supersonic beam.

For the hot clusters leaving the condensation zone and cooling afterwards by evaporating, *Klots* has coined the word of an "evaporative ensemble" [17]. It is with these still evaporating clusters that most cluster experiments have been performed so far. *C. Bréchignac* et al. have shown [18], how one can calculate internal energies and temperatures, provided the clusters start sufficiently hot, and the evaporation rate is known. The first condition is easy to fulfil experimentally, the second one is a big problem of contemporary cluster physics, as discussed in detail in Chapter 2.7. Using these ideas an estimate of the cluster temperature can be obtained the following way.

Assume that the clusters are formed very hot, like in an unseeded supersonic beam or are ionized in the focus of a high intensity laser beam. If cluster X_{n+1} has an internal energy $E^*(n+1)$, the energy of cluster X_n produced by evaporation of one atom $X_{n+1} \rightarrow X_n + X$ will be

$$E^*(n) = E^*(n+1) - D(n+1) - \varepsilon(n+1) \ , \tag{3.14}$$

where the dissociation energy $D(n+1)$ is needed to separate one atom from the remaining cluster, and $\varepsilon(n+1)$ is the energy of the recoiling products. At each evaporation step the cluster looses one atom and the amount of $D + \varepsilon$ on internal energy. If $E^*(n)$ is sufficiently high the cluster X_n will continue to decay and decay, until it is stable on the time scale of the experiment.

$$X_{n+1} \xrightarrow{k_{n+1}} X_n \xrightarrow{k_n} X_{n-1} \xrightarrow{k_{n-1}} X_{n-2} \xrightarrow{k_{n-2}} \tag{3.15}$$

If one would know the evaporation rates $k_n(E^*)$ of Eq. (3.15) one could calculate this evaporative process. Unluckily this is not the case. The simplest choice for it is given by the classical RRK expression:

$$k_n(E^*) = \nu g\left(1 - \frac{D(n)}{E^*}\right)^{s-1} \ . \tag{3.16}$$

Here ν is a typical vibrational frequency (10^{12} to 10^{13} Hz), $s = 3n - 6$ is the number of vibrational degrees of freedom, and g is a degeneracy factor, usually taken equal to the number of surface atoms. Figure 9 shows the lifetimes $\tau = 1/k_n(E^*)$. The rate depends strongly both on n and E^*. For a fixed experimental time window the clusters have different but relatively narrow excitation energies. There exist many different equations for the rates, as discussed by *Jarrold* in Chapter 2.7. They can differ by large factors, but there predictions of the internal energies differ typically by 30 to 50% and by a factor of 2 at most.

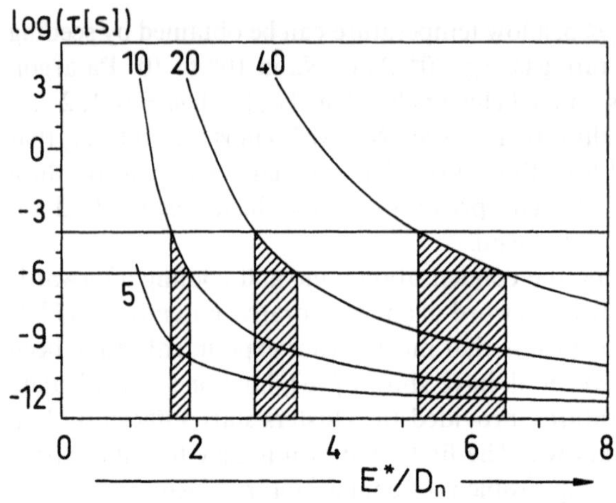

Fig. 9. Lifetimes of the $n = 5$, 10, 20, and 40 atom clusters as a function of the excitation energy E^* scaled by the dissociation energy. For a fixed time window 10^{-6} to 10^{-4} s the clusters have different but relatively narrow internal excitation energies

So far only internal energies have been calculated. How does one obtain temperatures? In the chapter by *R.S. Berry* (Ch. 2.8) the concept of a temperature for a finite system of particles is discussed in more detail. Here we define a temperature T as the internal energy per degree of freedom:

$$k_B T = E^*/s \ . \tag{3.17}$$

One obtains for the evaporation rate per surface atom

$$k_n(T) = \nu \left(1 - \frac{D(n)}{s k_B T}\right)^{s-1} \ . \tag{3.18}$$

From this expression one can deduce, that not the temperature, but the temperature scaled by the dissociation energy is the relevant parameter for an evaporative ensemble. The macroscopic limit of the RRK equation can be obtained analytically. Writing

$$1/x = -\frac{1}{s}\frac{D(n)}{k_B T}$$

one obtains

$$k_n(T) = \nu \left[\left(1 + \frac{1}{x}\right)^x\right]^{-D(n)/k_B T} \left(1 - \frac{D(n)}{s k_B T}\right)^{-1} \ . \tag{3.19}$$

In the limit of large clusters the last term converges to 1 and the term in square brackets gives:

$$\lim_{x \to \infty} \left(1 + \frac{1}{x}\right)^x = e \ ,$$

and one obtains the expected macroscopic behavior for the evaporation rate per surface atom:

$$k_n(T) = \nu \exp\left(-\frac{D(n)}{k_B T}\right) . \tag{3.20}$$

The deviation from the RRK result becomes large for $n \leq 10$ and negligible for $n \geq 20$. This makes this equation a convenient way to estimate the temperature of clusters. Note, however, that the temperature depends on the time window of observation and, in general, ν is not known.

The calculation of a temperature for a cluster can be done on three levels of increasing sophistication. One should always keep in mind that the equations given here and used below assume one specific rate constant for the decay. As the exact rate constant is not known, one has a systematic error of unknown magnitude. This error can be expected to vary slowly from one cluster size to the next. Therefore relative values of the temperature can be expected to be more accurate.

The simplest estimate uses Eq. (3.20), assuming that the clusters can be described as an evaporative ensemble. This is true if they are formed very hot, for example in a heavily condensed supersonic beam like in the experiments for Fig. 8. It takes a flight time of about 10^{-3} s for the cluster to go from the nozzle to the interaction zone. For the decay rate $k_n(T)$ one must have $k_n(T) \approx 10^3$ s. If $k_n(T)$ is larger the cluster has decayed already, if it is smaller the cluster has not been formed. Note that all clusters have started out very hot. Otherwise the argument is not tenable. Taking $\nu \approx 10^{-13}$ s one obtains

$$\frac{k_B T}{D(n)} = -\ln\left(\frac{\nu}{k_n}\right) = \ln(10^{10}) \approx 23 . \tag{3.21}$$

The data do indeed show this behavior.

A more refined estimate can be obtained by inverting Eq. (3.16)

$$E_n^{\max} = \frac{D_n}{1 - \sqrt[1-s]{\nu n t}} . \tag{3.22}$$

Here t is the lifetime since formation (or mass selection) and is taken again equal to the inverse of the decay rate. A cluster with larger energy would have decayed within the time t, one with lower energy not yet formed. An estimate of the lower bound to the clusters energy can be obtained from the formation process. The n-cluster has to be formed by evaporation from the $(n+1)$-cluster. From Eq. (3.14) one obtains:

$$E_n^{\min}(t) = E_{n+1}^{\max} - D_{n+1} - \varepsilon_{n+1} . \tag{3.23}$$

These energies are given in Fig. 10, using a $\nu = 10^{12}$ Hz, and $t = 23\sqrt{n}$ µs. The most accurate way is to simulate the decay chain given in Eq. (3.15). The results are similar to those given in Fig. 10. The distributions of the internal energy are flat topped at small cluster sizes, they become more rounded for increasing n.

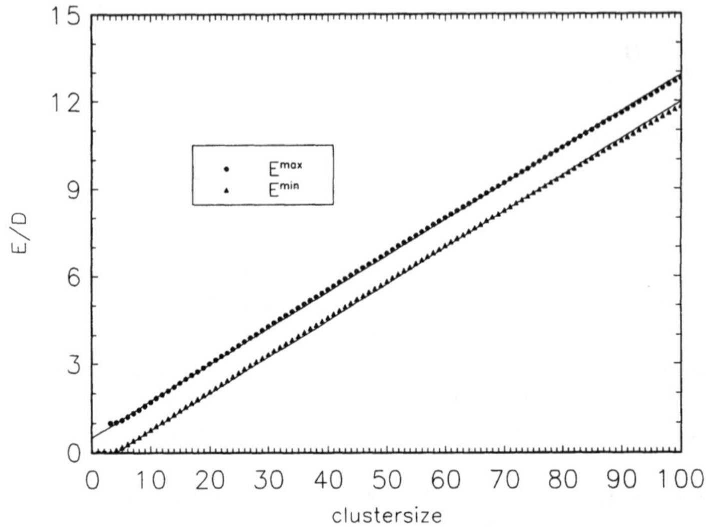

Fig. 10. Calculated internal energies of clusters, calculated from Eq. (3.22) and Eq. (3.23). Using Eq. (3.17) the internal energy can be converted to a temperature. A frequency factor of $\nu = 10^{12}$ Hz was assumed, and a time corresponding to a time-of-flight experiment: $t = 23\sqrt{n}$ μs

The ideas presented here have been spelled out directly for supersonic beams. This has been done to dispel the myth that clusters from supersonic beams are cold. Of course, the ideas have a much wider applicability. The ideas and equations given above are always applicable, provided the clusters form an evaporative ensemble [18].

The assumptions of a statistical distribution of internal energies and that of an evaporative ensemble have been very helpful in clarifying the ideas concerning cluster temperatures. But it becomes apparent that the assumption is too simple. Recent measurements show that the first atoms ejected after photoexcitation carry much more than their statistical energy share [19, 20, 21]. Also there is no guarantee that the very late parts of an evaporative cascade are statistical. This fact is well known for $(N_2)_n$ clusters. Here one N_2 molecule in the cluster can store several vibrational quanta, as the coupling between the molecules is very weak [22]. For rare gas cluster ions a similar behavior is probable [21].

Scaling Laws for Cluster Formation in Supersonic Jets. Cluster formation in supersonic jets has been studied widely in order to find a general scaling law. Put simply, one can say the higher the pressure, the larger the nozzle diameter, the smaller the cone angle of the nozzle opening, and the lower the temperature, the larger the clusters. The mean cluster sizes determined prior to 1985 should be regarded with suspicion because only after that time the problem of fragmentation of a cluster after ionization was recognised clearly. The mean cluster size of the ionized clusters is determined and not that of the neutral precursors. Nevertheless these older data still give valuable qualitative information. The only quantitative data for neutral clusters come from the group of *Buck* [23]. Using the method of preparing a neutral mass selected cluster beam as discussed in Section 3.4, cluster formation in supersonic jets could be studied up to $n = 10$. It is observed that the cluster density scales as $\varrho(\mathrm{Ar}_n) \sim (P_0)^n$, where P_0 is the inlet pressure. For very much larger clusters *Hagena* [24, 25] has derived scaling laws. The supersonic expansion can be characterized by three parameters: P_0, T_0, and D, which are taken as coordinates in a three-dimensional space. All points leading to a mean cluster size $\bar{N}$ in this space define a surface. Hagena could show experimentally and theoretically that this surface can be described as:

$$\begin{aligned} &P_0 \cdot D_{\mathrm{eff}}^{0.8} = \text{constant for } T_0 = \text{constant} \ , \\ &D_{\mathrm{eff}} \cdot T_0^{-2.9} = \text{constant for } P_0 = \text{constant} \ , \\ &P_0 \cdot T_0^{-2.35} = \text{constant for } D_{\mathrm{eff}} = \text{constant} \ , \end{aligned} \tag{3.24}$$

where D_{eff} is an effective nozzle diameter that takes into the account some special nozzle geometries. For standard convergent, or sonic nozzles as they are sometimes called, or for simple holes one has $D_{\mathrm{eff}} = D$. Hagena has obtained good agreement between this scaling law and experimental data. *Obert* [26] had extended the treatment to include divergent nozzles, having an opening angle 2Θ. The mean size of hydrogen clusters measured for different diameters, cone angles Θ, and temperature T varies over 2 orders of magnitude. Assuming $D_{\mathrm{eff}} = D/\tan\Theta$ one obtains the scaling law:

$$P_0 \cdot D_{\mathrm{eff}}^{1.5} \cdot T_0^{-2.4} = \text{constant} \ . \tag{3.25}$$

Mean hydrogen cluster sizes obey this scaling law. The agreement is surprisingly good regarding the crudeness of the models involved. Equation (3.25) seems to be more generally applicable as it describes sodium expansions quantitatively [27]. On a qualitative level the same behavior was observed by many authors.

Intensities. The intensity on axis of an unclustered supersonic jet from a sonic orifice of diameter D and with no attenuation due to background scattering, or skimmer interference effects is [4]:

$$I_0 = f(\gamma) n_0 D^2 \sqrt{\frac{2kT_0}{m}} \quad [\text{atoms or molecules/s} \cdot \text{sr}] \tag{3.26}$$

where $f(\gamma)$ depends weakly on the ratio of specific heats, and is roughly unity. To first order the intensity is given by the density before the expansion n_0, the nozzle area, and the velocity of sound at the stagnation temperature. This is the same behavior as observed for a normal effusive or Knudsen source. The higher intensity of a supersonic beam is mainly due to the much higher density n_0 before the expansion. The highest intensities one can expect without a skimmer and for a very large pumping speed, is about several 10^{19} atoms or molecules/s · sr. Due to insufficient pumping speed, skimmer interference, background scattering, imperfect alignment, etc. the experimental intensity can be much lower. Detailed formula and design parameters have been given by *Miller* [4]. The angular distribution of atoms or molecules coming from a sonic nozzle is very broad and most of the gas going through the nozzle is pumped away unused. The angular distribution of the gas is slightly more forward peaked than for an effusive source which has a cos θ distribution. This leads to a "peaking factor" of about 2 in Eq. (3.26). For conical or trumpet shaped nozzles this is very different. Figure 11 a shows angular distributions measured for hydrogen clusters coming from a source at $T_0 = 77$ K. A strong forward focusing is observed. This is even more pronounced in Fig. 11b, where the intensities and angular distributions of different nozzles are compared. Note

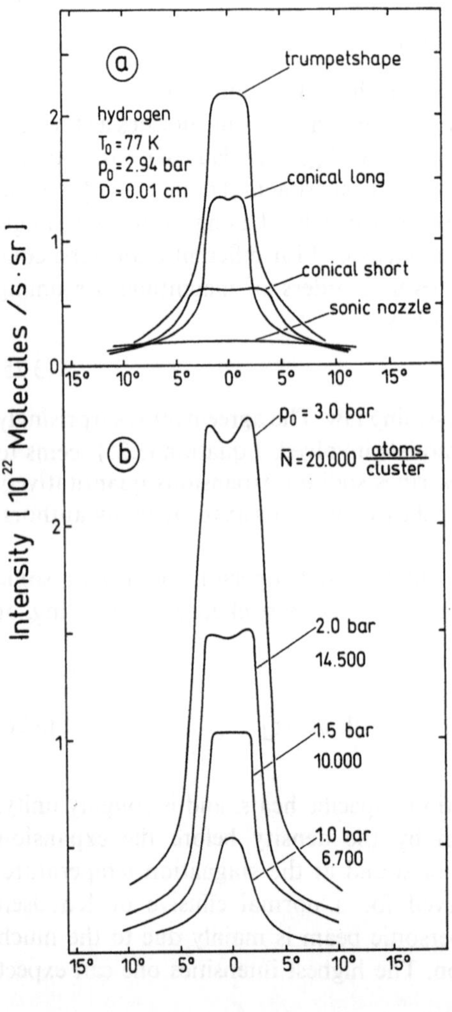

Fig. 11. a Angular distribution of a hydrogen cluster beam from a conical nozzle ($D = 0.35$ mm, nozzle length 60 mm, nozzle opening angle $2\theta = 14°$). The inlet pressure P_0 and the mean cluster size $\bar{N}$ are given as parameters. No skimmer was used. **b** Angular distribution of a hydrogen cluster beams from different nozzles. The stagnation pressure P_0 and the throat diameter D has been kept constant. Note the extreme forward peaking and high intensities compared to a simple hole nozzle

the tremendous increase compared to a sonic nozzle. The measured angular distributions are narrower than the cone angle of the nozzles, which seems to be due to boundary layer effects [26].

A small impurity content can increase the cluster size by a large factor. For the hydrogen clusters of Fig. 11 an impurity content of about 10^{-4} increases the mean cluster size by a factor of 2. The impurity acts as condensation germ so that cluster formation can start earlier than without it. Condensation phenomena in mixed beams have been studied by *Bartell* [28, 29] in a series of papers.

3.1.2 Gas Aggregation

Gas aggregation is a simple method to produce large clusters. A smoking fire or cloud and fog formation in nature are good examples. In all cases a liquid or solid is evaporated into a colder gas, which cools the evaporated atoms or molecules until they condense. Once their kinetic energy is low enough a stabilising three body collision can occur similar to that described in Eq. (3.9) earlier. Taking as an example the evaporation of copper into argon one has: $Cu + Cu + Ar \rightarrow Cu_2 + Ar$. The dimer will statistically be destroyed or grow until a cluster of critical radius r^* is formed where the value of r^* is given by Eq. (3.13). For larger radii supersaturation is attained and cluster formation becomes a very rapid process. Large clusters are efficient light scatterers as discussed in Chapter 6.5. This has led to the nickname of smoke source for gas aggregation cluster sources.

Figure 12 shows a smoke source for the production of carbon clusters (for references see Chapter 4.4). This source became famous as it allowed for the first time the production of macroscopic quantities of C_{60}. An electric arc is ignited between two carbon rods in a 10^4 Pa helium pressure. A black smoke coming from the arc rises because of the convection induced by the heat generated by the discharge. The soot sticking to the "smoke catcher" is dissolved in an organic solvent and large quantities of "bucky balls" can be isolated. Similar sources have been used to produce clusters for nanocrystalline materials as discussed in Chapter 3.7 of the second volume of this book [97].

Mainly very large clusters are generated by smoke sources. How can one reduce the cluster size? For a supersonic beam the rapid cluster growth is interrupted effectively by the freezing surface discussed in the paragraph after Eq. (3.8). The termination of the growing process is more difficult to achieve for an aggregation source. Figure 13 shows how to solve this problem. The gas is no longer stationary but streams over a container from which material is evaporated. Cluster growth stops effectively after the slits following the liquid nitrogen cooled condensation region.

The temperature of the liquid nitrogen cooled condensation cell has an influence, which is counter-intuitive at first glance. Smaller clusters are grown at lower gas temperatures. This can be qualitatively understood as: Consider two

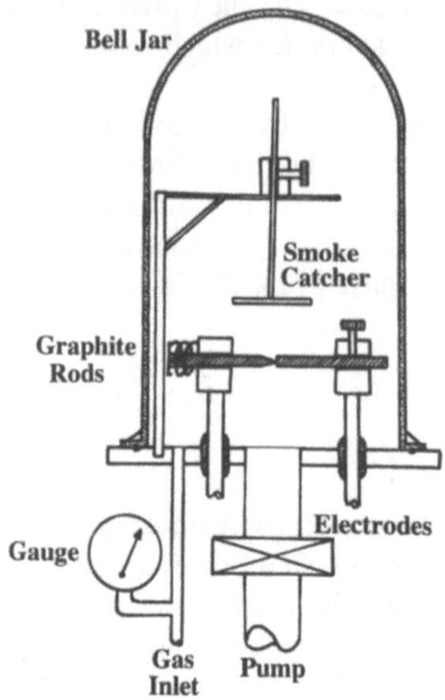

Fig. 12. A smoke source used for the production of C_{60}, C_{70}, and larger carbon clusters

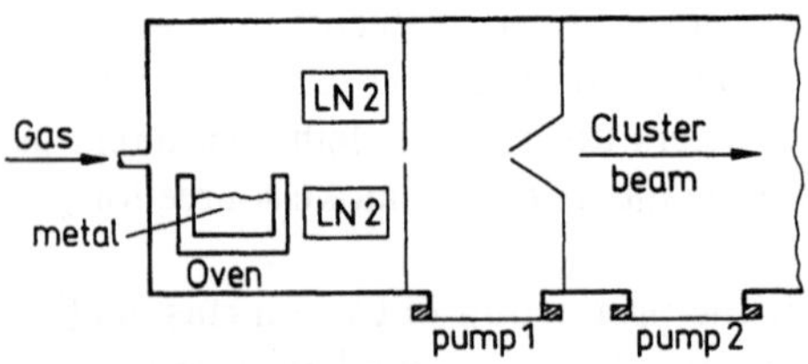

Fig. 13. Principle of a gas aggregation source. A metal is evaporated into a slow flow of argon of about 10 to 100 Pa. The metal atoms loose their kinetic energy by collisions with the gas, which also carries them into the aggregation cell. Cooling this part by liquid nitrogen reduces the mean cluster size considerably

otherwise identical sources to be at two temperatures $T_2 > T_1$. From Eqs. (3.11) to (3.13) one has for the critical radius r^* to reach supersaturation $r^*(T_1) < r^*(T_2)$. The clusters with $r < r^*$ are formed by a slow statistical process of decaying and growing. Only few clusters n_2 will reach the larger radius $r^*(T_2)$. Once supersaturation is reached at T_2 the few clusters grow rapidly until they have used up all the available material. A few ($\sim n_2$) large clusters have been grown. At the lower temperature T_1, $n_1 > n_2$ cluster arrive at the critical size $r^*(T_1)$. They compete for material while growing. If not enough material is available, the process stops with a smaller mean cluster size. A similar argument must be valid for small clusters, where classical condensation theory fails. If one wants to generate small clusters a liquid nitrogen cooled aggregation region is necessary.

The details of the condensation process are much more complicated than discussed above. For example, it is not understood why heavier atoms give smaller clusters than lighter ones under otherwise identical conditions. Other important parameters besides the temperature are the pressure and velocity of the gas, the geometry of the condensation cell, the diameters of the diaphragms used and the available pumping speed. Due to these many interdependent

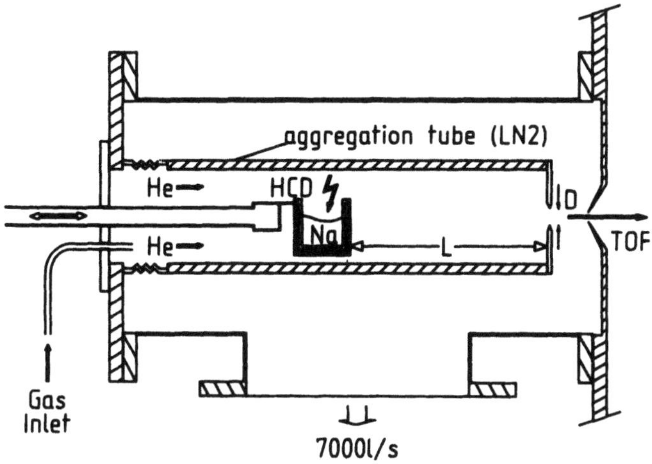

Fig. 14. A gas aggregation source for cold sodium cluster ions. An electric hollow cathode discharge (HCD) is ignited inside the liquid nitrogen cooled zone, the sodium container functioning as hollow cathode

parameters, the source is trickier to use and less well studied compared to supersonic beams. No scaling laws exist. The correlation of some experimental parameters with the intensity has been discussed by [30]. Other experimental details are explained in [31, 32, 33] and Fig. 13. A source working with a sputter instead of an evaporation source is described in [34, 35]. In Ref. [36] a gas aggregation source is described, which has an additional electric discharge to produce copious quantities of negatively charged clusters. A very simple source for the production of very large clusters has been described by *Satoh* and *Kimura* [37].

A source to produce cold cluster ions is shown in Fig. 14. An electric discharge is ignited inside the aggregation tube [38]. The source produces neutral and charged clusters of both polarities. If the distance L is large enough the clusters will be cooled to (nearly) the temperature of the helium gas. By changing its temperature, cluster ions of variable temperature can be produced. In this way the unknown heating effects of an ionizing laser or lamp can be avoided.

The clusters have the velocity of the streaming gas which is less than that of a supersonic beam. Sometimes these sources are used at so high a gas flow that a mild supersonic expansion occurs at the first diaphragm. If large clusters are produced their kinetic energies can become large. This is an important design parameter for the mass spectrometer separating and detecting large clusters.

3.1.3 Surface Erosion Sources

Another common method to produce clusters uses heavy particle impact, or intense laser radiation to remove material from a solid surface. In the first case an energetic ion beam (e.g. Xe^+ ions at 30 keV kinetic energy) hits a surface and atoms, molecules and clusters are ejected. This method works well with all solids and frozen gases and liquids [39, 40, 41, 42]. Or one can use the energetic ions near the cathode of a glow discharge for surface erosion ([34, 35]) or the violent processes on the electrodes of an electric arc. A more controlled but also more expensive method uses a high power laser pulse (10 to 20 mJ in 10 ns focused to an area of 1 mm^2). This gives an intense beam of small clusters for the elements C, Si or Ge. For the other elements mostly atomic ions are produced. The high intensity of the laser light ($\geq$ 100 MW/cm^2) erodes about 500 atomic layers per shot, leading to densities of $10^{18}/cm^3$. The trailing part of the pulse will ionize most of the material ejected by the leading part. A plasma of ten to twenty thousand Kelvin results. Focusing the laser beam not on the pure material but on a compound can give a large increase of the intensity. Using an organic polymer and an excimer laser a large flux of carbon clusters can be obtained [43] for example. The clusters from these sources are very hot, as they are produced by quite violent a process. They cool by evaporation of constituents. This leads to a well structured mass spectrum. Both these sources have been combined with supersonic beams or gas aggregation type sources in order to cool the clusters and manipulate the cluster sizes.

Laser Ablation Source. Smalley and coworkers were the first to combine a laser ablation source and a supersonic beam. In the early versions the laser ejected the material from a rotating and advancing rod, as can be seen in Fig. 15. Very large pumping speeds (up to 20,000 l/s) were used. Figure 16 shows the newest version [44]. A flat sample disk is used. A rod and pinion drive moves the disk, so that the fixed laser beam engraves a spiral groove into it. First the pulsed valve is opened, 60 microseconds later the laser is fired. The main improvement compared to the earlier design is the incorporation of a small "waiting room" for the clusters directly above the disk [45]. Here they make a turbulent motion and are continuously cooled by fresh, cool helium from the pulsed valve. The expansion

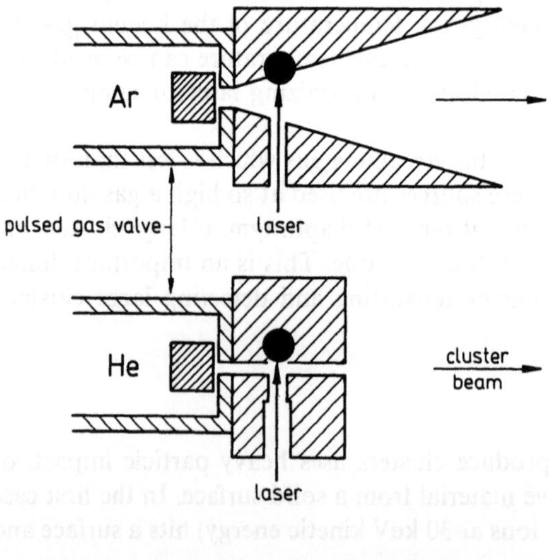

Fig. 15. Two variants of a laser ablation source combined with a pulsed supersonic beam. Material is laser ablated from a rotating and advancing rod, indicated as a black circle. It is entrained in the gas pulse, cooled and clusters are formed. The lower source can be used to produce pure metal clusters (e.g. Cu_n), while the upper one can produce clusters like $CuAr_n$

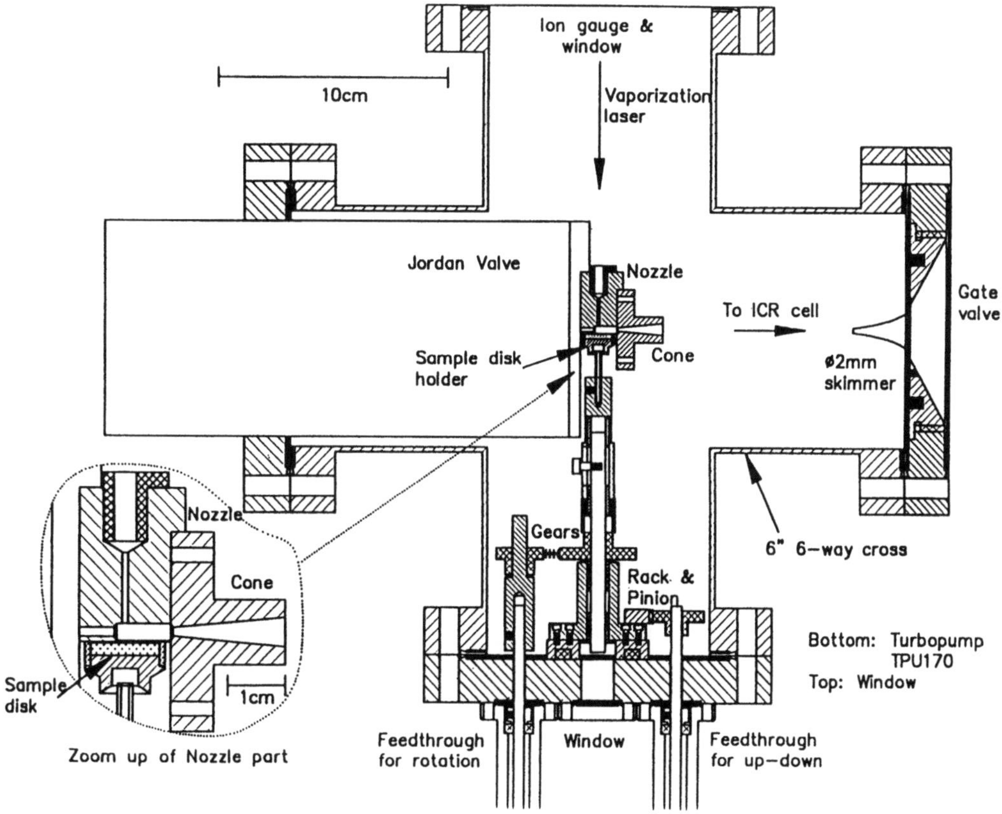

Fig. 16. A new version of the laser ablation/pulsed supersonic beam source

proceeds through a 2 cm long 10° cone. The setup is much smaller than the earlier one, and only a 170 l/s turbopump is used. It produces roughly the same cluster beam intensity as the earlier design, with far better control and reproducibility [44].

Another Surface Erosion Source: A source operating on a similar principle is shown in Fig. 17. The expensive laser is replaced by a high-current pulsed arc discharge which removes material from the electrodes. This is clustered in the following expansion. The authors call it a Pulsed Arc Cluster Ion Source (PACIS) [46].

3.1.4 Pick-up Sources

The principle of a pick-up source is shown in Fig. 18. Atoms, molecules or charges are "picked-up" either during the formation process of the clusters (at point A) or when they have cooled down (point B).

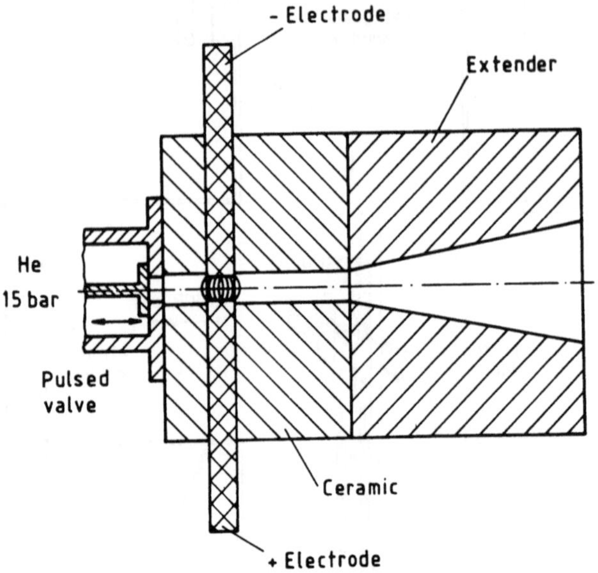

Fig. 17. Pulsed Arc Cluster Ion Source. The laser is replaced with a powerful electrical discharge

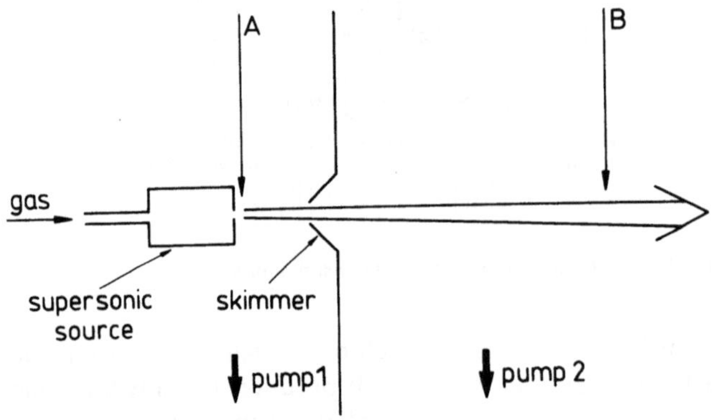

Fig. 18. Principle of a Pick-Up Cluster Source

Production of Mixed Neutral Clusters. In the experiments described in Chapter 2.2 of the second volume of this book [97] a supersonic argon cluster beam encounters a cloud of SF_6 or other molecules either at point A or B of Fig. 18. As a result the molecule can stick to the surface of the cluster which, because of its large mass, is deflected only negligibly from its original trajectory. If the cluster is large enough, the spectrum taken downstream of the pick-up point shows that the molecule remains on the cluster's surface. By contrast when the molecule is coexpanded with the clustering gas from the nozzle, it can become the nucleus of condensation and can be found spectroscopically inside the cluster. Pressures of the order of 10^{-6} Pa at point B through a length of a couple of centimeters are enough to dope a sizeable fraction of the clusters. Pick-up at point B is less efficient because of the pumping action of the jet which tends to defend the centerline of the beam from intrusion by the background gas. Since the pumping speed in the source chamber is normally quite large, a much larger consumption of the gas is needed to obtain pick-up at point A than B. See also Fig. 32. A pulsed variant of this source is described in Chapter 2.1 of the second volume of this book.

Production of Cold Cluster Ions. Cold cluster ions can be grown when electrons or ions are injected at point A of Fig. 18. The charges act as effective condensation germs reducing considerably the critical radius r^* of Eq. (3.13). Cluster ions can be grown this way, when this is not possible for neutrals. The solvated electron clusters discussed in Chapter 2.5 of the second volume of this book [97] have very often been grown this way. Crossing a charged particle beam at point at position B results nearly always in very hot clusters, which are prone to fragmentation.

Other Types of Sources. A large variety of other cluster sources exist, which do not fit too well into the classification scheme adopted here. The interested reader will find them in the more experimentally oriented sections of this book. A strong inhomogeneous field can suck cluster ions directly from a liquid surface [47]. This device has been termed Liquid-Metal-Ion-Source (LMIS). It is commercially available for some ions, as it is used in the semiconductor industry. Or a special electric discharge using a magnetic field (Penning discharge) has been used to generate intense beams of small cluster ions.

Problems with Cluster Sources. All known cluster sources have one severe disadvantage. They produce a broad distribution of clusters. The methods to select one cluster size only are reviewed in Chapter 3.4.

3.2 Detection of Cluster Ions

The detection of a slow neutral cluster seems to be impossible. The clusters have to be ionised for an efficient, mass selective detection. An atomic ion has a detection efficiency of near unity, if it is accelerated to some keV kinetic energy and impinges on a metal surface. The secondary electron emitted can be used for further amplification. The same method is used universally to detect clusters ions. One experiences a severe experimental problem, as can be seen from Fig. 19. Vanadium cluster ions, V_n^+, with $n = 1$ to 9 have been accelerated with energies between 15.5 and 25 keV and impinged on a stainless steel surface. Figure 19 shows the number of electrons η emitted per incident atom, as a function of the velocity of the impinging cluster. The points fall on a straight line with a threshold velocity of about $v_{th} \approx 50$ km/s. Below this value the detection probability becomes very small [48].

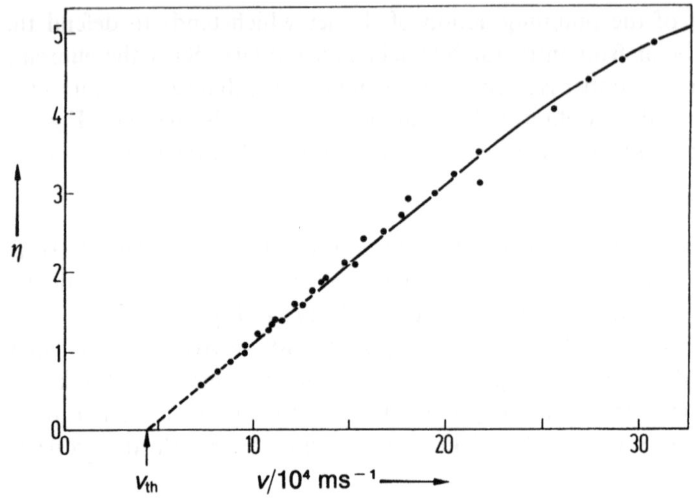

Fig. 19. Electron emission per impinging atom of mass selected vanadium cluster ions. Note that the emission probability scales linear with velocity. Only above a threshold of about 50 km/s the electron emission rises

A numerical example is instructive: A rather accurate equation for the velocity v is:

$$v\,[\mathrm{m/s}] = \sqrt{2E/m}\;10^4\,, \tag{3.27}$$

where E is the energy in electron volts and m the mass in atomic mass units ($^{12}C = 12\,\mathrm{amu}$). For $v \geq v_{th} = 50$ km/s one obtains for the threshold kinetic energy E_{th}:

$$E_{th}\,[\mathrm{eV}] \geq 12.5\,m\,. \tag{3.28}$$

One obtains for C_{60}: $E_{th} = 9$ keV, for Ar_{100}: $E_{th} = 50$ keV, and for Na_{2000}: $E_{th} = 525$ keV. Because of these large energies a small van de Graaf has been used to accelerate ion to a high kinetic energy before detection [49]. It helps, that Fig. 19 gives the probability per atom in the cluster. Even if η is very small, say 1%, one has for a $N = 100$ cluster still a 100% detection efficiency. The absolute thresholds have been determined by [43, 50].

A simple, effective, and robust detector for cluster ions is shown in Fig. 20. It is colloquially known as "Even-Cup" [51], and is especially convenient as detector for time-of-flight mass spectrometers. Positive ions are accelerated to -30 keV, and impinge a converter dynode which sits inside a cup-like structure. A recessed grid at its entrance prevents field penetration due to the high voltage. The ejected electrons are extracted by an electric field which penetrates through a hole. The electrons are accelerated by 30 keV to a grounded scintillator which is covered by a small Al film. The photons are transferred by a plexi-glass light pipe to a photomultiplier outside the vacuum system. For negative ions the converter has to be at $+30$ keV, and the scintillator at $+50$

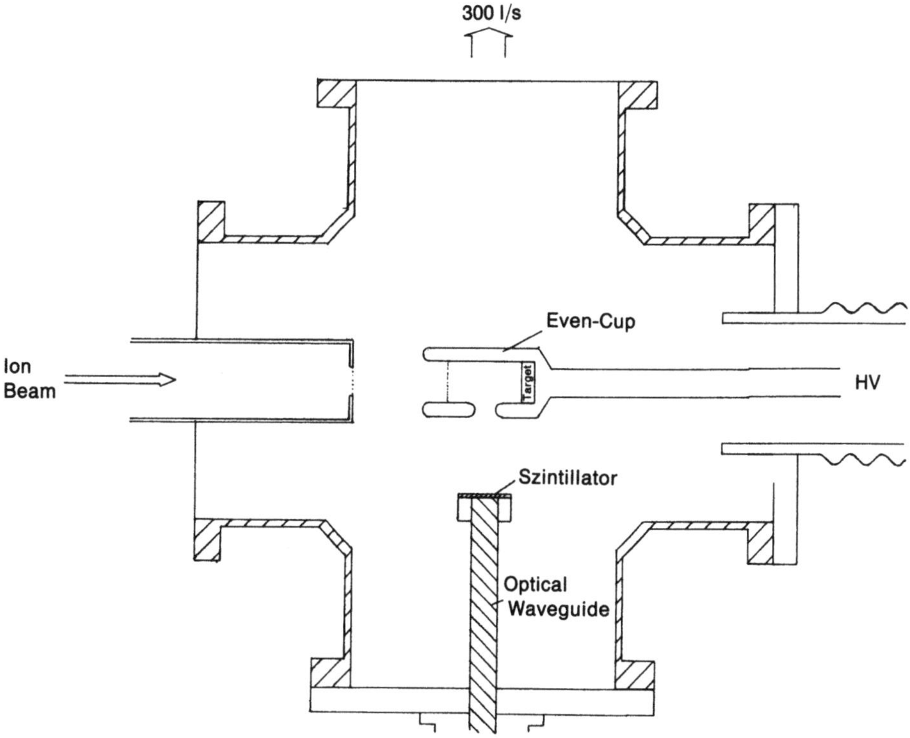

Fig. 20. The Even Cup is a simple and effective detector for cluster ions

to 60 keV. Also for positive ions it was found advantageous to electrically float the scintillator, as this can reduce noise. The SIMION software [52] was found to be very convenient in designing this detector and other ion and electro optics.

3.3 Electron Diffraction

A typical experiment consists of a well collimated electron beam with an energy of 30 to 50 keV crossing a supersonic cluster beam. The fast electrons are scattered from the atoms in the cluster, and the diffraction pattern is recorded mostly on photographic film. A series of diffraction rings around the position of the primary electron beam is recorded, analogous to the Debye–Scherrer rings observed from a macroscopic powder [53]. It takes only a few minutes to record the pattern, but a long time to interpret it. For the interpretation a geometrical model of the cluster is constructed and the diffraction pattern calculated. As explained in many a textbook the scattered intensity is the modulus of the Fourier Transform of the scatterers with respect to the momentum transfer. The agreement between the experimental and computed patterns are optimised in a trial and error fashion until an acceptable fit is obtained.

Three pieces of information can be extracted: 1) The mean geometry of the cluster, 2) the mean cluster size, and 3) the cluster temperature. Only a brief discussion of the underlying physics will be given here. More information can be found in [14, 28, 29]. The geometry is obtained directly from the fit of the assumed model to the experimental data. It is plausible that an icosahedron has a different Fourier Transform than a part of an fcc lattice, etc. The larger the cluster, the narrower are the diffraction rings. This is similar to light diffraction from a grating with few or many grooves. The more scatterers add their contributions coherently, the sharper is the resulting pattern. The physical idea of the derivation of the temperature is a little more involved. The position of an atom in the cluster is described by a vector:

$$\vec{r}(t) = \vec{r}_0 + \vec{u}(t)\,, \tag{3.29}$$

where $\vec{r}(t)$ is a constant vector pointing to an atom in the cluster and the time dependent vector $\vec{u}(t)$ describe the thermally activated vibrations. An average over the thermal motion [53] gives for the temperature dependence of the diffracted intensity:

$$\begin{aligned} I(T) &= I(T=0)\cdot\exp(-W) \\ W &= \frac{1}{3}\langle u^2\rangle k^2\,, \end{aligned} \tag{3.30}$$

where $\langle u^2\rangle$ is the mean square thermal motion of the atoms from Eq. (3.29), and k^2 is the square of the momentum transfer experienced by the electron beam upon scattering. The value of $\langle u^2\rangle$ increases with temperature. The diffraction lines do not broaden but decrease exponentially in intensity according to Eq. (3.30). The temperatures given in Fig. 8 have been obtained this way [14].

3.4 Methods for the Production of (Nearly) Mass Selected Neutral Cluster Beams

U. Buck

The production of cluster beams leads generally to a distribution of cluster sizes. By manipulation of the source conditions one can, at best, end up with a size distribution the halfwidth of which is about the average cluster size [54]. Such a result is by no means sufficient to carry out experiments with mass selected clusters. Therefore, a number of methods have been proposed to reach this goal [55]. One can, of course, use a mass spectrometer which is coupled to an ionization process which, in turn, provides size specific information of the neutral precursors. Such a process, however, works only if special methods are applied to avoid the fragmentation process. Examples are the two colour resonant two photon ionization of aromatic molecules with spectroscopic labelling [56] or the photoelectron-cluster ion coincidence technique for metals [57]. These are, however, no generally applicable procedures. Velocity analysis also does not help very much, since most of the clusters produced in adiabatic expansions or by gas aggregation have nearly the same velocity [58]. There are, of course, the well-known effects of the parallel velocity slip and the on-axis enrichment of the heavier species caused by variations in perpendicular velocities which both occur in seeded beams with large mass-mismatches [59]. They should also be observed with clusters under similar conditions, but they are again not very well suited for a general separation method. Deflections in inhomogeneous magnetic or electric fields did not prove to be very efficient for size selection, since the moments per atom depend only weakly on the cluster size [55]. There are only two methods left which have been explicitly used to produce clusters of one size: the deflection of the cluster beam by an atomic beam and the re-neutralization of cluster ions. The former method is based on the fact that in the scattering process larger clusters are scattered into smaller angles than the lighter ones. In this way mass distributions of large $(N_2)_n$ clusters ($n = 10{,}000$) were obtained by deflection from a crossed jet under multiple collision conditions [60]. In the high resolution version of this method, small Ar_n clusters ($n \leq 10$) were separated from each other by making use of their different angular and velocity distribution in a single collision experiment [61]. The method has been successfully applied to a series of measurements on the fragmentation by electron impact ionization and the infrared photodissociation of size-selected clusters [62, 63]. The problem with this method is the small intensity of the resulting beam after the scattering process. The second method starts with a well separated cluster ion which can easily be achieved in any type of mass filter. Then this ion is neutralized and the resulting neutral cluster can be used in subsequent experiments. Several methods have been proposed for the second step ranging from charge exchange of positive and negative ions to photodetachment of negative ions. First applications have been reported for S_n clusters [64] and Na_n clusters [65] using charge exchange. The problem with this method is that because of electronic excess energy and inelastic collisions fragmentation of the resulting neutral clusters cannot be excluded and as a result again a size distribution is found. In the case of Na_n clusters it was experimentally shown that mainly Na_n or Na_{n-1} clusters were produced [65].

In the present chapter the two methods are described in detail and the results are summarized at the end.

3.4.1 Scattering from Atomic Beams

The method of cluster separation in a scattering experiment with a secondary beam under single collision conditions is based on the fact that the heavier

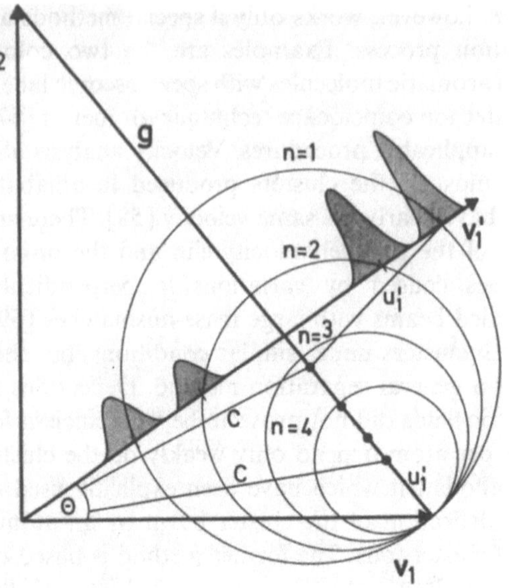

Fig. 21. Schematic Newton diagram for the scattering of a cluster beam X_n (velocity v_1) from an atomic beam (velocity v_2). The positions of elastically scattered monomers through tetramers are given by circles around the center of masses. g is the relative velocity, c is the velocity of the center of mass, and u_1' and v_1' are the final velocities in the cm and the lab system, respectively

clusters are scattered into smaller angles with smaller final velocities compared with the lighter clusters. This behavior is documented in a velocity vector (or Newton) diagram which is constructed on the basis of conservation of momentum and energy [66]. Such a schematic diagram is shown in Fig. 21, assuming that the cluster beam (velocity v_1, mass $n \cdot m_1$) is crossed by the atomic beam (v_2, m_2) at an intersection angle of 90°. The relative velocity is given by the difference vector $g = v_1 - v_2$. For elastic scattering all final center-of-mass (cm) velocities u_1' are restricted to end on a sphere around the center-of-mass with the radius

$$u_1^{(n)} = m_2 g/(nm_1 + m_2) \ . \quad (1)$$

In the case of inelastic scattering one has to take into account the transferred energy ΔE and the collision energy E according to $u_1' = u_1(1 - \Delta E/E)^{1/2}$.

Since the clusters have nearly the same velocity v_1, also the relative velocity is independent of the cluster size, but the final cm velocities $u_1^{(n)}$ are quite different. With increasing mass of the clusters they decrease and so do the radii of the circles in the diagram (see Eq. (1) and Fig. 21). For the laboratory(lab) angle shown in the diagram only dimers and monomers are detected, whereas the maximum scattering angle Θ_n for trimers and tetramers is smaller. It is immediately clear from this picture that larger clusters can be easily excluded by choosing the right angular range for detection. For the smaller clusters, two solutions are offered. The easiest way is to use a mass spectrometer to discriminate against the smaller masses. This procedure works only if, at least, a small fraction of the cluster X_n is detected at the corresponding mass of the ion X_n^+ or at a fragment mass X_m^+ which is larger than the next smaller cluster size X_{n-1}^+.

Such a case often occurs for hydrogen bonded systems, e.g. $(H_2O)_n$, which are detected at the protonated masses $(H_2O)_{n-1}H^+$. But even if the fragmentation is complete, the angular selection helps. We start the measurement at an angle $\Theta > \Theta_3$ so that only dimers contribute. With decreasing angle the contributions of the larger clusters appear size by size so that they can be sorted out.

The second solution is to use in addition to the angle also the final velocity as determining parameter. Depending on the experimental resolution, the intensity at a certain angle and velocity corresponds in a unique way to a certain cluster size as is demonstrated in Fig. 21 by the peaks belonging to $n = 1$ and $n = 2$. Since the masses of the clusters are usually heavier than the mass of the scattering partner, there are always two contributions to one cluster size which differ in the lab velocity v_1'. The procedure is general and completely independent of the subsequent detection method. Indeed, if a mechanical velocity selector is used for the specification of the velocity, a beam of one size is produced which can be used in subsequent experiments such like deposition onto surfaces or into matrices. If time of flight (TOF) analysis is used for the velocity specification, the answer is known only after the measurement. To summarize, there are mainly two modes of operation: (1) Measurements at fixed angle and mass and (2) measurements at fixed angle and velocity.

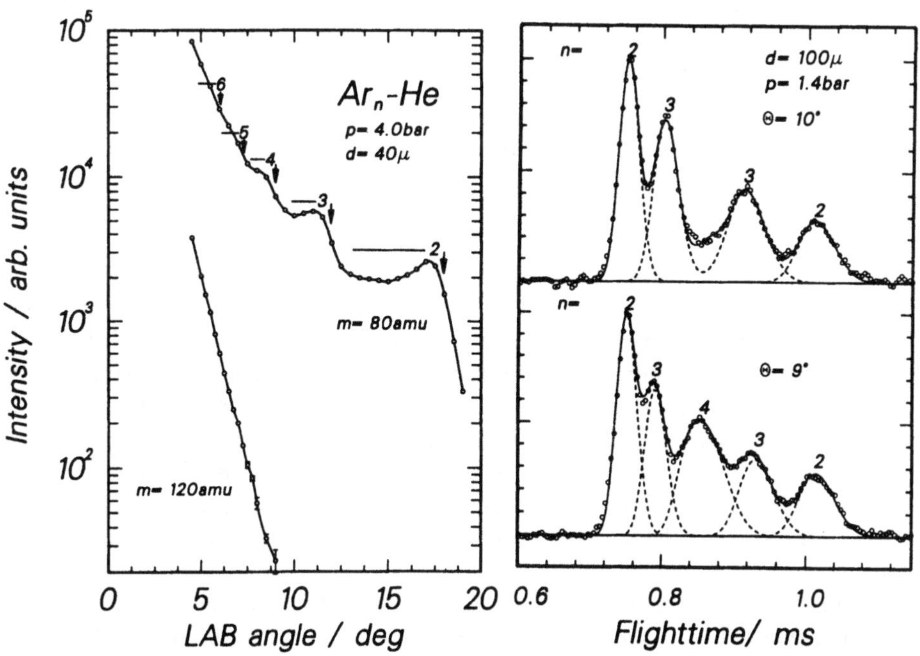

Fig. 22. Measured angular dependence (left panel) and time-of-flight spectra (right panel) for Ar_n–He scattering. The beam data (pressure *p*, nozzle diameter *d*) and the mass of the detector *m* are indicated. The TOF spectra are obtained at $m = 80$ amu [61]

Typical results are shown in Fig. 22 for small Ar_n clusters [61–63]. The Ar_n clusters are produced by adiabatic expansions and are then scattered from a He beam. The TOF spectra are taken using a mechanical pseudorandom chopper with a flight path of 449 mm. The calculated limiting angles for the different cluster sizes are $\Theta_2 = 18°$, $\Theta_3 = 12°$, $\Theta_4 = 9°$, and $\Theta_5 = 7°$. The results of the measured angular dependence are in good agreement with these predictions. At each limiting angle a rise of the intensity is observed indicating the contribution of the corresponding cluster size to the measured signal at the dimer mass ($m = 80$ amu). From these results it is immediately clear that extensive fragmentation occurs, because most of the signal under the given experimental conditions results from other clusters than dimers. The results obtained at $m = 120$ amu exhibit an extreme case. The first measurable intensity appears in the angular range which corresponds to pentamers so that there is no contribution of Ar_3 and Ar_4 to the ion signal Ar_3^+. Already these signals give a good qualitative overview about the cluster separation. For a quantitative analysis also the velocity has to be specified.

Two typical TOF spectra measured at different lab angles are also shown in Fig. 22. Since the quadrupole mass filter is operated at the dimer ion mass (80 amu), there are no contributions due to monomer scattering. Neutral trimers and dimers are found to contribute to the spectra at angles smaller than $\Theta = 12°$. At the maximum scattering angle for tetramers ($\Theta_4 = 9°$) another maximum is resolved. Note that the different peaks contain unambiguously the indicated cluster size.

We summarize the conditions for applying the scattering method for the production of single sizes clusters:

1. Collisional dissociation should be avoided. This condition is easily met by the appropriate choice of the collision energy (as low as possible) and the scattering partner (as light as possible, small m_2).
2. For the separation a high resolution molecular beam apparatus and intense cluster beams with good expansion conditions are required in order to minimize angular ($\Delta\Theta < 0.5°$) and velocity spread ($\Delta v/v < 0.05$) of the beams [67]. In the present experimental arrangement cluster sizes cannot be distinguished from each other, if their intensity peaks are less than 0.5° separated which happens for clusters larger than $n = 6$. This figure depends very much on the masses and velocities of the particles. Increasing m_1 and v_1 deteriorates the resolution, while increasing m_2 and v_2 improves it. We should, however, keep in mind that the intensity loss caused by the scattering process is about 10^{-6}. Nevertheless the method works and has been applied in many experiments.
3. During the scattering process, energy is transferred to the clusters which can be kept small, if m_2 is small. In principle, it does not disturb the size analysis unless the transfer is so large that the limiting angle of the next cluster is reached. On the other hand, by changing the collision conditions the amount of transferred energy can be varied on purpose.

3.4.2 Re-Neutralization of Ions

Since the production of cluster ions and their size selection in a mass filter is straightforward, there are many proposals to get monodispersed neutral clusters through the reneutralization of the selected ions. Negative ion clusters are neutralized by photodetachment [68–70], collisional detachment or charge exchange [71]. Positive ion clusters are neutralized by resonant or near resonant charge exchange [64, 65, 71–74]. Photodetachment is now a well established method to study the electronic properties of cluster anions [68, 70]. For Ag_n^- ions, however, fragmentation to smaller negatively charged daughter clusters was observed so that photofragmentation was found to compete well with photodetachment [68, 69]. In the charge exchange experiments

$$A_n^+ + B \rightarrow A_n + B^+ + \Delta E \tag{2}$$

the energy mismatch ΔE plays the key role both for the efficiency of the process and the competing fragmentation channels which, in turn, determine the distribution of produced neutrals. Experimental results have been reported for $S_n^+ + Zn$ [64, 72], $Na_n^+(K_n^+) + Cs$ [12, 21], $Rb_n^+ + Rb$ [20], $Si_n^- + SF_6$ [71], and $Pb^+ + Na$ [22]. If ΔE is made small, the collisional process involves little momentum transfer as was shown by translational spectroscopy in [72]. In the experiment with the negative cluster ions only negatively charged fragments of SF_6^- but no Si_m^- ($m < n$) were detected [71]. Therefore, the authors consider such a process to be a good candidate for the neutralization, since the excess energy is concentrated in the partner molecule SF_6. Since in one case only the neutral products have been analysed [65], we will describe this experiment in more detail.

The schematic diagram of the apparatus is shown in Fig. 23. The Na_n^+ cluster ions are generated by first producing neutral Na_n clusters in an adiabatic expansion through a 0.1 mm diameter conical nozzle and then ionizing them with a nitrogen laser. The ions are accelerated to about 1 keV and are selected according to their size in a time of flight mass spectrometer and then enter

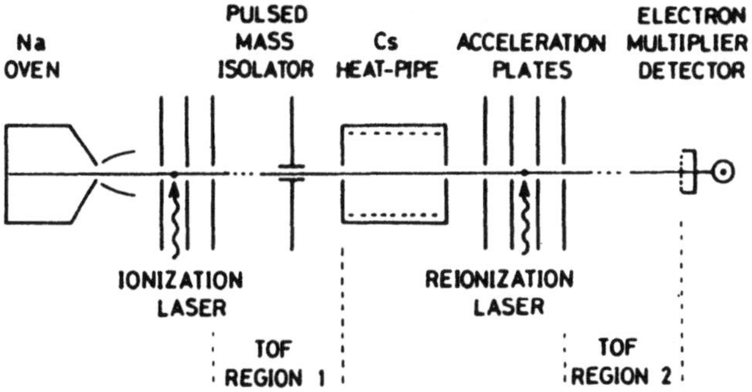

Fig. 23. Schematic diagram of the apparatus for the charge exchange after [65]

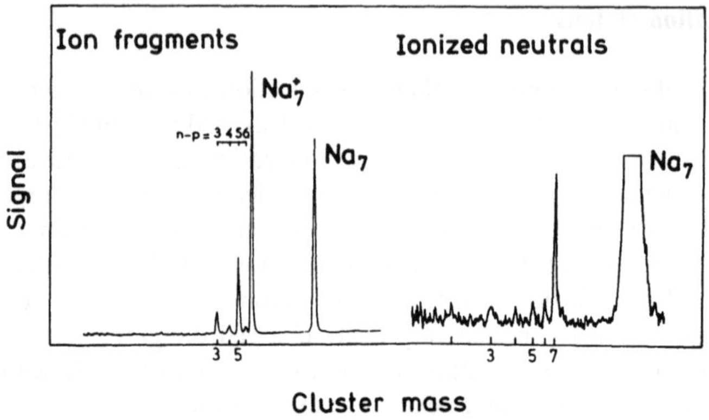

Fig. 24. Results of the charge exchange experiments for Na_7 after [65, 74]. From left to right we see the collision induced ion fragments, the parent ion, the neutral product, the reionized mass spectrum of the neutral product and the neutral product again

a 15 cm-long alkali metal heat pipe in which a uniform atomic density is maintained at low pressures so that only single collisions occur. Residual ions and product neutrals enter a second TOF mass spectrometer. Here the fragmentation of the surviving ion clusters, the fast neutrals and, after additional reionization with a second laser, the neutral product distributions are observed. The results for Na_7^+ are displayed in Fig. 24. The largest peak is that of the parent ions. To the left we find the fragmentation products caused by collisional dissociation

$$Na_n^+ + Cs \rightarrow Na_{n-p}^+ + pNa + Cs \ . \tag{3}$$

To the right the neutral product peak is seen which contains according to the reionized spectrum Na_7 and to a much lesser extend Na_6 and Na_5. In this case the energy mismatch $\Delta E = 0.13$ eV is very small so that a large neutral signal arises. For the reionization measurement the photon energy was chosen to lie above the ionization limit but below the dissociation energy so that any additional fragmentation results from the internal excitation of the charge transfer products which is sufficient to evaporate a monomer. Such processes are indeed observed for Na_7 when the photon energy is increased from 4.33 eV to 4.61 eV [76]. For all cases investigated the largest intensities are always found at Na_n and Na_{n-1} except for Na_9 and Na_{21} for which Na_{n-1} and Na_{n-2} are the larger contributions. In the latter two cases the tendency to evaporate an atom is probably due to the relative stability of the closed electron shells in Na_8 and Na_{20}. A detailed investigation of the charge transfer process revealed a strong dependence on the energy defect ΔE, the collision energy and the cluster size. In general the cross section are large and drop down from 40 Å^2 for the dimer to 10 Å^2 for $n = 20$ [74].

3.4.3 Summary

Out of the many proposed procedures only two methods have been actually shown to work, (1) the charge transfer of mass selected cluster ions and (2) the deflection of the cluster beam by an atomic beam. In the first method the critical parameter is the energy defect ΔE. If this value can be kept small, the cross sections are large and also the fragmentation into different neutral channels can be minimized. In the only investigated example, so far, the charge exchange of sodium clusters with cesium, $Na_n + Cs$, a very narrow final distribution was found of at most three masses Na_n, Na_{n-1} and Na_{n-2} with a most probable product of Na_n or Na_{n-1}. The advantage of the method is the easy extension to all cluster sizes with the disadvantage of a large velocity and a possibly high degree of internal excitation.

The second method is based on the momentum transfer from an atomic beam. In this manner, the different clusters of a supersonic nozzle beam are dispersed according to their size into different angles with different velocities. The method requires a high resolution molecular beam apparatus for separating the different contributions. The results of the preceding sections show that cluster sizes up to $n = 10$ could be easily separated from each other corresponding to an angular resolution of approximately 0.5°. An extension up to $n = 15$ should be possible for other scattering partners and better collimated beams. It is noted that the pure deflection of the cluster beam which is easier to handle than the full size selection, excludes the contribution of larger clusters successively.

3.5 Mass Spectrometers

As discussed above, it is difficult to sometimes impossible to mass select neutral clusters. Mass spectrometry has become important in cluster science because of this problem, and many methods do exist to mass select ions. The different types of mass spectrometers can be broadly classified into two categories: machines working with time-dependent or time-independent fields. Only the underlying principles will be discussed here. Several books and reviews are available on this subject.

Wien filters and magnetic as well as electric sector instruments belong to the second category. A wide variety of different combinations of static magnetic and electric fields can used to separate ions of different mass to charge ratio. In a magnetic field, particles with different momentum $\vec{p} = m\vec{v}$ follow identical trajectories, while in an electric field this is the case for particles of equal energy $E = mv^2/2$. This peculiar difference is deeply rooted in the different symmetry properties of electric and magnetic fields [77].

Mass spectrometers working with time–dependent fields are: quadrupoles, time-of-flight (TOF), and ion-cyclotron-resonance (ICR) machines. The latter two will be discussed briefly, a more detailed account can be found in [78].

Wien Filter. The force exerted on a moving ion in a magnetic field by the Lorentz force is compensated by an electric field in a Wien filter [79]. This is possible for one mass to charge ratio only. A great advantage of the Wien filter is its straight ion trajectory, a disadvantage its limited mass resolution.

Magnetic Sector Instruments. The mass spectrometer shown in Fig. 25 uses an electric and a magnetic sector field. Clusters are produced in a supersonic source, ionized by electron impact, and accelerated by lenses to the first slit S. After a first field free region the ions enter a homogeneous magnetic field. They have all been accelerated to the same kinetic energy ($E = mv^2/2$ = constant). Only one mass with the correct momentum can pass through the slit between magnetic and electric sector. The geometric width of the ion beam at this point is given by two contributions, the energy spread and the angular width of the ion beam. The energy dispersion of the electric sector can be so chosen, that it compensates that of the magnetic sector. One obtains a "double focusing" instrument of high resolving power, and a very good background suppression. The two field free regions can be used to study metastable decays, laser or

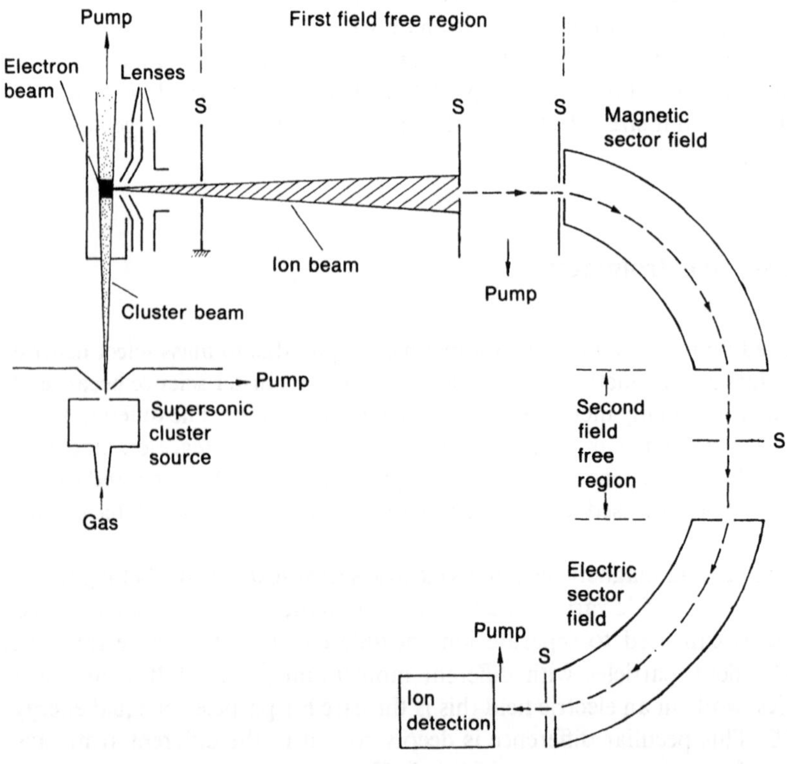

Fig. 25. Schematic of a double focusing magnetic mass spectrometer used by the Innsbruck group. Clusters are produced in supersonic expansion, ionized by electron impact, accelerated and mass analyzed

collision induced processes. Many experiments discussed in Chapters 2.6 and 2.7 of the second volume of this book [97] have been obtained using the machine shown in Fig. 25.

TOF-Mass Spectrometers. Time-of-flight (TOF) mass spectrometers have become popular in cluster science because of their large mass range, compactness, and relative cheap prize tag. The linear, single-focusing instrument described first by *Wiley* and *McLaren* [80] has a mass resolution not exceeding 300. Figure 23 shows two linear instruments in tandem. One mass is selected in TOF region 1. This mass undergoes a charge exchange collision in the heat-pipe. After reionization the products are analyzed in TOF region 2.

Mamyrin added an electrostatic mirror to the linear TOF, obtaining double focusing and a vastly improved resolution [81]. As the ions are reflected in the mirror these instruments are generally called "reflectrons" today. Resolution over 30,000 have been obtained in one very carefully designed instrument [81]. Low intensity is the prize one has to pay for this very high resolution. The ionization region can have a size of only a few mm^3. With large ionization regions resolutions in the 1000 to 3000 range are more typical.

Figure 26 shows a tandem TOF composed of a reflectron coupled to a small linear TOF-spectrometer [82]. Clusters are produced in a cluster source. They pass a skimmer, and are ionized between plates P_0 and P_1. The ions are accelerated by two different electric fields by applying potentials to plates P_0, P_1, and P_2. The ionizing beam has a certain geometric width δx. Therefore ions starting nearer to plate P_0 have a higher kinetic energy, but also a longer flight path than those starting near P_1. The fast ions overtake the slow ones at the first time focus P_3. At this point the *Wiley–McLaren* single focusing conditions are fulfilled [80]. A mass selector at this point allows to select ions of one mass only.

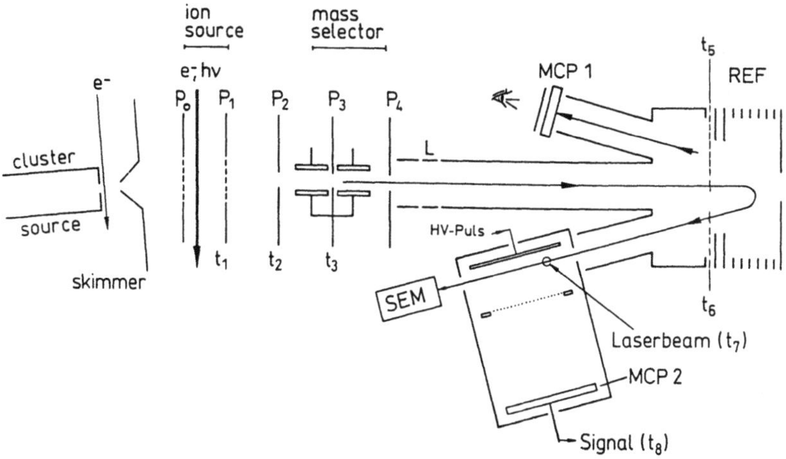

Fig. 26. A tandem TOF-mass spectrometer used for optical spectroscopy of mass selected cluster ions. A reflectron is combined with a small linear TOF. The reflector is moveable so that it can send the beam into the different arms of the machine

Once past P_3, the faster ions are ahead of the slower ones, but they have a longer flight path in the ion reflector, so on exiting the reflector, the slow ions are again ahead of the fast ones. The slow ones are again overtaken at the second time focus, which is the location of the multi-channel-plate 1 (MCP1) in Fig. 26. A phosphor screen behind MCP1 allows the visual inspection of the shape of the cluster beam. Choosing the electric fields properly, one can have time focusing in double or triple order [81]. The ion reflector is tiltable, so that the ion beam can be sent either on the MCP1 or into the small TOF in the lower part of the figure. The position of the second time focus intersects with that of a laser beam. The photofragments are recorded by applying a high voltage (HV) pulse and accelerating the photofragments onto MCP2. Note that there are no nets in the ion path, also the reflector is gridless. This gives a higher signal and lower background as the ions are not scattered and fragmented at the grids, the ion beam is narrower because of the lens action of the gridless reflector, and most important, the clusters are not heated on their way through the machine. A collection of review articles on time of flight machines can be found in Ref. [83].

ICR–Mass Spectrometers. An ion is moving on a circle in a homogeneous magnetic field, if its velocity is perpendicular to the magnetic field lines. The radius r of the circle is given by the condition that the Lorentz force evB is balanced by the centrifugal force mv^2/r. This gives for the radius $r = mv/eB$. The time for one revolution is $t = 2\pi r/v = 2\pi m/eB$. For the cyclotron frequency ω one obtains accordingly:

$$\omega = \frac{2\pi}{t} = \frac{eB}{m} . \tag{3.31}$$

Note that ω does not depend on the velocity of the ion. Particles with a higher v move on a larger radius r, which gives the same ω.

The ion-cyclotron-resonance (ICR) cell is a small box, having four isolated side plates, which can be seen in Fig. 27 and Fig. 28, and two isolated end plates. The unit is immersed in a homogeneous magnetic field. For cluster studies ions are generally injected into this cell, as shown in Fig. 28. The ions are excited by a radio-frequency (RF) pulse applied to one plate. The ion packets are bunched

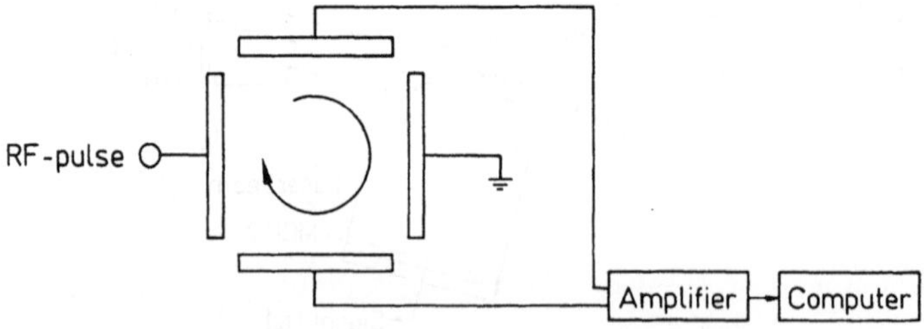

Fig. 27. Principle of an ICR mass spectrometer. Ions are excited by a radio-frequency (RF) pulse to circular orbits. The induced image currents are amplified. A computer is needed for Fourier transformation. The magnetic field is perpendicular to the paper

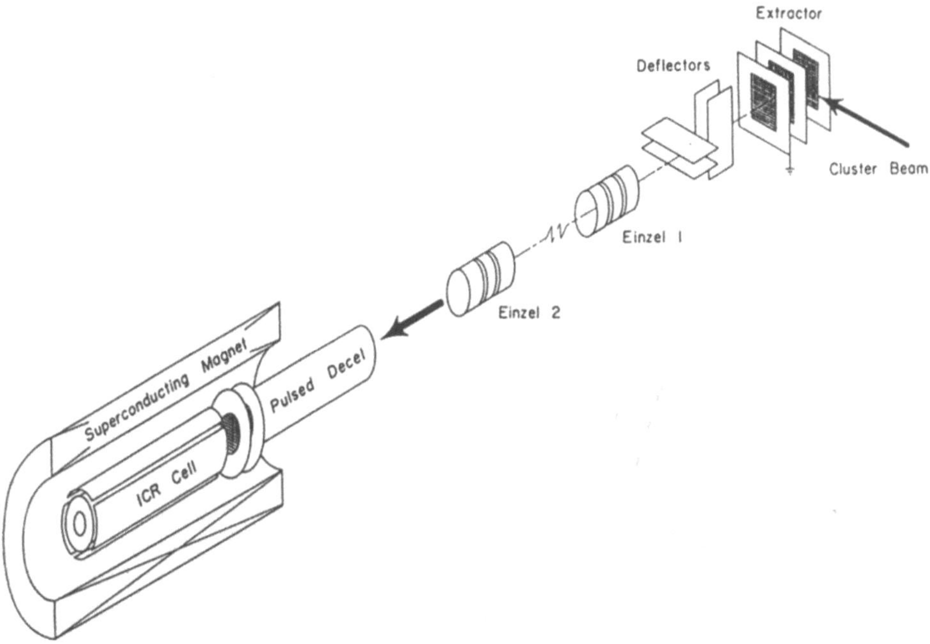

Fig. 28. Set-up used by the Rice group to inject cluster ions into an ICR cell. The ion cluster beam is steered by two deflectors and two einzel lenses. It is decelerated before being trapped in the ICR cell

and execute a circular motion. The image current induced on two opposite plates is amplified, and fed to a computer to do a Fourier transformation. If only one ion mass is in the ICR–cell the image current would show a single sine-curve. Its Fourier transform is one frequency only, from which one can calculate with Eq. (3.31) the mass. If one has a mixture of ions, the signal is a superposition of many sine-waves, and after Fourier transformation the complete mass spectrum can be obtained. The advantages of an ICR spectrometer are: simple mechanical structure, high transmission, simultaneous detection of all masses. By playing with the electronic parameters one can have low or high resolution, eject certain ions, while keeping other ones, etc. Its disadvantages are: the prize of the large magnet, the involved RF-pulse technology and the very fast computer needed.

3.6 Optical Spectroscopy

In one photon spectroscopy the cluster is directly excited or ionized by one photon. Two photon spectroscopy can probe the excited states by scanning the frequency ν_1 (compare Fig. 29), provided this state has a long enough life time.

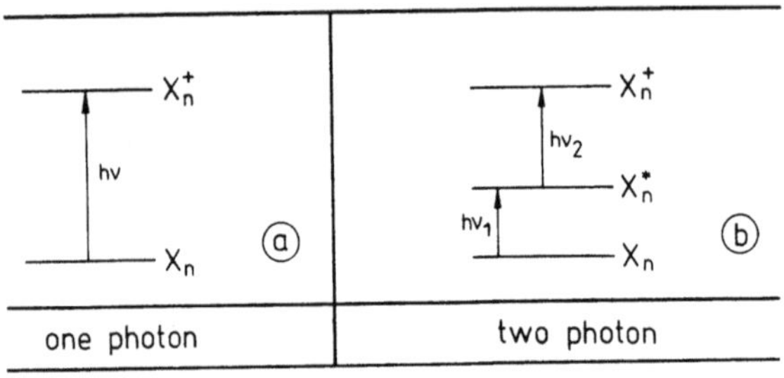

Fig. 29. Level scheme of one and two photon absorption. The experiment can be performed on neutral or charged clusters. Often the two frequencies of two photon spectroscopy are the same. If a high laser intensity is used one speaks of REMPI or Resonantly Enhanced Multi Photon Ionization.

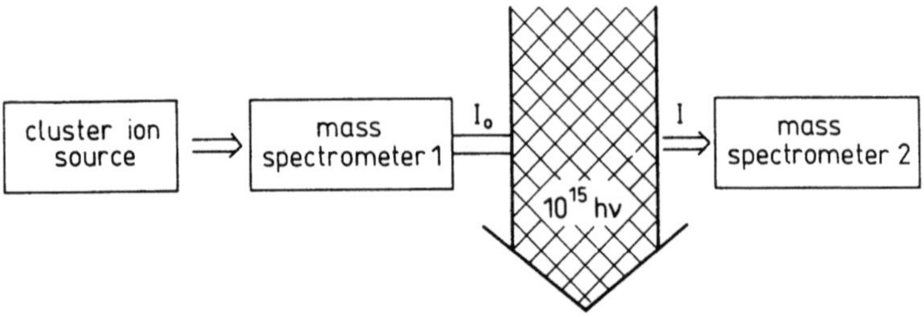

Fig. 30. The principle of depletion spectroscopy for cluster ions

This has been done successfully for diatomic molecules, Na_3 [16, 84], the solvation of aromatic molecules [85], and the spectroscopy of semiconductor clusters [86]. But for alkali clusters the intermediate states are very short lived. If one does not use lasers which have a pulse length in the 100 fs range the intermediate state has decayed long before the second photon is absorbed.

The fast decay of electronic into vibrational energy is typical for cluster science. It is more often quite a nuisance, but has been put to advantage in "depletion spectroscopy". Figure 30 shows its principle for cluster ions. Mass spectrometer 1 selects one mass, which has an intensity I_0. These ions are irradiated by an intense laser pulse, which leads to photo induced fragmentation. Only the intensity I is remaining of the originally selected cluster. If there would be perfect overlap between cluster and photon beam, one can calculate the cross section from the standard Lambert–Beer "absorption law":

$$I/I_0 = \exp(-\sigma\phi) \tag{3.32}$$

here ϕ is the photon fluence (which can be easily measured) and σ is the cross section for photo fragmentation. The overlap of the two beams is normally far from perfect and has to be determined experimentally [87].

If at least one atom is ejected in the time window of the experiment the photofragmentation cross section σ is equal to the desired photo absorption cross section. For large and cold clusters this is not true. This can be easily

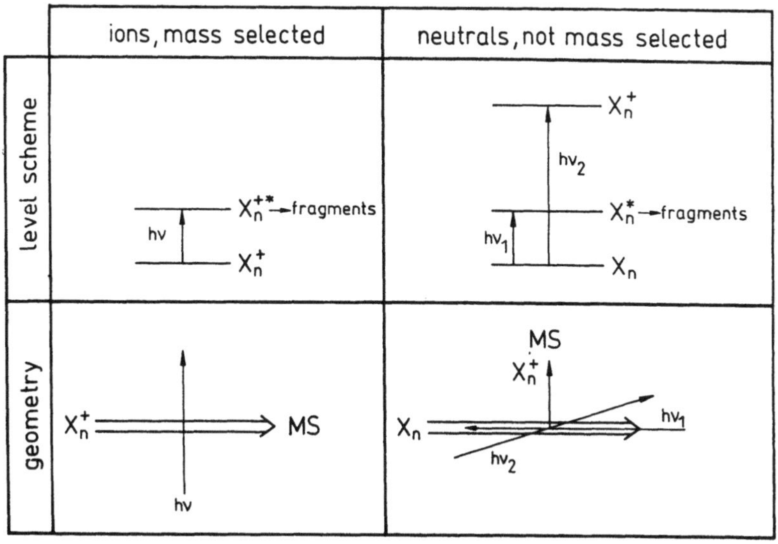

Fig. 31. The various possibilities of depletion spectroscopy. MS stands for mass spectrometer

understood from Fig. 9. For large and/or cold clusters the time scales for atom ejection become longer than the time window of the apparatus. A possible way out is multifragmentation [88]. Also it has been observed that the first atom ejected after photo excitation is ejected nonstatistically [21].

Normally one measures the absorption of a light beam by matter. Note that because of the extremely tenuous cluster beam the role is inverted. The photons play the role of an "absorber" in depletion spectroscopy. The technique is very sensitive, because of the high laser powers available today. For the spectrometer shown in Fig. 26 one needs at least one cluster ion per two to three laser pulses to record an optical spectrum.

Depletion spectroscopy for neutral and ionized clusters is compared in Fig. 31. While for ions the fragmentation can be monitored easily, one must make careful power and frequency dependent studies, and use a lot of physical arguments to rule out fragmentation for neutral depletion spectroscopy [86, 89, 90].

3.7 Infrared Spectroscopy

G. Scoles

Infrared spectroscopy of molecular beams was introduced by *Gough, Scoles* and *coworkers* more than a decade ago, and has been applied by them to the study of medium-large noble gas clusters seeded with infrared chromophores [91]. The early work demonstrated that information about the effects of the solvation on the guest molecule and the structural properties of the host cluster itself could be obtained as a function of the *average* cluster size, bridging the gap between molecular and bulk behavior. The existing vast literature on IR spectroscopy of molecules imbedded in a noble gas matrix with which the data obtained for the clusters can be compared, provides further motivation and support for this type of work. Some of the results are discussed in Section 2.2 of the second volume of this book [97].

Figure 32 shows a schematic of the experiment. The cluster beam can be seeded either by coexpanding a diluted mixture (typically one part in a few thousand) of the chromophore, or alternatively, by allowing the neat cluster beam to pass through a region (pick-up cell) in which the density of the chromophore can be controlled. As the clusters travel through the cell, they collide with the chromophore molecules and one or more of these may attach themselves to the cluster as described in Section 3.1.4. Spectroscopy is performed on the doped cluster beam, by intersecting it with a modulated CO_2 laser beam. The chromophore absorbs a photon from the laser, eventually relaxing into the cluster phonon modes. This excess energy released into the cluster is dissipated causing one or more atoms to evaporate. The laser induced signal is measured by one of the two bolometer detectors that are aligned with the molecular beam. The first, an annularly shaped Al_2O_3 disk with a little temperature sensor attached to it (AB), can collect fragments released from the clusters, at the same time allowing the main part of the beam to pass through the center. The fragments impart heat on to the sensor, leading to a sharp resistance change that is detected using lock-in techniques. This bolometer is used to record the spectra of Ar, Kr and Xe clusters. In contrast, a second bolometer (TB) directly on line with the molecular beam, detects the loss of fragments due to absorption of the photons and is used for the measurements on He, H_2 and D_2 clusters.

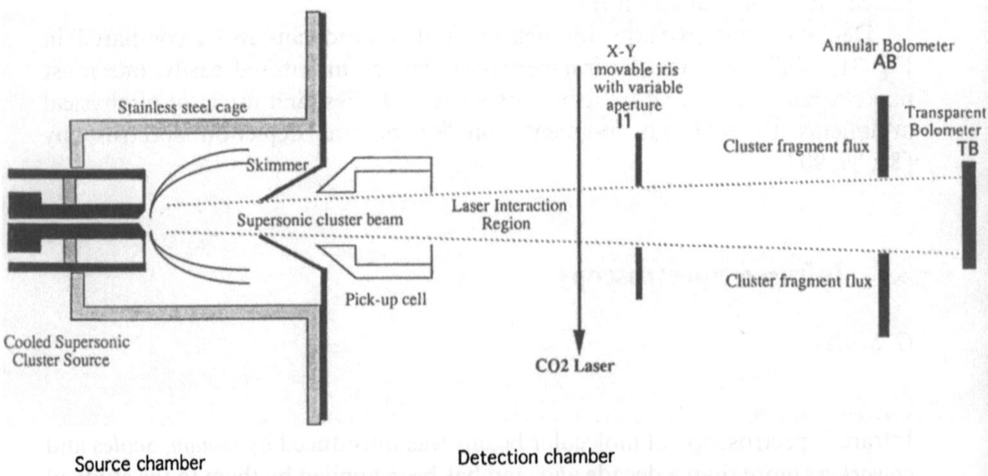

Fig. 32. A schematic of the Infrared photofragmentation spectrometer. The essential elements in the production of doped noble gas clusters and the bolometric detection scheme are shown

The signal is recorded for as many laser lines as can be produced in the absorption region of the chromophore using three isotopically substituted CO_2 lasers and one N_2O laser and the spectrum obtained by dividing the laser induced signal by the power. The resolution of the spectra is limited by the line tunability of the lasers, and it averages about 0.3 cm^{-1}. This method requires fairly large laser powers which has limited its application to the spectral region covered by the CO_2 lasers. In future the technique may be combined with continuously tunable narrow band lasers of higher power as these become available.

3.8 Photo Electron Spectroscopy

Photo electron spectroscopy (PES) measures the kinetic energy of photo ejected electrons. The energetic difference between charged (mostly negative so far) and neutral clusters can be measured this way. From an analysis of the experimental data the energy level structure of the neutral clusters can be obtained [92].

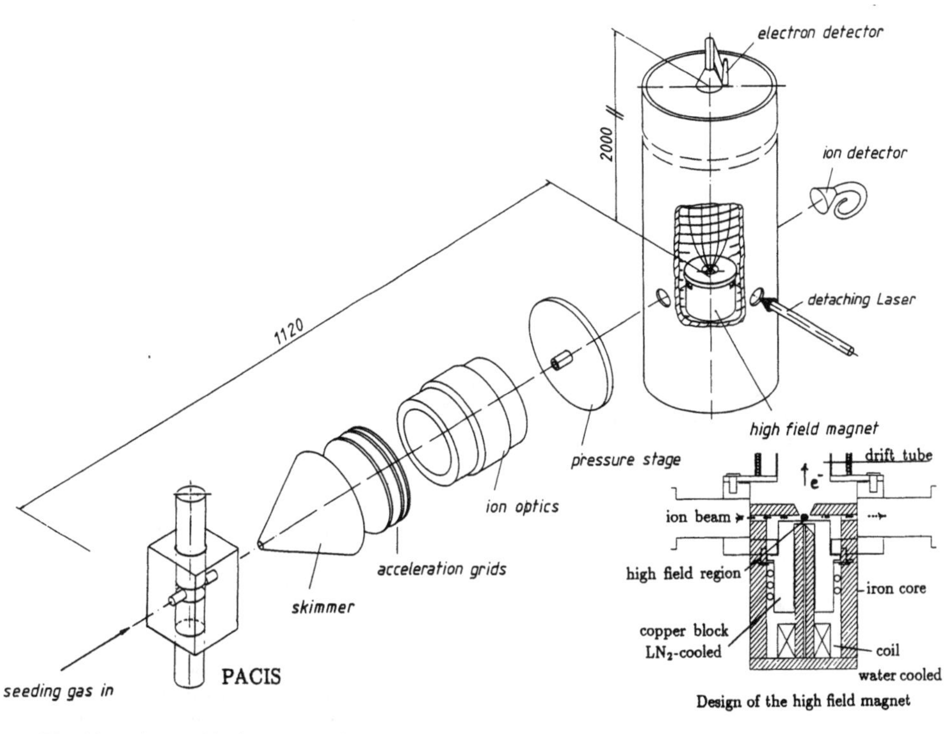

Fig. 33. A laser ablation source is coupled to the photo-electron-spectrometer of the Bielefeld group. Electrons are ejected from mass selected negatively charged clusters by the second laser. The flight times of the electrons are measured by the channeltron-detector. The electrons are trapped and guided by the inhomogeneous magnetic field

Figure 33 shows a typical experiment [93]. Negatively charged clusters are produced by a PACIS source (see page 226). They traverse a skimmer and are focused by a linear TOF. A second pulsed laser interacts with cluster ions of one mass only. Electrons are photo-detached, e.g. for Cu_{10}^-:

$$Cu_{10}^- + h\nu \rightarrow Cu_{10} + e^- \quad (3.33)$$

The electrons are photo ejected into a strong inhomogeneous magnetic field. They spiral around the field lines and are guided to the detector. The time between 2nd laser pulse and the electron's arrival time is measured. This is again a tandem time-of-flight (TOF) apparatus. The first TOF selects clusters of one mass, the second measures the flight times of the electrons. The machine in Fig. 33 works with a pulsed cluster and two pulsed laser beams. A cluster PES machine using continuous beams has been built at John Hopkins university [94]. The resolution of the electron spectrometers is typically 20 to 50 meV.

ZEKE-Spectroscopy. A much higher resolution can be obtained by the ZEKE-technique. The acronym stands for zero kinetic energy electrons. The techniques has been applied by their inventors to a variety of molecular systems [95]. Figure 34 shows an apparatus for ZEKE-spectroscopy on neutral clusters [96]. A neutral, not mass selected, sodium cluster beam is intersected by one or two narrow band, tunable laser beams. The laser frequency (or the sum of the laser

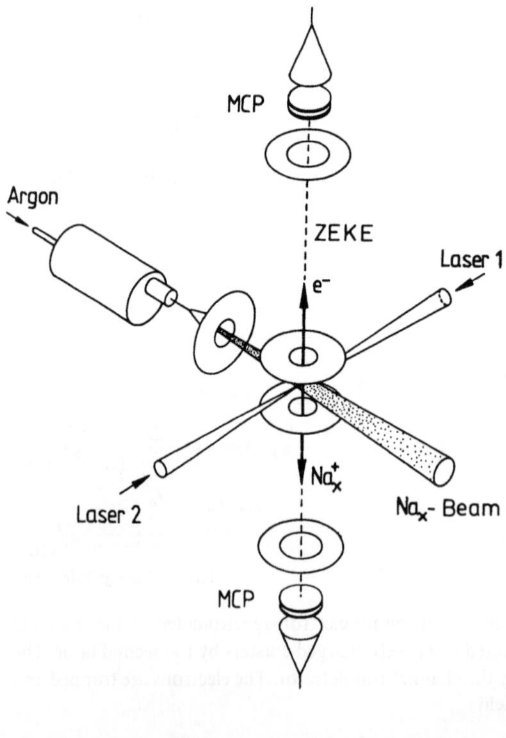

Fig. 34. Apparatus for ZEKE-spectroscopy of Na_n. One or two pulsed laser beams intersect a not mass selected sodium cluster beam. The sum of the laser frequencies is tuned to a threshold for electron production. The electron with nearly zero kinetic energy is detected by applying a weak voltage pulse to the plates above and below the laser/cluster beam interaction zone. Electrons and ions are registered in separate linear time-of-flight spectrometers equipped with multi-channel-plate (MCP) detectors

frequencies) is tuned very near to a threshold for electron production. Directly at threshold the electrons do indeed have zero kinetic energy. The intersection region of laser and cluster beam should be as free of electric fields as possible. A fast electron, which might also be produced by the laser, will leave the interaction region quickly. Only electrons with very low kinetic energy remain until about one micro second later a low voltage pulse (typically 1 to 5 V/cm) is applied. This accelerates these very low energy electrons into the electron detector. The energy resolution can be as low as 0.1 meV, which is two to three orders of magnitude better than conventional photo electron spectroscopy. The ZEKE-spectrum of Na_3 has been obtained this way [96]. For small clusters it is generally not necessary to measure electrons and ions in coincidence. The energetics of the system are well enough defined by the good energy resolution, that fragmentation processes can be avoided or reduced to a negligible level.

There exist two mechanisms to produce very low energy electrons. Either the direct ionization at threshold as discussed above. Alternatively a highly excited neutral cluster can be formed, which is field ionized by the low voltage pulse. For molecular systems and Na_3 the second route to ionization seems to be the dominant one.

3.9 New Developments

Cluster science is currently a very active and broad field, and only limited examples could be discussed above. Techniques which have not been described, and were a real progress has been made recently, are mentioned here very briefly: 1) A high pressure drift cell (≈ 500 Pa) can be used to discriminate between isomers of one cluster ion mass in a beam [98]. 2) Magnetic measurements on free [99, 100] and supported [101] clusters have been published. 3) Pump-probe experiments with ultrafast time resolution have become available [102–104]. 4) A pulsed supersonic beam is described in ref. [105] which produces clusters with variable vibrational temperature. 6) The electrospray source has been perfected. It has been used mainly to produce metal ions, which might be highly charged, solvated by organic molecules [106]. 7) Techniques used especially for metal clusters have been reviewed in [107]. For an up-to-date overview of current activities, a recent conference report is available [108].

References

1. K. Hilpert: Chemistry of inorganic vapors. Structure and Bonding **73**, 99 (1990)
2. M. Kappes, S. Leutwhyler: In: *Atomic and Molecular Beam Methods*, G. Scoles et al. (eds.), p. 381 (Oxford University Press, 1988)
3. L.D. Landau, E.M. Lifshitz: *Fluid Mechanics*. Pergamon Press, 1959

4. D.R. Miller, In: *Atomic and Molecular Beam Methods* G. Scoles et al. eds., p. 14 (Oxford University Press 1988)
5. H. Haberland, U. Buck, M. Tolle: Rev. Sci. Instrum. **56**, 1712 (1985)
6. O. Stern in his very first paper on the velocity distribution of an effusive atomic beam did not pay attention to this point. This was evidently remarked by A. Einstein: Z. Phys. **2**, 417 (1920)
7. R. Gentry: In: *Atomic and Molecular Beam Methods*, G. Scoles et al. (eds.), p. 54 (Oxford University Press, 1988)
8. K.L. Busarow, G.A. Blake, K.B. Laughlin, R.C. Cohen, Y.T. Lee, R.J. Saykally: J. Chem. Phys. **89**, 1268 (1988)
9. A. McIlroy, D.J. Nesbitt: J. Chem. Phys. **91**, 104 (1989)
10. D. Lippman, W.C. Schieve, C. Cantero: J. Chem. Phys. 4969 (1984)
11. F.F. Abraham: Rep. Prog. Physics **45**, 1113 (1982)
12. A.C. Zettlemoyer: *Nucleation Phenomena.* Wiley, New York, 1977
13. J.M. Soler, N. Garcia, O. Echt, K. Sattler, E. Recknagel: Phys. Rev. Lett. **49**, 49 (1992)
14. J. Farges, M.F. de Feraudy, B. Raoult, G. Torchet: Surf. Science **106**, 95 (1981)
15. J. Gspann: Z. f. Phys. **D3**, 43 (1986)
16. G. Delacrétaz, L. Wöste: Chem. Phys. Lett. **120**, 342 (1985)
17. C.E. Klots: J. Phys. Chem. **92**, 5864 (1988)
18. J. Leygnier, C. Bréchignac, P. Cahuzac, J. Weiner: J. Chem. Phys. **90**, 1493 (1989)
19. G. Gerber: priv. communication (1992)
20. N.G. Gotts, R. Hallett, J.A. Smith, A.J. Stace: Chem. Phys. Lett. 491, 1991; J. Chem. Phys. submitted for publication (1992)
21. H. Haberland, Ch. Ellert, B.V. Issendorff, Th. Reiners, M. Schmidt: unpublished results
22. T. Leisner, O. Echt, Xue Jian Yan, E. Recknagel: Chem. Phys. Lett. 386 (1988)
23. U. Buck, H. Meyer: Surf. Sci. **156**, 275 (1985)
24. O.F. Hagena: Surf. Sci. **106**, 101 (1981)
25. O.F. Hagena: Z. Phys. **D4**, 291 (1987)
26. W. Obert: In R. Campargue, ed., *11. Symposium on Rarefied Gas Dynamics.* 1181 (1979)
27. U. Buck: private communication (1990)
28. L.S. Bartell: Chem. Rev. **86**, 491 (1986)
29. L.S. Bartell, Y.Z. Barshad: J. Phys. Chem. **91**, 2890 (1987)
30. H. Schaber, T.P. Martin: Surf. Sci. **156**, 64 (1985)
31. J. Mühlbach, K. Sattler, E. Recknagel: Surf. Sci. **106**, 188 (1981)
32. F. Frank, W. Schulze, B. Tesche, J. Urban, B. Winter: Surf. Sci. **156**, 90 (1985)
33. C. Bréchignac, P. Cahuzac: Z. Phys. D (1990)
34. H. Haberland, M. Karrais, M. Mall: Z. Phys. D 413 (1990)
35. H. Haberland, M. Karrais, M. Malland, Y. Thurner: *J. Vacuum Science and Technology*, **A10**, 3266 (1992).
36. K.M. Mc.Hugh, H.W. Sarkas, J.G. Eaton, C.R. Westgate, K.H. Bowen: Z. Phys. **D12**, 3 (89)
37. N. Sato, K. Kimura: J. Am. Chem. Soc. **112**, 4688 (1990)
38. Th. Reiners: Ph.D-thesis, Freiburg 1992, unpublished
39. I. Katakuse: Microclusters, *Springer Series in Material Science* **4**, 10 (1987)
40. W. Begemann, S. Dreihöfer, K.H. Meiwes-Broer: Z. Phys. D **3**, 183 (1986)
41. T.F. Magnera, D.E. David, J. Michl: J. Chem. Soc. Faraday Trans. **86**, 2427 (1990)
42. P. Fayet, L. Wöste: Surf. Sci. **156**, 134 (1985)
43. E.E.B. Campbell, H.G. Ulmer, B. Hasselberger, H.G. Busmann, I.V. Hertel: J. Chem. Phys. **93**, 6900 (1990)
44. S. Maruyama, L.R. Anderson, R.E. Smalley: Rev. Sci. Instrum. **61**, 3686 (1990)
45. P. Milani, W.A. deHeer: Rev. Sci. Instrum. **61**, 1835 (1990)
46. G. Ganteför, H.R. Siekman, H.O. Lutz, K.H. Meiwes-Broer: Chem. Phys. Lett. **165**, 293 (1990)
47. N.D. Baskar, R.F. Frueholz, C.M. Klimack, R.A. Look: Phys. Rev. **B36**, 4418 (1987)
48. F. Thum, W.O. Hofer: Sur. Sci. **90**, 331 (1979)
49. L. Friedman, G.H. Vineyard: Comments At. Mol. Phys. **15**, 251 (1984)
50. P.W. Geno, R.D. MacFarlane: Int. J. Mass Spectrom. Ion Proc. **92**, 195 (1989)

51. D. Bahat, O. Chesnovsky, U.C. Even, N. Lavie, Y. Magen: J. Phys. Chem. **91**, 2460 (1987)
52. D.A. Dahl, J.E. Delmore, A.D. Appelhaus: Rev. Sci. Instrum. **61**, 607 (1990)
53. C. Kittel: *Introduction to Solid State Physics*. John Wiley, New York (1976)
54. M. Kappes, S. Leutwyler: In: *Atomic and Molecular Beam Methods*, G. Scoles (ed.), Chap. 15, p. 380 (Oxford University Press, New York, 1988)
55. M. Kappes: Chem. Rev. **88**, 373 (1988)
56. S. Leutwyler, J. Bösinger: Chem. Rev. **90**, 89 (1989)
57. K. Rademann, T. Rech, B. Kaiser, U. Even, F. Hensel: Rev. Sci. Instrum. **62**, 1932 (1991)
58. T. Pflaum, K. Sattler, E. Recknagel: Z. Phys. D **1**, 131 (1986)
59. D. Miller: In: *Atomic and Molecular Beam Methods*, G. Scoles (ed.), Chap. 2, p. 14 (Oxford University Press, New York, 1988)
60. J. Gspann, H. Vollmer: *Rarefied Gas Dynamics*, K. Karamcheti (ed.), p. 261. (Academic Press, New York, 1974)
61. U. Buck, H. Meyer: Phys. Rev. Lett. **52**, 109 (1984); J. Chem. Phys. **84**, 4854 (1986)
62. U. Buck: J. Phys. Chem. **92**, 1023 (1988)
63. U. Buck: In: *The Chemical Physics of Atomic and Molecular Clusters*, Enrico Fermi School, G. Scoles (ed.), p. 543 (North Holland, Amsterdam, 1990)
64. M. Arnold, J. Kowalski, G. zu Putlitz, T. Stehlin, F. Träger: Z. Phys. **A 322**, 179 (1985)
65. C. Bréchigniac, Ph. Cahuzac, J. Leynier, R. Pflaum, J. Weiner: Phys. Rev. Lett. **61**, 314 (1988)
66. U. Buck: In: *Atomic and Molecular Beam Methods*, G. Scoles (ed.), Chap. 18, p. 449 (Oxford University Press, New York, 1988)
67. U. Buck, F. Huisken, J. Schleusener, J. Schaefer: J. Chem. Phys. **72**, 1512 (1980)
68. L.-S. Zeng, C.M. Karner, P.J. Brucat, S.H. Yang, C.L. Pettiette, M.J. Craycraft, R.E. Smalley: J. Chem. Phys. **85**, 1681 (1986)
69. P. Fayet, L. Wöste: Z. Phys. D **3**, 177 (1986)
70. G. Ganteför, M. Gausa, K.H. Meiwes-Broer, H.D. Lutz: Z. Phys. D **9**, 253 (1988)
71. W. Begemann, R. Hector, Y.Y. Liu, J. Tiggesbäumker, K.H. Meiwes-Broer, H.O. Lutz: Z. Phys. D **12**, 229 (1989)
72. M. Abshagen, J. Kowalski, M. Meyberg, G. zu Putlitz, F. Träger, J. Well: Europhys. Lett. **5**, 13 (1988)
73. N.D. Bhaskar, R.P. Fruhholz, C.M. Khincak, R.A. Cook: Chem. Phys. Lett. **154**, 175 (1989)
74. C. Bréchigniac: Ph. Cahuzac, F. Calier, J. Leygnier, I.V. Hertel: Z. Phys. D **17**, 61 (1990)
75. M. Abshagen, J. Kowalski, M. Meyberg, G. zu Putlitz, J. Slaby, F. Träger: Chem. Phys. Letters **174**, 455 (1990)
76. C. Bréchigniac, Ph. Cahuzac: In: *The Chemical Physics of Atomic and Molecular Clusters*, Enrico Fermi School, G. Scoles (ed.), p. 91 (North Holland, Amsterdam, 1990)
77. A. Messiah: *Quantum Mechanics II, chapter XV*, paragraph 14. North Holland, 1964 (1964)
78. D. Price, J.F.J. Todd (eds.): Dynamic Mass Spectrometry 1980 and 1981. Heyden & son Ltd. London
79. W. Aberth: Int. J. Mass Spectrom. Ion Proc., 209 (1986)
80. W.C. Wiley, I.H. McLaren: Rev. Sci. Instrum. **26**, 1150 (1955)
81. T.P. Martin, H. Schaber: Rev. Sci. Instrum. **60**, 347 (1989)
82. H. Haberland, H. Kornmeier, Ch. Ludewigt, A. Risch, M. Schmidt: Rev. Sci. Instrum. 2621 (1991)
83. E.W. Schlag (ed.): Time-of-flight mass spectrometry and applications. Int. J. Mass Spectrometry and Ion Physics, 1993, to be published
84. H.-J. Foth, J.M. Gress, C. Hertzler, W. Demtröder: J. Phys. **D18**, 257 (1991)
85. T. Troxler, S. Letwyler: J. Chem. Phys. **95**, 4010 (1991)
86. K.D. Kolenbrander, M.L. Mandich: J. Chem. Phys. **92**, 4759 (1990)
87. C. Bréchignac, Ph. Cahuzac, F. Carlier, M. de Frutos, J. Leygnier: Chem. Phys. Lett. **189**, 28 (1992)
88. C. Bréchignac, Ph. Cahuzac, N. Kebaïli, J. Leygnier, A. Sarfati: Phys. Rev. Lett. **68**, 3916 (1992)
89. S. Pollack, C.R.C. Wang, M.M. Kappes: J. Chem Phys. **94**, 2496 (1991)
90. H. Fallgreen, T.P. Martin: Chem. Phys. Lett. **168**, 233 (1990)

91. T.E. Gough, D.G. Knight, G. Scoles: Chem. Phys. Lett. **97**, 155 (1983)
92. O. Cheshnovsky, K.J. Taylor, J. Conceicao, R. Smalley: Phys. Rev. Lett. **64**, 1785 (1990)
93. M. Seidl, K.-H. Meiwes-Broer, M. Brack: J. Chem. Phys. **95**, 8251 (1991)
94. J.V. Coe, J.T. Snodgrass, C.B. Freidhoff, K.M. McHugh, K.H. Bowen: J. Chem. Phys. **84**, 618 (1986)
95. G. Reiser, W. Habenicht, K. Müller-Dethlefs, E.W. Schlag: Chem. Phys. Lett. **152**, 119 (1989)
96. R. Thalweiser, S. Vogler, G. Gerber: SPIE Proceedings, Vol. 1858, page 196 (1993)
97. F.G. Amar, S. Goyal, D.J. Levandier, L. Pereva, G. Scoles: In: *Springer Series in Chemical Physics*, **Vol. 56**, H. Haberland (Ed.). Springer-Verlag, Berlin, Heidelberg (1994) in press
98. G. van Helden, M.T. Hsu, N.G. Gotts, P.R. Kemper, M.T. Bowers: Chem. Phys. Lett. **204**, 15 (1993)
99. W.A. de Heer, P. Milani, A. Châtelain: Phys. Rev. Lett. **65**, 488 (1990)
100. J.P. Bucher, D.C. Douglas, L.A. Bloomfield: Phys. Rev. Lett. **66**, 3052 (1992)
101. L.J. de Jong: Physica **155B**, 289 (1989)
102. J.M. Papanikolas, V. Vorsa, M.E. Nadal, P.J. Campagnola, W.C. Lineberger: J. Chem. Phys. **97**, 7002 (1992)
103. T. Baumert, C. Röttgermann, C. Rothenfusser, R. Thalweiser, V. Weiss, G. Gerber: Phys. Rev. Lett. **60**, 1512 (1992)
104. J.A. Syage: Chem. Phys. Lett. **202**, 227 (1993)
105. J.P. Bucher, D.C. Douglas, L.A. Bloomfield: J. Chem. Phys. **63**, 5667 (1992)
106. M.H. Allen, M.L. Vestal: J. Am. Soc. Mass Spectrom. **3**, 18 (1992)
107. W.A. de Heer: Rev. Mod. Phys. **65**, 611 (1993)
108. *Proceedings of the Sixth International Meeting on Small Particles and Inorganic Clusters*, Z. Phys. **D28** (1993) and supplement to it

4 Across the Periodic Table

4.1 Alkali Clusters

C. Bréchignac

4.1.1 Introduction

Because of the freely mobile electrons the alkali metals are often considered as prototypes of metals well described with the free electron methods. In connection with that alkali clusters are also considered as prototypes of metals clusters. However the applicability of methods based on concepts well proven in solid state physics can start to be questionable when a cluster is very small. Going from atom to the bulk the question is from what cluster size the electronic properties of alkali clusters can be described with the free electron model? Moreover how does the quantum size effect affect their electronic structure? On the other hand how can notions such as surface tension or surface curvature be employed to interpret deviation from bulk properties concerning ionization potentials or binding energies per atom?

In this chapter comparison will be done essentially between the properties of alkali clusters and classical or quantum metallic droplet. Such a similarity between alkali cluster and metallic droplet is interesting to be proven since the metallic droplet, which involved very simple notion, can be used as reference for other clusters made of metal atoms.

4.1.2 Ionization of s^1 Clusters

4.1.2.1 Ionization Potentials of s^1 Metal Clusters

4.1.2.1a) First Ionization Potential

The simplest idea of a metal cluster is a metal droplet. In that respect the comparison between ionization potential (IP) of free clusters and the work function of an isolated metallic droplet can be used as a criterium to define the metallic character of small particles made of metal atoms.

The ionization potential of small alkali clusters was first recorded [1] about 25 years ago during the early development of beam techniques for making clusters. Latter measurements were done for Na clusters at higher resolution

[2–3] and extended to potassium and mixed NaK clusters [4–5]. Recently IP measurements were also obtained for group Ib clusters like silver [6].

In its principle the experimental arrangement, to measure the first ionization potential IP^0 which is the minimum energy required for the process:

$$X_n + h\nu \rightarrow X_n^+ + \bar{e} \tag{1}$$

is simple. Alkali cluster beam is formed either by pure or seeded adiabatic expansion, silver clusters are obtained by gas aggregation techniques. Once formed the clusters enter an ionizing region where they are gently one phton ionized. The usual sources for ionization are either a selected narrow range from a broad-band light source or a tunable laser either pulsed or continuous. Photoions are then mass selected using a mass spectrometer (quadrupole, magnet, time-of-flight). The experiment then consists of following a mass selected ion intensity as the photon energy is varied. Photoionization efficiency (PIE) curves are measured for every component of the beam, they can be made simultaneously for all clusters present with one scan of wavelengths. For clusters larger than $n = 2$, PIE curves show slowly rising shapes and there is no theoretical basis to date on how to extract either vertical or adiabatic ionization potential. The best discussion on the interpretation of PIE curves for clusters was done by *M. Kappes* and *E. Schumacher* [7]. The appearance potential which corresponds to the extrapolated energy for vanished ion signal is usually classed as IP.

The general trends is a decrease of the IP as cluster size increases. This evolution is expected for all elements since IP of a single atom is always larger than the work function of the bulk. An interesting point may be reached from the decreasing law, which must depend on the nature of the atoms. For alkalies, most of the experiments have been done on sodium and potassium [1–5, 7] and recently on lithium. It has to be noted that the ratio between the IPs of sodium and potassium clusters having the same size is roughly independent of the cluster size and remains equal to 1.2 from atom to the bulk. Few experiments deal with the IP of mixed Na_xK_y clusters [5, 8]. However, as E. Schumacher pointed out in this earlier paper [8] that the IPs of mixed clusters is a simple combination of the IPs of the corresponding homonuclear species. They can be found with the simple rule:

$$\mathrm{IP}(\mathrm{Na}_p\mathrm{K}_{n-p}) = \frac{p}{n}\,\mathrm{IP}(\mathrm{Na}_n) + \frac{n-p}{n}\,\mathrm{IP}(K_n) \tag{2}$$

which indicate a total miscibility of the two elements

4.1.2.1b) Multiionization Potentials

Ionization of cluster ion, having q charges X_n^{q+} corresponding to the process

$$X_n^{q+} + h\nu \rightarrow X_n^{q+1} + \bar{e} \tag{3}$$

also probe the "metallicity" of a cluster. If IP^q is the minimum energy required for ionization of X_n^{q+} the energy to extract q electrons from the neutral X_n cluster is: $\sum_{k=0}^{q-1} \mathrm{IP}^k$.

This quantity is not proportional to the number of ejected electrons as it is the case for the bulk due to the remaining charge located in the finite cluster volume. The question of the delocalization of the charge all over the cluster is an important point for the concept of the "metallicity".

From an experimental point of view, the set up for measuring multiionization potentials is more sophisticated than that for the first ionization potentials, since it concerns ionization of mass selected ions. The experimental arrangement used for potassium clusters is represented on Fig. 1a. After ionization by a pulsed laser of a neutral cluster distribution formed from a pure adiabatic expansion, the cluster ions enter a tandem time-of-flight mass spectrometer. A gate located in the first part of the drift tube allows a size-selected mass packet to proceed further. A second laser pulse delayed from the ionizing laser reionized the cluster ions which are then mass dispersed in a second time-of-flight. The photon energy ranges between 3.5 and 6.17 eV in order to obtain the second, third and fourth ionization potentials. Figure 1b shows ionization of K_{300}^+ and K_{600}^+ mass selected in the same packet.

However for a given charge $+qe$, there is a critical size $n_c(q)$ below which the cluster X_n^{q+} cannot be obtained by direct photoionization of $X_n^{(q-1)+}$ because of

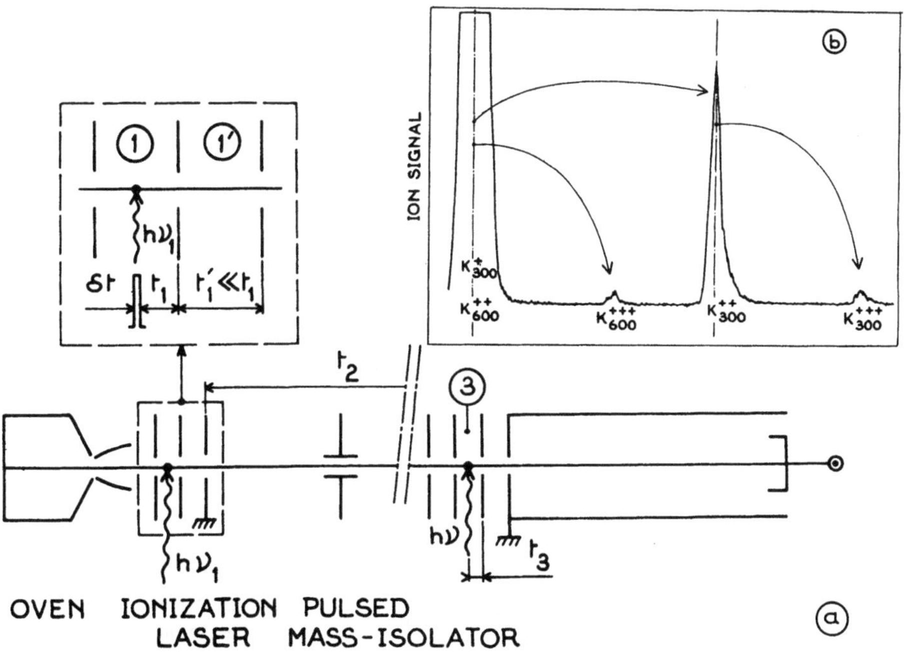

Fig. 1. a) Experimental set up, **b)** mass spectrum of multiionized mass selected Na_{300}^+ and Na_{600}^+ with a photon energy $h\nu = 4.66$ eV

the Coulombic repulsion forces [9]. The stability of multicharged clusters will be discussed later and we first focus on the ionization potential thresholds.

4.1.2.2 Work Function of a Metallic Droplet

4.1.2.2a) Classical Calculation

The work function of a metallic droplet is the energy required to remove an electron from the droplet to infinity. It can be calculated as the work ϕ done against the image force, which is the principal interaction beyond a few angstroms above the surface. For better understanding the image potential, it is significant to decompose the three stages of the calculation and to compare the result to that of a metallic plane as schematically drawn in Fig. 2.

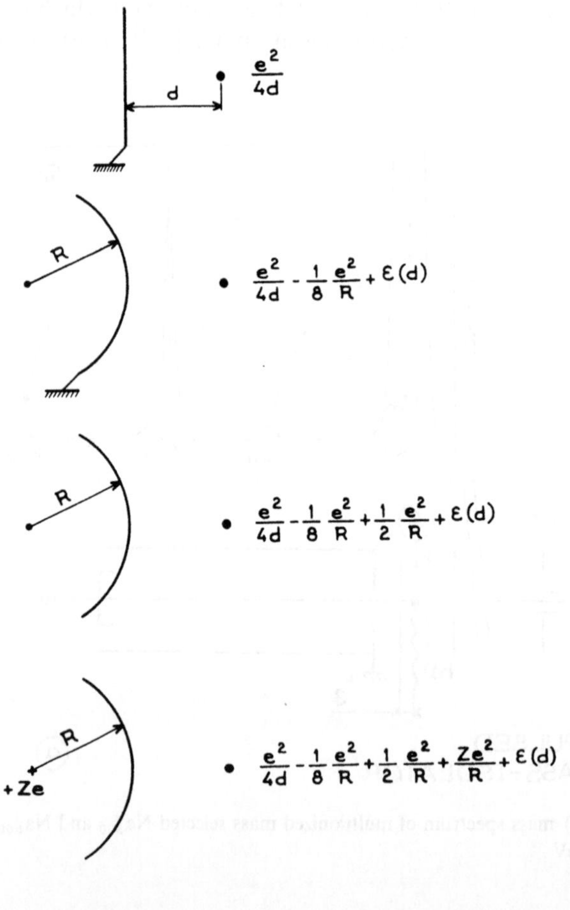

Fig. 2. Potential energy of a system formed by a piece of metal and an electron. *d* is the distance from the electron to the metallic surface

In the case of a planar surface maintained at zero potential the work done against the image force [10] is:

$$\phi_p = \frac{e^2}{4d} , \tag{4}$$

where d is the distance of the charge above the surface.

For a sphere of radius R maintained at *zero potential* [11]

$$\phi_s = \frac{e^2 R}{2d(2R+d)} = \frac{e^2}{4d}\left(\frac{1}{1+d/2R}\right) \tag{5}$$

$$\phi_s = \frac{e^2 R}{2d(2R+d)} = \frac{e^2}{4d} - \frac{1}{8}\frac{e^2}{R} + \varepsilon(d) \quad \text{for } R \gg d . \tag{6}$$

It is thus interesting to note that the image potential for a *grounded metallic* sphere of radius R is *lower* by $(1/8)(e^2/R)$ than that for a metallic plane. Analogous considerations have to be done for a real surface for which the local curvatures introduce corrections to the work function of an ideal plane.

For an *isolated* metallic sphere, the well known energy $(1/2)(e^2/R)$ [10] to charge the sphere must be added, leading for the image potential:

$$\phi(R) = \frac{e^2}{4d} - \frac{1}{8}\frac{e^2}{R} + \frac{1}{2}\frac{e^2}{R} + \varepsilon(d) . \tag{7}$$

Hence the image potential of an *isolated sphere* will be *increased* by $(3/8)(e^2/R)$ relative to the value of a grounded plane [12].

For a Ze charged metallic sphere the image potential is:

$$\phi(R, Z) = \frac{e^2}{4d} - \frac{1}{8}\frac{e^2}{R} + \frac{1}{2}\frac{e^2}{R} + \frac{Ze^2}{R} + \varepsilon(d) . \tag{8}$$

When d tends to the radius of the electron the term $e^2/4d$ tends to the work function of a planar surface. With the same approximation the work function of a metallic sphere having Ze charges is

$$W(R, Z) = W_\infty + \left(Z + \frac{3}{8}\right)\frac{e^2}{R} . \tag{9}$$

It has to be noted that the misunderstanding which exists in the literature originates from the difference between isolated and grounded sphere. Some authors [13–14] assimilate the work function of a metallic grounded sphere to that of a grounded plane they neglect in that case the curvature effect of a piece of metal. Within this approximation

$$W(R, Z) = W_\infty + \left(Z + \frac{1}{2}\right)\frac{e^2}{R} . \tag{10}$$

Recent jellium calculation taking into account curvature energy [15] are in better agreement with Eq. (9).

4.1.2.2b) Comparison with the Metallic Droplet

In order to compare the IP of clusters with the work function of an isolated metallic droplet, IPs of clusters are plotted against I/R where R is the radius assigned to a cluster of n atoms which is supposed to be spherical. Assuming that the volume of the cluster is the volume of one atom multiplied by n, the cluster radius is $R = r_s n^{1/3}$ where r_s is the Wigner Seitz radius [16]. Figure 3 shows the ionization potentials for neutral [4], singly [17] and doubly [17] potassium clusters versus $n^{-1/3}$. Also are drawn the straight lines representing the work function of the corresponding metallic sphere having zero, one and two charges, up to the critical size of stability. W_∞ is taken as the polycrystalline work function [16].

The experimental measurements follow remarkably well the behavior of a conductive droplet even for small sizes. The same is obtained for Na and Ag clusters. On the opposite, significant deviation are observed in the general decrease of IP versus cluster size for other elements like Hg, Sb . . . The way how IPs converge to the behavior of a macroscopic metallic droplet can be used as a first criterium of metallic properties.

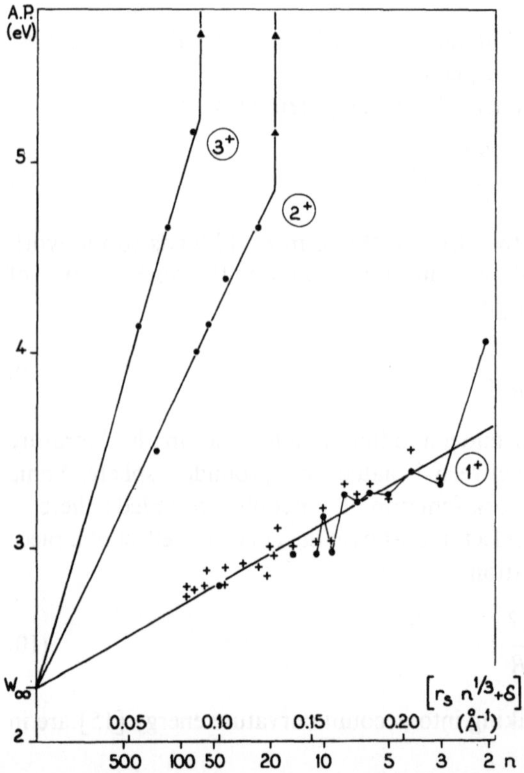

Fig. 3. Appearance potentials (A.P.) for single 1^+, doubly 2^+ and triply 3^+ charged potassium clusters. The respective straight lines coincide with those predicted by Eq. (9) merging to $W_\infty = 2.28$ eV

4.1.2.2c) Quantum Metallic Droplet

Even if the general trends of IP of alkali cluster with size is the one of a macroscopic metallic droplet it is unrealistic to consider a three-atom-cluster as a classical metallic droplet. For such a small cluster the quantum size effects must be considered [18]. The quantum metallic droplet is treated by jellium approximation. The electrons move in an effective potential created by the other electrons and ions. The effect of ionic cores is simulated by a uniformly positive charged background. The many body effects are replaced by an effective one-electron potential which significantly simplifies the problem. When the one electron equations are solved self consistently in spherical coordinates, the eigenstates are labeled by the principal quantum number N, where $N - 1$ is equal to the number of modes of the radial wavefunction and by the angular momentum number 1. The high degeneracy $2(2l + 1)$ of an energy level result from the spherical symmetry. This rough description of a quantum metallic droplet was first employed for clusters by *W. Ekardt* [13] and *W. Knight* [19]. It predicts a shell electronic structure with shell closing for a total number of electrons $n = 2, 8, 18, 20, 34, 40, 58, 68, 70, 92, 106, 112, 138, 156, 166, 168, 198 \ldots$ associated to $1s, 1p, 1d, 2s, 1f, 2p, 1g, 2d, 3s, 1h, 2f, 3p, 1i, 2g, 3d, 4s, 1j \ldots$ For small masses most of the shells are observed. For large masses only those with the first quantum number equal to 1 have been seen.

Direct evidence for shell structure and the associated energy gaps must appear in the variation of IP with cluster size. For example the IP for a closed-shell cluster is expected to be significantly larger than for open shell larger clusters. This was first observed for potassium clusters between 2 and 30 and

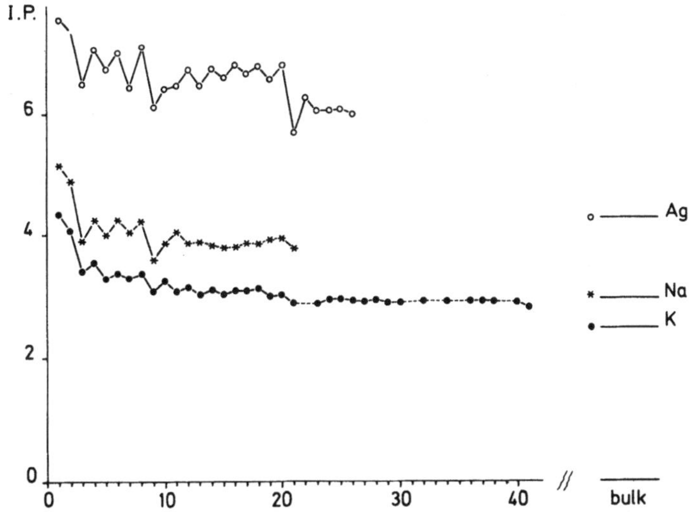

Fig. 4. Ionization potential of silver (○), sodium (*) and potassium (●) clusters as a function of size

extended to larger clusters beyond the shell closing at 92. Measurements for sodium [7] recently for silver clusters [6] are also understood in terms of electronic shell structure (Fig. 4). Shell structure is more pronounced for silver, than for sodium and potassium due to deeper potential well for silver than for sodium and potassium.

If the variation of IP with cluster size is qualitatively described with electronic shell model there are quantitative discrepancies between measurements and spherical jellium calculations. Improvement were obtained using non spherical cluster model [20]. Moreover the asymptotic limit of jellium model for large sizes must be taken considering the curvature of the grounded metallic sphere [15]. However in order to make the link between dimer and cluster it is very important to discuss the bonds in alkali metal clusters from the molecular point of view. This has been carefully done [21] by CI calculations where the topology of the cluster is taken into account. Such ab initio calculations can be done only for small masses $n < 20$ in good agreement with experimental measurements.

4.1.2.3 Photodetachment of Negative Cluster Ions

The recent development of intense negatively charged cluster beams and their subsequent mass analysis has allowed to apply the technique of photoelectron spectroscopy (PES) to clusters of a defined mass and to gain insight into the electronic level structure of these systems. Negative ion photoelectron spectroscopy is conducted by crossing a mass selected beam of netative ions with a fixed frequency laser beam and energy analyzing the resultant photodetached electrons. The photoelectron spectra of Na_n^-, K_n^-, $n < 7$ [22] and Ag_n^-, Cu_n^-, $n < 21$ [23] are highly structured. The peaks correspond to photodetachment transitions between the ground electronic states of the cluster anions and the ground and excited electronic states of their corresponding neutral clusters. Photoelectron spectra provide also the photodetachment threshold energies as estimates of the upper bonds of electron affinity (EA). Since photoelectron spectra show a slowly rising shape threshold behavior, as for photoionization spectra, there is an uncertaincy in the onset determination.

If the cluster is assumed to be a conducting sphere, the simple classical metallic droplet model may be used to describe the trends of EA.

$$\mathrm{EA}(R) = W_\infty - \frac{5}{8}\frac{e^2}{R} \,. \tag{11}$$

Apparently, as for ionization potentials, the electron affinity of group I clusters follows roughly the electrostatic curve.

Quantum chemical calculations performed on small sodium clusters yield an explanation for Na_n^-, $n = 2$–5 in terms of molecular electronic states, which can be extended to other alkali and alkali-like clusters. For larger clusters jellium, model reproduce strong discontinuities in the gap of the two first peaks of the PESs for $n = 8$–9 and $n = 20$–21, in agreement with shell structure.

4.1.2.4 Conclusion

Photoionization and photodetachment spectroscopy measure the *binding energy of one electron* and tests the electronic structure. The general trends of the energy required to extract one electron from cluster versus cluster size is described by the classical metallic sphere model in agreement with electron delocalization over the cluster volume. However the electron confinement to small volume introduces discrete energy states which is called the quantum size effect [18]. The essential idea, which has been pointed out by Kubo, is that the spacing between adjacent energy states decreases as cluster size increases. When the spacing between two successive occupied electronic states is lower than kT the small particle presents the metallic character of conducting sphere. Then an interesting question may be asked: up to what cluster size electronic shell structure can be observed? It seems that the cooler the cluster is the better electronic structure can be seen. However for cold cluster the architecture of ionic cores must be taken into account that can notably modify the jellium shell predictions.

4.1.3 Stability of s^1 Clusters

4.1.3.1 Unimolecular Dissociation

To the question: what is the energy stored into a cluster without it fragmentates? The answer is not obvious because the time during which the energy is stored may be considered. Starting from a "metastable cluster" in which the internal energy E^* is randomly distributed among the $3n - 6$ vibrational modes the probability of localizing enough energy in a single mode so as to overcome to obtain the dissociation associated to the process:

$$X_n \rightarrow X_{n-p} + X_p \ . \tag{12}$$

is calculated using statistical mechanics (see the chapter on Introduction to Statistical Reaction Rate theories).

It has to be noted that for small systems as well as for molecular clusters in which soft and hard modes coexist the simplified versions of statistical theories are inadequate. However for homogeneous systems X_n with $n > 10$ these simplified thories allow to predict the behavior of unimolecular evaporation. Using RRK [24] theory for example the dissociation rate $k(E^*)$ for X_n is estimated to

$$k(E^*) = \nu g\left(1 - \frac{B_{n,p}}{E^*}\right)^{3n-6} , \tag{13}$$

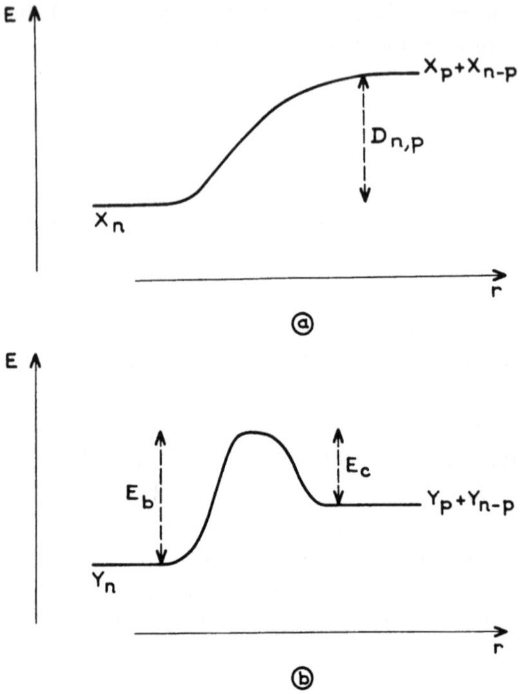

Fig. 5. Potential curves of cluster dissociation versus the distance r of the two fragments. Trace a represents the dissociation without barrier (neutral and simply ionized parent). Trace b represents the dissociation with barrier (doubly charged cluster)

where ν is the mean value of the vibrational frequency, g is the degeneracy factor and $B_{n,p}$ is the energy to surpass to obtain dissociation. For a dissociation channel without barrier the energy to surpass to obtain dissociation is the dissociation energy $D_{n,p}$ which is the difference between the atomization energies of initial and final states (Fig. 5a)

$$D_{n,p} = E(X_{n-p}) + E(X_p) - E(X_n) \ , \tag{14}$$

where $-E(X_n)$ is the energy associated to the process:

$$X_n \rightarrow nX \ . \tag{15}$$

For a dissociation channel with barrier the energy to surpass to promote dissociation is the energy barrier E_b at the transition state (Fig. 5b).

In all cases the unimolecular dissociation favors the dissociation channel which minimizes the energy cost. This point essentially concerns several competing channels.

4.1.3.2 Single Step in Unimolecular Dissociation

4.1.3.2a) Experimental Procedure

The experimental set up is a slightly modified version of one schematically illustrated Fig. 1a. The neutral cluster distribution is photoionized with a pulsed laser between the first two plates of an ion focusing and accelerating system. At high laser fluence ionized clusters sequentially absorb several photons which either lead to multiionized clusters and rapid sequential evaporative cooling process. After acceleration to the final ion beam energy of few keV, the ion clusters enter a field free tube where they are spatially resolved into separate mass packets. The packets pass between a gate deflection plates which remove them from the beam path except the packet of interest which proceed further. The time-of-flight τ in this first drift tube ranges from 10 to 200 μs for $n = 3$ to 900. During this drift period "metastable" cluster ions undergo further evaporation with parent and fragment particles proceeding at the center of mass velocity. A decelerating region slows differently parent and fragment ions which are then spatially dispersed in a second drift region before detection. The fragmentation pattern points out unimolecular dissociation during the time-of-flight τ.

4.1.3.2b) Unimolecular Dissociation of Singly Charged Clusters

As has been mentioned above, the dissociation events are governed by the time to localize enough energy on one dissociation mode. If unimolecular dissociation occurs within experimental time window τ, the time to localize the energy is smaller than τ. Figure 6 shows three unimolecular dissociation spectra of mass selected sodium cluster ions (dashed peaks).

The potential curve of two particles leaving each other is identical to the potential curve of two particles approaching each other and strongly depends on the charge state of the particles. When a cluster is dissociating, if one of the two fragments is neutral the potential curve smoothly decreases as the distance between the two fragments decreases and does not present any barrier. The

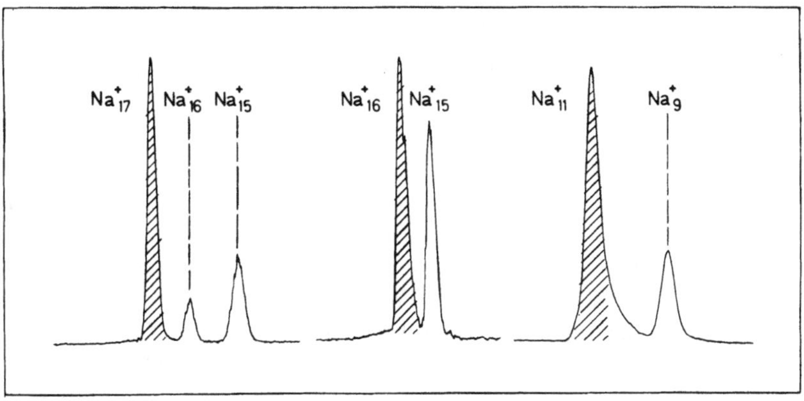

Fig. 6. Unimolecular dissociation spectra of three mass-selected parents Na_{17}^+, Na_{16}^+ and Na_{11}^+. The dissociation takes place within few tenth of microseconds

energy difference between initial and final states is the dissociation energy (Fig. 5a). As predicted by statistical model the dissociation patterns well follow the ordering of the dissociation energies. For Na_{17}^+ the dissociation energies $D_{17,1}^+$ and $D_{17,2}^+$, corresponding to the ejection of monomer and dimer, are close in energy. The two channels are present in unimolecular dissociation. On the contrary for Na_{16}^+ and Na_{11}^+ one dissociation channel dominates.

All results obtained on alkali clusters show [25–26] that for neutral and singly charged clusters only the two dissociative channels associated to the ejection of atom and neutral dimer are observed

$$X_n^+ \begin{cases} \nearrow X_{n-1}^+ + X & D_{n,1}^+ \\ \searrow X_{n-2}^+ + X_2 & D_{n,2}^+ \end{cases} . \tag{16}$$

Moreover even-numbered clusters present only the evaporation of atom whereas odd numbered clusters dissociate along two pathways, evaporating either a monomer or a dimer. At higher masses, however, monomer ejection tends to predominate in agreement with the energetic considerations. The energy balance of the following cycle

$$X_n^+ \begin{cases} \nearrow X_{n-1}^+ + X \searrow \\ \searrow X_{n-2}^+ + X_2 \nearrow \end{cases} X_{n-2}^+ + 2X \tag{17}$$

leads to

$$D_{n,1}^+ + D_{n-1,1}^+ = D_{n,2}^+ + D_2 \ . \tag{18}$$

For large clusters the dissociation energies D_{k1}^+ tends to the bulk values which are larger than the binding energies D_2 of dimers (Table 1). To verify Eq. (18), $D_{n,1}^+$ must be smaller than $D_{n,2}^+$ which indicates that monomer ejection prevails.

The most important application of the unimolecular dissociation is in the determination of cluster dissociation energies. This can be done from unimolecular dissociation rates. However if the dissociation rate is measured only for one cluster the binding energy which is deduced from the dissociation rate strongly depends on statistical method used to determine the unimolecular time scale, so the determination of the binding energy is not reliable. To minimize the uncertainty due to statistical method, relative binding energies can be obtained from the unimolecular evaporation of an "evaporative ensemble" in which every cluster has undergone at least one evaporation before the measurements. The relative binding energies are obtained by a recursive procedure from one size to the next one [26] which is not sensitive to the absolute value of the dissociative probability, but to their relative ratios. The absolute energies are calibrated by Eq. (18). Figure 7 shows the dissociation energies of the lowest dissociation channel for Na_n^+ and K_n^+ with $n \leq 25$. They present similar sawtooth behavior well interpreted in the framework of electron shell model [27]. The sharp drops after ion cluster masses 9 and 21 correspond to electron configuration shell closing with 8 and 20 electrons. The odd even alternation showing more stable cluster with even number of electrons is attributed to pairing effect. It should be noted that such behavior is less pronounced for K_n^+ than for Na_n^+. Since the

Table 1.

	Binding energy (eV)	Equilibrium distance (Å)
Li_2	1.05	2.67
Li_2^+	1.44	
bulk	1.63	3.02
Na_2	0.72	3.08
Na_2^+	0.96	3.54
bulk	1.13	3.66
K_2	0.51	3.91
K_2^+	0.85	4.11
bulk	0.93	4.53
Rb_2	0.49	
Rb_2^+	0.72	3.94
bulk	0.85	4.84
Cs_2	0.39	4.47
Cs_2^+	0.61	4.44
bulk	0.80	5.25
Cu_2	2.03	2.22
bulk	3.49	2.56
Ag_2	1.66	
bulk	2.95	2.89
Au_2	2.30	2.47
bulk	3.81	2.88

potential well in which electrons are moving is less pronounced for potassium than for sodium, the electronic levels are less spaced in potassium than in sodium.

4.1.3.2c) Unimolecular Dissociation of Doubly Charged Clusters: Evaporation Against Fission

The unimolecular dissociation of doubly charged clusters is more complex than that of singly charges ones – Two possibilities occur; either the two charges stay on one fragment, or each fragment takes one charge

$$X_n^{++} \begin{cases} \rightarrow X_{n-p}^{++} + X_p \\ \rightarrow X_{n-q}^{+} + X_q^{+} \end{cases} . \tag{19}$$

If the two charges stay on one fragment the potential curve does not present any barrier. The kinetic energy of the fragments is roughly kT, where T is the temperature of X_n^{++}. If each fragment takes one charge the potential curve presents a Coulombic barrier due to the repulsive energy of X_{n-q}^+ and X_q^+. The fragments separate at the transition state with the Coulombic kinetic energy. Such a last process is often call Coulombic fission.

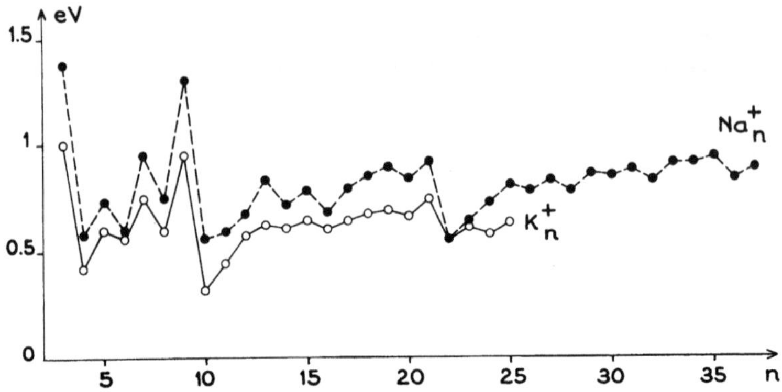

Fig. 7. Dissociation energies of the lowest dissociation channel for sodium (●) and potassium (○) cluster ions

Small doubly charged clusters X_n^{++} are often unobservable in mass spectra unless their size exceeds a well defined critical size n_c [9]. It is commonly admitted that small doubly charged clusters below the critical size are not stable because the Coulombic repulsive energy between positive holes exceeds the binding energy. In that respect the critical size must be dictated by the dynamics of the unimolecular dissociation. Figure 8 shows unimolecular dissociation spectra for several mass to charge ratios near the critical size. Selecting half integer size to charge ratios makes it possible to focus on unimolecular dissociation of doubly charged Na_n^{++} exclusively, whereas by selecting an integer size to charge ratio unimolecular dissociations of both Na_n^{++} and $Na_{n/2}^{+}$ are present. The unimolecular fragments on the right-hand side of the parent (dashed peak) correspond to neutral ejection whereas the unimolecular fragments on the left-hand side of the parent show the heaviest singly charged fragment from asymmetric Coulombic fission. For large cluster size $n > 30$ evaporation of neutral atom prevails, the Coulombic barrier is higher than the dissociation energy. For intermediate sizes $26 < n < 30$ both evaporation and fission coexist indicating a Coulombic barrier height which roughly equates the binding energy. The discrimination which exists between the intensities of fragments coming from evaporation and fission lies in the kinetic energies of the fragments. Evaporation imparts a kinetic energy of about 0.005 eV to the fragments whereas fission impart roughly 1 eV which is the energy difference between the top of the barrier and the emerging fragments. In time-of-flight experiment the recoil velocity isotropically adds to the center of mass velocity decreases the intensity of the fission fragments more than the intensity of evaporating fragments. For small cluster sizes $n < 24$ the height of Coulombic barrier is smaller than the binding energy and the doubly charged clusters vanish. The critical size below which doubly charged group I clusters are unstable versus Coulombic fission are 24, 17, 19, 3, 5 for Na, K, Cs, Au and Ag respectively [28–33].

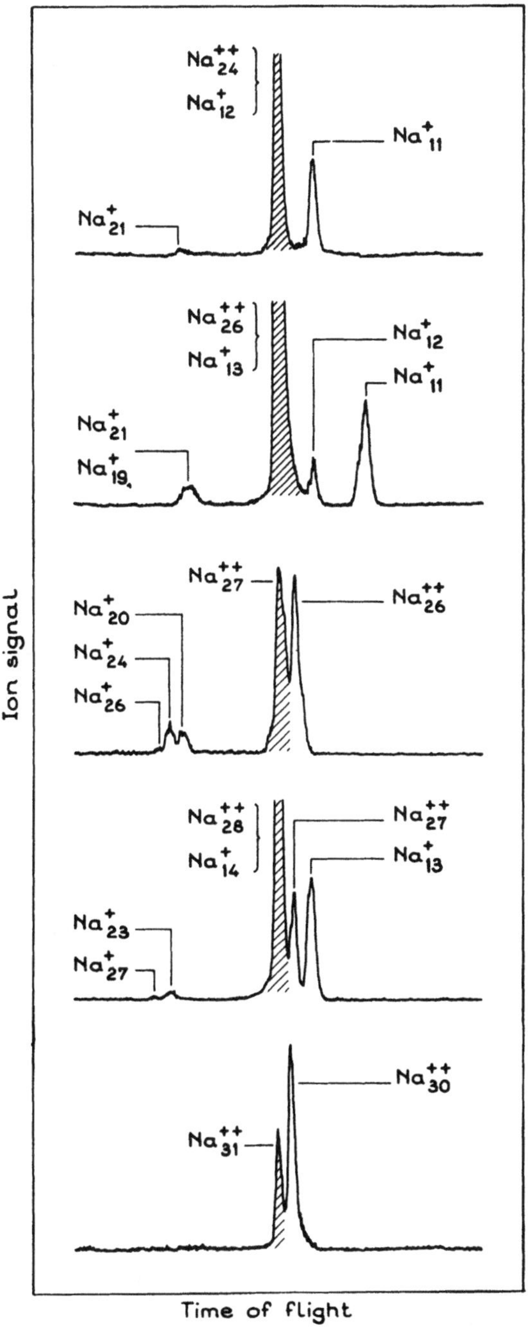

Fig. 8. unimolecular dissociation spectra of doubly charged sodium clusters within 50 μs

4.1.3.3 Evaporation Cascades: Atomisation Energies

4.1.3.3a) Multistep Evaporation

When the internal energy deposited into the cluster is large enough sequential evaporations can be observed within experimental time window. Considering the sequential evaporation following the photon absorption of mass selected cluster ion X_n^+. For a cluster large enough so that only monomer evaporations are concerned:

$$X_n^+ + h\nu \rightarrow X_{n-p}^+ + pX \ . \tag{20}$$

It has to be noted that the number of atoms lost depend on the photon energy on the observational time window and on the initial cluster temperature. Starting with a cluster of an "evaporative ensemble" the temperature of the selected cluster and so its internal energy is defined by the evaporating time of its last evaporation t_2 (Fig. 1). The average value of its internal energy E^* in dissociation energy units varies almost linearly with cluster sizes n as numerically obtained after inverting Eq. (13)

$$\frac{E^*(t, n)}{D_{n,1}^+} = \mu(t)n + a \ . \tag{21}$$

If the last evaporation has taken place within t_2 the internal energy before absorbing a photon is $E^*(t_2, n)$. The photoinduced evaporation within the time t_3 leads to and energy balance

$$E^*(t_2, n) + h\nu = E^*(t_3, n - p) + \sum_{k=n-p+1}^{n} (D_k^+ + \varepsilon_k) \ . \tag{22}$$

Neglecting the kinetic energy release ε_k with respect to the dissociate energy D_k^+ and assuming that the dissociation energy smoothly vary along one sequence of evaporation, which is the case for large clusters Eqs. (21) and (22) lead to

$$n\mu(t_1) + \frac{h\nu}{D^+} = (n - p)\mu(t_3) + p \ , \tag{23}$$

$$p = \frac{h\nu}{D^+[1 - \mu(t_3)]} - n\frac{\mu(t_3) - \mu(t_2)}{1 - \mu(t_3)} \ . \tag{24}$$

Experimental data plotted in Fig. 9 show the number of lost atoms after the photoexcitation of three different parents K_{21}^+, K_{53}^+, K_{100}^+ within the time window t_3 versus the photon energy. For each mass the experimental points follow a straight line which does not pass through zero in agreement with the Eq. (24). Since the time window t_2 is larger than the time widnow t_3 the parent are colder than the fragments (because they had more time to be cooled). The

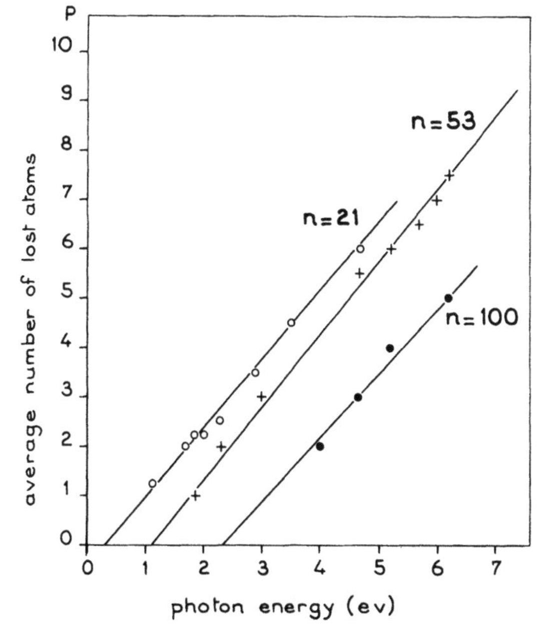

Fig. 9. Average number of lost atoms after photoexcitation of mass selected clusters K_n^+ with $n = 21$, 53, 100, versus photon energy. The evaporation takes place within 1 μs. Notice that the straight lines do not pass through zero. The interception with abscissa axis is related to the initial temperature of the parent as compared to temperature of the fragments

energy $h\nu_{min}$ defined as:

$$h\nu_{min} = nD_n^+ [\mu(t_3) - \mu(t_2)] \tag{25}$$

is the energy needed to heat the parent from its initial temperature associated to the time window t_2 to the temperature associated to the time window t_3.

The mean value of the dissociation energy over few masses is obtained from the slope of the straight lines.

4.1.3.3b) Atomization Energies of Neutral and Singly Charged Ions

The mean values of the dissociation energies obtained from photoevaporation experiments combined with the dissociation energies obtained from unimolecular dissociation permit to determine the atomization energy ΔE_n^+

$$\Delta E_n^+ = \sum_{k=2}^{n} D_{n,1}^+ \tag{26}$$

associated to the reaction

$$X_n^+ \rightarrow (n-1)X + X^+ \ . \tag{27}$$

Using the thermodynamic Born–Harber cycle

$$\begin{array}{l} X_n \rightarrow X_n^+ + \vec{e} \\ \downarrow \qquad \quad \downarrow \\ nX \rightarrow (n-1)X + X^+ + e^- \end{array} \tag{28}$$

the atomization energy of neutral cluster X_n associated to the energy charge

$$X_n \rightarrow nX \tag{29}$$

is

$$\Delta E_n = \Delta E_n^+ + \mathrm{IP}(X_n) - \mathrm{IP}(X) \ , \tag{30}$$

where $\mathrm{IP}(X_n)$ and $\mathrm{IP}(X)$ are the ionization potentials of X_n and X respectively.

Taking the measured values of the ionization potentials it is interesting to compare the atomization energies per atom $\Delta E_n^+/n$ and $\Delta E_n/n$ for ionic and neutral clusters.

Figure 10 shows the atomization energy per atom as a function of $n^{-1/3}$ for sodium and potassium clusters. For *neutral* clusters the atomization energy per atom follows the classical drop model equation

$$\frac{\Delta E_n}{n} = a_v - a_s n^{-1/3} \ , \tag{31}$$

where a_v represents a volume energy and a_s is referred to a surface energy. The values of a_v and a_s deduced from the best fit of the classical drop model are given in Table 2 together with the bulk values. For both sodium and potassium clusters a_v coincides with the cohesive energy of the bulk. The surface coefficient a_s is larger for clusters than the surface tension of the bulk. This difference should be explained by surface stress. The fact that the surface coefficient is equal to the volume coefficient indicates that the binding energy of a surface atom is roughly equal to the binding energy of a volume atom. This property comes from the nature of the "metallic bond" for which the cohesion is maintained by the kinetic energy of the delocalized electrons. In the case of either directive bonds or dispersive forces the surface coefficient might be lower than the volume term. Moreover it has to be noted that the comparable values for a_v and a_s lead to zero energy for the atom taken here as the energy reference.

For ionic clusters satisfactory fit is obtained using classical drop model for the atomization energies per atom of large cluster ions. For small cluster ions $n < 40$ atomization energies deviate from the drop model. This deviation shows that the cluster cations are more strongly bound than the corresponding neutral clusters. This difference vanishes progressively with increasing cluster sizes and tends to zero since the atomization energy of the bulk with more or less one electron should be constant. The deviation of the atomization energies per atom from the classical drop behavior reflects the solvation energy of the charge inside the cluster. In fact the reaction

$$X_n^+ \rightarrow (n-1)X + X^+ \tag{27}$$

entails not only bond breaking but also significant relocalization of the single charge which was delocalized over the cluster [34].

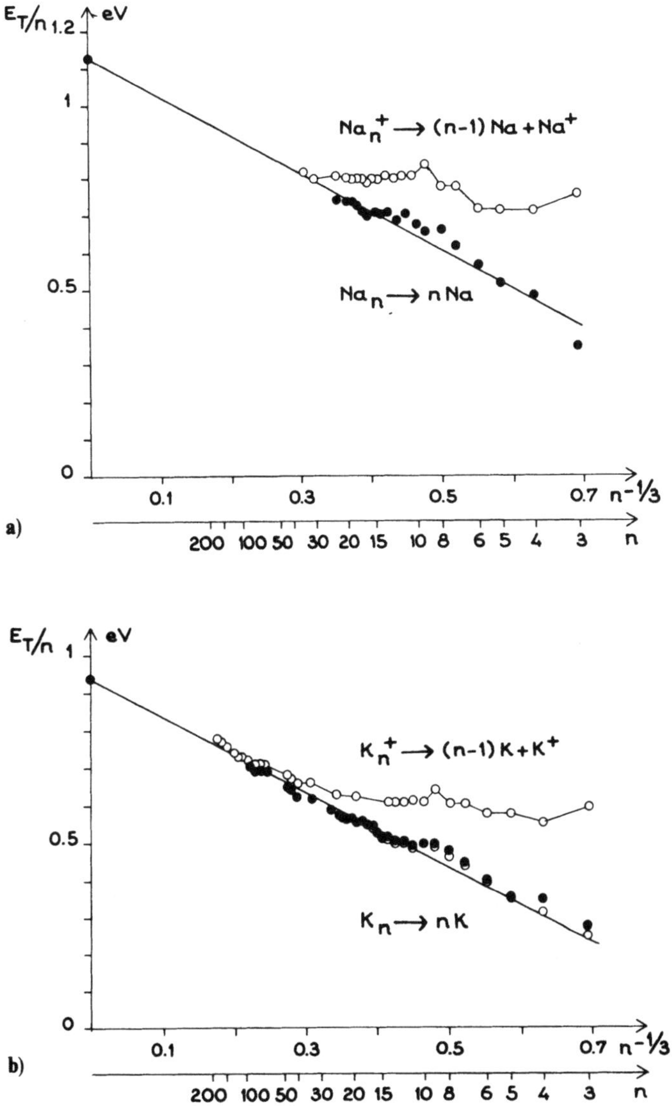

Fig. 10. Cohesive energies for neutral (●) and ionized (○) sodium (**a**) and potassium (**b**) clusters versus $n^{-1/3}$ deduced from photoevaporation experiments

Table 2.

	Cluster		Bulk	
	a_v (eV)	a_s (eV)	a_v (eV)	a_s (eV)
Na	1.12	1.02	1.13	0.89
K	0.94	0.98	0.94	0.75

4.1.3.4 Drop Model and Shell Corrections

4.1.3.4a) Energy of a Metallic Droplet

It is easy to evaluate the atomization energy $E(X_n^{z+})$ of a Z-charged metallic droplet (X_n^{z+}) from the following thermodynamic cycle

$$\begin{array}{l} X_n^{z+} \rightarrow zX^{+} + (n-z)X \\ \uparrow \qquad \uparrow \\ X_n \rightarrow nX \end{array} \tag{32}$$

leading to the equation:

$$E(X_n^{z+}) + \mathrm{IP}^z(X_n) = E(X_n) + z\,\mathrm{IP}(X) \; . \tag{33}$$

Using the expression of the atomization energy of a neutral drop (31) and the ionization potential of a metallic droplet (10)

$$E(X_n^{z+}) = a_v n - a_s n^{2/3} + z\,\mathrm{IP}(X) - \sum_{q=0}^{q=z-1} W(q, R) \tag{34}$$

$$E(X_n^{z+}) = a_v n - a_s n^{2/3} + z\left(\mathrm{IP}(X) - W_\infty + \frac{1}{8}\frac{e^2}{r_s n^{1/3}}\right) - \frac{z^2}{2}\frac{e^2}{r_s n^{1/3}} \tag{35}$$

$$E(X_n^{z+}) = a_v n - a_s n^{2/3} + z\beta - a_c\left(z^2 - \frac{z}{4}\right)n^{-1/3} \tag{36}$$

where a_c is the Coulombic coefficient

$$a_c = \frac{e^2}{2r_s} \tag{37}$$

and β is the difference between the ionization potential of the atom and the work function of the bulk

$$\beta = \mathrm{IP}(X) - W_\infty \; . \tag{38}$$

4.1.3.4b) Dissociation of Neutral Clusters

The dissociation energy of a metallic neutral droplet in two fragments, associated to the reaction

$$X_n \rightarrow X_{n-p} + X_p \tag{39}$$

$$D_{n,p} = E(X_n) - (X_{n-p}) - (X_p) \; . \tag{40}$$

$$D_{n,p} = a_s[p^{2/3} + (n-p)^{2/3} - n^{2/3}] \tag{41}$$

which is a surface energy. In Fig. 11a are plotted, as an example, the dissociation energies obtained from drop model versus the fragment size p, for a given parent Na_{13}. These energies are compared to the dissociation energies from Hückel

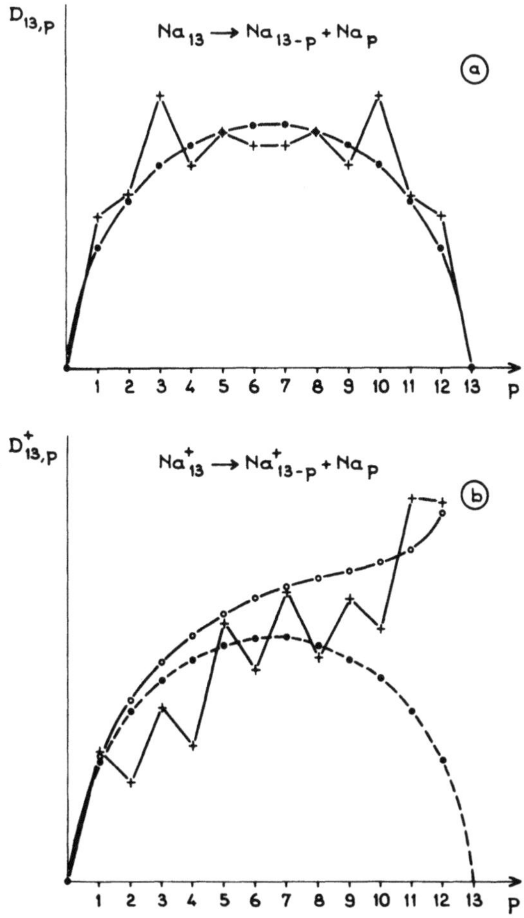

Fig. 11. Calculated dissociation energies $D_{n,p}$ and $D^+_{n,p}$ for $n = 13$. The points correspond to the drop model, the crosses to Hückel calculations [34]

calculations [34]. The smooth behavior of the drop model coincides with the mean value of the sawtooth behavior of the Hückel model. In both cases the lowest dissociation energy corresponds to asymmetric dissociation which minimizes the variation of the surface energy. It has to be noted that from the drop model the evaporation of the monomer always prevails for all parents whereas taking into account pairing effect, dimer ejection can prevails in some cases for small even numbered clusters.

4.1.3.4c) Dissociation of Singly Charged Clusters

The dissociation energy of a singly charged metallic droplet associated to the dissociation

$$X_n^+ \rightarrow X_{n-p}^+ + X_p \tag{42}$$

is

$$D^+_{n,p} = E(X^+_n) - E(X^+_{n-p}) - E(X_p) \quad (43)$$

$$D^+_{n,p} = a_s[p^{2/3} + (n-p)^{2/3} - n^{2/3}] + \frac{3}{4}a_c[(n-p)^{-1/3} - n^{-1/3}] \; . \quad (44)$$

The dissociation energy is the sum of two terms: a surface and a Coulombic term. As shown in Fig. 11b the Coulombic term increases the dissociation energy as the size of ionic fragment decreases. As for neutral clusters, the smooth behavior of the drop model, coincides with the sawtooth behavior of Hückel calculations [34]. These models well explain the experimental results. The lowest dissociation channel which is observed from unimolecular dissociation is an asymmetric fragmentation in agreement with a large surface tension. The charge stays on the highest fragment because the Coulombic contribution minimizes the charge solvation. Moreover the pairing and shell effects which are included in Hückel and jellium calculations can induced dimer ejection instead of the monomer ejection predicted by the droplet model, as it is the case for Na^+_{13}.

4.1.3.4d) Dissociation of Doubly Charged Clusters

The dissociation of doubly charged clusters when the two charges stay on one fragments behaves like singly charged clusters. The dissociation channels do not present any barrier and the lowest dissociation channel correspond to an atom ejection in agreement with the energetics. When the two charges are shared with the two fragments

$$X^{++}_n \rightarrow X^+_{n-p} + X^+_n \quad (45)$$

the energy barrier, which has to overcome to produce fragmentation (Fig. 5b), can be evaluated from the total energies of the fragments and the repulsive energy of the two singly charged fragments E_c.

$$B^{++}_{n,p} = E(X^+_{n-p}) + E(X^+_p) + E_c - E(X^{++}_n) \quad (46)$$

The repulsive Coulombic energy E_c is calculated from the work made against the Coulomb forces when the charges move from infinite distance to the contact distance d to form the parent.

$$E_c = \frac{e^2}{d} \quad (47)$$

d is the sum of the radii of the two fragments corrected by the spill out factor δ which take into account the extension of the electron density over the cluster

$$d = r_s((n-p)^{1/3} + p^{1/3}) + \delta \; . \quad (48)$$

Using the drop model to evaluate the total energies

$$B_{n,p}^{++} = a_s(p^{2/3} + (n-p)^{2/3} - n^{2/3}) + \frac{3}{4}a_c(p^{1/3} + (n-p)^{1/3} - n^{1/3})$$
$$- \frac{11}{4}a_c n^{1/3} + 2a_c((n-p)^{1/3} + p^{1/3} + \delta)^{-1} . \quad (49)$$

For a given parent the surface term increases from asymmetric $p = 1$ to symmetric fission $p = n/2$, whereas the Coulombic term decreases from asymmetric $p = 1$ to symmetric fission $p = n/2$. So, if the surface tension dominates asymmetric fission will prevail, if the Coulombic term dominates symmetric fission will impose the dissociation channel. Figure 12 shows the results obtained for Na_{23}^{++}. It is clearly seen, from drop model, that the lower barrier stands for $p = 3$ to 5. The shell corrections will favor the close shell fragment and odd numbered clusters, in agreement with experimental results.

The competition between the fission in two singly charged fragments and the evaporation depends on the relative values between the Coulombic barrier $B_{n,p}^{++}$ and the dissociation energy $D_{n,1}^{++}$. The critical size below which doubly charged clusters are unstable is the size for which $B_{n,p}^{++} \leq D_{n,1}^{++}$. Its estimation is compatible with experimental results. To go further in the dynamics of the

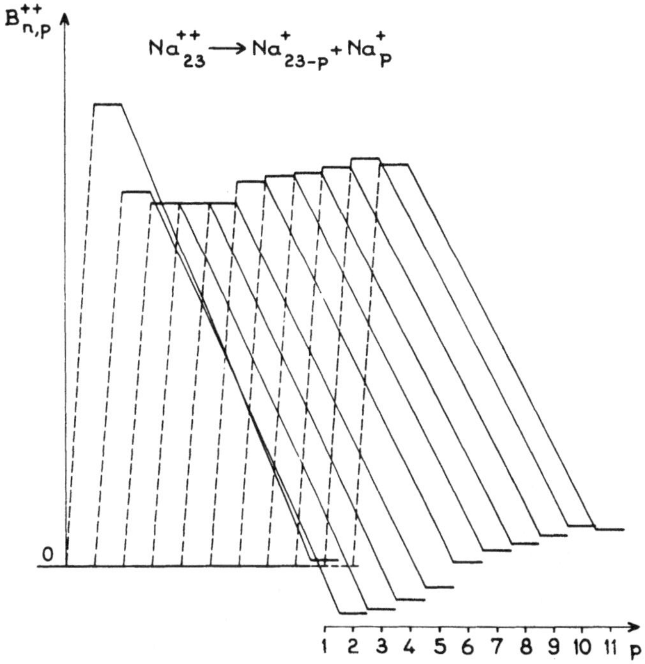

Fig. 12. Calculated barrier of doubly charged cluster Na_{23}^{+} from drop model equation (49). The smallest barrier corresponds to asymmetric fission due to the predominance of the surface term compared to the Coulombic term

dissociation the knowledge of the multidimensional energy surface from more refined calculations would be useful [35].

4.1.3.5 Conclusion

Unimolecular dissociation for which the storaged energy provides evaporation or fission of a mass selected cluster is the key experiment for studying cluster stability. For s^1 clusters the dissociation energies show that the cluster stabilities follow their electronic shell structures. The atomization energy per atom, mean value of the dissociation energies over lower masses, well obeys to metallic drop model. This simple model gives the physical understanding of the dissociation behavior and interpret the special abundance in cluster mass distribution (Fig. 13).

4.1.4 Optical Response of s^1 Clusters

The optical response of free metal clusters must be connected to the photoexcitation of small metal particles (diameters in the range 2–150 nm) which has been investigated for a considerable time either on free [36] or on supported particles

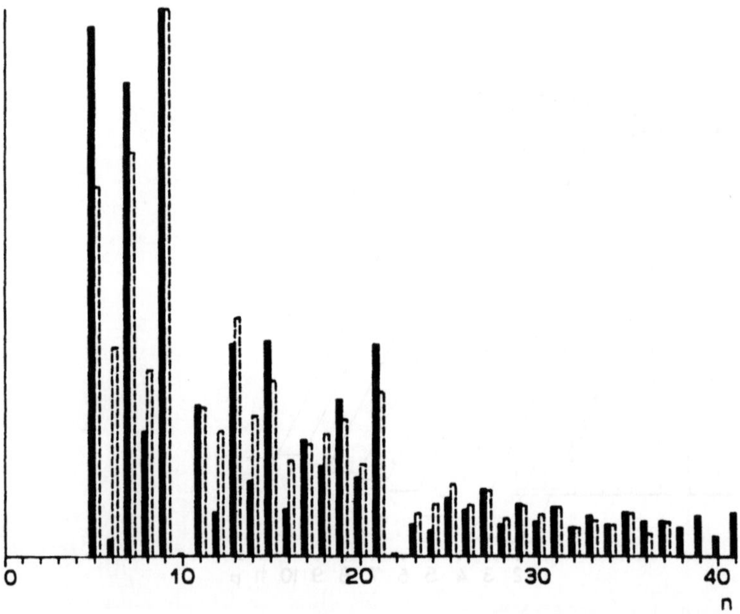

Fig. 13. Experimental spectra of sodium clusters, solid lines, showing ionic shell structure, compared to calculated one, dashed lines, from the evaporative ensemble model

[37–41]. The *Mie* solution [42] for scattering and absorption of light by a single homogeneous sphere provides a good description of the optical properties of isolated spherical particles. An extended Mie theory for both prolate and oblate ellipsoids of revolution has been derived by *Gans* in 1915 [43] and applied to small prolate spheroids of silver particles embedded in gelatin. It is interesting to know down to which cluster size the absorption can be described by classical Mie model.

For very small clusters, quantum size effects are predicted. The absorption spectra should exhibit discrete structure and the familiar Mie sphere resonance is expected to be considerably broadened as a consequence of Landau damping in a discrete system [18]. In particular it is an open question as to how important the contribution to the linewidth due to the direct coupling with vibration is. Moreover the main problem of the giant resonance in small systems either in nuclear physics [44–45] or in atomic cluster physics [46] is to explain their nature and find out the principal factors which determine their positions and widths. Two different approaches are used in theoretical investigations of the giant multipole resonances. The single particle approach in which the giant resonance is a coherent superposition of a definite number of particle-like and hole-like excitations and the collective approach involving collective electron motion. In this respect obtaining the photoexcitation spectra of small and intermediate cluster size is of fundamental interest.

4.1.4.1 Classical Mie Theory

4.1.4.1a) Spherical Particle

Let us consider the collective motion of an electron gas, in the potential well created by the ion cores and the electrons themselves, excited by an electromagnetic field. The equation of motion of one electron is:

$$\ddot{\mathbf{r}} + \Gamma\dot{\mathbf{r}} + \omega_0^2\mathbf{r} = -\frac{e}{m}\mathbf{E}_0 e^{-i\omega t} \tag{50}$$

$\mathbf{r}$ is the deviation of the position of the electron against its equilibrium distance, Γ is a damping coefficient, $\omega_0^2\mathbf{r}$ is the restoring force due to ionic cores screened by the electrons, m is the electron mass, and $\mathbf{E}_0 e^{-i\omega t}$ the electromagnetic field. The dipole moment of the n electron system is

$$\mathbf{P} = [\chi]\mathbf{E} = ne\,\mathbf{r}\ , \tag{51}$$

where $[\chi]$ is the polarizability tensor. For spherical symmetry the polarizability of the system is isotropic:

$$\chi = \frac{ne^2}{m\varepsilon_0}\frac{1}{\omega_0^2 - \omega^2 - i\omega\Gamma}\ . \tag{52}$$

In the case of a static field ($\omega = 0$), ω_0^2 is related to the static polarizability α.

$$\alpha = \frac{ne^2}{m\varepsilon_0} \omega_0^2 \ . \tag{53}$$

When the system is excited by an electromagnetic field the dynamic polarizability of Eq. (51) leads to an optical absorption with an absorption cross section:

$$\sigma = \frac{ne^2}{mc\Gamma\varepsilon_0} \frac{\omega^2}{(\omega^2 - \omega_0^2)^2 + (\omega\Gamma)^2} \tag{54}$$

which is the well known Mie formula.

For a classical monovalent metal sphere, the Gauss theorem leads to a restoring force $\mathbf{F} = (e^2/4\pi\varepsilon_0)(\mathbf{r}/r_s^3)$ where r_s is the Wigner Seitz radius. Its resonance frequency is then $\omega_0^2 = e^2/4\pi\varepsilon_0 r_s^3$. The corresponding plasma frequency of the bulk is given by $\omega_p^2 = e^2/4\pi\varepsilon_0 v$ where v is the volume of one electron $v = \frac{4}{3}\pi r_s^3$. So the resonance frequency of the metallic sphere is expected to be

$$\omega_0 = \frac{\omega_p}{\sqrt{3}} \ . \tag{55}$$

4.1.4.1b) Ellipsoidal Particle

The polarizability of a metal ellipsoid has a different value α_i along each of the three principal axes, and the photoabsorption cross section depends on the orientation of the ellipsoid with respect to the photon polarization. For randomly oriented ellipsoidal clusters the photoabsorption cross section is the sum of three terms of equal weight. Equation (54) is generalized to give

$$\sigma = \frac{ne^2}{3mc\Gamma\varepsilon_0} \sum_{i=1}^{3} \frac{\omega^2 \Gamma_i}{(\omega^2 - \omega_{0i}^2)^2 + (\omega\Gamma_i)^2} \ . \tag{56}$$

Γ_i represent the damping along the i axis associated to the resonance ω_{0i} related to the polarizability α_i:

$$\alpha_i = \frac{ne^2}{m\varepsilon_0 \omega_{0i}^2} \ . \tag{57}$$

Moreover for an homogeneous ellipsoid the three resonances frequencies are related to the resonance frequency ω_0 of the corresponding sphere having the same volume by the relation

$$\omega_{01}\omega_{02}\omega_{03} = \omega_0^3 \ . \tag{58}$$

4.1.4.2 Static Polarizabilities of Clusters

4.1.4.2a) Theoretical Predictions

Early attempts to calculate the polarizabilities of metal clusters assumed them to be spherical and used the classical solution for the polarizability of a metallic sphere which is proportional to R^3 where R is the radius of the sphere. A rough approximation leads to a polarizability per atom which is independent of a cluster size giving a resonance frequency $\omega_0 = \omega_p/\sqrt{3}$.

The application of a jellium model to small spheres [47–48] evaluate the electron density using a gradient expansion approximation. The resulting polarizability was expressed in a form

$$\alpha = a(R + \delta)^3 \;, \tag{59}$$

where δ is the analog of the Lang Kohn coefficient indicating that the electron density of the clusters spills out beyond the edge of the ionic core density. The calculated polarizabilities of small clusters can be 30% larger than the bulk sphere.

Subsequent calculations are carried out using RPA method [49] or CI calculations [50], all showing a spill out factor which decreases with cluster size.

4.1.4.2b) Experimental Results

The experimental method involves molecular-beam deflection in an inhomogeneous electric field [51]. After deflection the neutral cluster beam is then ionized in order to achieve mass selection. From Knight's results the polarizabilities per atom, normalized to the atomic polarizability are plotted for sodium and potassium clusters in Fig. 14. The experimental values for Na and K are the same within experimental error. They follow a general decrease toward the bulk value as cluster size increases in agreement with theoretical trends. However the calculated values are significantly lower than the experimental determinations. Several effects may contribute to this discrepancy. One of them may be the determination of Wigner Seitz radius for cluster.

4.1.4.3 Absorption Cross-Section

4.1.4.3a) Experimental Procedures

Different experimental procedures are used to obtain absorption spectra of clusters.

For every small ones $n \leq 4$, two step ionization through intermediate state gives the absorption spectra when the photon energy is varied. Such a procedure

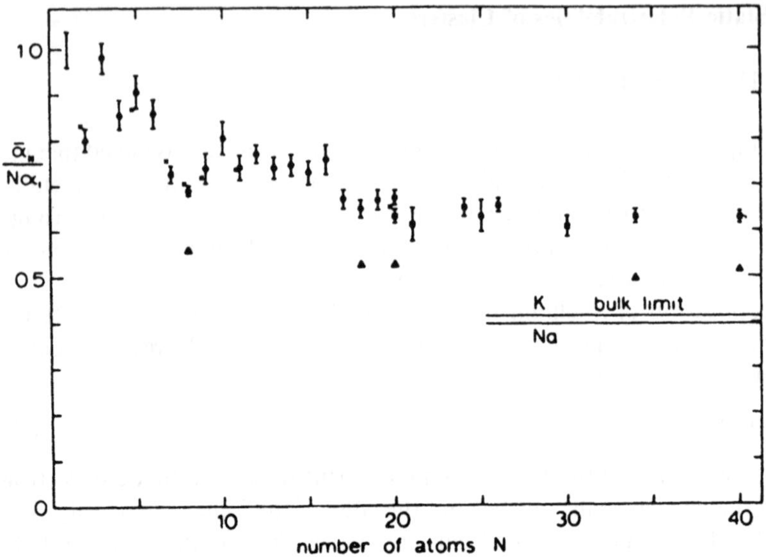

Fig. 14. Cluster polarizability versus cluster size from Ref. [51]

can be achieved if the lifetime of the intermediate state is longer than the delay between the two successive absorptions.

For larger cluster sizes $n > 5$ and the highly excited states of small ones, the relaxation of the electronic excitation into vibrational excitation heats the cluster, causing evaporation of atoms (or dimers). For neutral clusters flighting with thermal kinetic energy, the transverse recoil of the daughter cluster removes it from the collimated beam. The absorption cross section is then monitored by beam depletion. For ionic clusters which can be accelerated to a few of KeV photoevaporation spectroscopy, previously described, is used to measure absorption cross section on mass selected cluster in a cross beam configuration (Fig. 1a).

4.1.4.3b) Giant Resonances in Absorption Spectra of Free Clusters

The absorption spectra of dimers, trimers [8, 52] and tetramers [53] shows the presence of electronic states with well resolved vibrational progressions which are understood by conventional molecular picture. As cluster size increases the density of states increases, and the description of the cluster by quantum chemistry is difficultly extended to large clusters. Moreover a pertinent description of collective effects can be only achieved from collective models. Such effects were observed in alkali clusters for size $n > 8$, the intermediate region $n = 5, 6, 7$ is more or less chaotic.

Figure 15 shows the photoabsorption spectra of 8 electron closed shell clusters. For each element absorption spectrum is dominated by one resonance

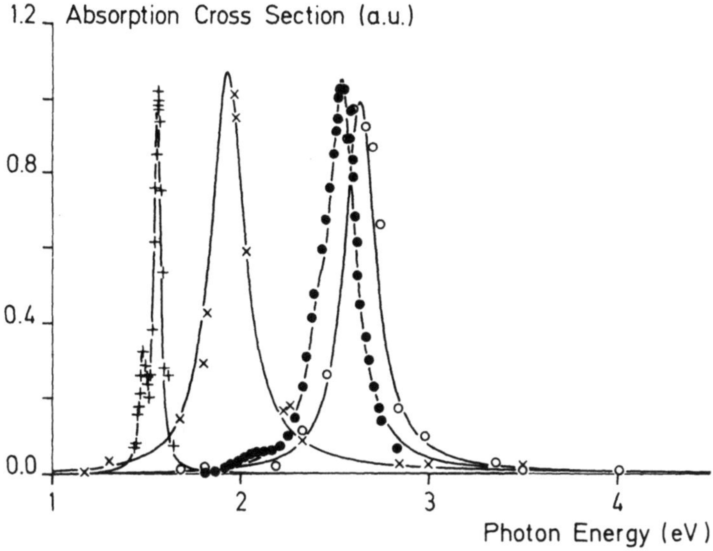

Fig. 15. Giant resonances for 8-electron clusters (+) Cs_8, (×) K_9^+, (●) Na_8, (○) Na_9^+. The single peak put in evidence spherical symmetry

Table 3.

		Li	Na	K	Rb	Cs
$\hbar\omega_0$	X_n	2.58 [53]	2.52 [46, 56]			1.56 [54]
(eV)	X_{n+1}^+		2.62 [55]	1.93 [55]		
$\alpha/8$	(Å$_3$)	15	16	27		41
α_a	(Å$_3$)	22	24	45	49	63
$\omega_p/\sqrt{3}$	(eV)	4.62	3.4	2.4		1.63
α_b	(Å$_3$)	4.7	9	18		38

Resonance energies and corresponding polarizabilities for atom, 8-electron clusters, and bulk. α_a is taken from Ref. [57], and α_b from Ref. [58]

easily fitted by Eq. (54). The blue shift of the resonance from Cs_8 to Na_8 [46, 54–56] agrees with an increase of the polarizability from sodium to cesium. Table 3 regroups the resonance energies $\hbar\omega_0$ for several 8 electron clusters. The cluster polarizabilities are deduced from Eq. (53). The polarizability per electron α_8/n is compared to the atomic [57] and bulk [58] polarizabilities. It is clearly seen that the ratio between the polarizability per electron of the 8 electron cluster and the atomic polarizability is the same whatever the nature of the cluster is. However the evolution to the bulk cannot be deduced from an universal scaling law. If the polarizability of C_8 is close to the bulk value than to the atomic value, the reverse prevails for lithium indicating different evolution of the screening effect with the cluster nature.

For the same element the two resonances Na_8 and Na_9^+ are close in shape and energy indicating that the main behavior of the photoabsorption giant resonance is defined by the number of valence electrons. The small blue shift of Na_9^+ with respect to Na_8 is understood as a smaller polarizability of Na_9^+ than Na_8. The 9-ions of Na_9^+ create stronger restoring force on the 8-valence electron cloud than the 8-ions of Na_8. The difference between neutrals and ions having the same number of electrons must vanish as cluster size increases.

Figure 16 shows the photoabsorption spectra of an open shell cluster. In that case the giant resonance is splitted, showing two components well understood by Eq. (55) where $\omega_1 = \omega_2 = \omega_\perp$ and $\omega_3 = \omega''$. The resonance energy of the corresponding theoretical spherical cluster is $\omega_0 = [\omega, \omega_\perp^2]^{1/3}$. It is very close to the resonance energy of the nearest in size closed shell cluster. Moreover there is a great similarity in the resonance shapes of clusters having the same number of electrons as Na_{10} [59] and Na_{11}^+ [60] for example.

These similarities between resonance shapes of clusters having the same number of electrons as well as their interpretation with spherical and ellipsoidal classical drop model point out that the geometrical shape of an alkali cluster is dictated by the symmetry of its electron gas and reflect the collective excitation of the electron cloud with respect to ionic cores.

The observed giant dipole resonances in nuclei [61], which can be described as the collective motion of the protons with respect to the neutrons, show many

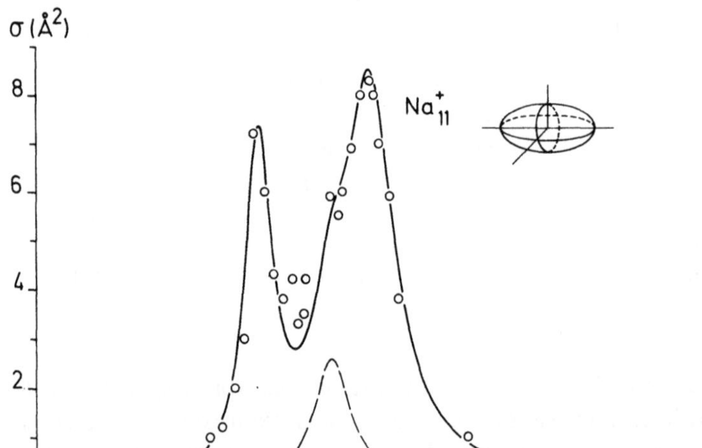

Fig. 16. Giant resonance for the ellipsoidal Na_{11}^+ cluster. The law energy peak reflects the oscillation along the long axis, the high energy peak reflects the oscillation along the short axis. Dashed curve represents the excitation of Na_9^+ which is present in the selected packet Na_{11}^+

similarities to giant resonances in alkali clusters. These nuclear resonances show single peak for spherical nuclei or double peak for ellipsoidal ones and are seen in nuclei as small as ^{4}He. Studies of the damping of nuclear giant dipole resonances find that Landau damping and coupling to nuclear shape oscillations are important [44], giving support to the predictions that these damping mechanisms exist in clusters. Moreover the direct coupling between collective excitation and ionic core motion provides a major contribution to the width of the giant resonance [44]. All the theoretical models as jellium with RPA, are easily adapted to clusters.

4.1.4.3c) Collective Mode Resonance in Small Absorbed Clusters

It is interesting to compare the absorption spectra of free clusters to supported ones or clusters embedded in matrix environments. Extensive results were obtained by *Parks* and *Mc Donald* [62] for sodium microclusters adsorbed on pyrolytic boron nitride surfaces for a range of average cluster size from $\langle n \rangle = 10$ to 300 atoms per cluster. These spectra exhibit line shapes dominated by the dipolar collective mode resonance of the valence electrons making a link between giant resonance and surface plasmon excitation. In fact the resonance is observed to shift from a lower energy of about 2.4 eV at the smallest adsorption, in agreement with the resonance peak of free sodium clusters, to near the bulk limit of 3.4 eV as cluster size increases (Fig. 17). However the very recent measurement on giant resonances of free large clusters (few hundred of atoms) exhibit a shift between free and supported clusters which may be due to cluster surface interaction [63]. Moreover new results on large free lithium clusters are beyond the jellium model [64].

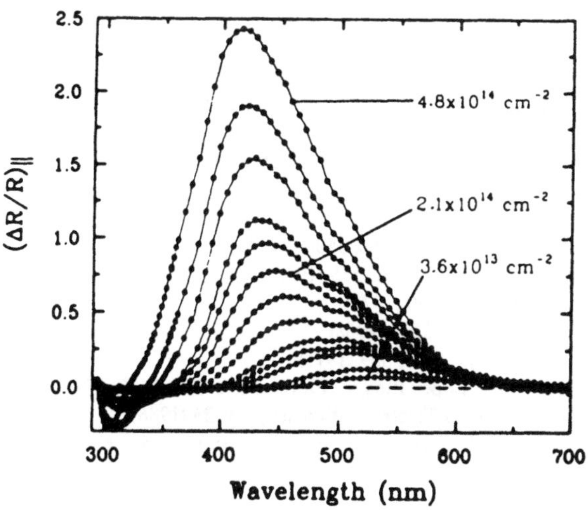

Fig. 17. Giant resonance of supported sodium clusters taken from Ref. [62]

References

1. E.J. Robins, R.E. Leckenby, P. Willis: Adv. Phys. **16**, 739 (1967)
2. A. Herrmann, E. Schumacher, L. Wöste: J. Chem. Phys. **68**, 2327 (1978)
3. K.I. Petersen, P.D. Dao, R.W. Farley, A.W. Castleman Jr: J. Chem. Phys. **80**, 1780 (1984)
4. W.A. Saunders, K. Clemenger, W. de Heer, W. Knight: Phys. Rev. B **32**, 1366 (1985); W.A. Saunders: Ph.D. thesis, University of California, Berkeley (1986)
5. C. Bréchignac, Ph. Cahuzac, J.Ph. Roux: Chem. Phys. Lett. **127**, 445 (1986)
6. W. Shulze: to be published
7. M.M. Kappes, M. Schär, U. Röthlisberger, C. Yeretzian, E. Schumacher: Chem. Phys. Lett. **143**, 251 (1988)
8. A. Herrmann, S. Leutwyler, E. Schumacher, L. Wöste: Helv. Chem. Acta **61**, 453 (1978)
9. O. Echt: in *Physics and Chemistry of Small Clusters*, ed. by P. Jena, B.K. Rao, S.N. Khanna (Plenum, New York, 1987) p. 623 and references therein
10. W.R. Smythe: *Static and Dynamic Electricity*, 3rd ed. (Mc Graw Hill, New York, 1968) p. 128
11. L.D. Landau, E.M. Lifshitz: *Electrodynamics of Continuous Media* (Pergamon, New York, 1960) p. 11
12. D.M. Wood: Phys. Rev. Lett. **46**, 749 (1981)
13. W. Ekardt: Phys. Rev. B **29**, 1558 (1984)
14. J.P. Perdew: Phys. Rev. B **37**, 6175 (1988)
15. E. Engel, J.P. Perdew: Phys. Rev. B **43**, 1331 (1991)
16. N.W. Ashcroft, N.D. Mermin: *Solid State Physics* (Holt Sauders, Tokyo, 1976)
17. C. Bréchignac, Ph. Cahuzac, F. Carlier, J. Leygnier: Phys. Rev. Lett. **63**, 1368 (1989)
18. D.M. Wood, N.N. Ashcroft: Phys. Rev. B **25**, 6255 (1982)
19. W.D. Knight, K. Clemenger, N.A. de Heer, W.A. Saunders, M.Y. Chou, M.L. Cohen: Phys. Rev. Lett. **52**, 2141 (1984)
20. K. Clemenger: Phys. Rev. B **32**, 1359 (1985)
21. J. Koutecky, P. Fantucci: Chem. Rev. **86**, 539 (1986)
22. K.M. Mc Hugh, J.G. Eaton, G.H. Lee, H.W. Sarkas, L.H. Kidder, J.J. Snodgrass, M.R. Manaa, K.H. Browen: J. Chem. Phys. (1990)
23. G. Ganteför, M. Gausa, K.H. Meiwes Broer, H.D. Lutz: Faraday Discuss. Chem. Soc. **86**, 197 (1988)
24. L.S. Kassel: J. Chem. Phys. **32**, 225 (1928)
25. C. Bréchignac, Ph. Cahuzac, J. Leygnier, J. Weiner: J. Chem. Phys. **90**, 1492 (1989)
26. C. Bréchignac, Ph. Cahuzac, F. Carlier, M. de Frutos, J. Leygnier: J. Chem. Phys. **93**, 7449 (1990)
27. Z. Penzar, W. Ekardt: Z. Phys. D **17**, 69 (1990)
28. C. Bréchignac, Ph. Cahuzac, F. Carlier, M. de Frutos: Phys. Rev. Lett. **64**, 2893 (1990)
29. C. Bréchignac, Ph. Cahuzac, J. Leygnier, J. Weiner: J. Chem. Phys. **90**, 1492 (1989)
30. T.P. Martin: J. Chem. Phys. **81**, 4426 (1984)
31. P. Sudraud, C. Colliex, J. van der Walle: J. de Physique **40**, 207 (1979)
32. W. Saunders: Phys. Rev. Lett. **64**, 3046 (1990)
33. I. Katakuze, H. Ichihara: Int. J. Mass spectrum. Ion Proc. **97**, 47 (1990)
34. D.M. Lindsay, Y. Wang, T. George: J. Chem. Phys. **86**, 3500 (1987)
35. S.N. Khanna, F. Reuse, J. Buttet: Phys. Rev. Lett. **61**, 535 (1988)
36. D.M. Mann, H.P. Broida: J. Appl. Phys. **44**, 4950 (1973)
37. D.C. Skillman, C.R. Berry: J. Chem. Phys. **48**, 3297 (1968)
38. L. Genzel, T.P. Martin, U. Kreibig: Z. Phys. B **21**, 339 (1975)
39. Y. Borensztein, P. Andres, R. Montreal, T. Lopez Rios, F. Flores: Phys. Rev. B **33**, 2828 (1986)
40. M. Acheche, C. Colliex, P. Trebbia: Scanning Electron Microscopy 1, **25** (1986)
41. W. Hoheisel, K. Jungmann, M. Wollmer, R. Weidenauer, F. Trager: Phys. Rev. Lett. **60**, 1649 (1988)
42. G. Mie: Ann. Physik **25**, 377 (1908)
43. R. Gan: Ann. Physik **47**, 270 (1915)

44. G.F. Bertsch, P.F. Bortignon, R.A. Broglia: Rev. Mod. Phys. **55**, 287 (1983)
45. G.F. Filipov, V.S. Vasilevskii, S.P. Kruchinin, L.L. Choposkii: Soviet J. Nucl. Phys. **43**, 536 (1986)
46. V. de Heer, K. Selby, V. Kresin, J. Masui, M. Vollmer, A. Chatelain, W.D. Knight: Phys. Rev. Lett. **59**, 1805 (1987)
47. W. Ekardt: Phys. Rev. B **31**, 6360 (1985)
48. D.E. Beck: Phys. Rev. B **35**, 7325 (1987)
49. M. Brack: Phys. Rev. B **39**, 3533 (1989)
50. V. Bonacic Koutecky, P. Fantucci, J. Koutecky: to be published
51. W.D. Knight, K. Clemenger, W.A. de Heer, W.A. Saunders: Phys. Rev. B **31**, 2539 (1985)
52. G. Delacretaz, E.R. Grant, R. Whetten, L. Wöste, J. Zwanziger: Phys. Rev. Lett. **56**, 2598 (1986); M. Broyer, G. Delacretaz, P. Labastie, J.P. Wolf, L. Wöste: Phys. Rev. Lett. **57**, 1851 (1986)
53. M. Broyer, J. Chevaleyre, P. Dugourd, J. Wolf, L. Wöste: Phys. Rev. A **42**, 6954 (1990).
54. H. Fallfren, T.P. Martin: Chem. Phys. Lett. **168**, 233 (1990)
55. C. Bréchignac, Ph. Cahuzac, F. Carlier, J. Leygnier: Chem. Phys. Lett. **164**, 433 (1989); C. Bréchignac, Ph. Cahuzac, F. Carlier, M. de Frutos, J. Leygnier: Z Phys. D (1991)
56. C.R.C. Wang, S. Pollack, M. Kappes: Chem. Phys. Lett. **166**, 26 (1990)
57. W.D. Hall, J.C. Zorn: Phys. Rev. A **10**, 1141 (1974)
58. C. Kunz: Phys. Lett. **15**, 312 (1965)
59. K. Selby, M. Wollmer, J. Masui, V. Kresin, W. de Heer, W. Knight: Phys. Rev. B **40**, 5417 (1989)
60. C. Bréchignac, Ph. Cahuzac, F. Carlier, M. de Frutos, J. Leygnier: Chem. Phys. Lett. **189**, 28 (1992)
61. B.L. Berman, S.C. Fultz: Rev. Mod. Phys. **47**, 713 (1975)
62. J. Parks, S. Mc Donald: Phys. Rev. Lett **62**, 2301 (1989)
63. C. Bréchignac, Ph Cahuzac, N. Kebaili, J. Leygnier, A. Sargati: Phys. Rev. Lett. **68**, 3916 (1992)
64. C. Bréchignac, Ph. Cahuzac, J. Leygnier, A. Sargati: Phys. Rev. Lett. **70**, 2036 (1993)

4.2 Clusters of s^2p^1 Metals and Semiconductors

M.F. Jarrold

4.2.1 Introduction

The elements of group 13 of the periodic table: boron, aluminum, gallium, indium, and thallium; have an s^2p^1 valence electron configuration. As is the rule, the properties of the first member of this group are quite different from those of the others. The ionization energies of boron are large, so covalent bond formation is important in its chemistry. Pure boron is a large band-gap semiconductor. Three allotropic forms of bulk boron, consisting of different arrangements of hollow B_{12} icosahedra, have been characterized. The next member of group 13, moving down the periodic table, is aluminum. While pure aluminum is an almost ideal free electron metal, many of its compounds have considerable covalent character, as do many of the compounds of Ga, In, and Tl. Trivalent behavior is dominant for all of the group 13 elements, except for thallium, where the monovalent state is also important.

Of the members of group 13, aluminum clusters have been by far the most widely studied. There have been a few studies of the properties of boron clusters, but very little work has been performed on clusters of the other elements in this group. We will start by considering boron clusters and then move down the periodic table.

4.2.2 Boron Clusters

In the solid state many boron compounds consist of three-dimensional networks of boron clusters. For example, metal borides with formulae MB_6 and MB_{12} are well known. The MB_6 compounds consist of B_6 octahedra packed in a cubic array, and the MB_{12} compounds generally have B_{12} cubo-octahedra. Given the importance of boron clusters, and in particular the B_{12} unit, in the solid state chemistry of boron, one might anticipate some interesting trends in the properties of the isolated clusters. However, there have been only a few studies of the properties of isolated boron clusters. This is unfortunate, not only for the reasons mentioned above, but also because it should be possible to perform reasonably reliable theoretical calculations on quite large boron clusters.

Most of the studies of isolated boron clusters have been performed by *Anderson* and coworkers [1–4]. They have investigated the collision induced dissociation [1] of boron cluster ions, B_n^+ ($n = 2$–13), and studied the chemical reactions of the clusters with O_2 [2], D_2 [3], and D_2O [4]. They have also performed ab initio molecular orbital studies of boron clusters containing up to 6 atoms [1]. There have been a few other theoretical studies of boron clusters [5].

The clusters for the experimental work of *Anderson* and coworkers were generated by pulsed laser vaporization of solid boron [1]. *Chupka* and *Berkowitz* [6] reported the observation of small boron cluster ions from ruby laser vaporization of the bulk solid as long ago as 1964. The laser vaporization source of *Anderson* and coworkers [1] is unique in that the cluster ions are trapped and cooled in an octopole ion guide. After cooling a particular cluster size is selected by a Wien filter and injected into a second octopole ion guide. In the collision induced dissociation experiments the clusters are then collided with xenon and the products analyzed by a quadrupole mass spectrometer. Measurement of the collision induced dissociation cross sections as a function of collision energy can provide information on bond dissociation energies, and the product distribution can provide qualitative information on the cluster ionization energies. Clusters with 5, 10, 11, and 13 atoms were found to be particularly abundant in the mass spectrum of clusters produced by the source. However, as described below, while the collision induced dissociation data indicates that the 5 and 13 atom clusters are particularly stable, B_{10}^+ and B_{11}^+ appear to be no more stable than B_{12}^+ (which occurs in the mass spectrum with low intensity). Clearly, "magic numbers" alone are not a reliable indication of stability. At low collision energies (10 eV) two product channels are observed in the collision induced dissociation of B_{2-13}^+: loss of an atom to give B_{n-1}^+ and formation of B^+. At higher collision energies B_{n-2}^+ and B_{n-3}^+ are also observed as products from some of the larger clusters. Figure 1 shows a plot of the collision induced dissociation cross sections recorded as a function of collision energy for B_8^+. The line drawn through the experimental data is the result of a simulation, taking into account the energy spread of the ion beam and thermal motion of the target gas, to determine the true dissociation threshold. The thresholds determined from this analysis are summarized in Table 1 and the dissociation energies are plotted in Fig. 2. Clusters with 5, 7, 10, and 13 atoms appear to be somewhat more stable than their neighbors. The dissociation energies can be compared with the bulk cohesive energy of boron which is 5.78 eV/atom. The results suggest that the dissociation energies of boron clusters approach the bulk value for quite small clusters. There is no other experimental data available on the dissociation energies of these small boron cluster ions. The dissociation energy measured for the dimer ion (0.8 ± 0.5 eV) appears to be more than 1.0 eV smaller than the value recommended by *Bruna* and *Wright* [7] from their theoretical calculations (2.25 eV). It seems likely that there is a problem with the experimental measurement of the dimer dissociation energy. Anderson and coworkers have suggested that this cluster may not be efficiently cooled in their cooling trap before undergoing collision induced dissociation.

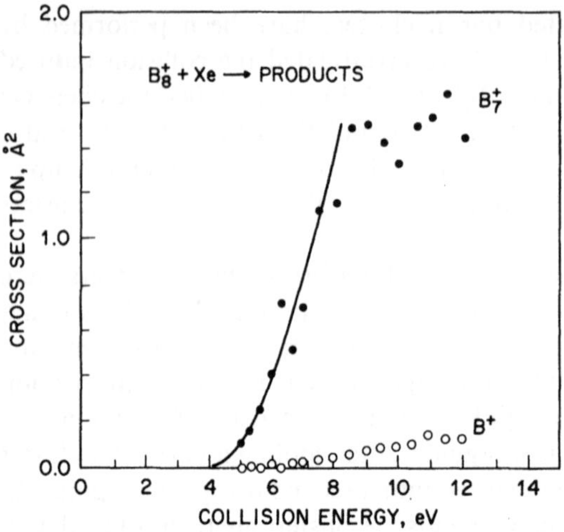

Fig. 1. Cross sections for the collision induced dissociation of B_8^+. The points are the experimental data and the line is the result of a simulation to determine the dissociation threshold. Adapted from Ref. [1]

Table 1. Appearance energies for $B_n^+ + Xe \rightarrow B^+, B_{n-1}^+$

B_n^+	B^+, eV	B_{n-1}^+, eV
2	0.8 ± 0.6	—
3	2.3 ± 0.6	4.3 ± 0.7
4	2.4 ± 0.6	8.0 ± 1.5
5	3.6 ± 0.6	7.1 ± 0.6
6	3.2 ± 0.7	2.7 ± 0.6
7	4.8 ± 0.5	5.5 ± 0.7
8	6.2 ± 0.7	4.2 ± 0.5
9	4.3 ± 0.7	4.0 ± 0.5
10	5.6 ± 1.0	5.4 ± 0.5
11	6.5 ± 0.8	5.6 ± 0.8
12	7.0 ± 1.5	5.5 ± 0.5
13	7.8 ± 0.9	8.0 ± 1.5

The two main products in the dissociation of boron cluster ions ($B_{n-1}^+ + B$ and $B^+ + B_{n-1}$) differ only by the location of the charge. Information on the clusters' ionization energies can be obtained from the collision induced dissociation product distributions using two different methods. Unfortunately, neither method is very reliable. In the first method, the ionization energies are obtained from the difference between the measured thresholds for the $B^+ + B_{n-1}$ and $B_{n-1}^+ + B$ product channels. The ionization energy of the boron atom is of course known, so the ionization energy of the B_{n-1} cluster is given by

$$IE_{B_{n-1}} = T(B_{n-1}^+) - T(B^+) + IE_B \ , \qquad (1)$$

where $T(B_j^+)$ are the thresholds. The problem with this method is that the energy of the threshold for the higher energy channel is usually overestimated

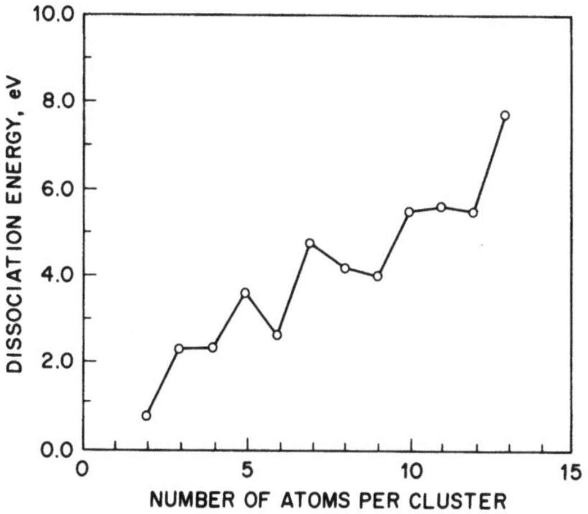

Fig. 2. Dissociation energies of B_n^+ ($n = 2 - 13$) determined from collision induced dissociation studies. Adaped from Ref. [1]

because it must compete with the lower energy one, so this method tends to exaggerate the differences between the ionization energies of the clusters and the atom. The second method for detemining ionization energies uses a pseudo-equilibrium approach [8] with the cluster ionization energies given by

$$IE_{B_{n-1}} = k_B T \ln\left(\frac{I(B^+)}{I(B_{n-1}^+)}\right) + IE_B . \quad (2)$$

In this expression $I(B_j)$ is the intensity of fragment ion j, and k_B is the Boltzmann constant. The problem with this method is that it is necessary to assign a temperature to the dissociating clusters. Ionization energies determined for B_2–B_{12} using these two methods are plotted in Fig. 3. As might be expected the two methods give somewhat different values for the ionization energies. However, both methods predict that the ionization energies of the smaller clusters are larger than the ionization energies of the atom. The ionization energies drop below the atomic value for clusters with 5–7 atoms and then continue to fall with increasing cluster size, to ultimately approach the bulk work function (4.45 eV).

The observation of that the dissociation energies of small boron clusters increase rapidly with size suggests that the number of bonds being broken increases with cluster size. This in turn suggests that the clusters have compact three-dimensional structures. The observation that the lower energy fragmentation channels arise from loss of a boron atom (either as B or B^+) supports this conclusion. Theoretical calculations for B_{2-6} and B_{2-6}^+ also suggest that the clusters have close packed structures. The calculated structures for B_n and B_n^+ are very similar. According to the calculations B_3 is triangular, B_4 is a planar rhombus, B_5 is a trigonal bipyramid, and B_6 is a capped pentagon. The dissociation energies from the theoretical calculations of Anderson and coworkers for B_3^+–B_6^+ are $\sim$ 60–80% of the measured values. For the dimer the calculated dissociation energy is larger than the measured value. Theoretical

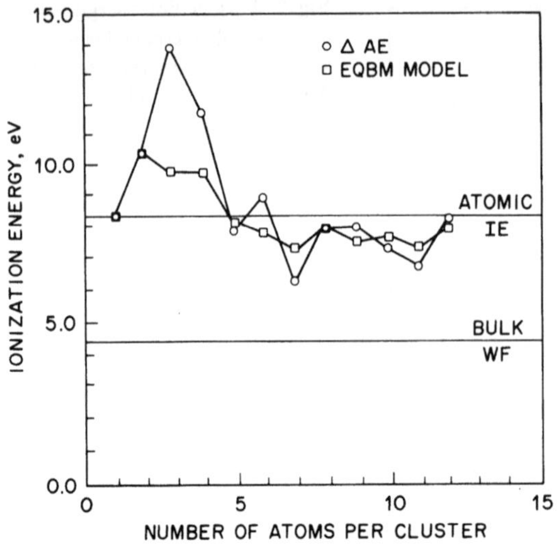

Fig. 3. Ionization energies of B_2–B_{12} estimated from collision induced dissociation measurements. Ionization energies estimated using two different approaches are shown (see text). The agreement between the values obtained by the different methods is, in some cases, not very good. Adapted from Ref. [1]

calculations have not yet been performed for larger bare boron clusters. However, it is tempting to speculate that the particularly stable B_{13}^+ cluster is icosahedral or cubo-octahedral, with the extra atom inside a hollow B_{12} unit.

Anderson and coworkers have also reported studies of the chemical reactions of the boron clusters with O_2 [2], D_2 [3], and D_2O [4]. The reactions with oxygen result in fragmentation of the clusters to yield a large number of B_x^+ and B_yO^+ products ($x = 1, n-1, n-2, n-3$; and $y = n, n-1, n-2$). Clusters with $n < 6$ give mainly B^+ over the 0.25–10.0 eV collision energy range investigated, and clusters with $n > 6$ give mainly B_{n-2}^+. For the larger clusters the $B_{n-1}O^+$ product is a minor one over the collision energy range studied, but the cross sections for this product (unlike the other products) generally increase at the lower collision energies, suggesting that $B_{n-1}O^+$ may dominate at thermal energies. The main products observed in the reactions of the clusters with D_2 are $B_nD_2^+$, B_nD^+, B_{n-1}^+, and B^+. For most clusters $B_nD_2^+$ adduct formation dominates at the lower collision energies. This process occurs without an activation barrier for small clusters ($n < 10$), but collision energy thresholds were observed for the larger clusters ($n = 10$–16). With D_2O the smaller clusters give a wide range of different products, but for the larger clusters ($n > 6$) the product distributions at low energy are dominated by the reaction:

$$B_n^+ + D_2O \rightarrow B_{n-1}D^+ + DBO . \qquad (3)$$

In this reaction a B atom in the cluster has been replaced by a D atom. The $B_{n-1}D^+$ products react in a similar way, and so a series of sequential exchange reactions can occur. For example, for B_{12}^+ sequential exchange all the way to $B_5D_7^+$ was observed. A more detailed discussion of the chemistry of boron cluster ions can be found in a separate chapter.

4.2.3 Aluminum Clusters

In contrast to boron, which is a large band gap semiconductor, aluminum is an almost ideal free electron metal. Aluminum clusters have been quite widely studied and have emerged as one of the model systems for testing and developing ideas about metal clusters. There are several reasons for this. Aluminum is widely used commercially, and there is an extensive literature on the chemical properties of aluminum surfaces. While not as simple, electronically, as boron it is still possible to perform reasonably reliable theoretical calculations on aluminum clusters. Finally, aluminum is light and only has one isotope so it is attractive for experimental studies as well.

The dissociation of aluminum cluster ions has been investigated by several groups. *Begemann, Meiwes-Broer,* and *Lutz* [9] have investigated the metastable dissociation (spontaneous unimolecular dissociation) of aluminum cluster ions generated by sputtering. Collision induced dissociation of aluminum cluster ions has been studied by *Meiwes-Broer* and coworkers [10], *Jarrold, Bower,* and *Kraus* [8], and *Hanley, Ruatta,* and *Anderson* [11]. *Woste* and coworkers [12] and *Jarrold* and coworkers [13, 14] have investigated the photodissociation of aluminum cluster ions. The smaller cluster ions ($n < 8$) dissociate to give mainly Al^+ and the larger ones ($n > 13$) give mainly Al_{n-1}^+. There are some minor differences in the product branching ratios measured by different groups. While some of these differences may reflect the different methods of exciting the clusters, it is likely that at least some of the discrepancies arise from experimental difficulties in detecting products with widely different masses with equal efficiency.

The products observed in the dissociation of aluminum cluster ions are similar to those observed in the dissociation of boron cluster ions: loss of an atom as either M^+ or M. It appears that metal clusters generally (but not always) dissociate by evaporation rather than fission into two cluster fragments. Studies of the photodissociation [15] and collision induced dissociation [16] of transition metal clusters show that the dominant dissociation process is evaporation. All the available evidence indicates that the dissociation of these metal cluster systems occurs on the ground electronic state potential surface by statistical unimolecular dissociation. In which case the dominance of the evaporation channel simply occurs because it is a lower energy process than cluster fission. This is expected for metallic systems where the binding energy per atom increases relatively smoothly with cluster size. For covalently bonded clusters such as silicon or carbon cluster ions "magic fragments" are observed. Small carbon cluster ions dissociate by loss of C_3 units [17] and silicon cluster ions dissociate preferentially to give products with 6, 7, 10, and 11 atoms [18]. Thus it is somewhat surprising that boron clusters behave like aluminum and other metal clusters, because bulk boron is more like bulk silicon than bulk aluminum.

It was apparent from the studies of the dissociation of aluminum cluster ions that clusters with $n = 7$ and 13–14 were somewhat more stable than their neighbors. Al_7^+ and Al_{14}^+ appear as "magic numbers" in some mass spectra of aluminum clusters. Though the "magic numbers" in aluminum are not as well defined or as reproducible as in the alkali metals (see Chapter 4.1). The enhanced stability of clusters with $n = 7$ and 13–14 can be accounted for by the electronic shell model or jellium model [19]. This model has been described in detail in Chapter 4.1. It is an approximate model in which the clusters are treated as metallic. The ionic cores are replaced by a uniformly positively charged background, and then the electronic energies are calculated self-consistently to obtain the energy levels. The energy levels show a shell structure, due to electronic angular momentum, similar to that which is well established for electronic and nuclear energy levels. The location of the shell closings depends on the nature of the potential. For a square well potential, shell closings occur with 8, 18, 20, 34, 40, 58, 68, and 92 valence electrons corresponding to the $1p$, $1d$, $2s$, $1f$, $2p$, $1g$, $2d$, and $3s$ shells. Shell closings occur with 8, 20, 40, and 70 valence electrons for a harmonic oscillator potential. *Chou* and *Cohen* have performed electronic shell model calculations for neutral aluminum clusters [20] using a local density-functional scheme and found shell closings with 18, 20, 40, 58, 70, and 92 electrons. Aluminum is trivalent and the shell closings generally do not occur with an integral number of three valence electrons so the effect of the shell closings are smeared out over several clusters. The calculations of *Chou* and *Cohen* [20] predict that the shell closings with 20 and 40 valence electrons will be particularly prominent. These correspond to the "magic numbers" Al_7^+ (with 20 valence electrons: $7 \times 3 - 1$ (for the charge) = 20) and Al_{13}^+–Al_{14}^+ (38 and 41 valence electrons). However, the application of the electronic shell model to aluminum clusters has not been universally accepted. One obvious conceptual problem is that not all the valence electrons in the aluminum atom are equivalent. There is a 3.6 eV gap between the filled $3s$ level and the partly filled $3p$ levels in the atom. In addition the core is less well shielded in high valence elements like aluminum, so the interaction of the electrons with the core will be stronger. The most recent theoretical results [21], which will be discussed in more detail below, suggest that the large extra stability conferred by a closed shell configuration makes the clusters conform to the electronic shell model at the shell closings, but the electronic structure may depart from the predictions of the shell model away from the shell closings. Before leaving "magic numbers" and the electronic shell model it is worth mentioning some experiments performed on the collision induced dissociation of oxidized aluminum cluster ions, $Al_nO_m^+$ ($m = 1, 2$) [22]. The main product from the collision induced dissociation of these species arises from loss of Al_2O, which is a particularly stable molecule. As noted above bare clusters with $n = 7$ and 13, 14 were found to be particularly stable. With the oxidized clusters the particularly stable clusters shifted to larger values of n. It appears that the particularly stable oxidized clusters can be accounted for by the electronic shell model using simple chemical valence ideas where binding an oxygen atom (either as O^{2-} or as $O{=}Al$) costs the cluster two valence electrons. Thus performing an electron count for the $Al_{15}O_2^+$ cluster (which was found to be particularly stable) shows that this cluster has $3 \times 15 - 1$

(for the charge) − 4 (2 per oxygen) = 40 valence electrons, and so has a closed electronic shell configuration according to the electronic shell model.

The observation that Al^+ is the main product from the dissociation of the smaller clusters and Al_{n-1}^+ dominates for the larger ones, indicates that, like boron, the ionization energies of the smaller clusters are above the atomic value. A number of groups have derived approximate values for the ionization energies of aluminum clusters from the product distributions in dissociation experiments (using the approaches described above for boron). However, for aluminum more direct measurements using laser photoionization have been performed. The laser photoionization measurements should generally be more reliable than the more indirect approaches. The first attempt to measure the ionization energies of aluminum clusters by laser photoionization was by *Cox* and coworkers [23]. More recently *Whetten* and coworkers have described photoionization measurements for aluminum clusters with up to 70 atoms [24]. For these experiments the clusters were generated using pulsed laser vaporization with a pulsed valve, but a Laval type conical nozzel was employed to optimize the cooling of the clusters. In photoionization experiments it is important that the clusters be cold since the photoionization threshold will be shifted to lower energy by the internal energy in the cluster, and so the ionization energies will be underestimated. It is very difficult to determine the temperatures of clusters in the gas phase. *Whetten* and coworkers showed that their aluminum clusters were sufficiently cold that Ar (added to the He buffer gas) will condense on to them to give Al_nAr_m species. Figure 4 shows the photoionization threshold energies for aluminum clusters as a function of size. The points are the experimental data and the dashed line is the prediction of the classical metallic sphere model [25]. According to this model the ionization energies are given by

$$IE_n = WF + \frac{3}{8}\frac{e^2}{R}, \tag{4}$$

where WF is the bulk work function and R is the cluster radius. There has been a lot of discussion over whether 3/8 or 1/2 is the correct value in this expression.

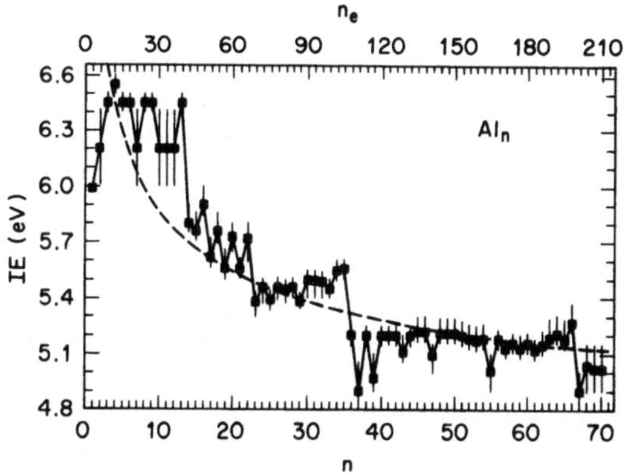

Fig. 4. Ionization energies of aluminum clusters obtained from photoionization threshod measurements on cold clusters. The dashed curve is the predictions of the classical metallic sphere model (see text). Adapted from Ref. [24]

The simple classical model (with 3/8) seems to fit the experimental data on a wide range of different metal cluster systems [26].

Ionization energies estimated from aluminum cluster dissociation experiments are in reasonable agreement with the values obtained by photoionization, except that the dissociation experiments suggest that clusters in the 7–13 atom size regime have lower ionization energies (closer to the atomic value) than obtained from photoionization. For example, photoionization experiments yield a value of 6.0–6.4 eV for the ionization energy of Al_7, but in photodissociation experiments on Al_8^+ [14] the threshold for production of $Al_7^+ + Al$ is ~ 0.4 eV lower in energy than the threshold for $Al^+ + Al_7$, suggesting that the ionization energy of Al_7 is significantly below the atomic value (5.998 eV). Similarly, the dissociation experiments suggest that the ionization energy of Al_{13} is closer to the atomic value than the > 6.4 eV value obtained from photoionization. It is possible that there are small activation barriers associated with the dissociation of Al_n^+ to $Al_{n-1}^+ + Al$. However, a more likely explanation for the discrepancy is that the true threshold was not observed in the photoionization experiments because of poor Franck–Condon factors (a significant geometry change upon ionization). There has been considerable interest in the ionization energy of Al_2. Collision induced dissociation experiments [8, 11] provide estimates which range from 5.2 eV to 6.2 eV depending on how the data is analyzed. Photoionization experiments by *Cox* and coworkers [23] and *Whetten* and coworkers [24] yield brackets of 6.0–6.4 eV. Theoretical calculations [27] indicate that a substantial geometry change occurs upon ionization so it is difficult to observe the true adiabatic ionization threshold. Recent zero kinetic energy threshold photoelectron spectroscopy measurements by *Harrington* and *Weisshaar* [28] appear to resolve this issue and provide a value of 5.989 ± 0.002 eV for the adiabatic ionization energy of Al_2.

As can be seen from Fig. 4 there are sharp drops in the ionization energies for clusters with 7 (21), 14 (42), 23 (69), 36 (108), and 67 (201) atoms (the number of valence electrons are shown in brackets). According to the electronic shell model, shell closings occur with 20, 40, 58, 70, 92, 138, and 198 electrons for aluminum [20]. Drops in the ionization energies are expected for neutral clusters with one or two electrons over the number required for a closed shell. The agreement is not striking. While the observed drops in the ionization energies at $n = 7$ and 14 are in agreement with the predictions of the electronic shell model, for larger clusters the discontinuities do not agree with this model. The low ionization energy of Al_{23} (with 69 valence electrons) rather than at Al_{24} (with 72 valence electrons) may indicate that there is a shell closing with 68 electrons (expected for a square well potential) rather than 70 electrons (predicted by the local density-functional theory calculations of *Chou* and *Cohen* [20]). The observation that the smaller clusters agree with the predictions of the shell model and the larger clusters do not is rather puzzling because the electronic shell model would expected to be better for the larger clusters.

With alkali metal clusters the ionization energies drop fairly continuously from the atomic value towards the bulk work function (except for small discontinuities at the shell closings). As described above the ionization energies of both boron and aluminum clusters rise above the atomic value for small clusters. This behavior was first predicted by Upton in his theoretical calculations on small aluminum clusters (n up to 6) [29]. His explanation for this phenomena is illustrated schematically in Fig. 5. For the alkali metal clusters with a half filled s band of orbitals the ionization energy decreases smoothly with increasing cluster size. The group 13 elements have a 1/6 filled p band of orbitals and the energy of the highest occupied molecular orbital initially decreases with increasing cluster size. Whetten and coworkers have argued that as the s and p bands overlap the ionization energies should approach the predictions of the classical metallic droplet, and suggest (on the basis of the data shown in Fig. 4) that this occurs for clusters with 5–25 atoms.

Photoelectron spectroscopy can provide information on the band structure of clusters. Furthermore, it is a technique which is applicable to both molecules and bulk materials and so has the potential for showing how the band structure develops with cluster size. Since it is important to do these experiments on size selected clusters, measurements on isolated clusters in the gas phase are generally performed on cluster anions (so they can be mass analyzed before photoionization). An additional advantage of performing these experiments on cluster anions is that their electron affinities are fairly small so photoelectron spectra can be recorded using UV lasers. The high photon fluxes available from these lasers (versus UV lamps) partly compensates for the low cluster anion densities. Another advantage of working with the anions in these experiments is that the

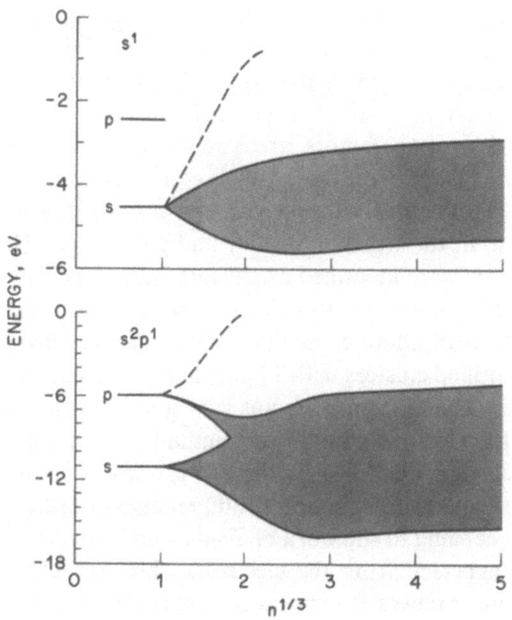

Fig. 5. Diagram showing the evolution of the band structure in s^1 metals and s^2p^1 metals. Adapted from Ref. [24]

photoelectron spectrum yields information on the neutral excited states (rather than the cation states as in conventional photoelectron spectroscopy). However, this information is for the neutral excited states in the geometry of the anion, and there may be a significant geometry change on going from the anion to the neutral cluster.

Photoelectron spectra of size selected aluminum cluster anions have been recorded by two groups [30, 31]. The experimental approaches employed were similar. The cluster anions were generated by pulsed laser vaporization, and then size selected by time of flight mass analysis. The size selected cluster anions are then irradiated with a pulsed UV laser. Since the number of cluster anions present in the UV laser interaction region is small, the photoelectron spectra were recorded using magnetic bottle spectrometers which collect photoelectrons emitted in virtually all directions (conventional photoelectron spectrometers generally only collect photoelectrons emitted into $\sim 10^{-4}$ sterad). The magnetic bottle spectrometer consists of two regions: a high field region where photodetachment occurs, the electrons are then guided adiabatically along the field lines to the second region, a long low field region where the electrons are ultimately guided to the detector. Photoelectron spectra are recorded by time of flight. The ability to detect nearly all the emitted photoelectrons has a price: the energy resolution (which depends on the ratio of the magnetic fields in the low and high field regions) is quite low, and much less than can be achieved using more traditional electrostatic analyzers.

Figure 6 shows photoelectron spectra of aluminum cluster anions, with 3–32 atoms, recorded with an ArF excimer laser (6.4 eV photons). According to convention the data are plotted against electron binding energy (photon energy minus electron kinetic energy). The first piece of information that can be obtained from these results is the electron affinities (which is the energy required to remove an electron from the cluster anions). These can be derived from the photoelectron thresholds, and are represented by the arrows in the figure. The electron affinities cannot be derived precisely because neither the clusters' temperatures nor the vibrational overlap factors are known. Low electron affinities are expected for neutral clusters with closed shell configuration, and high electron affinities are expected for neutral clusters with one or two electrons under a closed shell configuration. In the work of *Smalley* and coworkers [31] clusters with 6, 13, 19, and 23 atoms were identified as clusters with "especially high" electron affinities. Clusters with 4, 6, 9, 19, and 22 atoms were identified as clusters with particularly high electron affinities in the work of *Meiwes-Broer* and coworkers [30], they also identified clusters with 14 and 23 atoms as having particularly low electron affinities. The agreement is not very good. As can be seen from Fig. 6 the changes in the electron affinities are actually rather small and the differences between the clusters identified by the two groups as having "especially high" electron affinities could simply reflect the different assumptions used to identify the thresholds. According to the work of *Smalley* and coworkers [31] high electron affinities are expected from the electronic shell model for clusters with 6, 13, 19, and 23 atoms. Meiwes-Broer and coworkers showed that

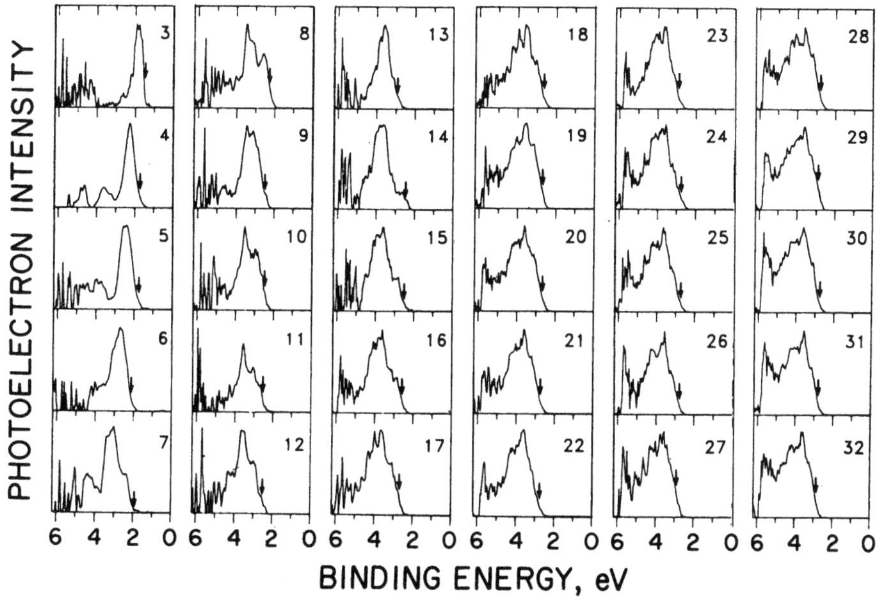

Fig. 6. Photoelectron spectra of Al_3^-–Al_{32}^- recorded using an ArF excimer laser (6.4 eV photons). The arrows mark the assigned thresholds which are used to derive the electron affinities. Adapted from Ref. [31]

the electron affinities approximately followed the

$$EA_n = WF - \frac{5}{8}\frac{e^2}{R} \tag{5}$$

dependence expected from the classical metallic sphere model [25].

As can be seen in Fig. 6 a number of poorly resolved features are apparent in the photoelectron spectra. Unfortunately, it is not possible to say much about these features without the aid of detailed theoretical calculations (which are not yet available). The behavior around Al_{13}^- (which presumably has a closed shell configuration with 40 valence electrons), however, deserves further comment. The threshold for Al_{13}^- is sharp. That for Al_{14}^- has a foot or step which in the work of *Meiwes-Broer* and coworkers [30] is resolved into a separate peak. This separate low intensity peak could arise from photoelectrons from the newly opened shell of Al_{14}^-. An obvious question in the photoelectron spectroscopy of aluminum cluster anions is at what point do the spectra resemble the bulk. In the bulk the *s* and *p* bands (which are separated by 3.6 eV in the atom) have merged and are indistinguishable. The higher binding energy feature which appears to grow in for clusters with $n > 20$ (see Fig. 6) could well be the 3*p* band beginning to overlap with the 3*s* band.

Cox and coworkers [32] have investigated the magnetic properties of aluminum clusters using a Stern–Gerlach experiment. In these experiments a beam of neutral aluminum clusters, generated by pulsed laser vaporization, is collimated and passed through a Stern–Gerlach magnet which generates a magnetic field gradient. Species passing through the field are deflected by an amount proportional to their magnetic moment. After passing through the magnet the clusters are ionized by a laser. The cluster beam profile is measured by translating the ionizing laser beam across the width of the cluster beam. Since aluminum has an odd number of electrons the even-numbered clusters have ground states which have $S = 0, 1, 2, \ldots$ (singlet, or triplet, or quintet, etc.); and the odd-numbered clusters have $S = 1/2, 3/2, \ldots$ (doublet, or quartet, etc.). The spin is quantized along the magnetic field direction with $2S + 1$ values of M_s which range from $+S$ to $-S$. Since the clusters are generated by a supersonic expansion, all clusters have approximately the same velocity, regardless of size. Thus the deflection for a given total magnetic moment will decrease with increasing cluster size. Experimentally, it was found that the deflection decreases with increasing cluster size and becomes virtually unmeasurable for clusters with $n > 12$. The results were consistent with the odd atom clusters having doublet ground states, and even atom clusters having either singlet or triplet ground states. Al_2, Al_6, and Al_8 were identified as triplets. Theoretical calculations for Al_2 [31] suggest that the ground state is a triplet, though the first singlet excited state lies only $\sim 2000\ cm^{-1}$ higher in energy.

In experiments that are closely related to the magnetic deflection experiments described above, *de Heer* and coworkers [33] have measured the polarizabilities of aluminum clusters containing 15–61 atoms. To measure the polarizabilities a collimated aluminum cluster beam is passed through an inhomogeneous electric field. The deflection in this field is proportional to the clusters polarizability and inversely proportional to its mass. The measurements performed by de Heer and coworkers were performed using a position sensitive technique which avoids scanning the position of the laser beam. The polarizabilities per atom of Al_{15}–Al_{61} are shown in Fig. 7. For a metal sphere the polarizability is R^3. For a small metal cluster the polarizabilities are enhanced because the electrons spill out over the edge of the cluster and the polarizabilities per atom can be written

$$\alpha_n = \frac{(R + \delta)^3}{n}, \tag{6}$$

where δ is related to the electronic spillout. The dashed line in Fig. 7 shows polarizabilities per atom calculated for jellium spheres using the Thomas–Fermi approximation. The polarizabilities for most clusters with $n < 40$ are considerably below the jellium model predictions. In contrast, for alkali metal clusters the measured polarizabilities are in good agreement with the predictions of this jellium model. *de Heer* and coworkers [33] have suggested that their measurements indicate that a non-jellium to jellium transition occurs for aluminum clusters with ~ 40 atoms. They suggest that the smaller clusters have polarizabilities which depart from the predictions of the jellium model because of perturbations by the ionic core potentials. *Whetten* and coworkers [24] have

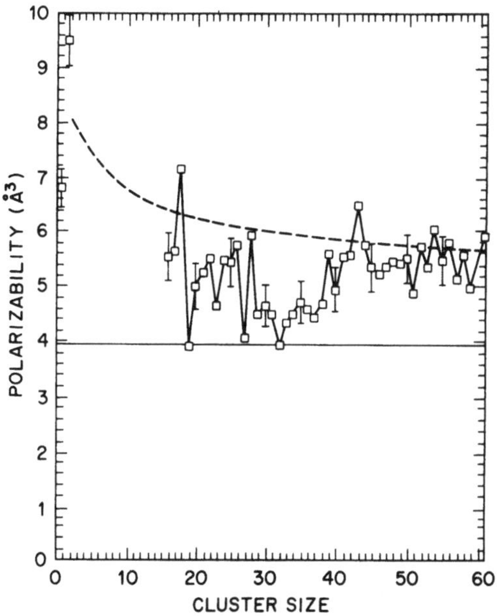

Fig. 7. Polarizabilities per atom of $Al_{15} - Al_{61}$. The dashed line shows predictions for a jellium sphere using the Thomas–Fermi approximation. Adapted from Ref. [33]

proposed an alternative explanation. They suggest that the low polarizabilities of the smaller clusters may be due to the suppressed response of the *s*-band electrons, since this is the size regime where the *s* and *p* bands are expected to merge.

Our experimental knowledge of the geometric and electronic structure of molecules is derived mainly from spectroscopy. Unfortunately, attempts to apply these techniques to metal clusters have not been particularly fruitful. There are several reasons for this. Since there are generally a broad range of cluster sizes present, it is difficult to identify the carrier of a particular absorption. This is a serious problem in matrix isolation experiments, and in direct absorption experiments in the gas phase. The technique of resonant two photon ionization overcomes this problem. Here the first photon excites the cluster to a particular excited state and the second photon ionizes it. The ions are then analyzed by time of flight mass spectrometry and spectra are recorded by scanning the wavelength of the first photon. The energy of the second photon is kept sufficiently small that it cannot ionize the cluster unless it has already adsorbed the first photon and is in an excited electronic state. Unfortunately, even this technique has had limited success. It appears that the excited state lifetimes of small ($n > 3$) metal clusters are extremely short. As a consequence, after absorbing the first photon the excited electronic state rapidly internally converts to the ground state (the electronic excitation is degraded into vibrational excitation), and the excited cluster is not efficiently ionized by the second photon.

Al_3 is the largest neutral aluminum cluster for which a spectrum has been measured using resonant two photon ionization [34]. The spectrum, recorded by Morse and coworkers, shows a discrete band system in the 520–610 nm region consisting of an extended vibrational progression in a single vibrational mode ($\omega'_1 = 273.2\ cm^{-1}$). Hot bands and combination bands were also observed. A detailed assignment of the ground and excited states has not yet been performed and requires high resolution studies to resolve the rotational structure of the transitions. The vibrational progression sits on top of a continuum which gradually grows in with decreasing wavelength. Both the continuum and the vibrational progression appear to abruptly terminate at 516 nm. Morse and coworkers suggested that this marks the dissociation threshold of Al_3, in which case the dissociation energy would be 2.40 eV.

One approach to measuring spectroscopic information for metal clusters that avoids the problem of the intermediate state lifetime is photodissociation spectroscopy. Here after absorbing a photon the cluster subsequently dissociates, and the excited state lifetime is unimportant (except, obviously it may influence the widths of the transitions). Spectra are recorded by scanning the wavelength and measuring the fraction which dissociates. The main limitation of this approach is that it is necessary to work with photon energies above the dissociation energy of the cluster, though two-photon dissociation spectroscopy is possible. These measurements can be performed with neutral and charged clusters. With neutral clusters only depletion measurements can be made. Here the laser beam counter-propagates along a collimated neutral beam and upon dissociation the kinetic energy between the neutral fragments causes them to leave the beam. With ions it is possible to size select the clusters before photodissociation and so the products and undissociated ions can be directly measured. This approach has definite advantages, particularly if the absorptions are weak. *Jarrold* and coworkers [13] have used photodissociation spectroscopy to study small aluminum cluster ions, Al_3^+–Al_6^+. The photodissociation spectra are shown in Fig. 8, they are plots of the relative abundance of the observed photoproducts against wavelength. Note that two photoproducts were observed in each case: Al^+ and Al_{n-1}^+. Unforunately, the spectra do not show any vibrational structure. One factor which probably contributes to the absence of vibrational structure is that the clusters in these experiments were probably not cold, even though they were believed to have undergone many collisions with the He buffer gas in the source. The ionization energies of small aluminum clusters are larger than the atomic value so in the source the small clusters presumably undergo charge transfer reactions with neutral aluminum atoms. The small aluminum clusters which do make it out of the source are probably formed close to the source exit and may not be completely thermalized. A number of features are apparent in the spectra. Al_3^+ has an absorption centered around 500 nm and then absorbs strongly towards the UV. Al_4^+ only weakly absorbs throughout most of the visible region of the spectrum, though an absorption feature grows in towards the UV. For both Al_5^+ and Al_6^+ broad absorption features, apparently composed of several overlapping transitions are observed. These clusters also show sharp thresholds towards the red end of the spectra. Measurements of the fluence dependence of the photoproducts around

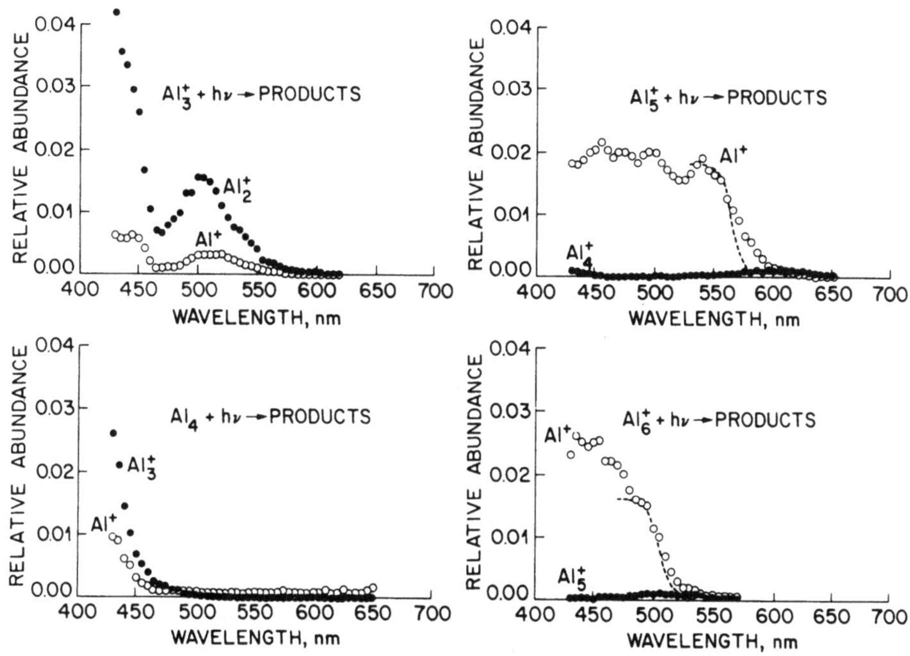

Fig. 8. Photodissociation spectra recorded for Al_3^+–Al_6^+. The spectra are plots of the relative abundance of the observed products against wavelength. The data were recorded using a constant laser fluence of 3.3 mJ/cm^2. Adapted from Ref. [13]

the threshold region indicated that the thresholds observed in the photodissociation spectra are due to the dissociation thresholds of the clusters. For wavelengths longer than the thresholds, dissociation requires two photons. From this data, values for the dissociation energies of 2.20 eV and 2.35 eV were determined for Al_5^+ and Al_6^+ respectively. The features observed in the photodissociation spectra of Al_3^+–Al_6^+ have not yet been assigned. However, they are not like the strong absorption features, assigned to surface plasmons, that have been observed for alkali metal clusters in the visible region of the spectrum [35]. Based on the bulk surface plasma frequency the surface plasmon for these aluminum clusters should occur deep in the UV region of the spectrum. The photodissociation of Al_2^+–Al_8^+ and Al_2^-–Al_8^- has been studied by *Woste* and coworkers [36]. The results obtained for the positive cluster ions were qualitatively similar to those described above. For the negative cluster ions the only product observed was neutral atom evaporation to give Al_{n-1}^-. This result is consistent with the electron affinity measurements described above. However, the photodissociation cross sections for the negative cluster ions were 20–100 times larger than those of the positively charged clusters. The origin of this large difference in the photodissociation cross sections is not yet understood.

The dissociation energies of aluminum cluster ions have been measured by *Hanley, Ruatta,* and *Anderson* [11] and by *Ray, Jarrold, Bower,* and *Kraus*

[13, 14]. Anderson and coworkers employed collision induced dissociation as described above for boron cluster ions. The aluminum cluster ions were generated by sputtering and cooled in an rf trap. Jarrold and coworkers used photodissociation with the clusters generated by pulsed laser vaporization. Both methods depend on preparing the clusters with a known amount of energy and noting whether dissociation occurs. Clusters containing more than a few atoms contain many internal degrees of freedom. After the clusters are excited the energy is distributed among these degrees of freedom and it takes time for the energy to become localized in a particular mode and cause dissociation. The amount of time required increases as the number of atoms increase, and the larger clusters require an energy considerably above the dissociation threshold (often called a kinetic shift) in order for dissociation to occur on a reasonable timescale. These aspects of cluster dissociation are discussed in more detail in Chapter 2.7. For the dissociation of aluminum cluster ions the kinetic shift becomes important for clusters with $n > 6$. Anderson and coworkers studied aluminum clusters with up to seven atoms and so they avoided the kinetic shift problem. To measure dissociation energies for larger clusters *Jarrold* and coworkers [14] adopted a different approach. Instead of attempting to measure the dissociation thresholds, they directly measured the lifetimes of photoexcited clusters at several different photon energies and then fit the lifetime data using an RRKM model (see Chapter 2.7) to deduce the dissociation energies. Dissociation energies available for aluminum from a number of different sources are summarized in Fig. 9. A value for the dissociation energy of Al_2^+ can be derived from the accepted value for the dissociation energy of Al_2 and the ionization energies of Al_2 and Al. Similarly, a value for the dissociation energy of Al_3^+ can be derived from the dissociation energy of Al_3 deduced by *Morse* and coworkers [34] and the ionization energies of Al_3 and Al_2. The results of Anderson and

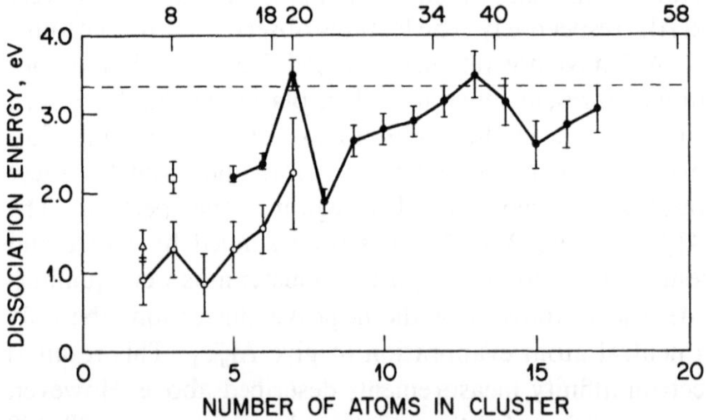

Fig. 9. Plot of the dissociation energies of aluminum cluster ions containing up to 17 atoms: (●) *Jarrold* and coworkers [13, 14]. (○) CID measurements of *Anderson* and coworkers [11]; (□) *Morse* and coworkers [34] (see text); and (△) (see text). The numbers along the top of the figure show the locations of the shell closings according to the simple electronic shell model. Adapted from Ref. [14]

coworkers show the same overall trends, but are significantly below the other measurements. A possible explanation for this result is that the clusters in their experiments were not cooled as efficiently as anticipated in their cooling trap. It is apparent from the data shown in Fig. 9 that there are maxima in the dissociation energies for clusters with seven and 13 atoms as predicted by the electronic shell model. The dashed line in Fig. 9 shows the bulk cohesive energy (3.358 eV atom). The dissociation energies show an overall increase with increasing cluster size, but a more convenient way to follow the development of the bulk cohesive energy is to plot cohesive energies per atom of neutral clusters against $n^{-0.33}$. This is done in Fig. 10, where the experimental measurements are compared with the results of a number of theoretical calculations. The theoretical results nearly all lie below the experimental measurements. The dashed line in Fig. 10 shows the prediction of a simple model, based on bulk properties, originally due to *Miedema* and *Gingerich* [40]. According to this model the cluster cohesive energies per atom are given by

$$E_c(n) = \Delta H^0_{\text{vap}} - \left[\frac{36\pi V_a^2 \gamma^3}{n}\right]^{0.33}, \tag{7}$$

where ΔH^0_{vap} is the heat of vaporization of the bulk at 0 K, V_a is the volume occupied by an atom in the bulk, and γ is the surface energy of the solid metal at 0 K. As can be seen from Fig. 10, the experimental values approach the predictions of this model for clusters with $n > 6$. This may be related to the assumption that the clusters are spherical, which is made in deriving Eq. (7). However, it is remarkable that such a simple model (employing bulk properties) is in such good agreement with the cohesive energies of clusters with as few as seven atoms.

The chemistry of aluminum clusters has been studied by several groups. *Cox* and coworkers [23] have investigated the reactions of neutral aluminum clusters

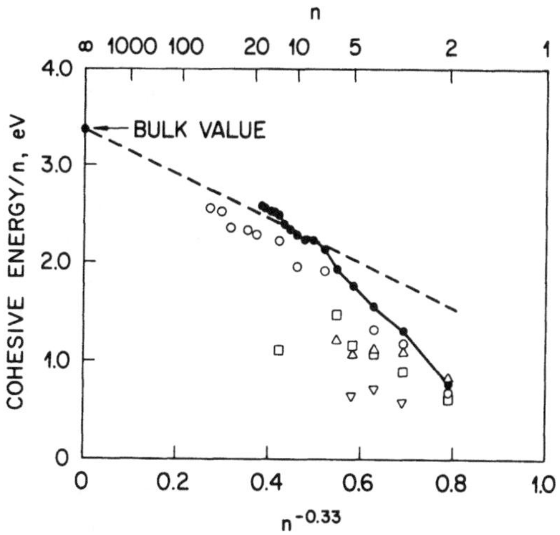

Fig. 10. Plot of cohesive energies per atom for neutral clusters against $n^{-0.33}$. The solid points (●) are the experimental results and the other points are the results of theoretical calculations: (▽) *Pacchioni* and *Koutecky* [37]; (△) Upton [29]; (□) *Pettersson, Bauschlicher*, and *Halicioglu* [38]; and (○) represents the semi-empirical calculations of *Pacchioni* and *Fantucci* [39]. The dashed line shows the predictions of the simple model (discussed in the text) based on the bulk cohesive energy and the cluster surface energy. Adapted from Ref. [14]

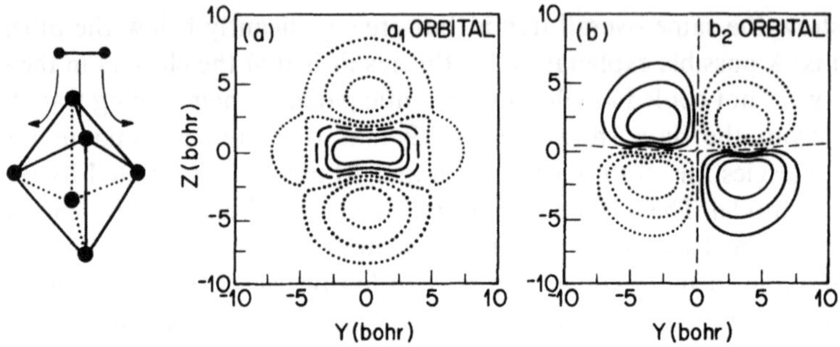

Fig. 11. A diagram showing the interaction geometry between Al_6 and H_2, and a plot of the two highest occupied molecular orbitals. At large Al_6–H_2 separations the Al_6 a_1 orbital is populated. At the transition state the electrons have shifted into the b_2 orbital. Adapted from Ref. [41]

with a number of different molecules. With D_2 the smallest cluster that reacts is Al_6, and for $n > 6$ the reaction rates rapidly decrease. Upton has performed theoretical calculations on the chemisorption of hydrogen on small aluminum clusters [29, 41]. He found that for clusters with $n < 6$ dissociative chemisorption is endothermic. Dissociative chemisorption of H_2 on Al_6 was exothermic but according to his calculations there is a significant activation barrier. Figure 11 shows the molecular orbitals important in the reaction. At large Al_6–H_2 separations the a_1 orbital on Al_6 is filled and the b_2 orbital is empty. As the H_2 approaches the filled a_1 orbital is destabilized by exchange interactions with the filled σ orbital of H_2. This repulsive interaction between the filled a_1 orbital and the filled σ orbital causes the activation barrier. Eventually, electrons in the a_1 orbital shift into the b_2 orbital. The interaction between the Al_6 a_1 orbital and the H_2 σ^* orbital, and between the Al_6 b_2 orbital and the H_2 σ orbital is now favorable and ultimately leads to Al–H bond formation. Why the reactivity drops dramatically for clusters larger than Al_7 has not yet been explained. The reactions of aluminum cluster ions with D_2 have been studied by *Jarrold* and *Bower* [42] using low energy ion beam techniques. Activation barriers for chemisorption of D_2 where measured for clusters with 10–27 atoms. The activation barriers increase from ~ 1.2 eV to ~ 2.0 eV with increasing cluster size and then appear to level off. D_2 does not chemisorb on clean aluminum surfaces at room temperature. Unfortunately, the activation barrier for the bulk surface has not been measured but it is probably ~ 1.0–1.5 eV. Thus the activation barriers must decrease for clusters with $n > 27$ in order to approach the bulk value. *Jarrold* and *Bower* [43] also measured activation barriers for chemisorption of D_2 on aluminum clusters with preadsorbed oxygen, $Al_nO_m^+$ ($n = 10$–27, $m = 1, 2$). Oxygen preadsorption was found to have a dramatic effect on the activation barriers for some of the smaller clusters. For example, preadsorbing one oxygen on Al_{15}^+ caused the activation barrier to increase by 0.6 eV.

The reactions of aluminum clusters with oxygen have been studied by a number of different groups. For the neutral clusters (studied using a fast flow reactor) the products arise from adsorption to yield Al_nO_2 species [23]. The atom and dimer react rapidly, then there is a sudden drop in reactivity followed by an almost monotonic increase of the reaction rate with cluster size. The oxidation of positively charged aluminum clusters has been studied by *Anderson* and coworkers [44] and *Jarrold* and *Bower* [45] using low energy ion beam techniques. For clusters with $n < 7$ Al^+ and Al_2O^+ are important products. For larger clusters etching reactions such as

$$Al_n^+ + O_2 \rightarrow Al_{n-4}^+ + 2Al_2O \quad (8)$$

$$Al_n^+ + O_2 \rightarrow Al_{n-5}^+ + 2Al_2O + Al \quad (9)$$

occur. Reaction thresholds of up to a few tenths of an electron-volt were measured for clusters with 3–9 and 25 atoms [42, 46] (the intervening clusters have not yet been studied). Bulk aluminum reacts rapidly with oxygen and on the bulk surface there is no significant activation barrier, so it appears that the clusters are less reactive than the bulk. The reactions of negatively charged aluminum clusters with oxygen have been studied by *Leuchtner*, *Harms*, and *Castleman* [47] using a flow tube. A series of sequential etching reactions occur. These experiments were not performed on size selected clusters so the products of the individual reactions could not be identified. However, after exposure to excess oxyen it is apparent that product builds up at clusters with 13, 23, and 37 atoms, and that these clusters are inert towards oxygen. Al_{13}^-, Al_{23}^-, and Al_{37}^- with 40, 70, and 112 valence electrons respectively, have closed shell configurations assuming a harmonic oscillator potential. *Castleman* and coworkers have also investigated the reactions of aluminum alloy clusters ions with oxygen [48]. They found Al_4Nb^- and Al_6V^- to be inert. Nb has a $4s^24d^3$ valence electronic configuration so if all these electrons are counted Al_4Nb^- has 18 electrons, which is a jellium model shell closing. On the otherhand, if all the electrons are counted for Al_6V^- (V also has an s^2d^3 configuration) it has 24 valence electrons, which is not a shell closing.

A number of other reactions have been studied but these will not be discussed in detail here. *Cox* and coworkers [23] have studied the reactions of neutral aluminum clusters with CH_3OH, CO, D_2O, and CH_4 and reported that the overall reactivity was roughly ordered as $O_2 > CH_3OH > CO > D_2O > D_2 > CH_4$. It was suggested that this reactivity could be correlated with the HOMO/LUMO energies of the reactants assuming that CH_3OH and D_2O acted as donor molecules and D_2, CO, and CH_4 acted as acceptors. *Jarrold* and *Bower* measured activation barriers for chemisorption of CO, N_2, CH_4, and C_2H_4 on Al_{25}^+ [46, 49] and found a correlation with the HOMO/LUMO energies of the reagents. Finally, *Anderson* and coworkers [44] have studied the reactions of aluminum cluster ions with N_2O.

Several groups have performed theoretical calculations on aluminum clusters. Some of this work has been mentioned above, where the theoretical results were compared with experiment. Here we consider the theoretical results with emphasis on the geometric structure of the clusters. Geometric structure is an extremely important topic, for which there is virtually no experimental data available. So the only insight we have to cluster geometry is from theory. *Pacchioni* and coworkers [37, 50] have performed calculations on neutral aluminum cluster containing up to five atoms. They found the lowest energy structures for Al_3–Al_5 to be linear, a rhombus, and a square pyramid respectively. These structures can be considered deformed sections of the fcc lattice. The ground states were found to have high spin multiplicity. The ground state structures found by *Upton* [29] are slightly different from those found by Pacchioni and coworkers [37, 50]. For Al_3–Al_6 Upton found an isosceles triangle, a distorted rhombus, a square pyramid, and an octahedron. In these calculations the ground states were found to be low spin, except for Al_2. This is in better agreement with the magnetic deflection experiments [32] described above. *Bauschlicher* and coworkers [38, 51] have reported several studies of aluminum clusters. They have studied Al_2–Al_6 and Al_{13} employing both ab initio techniques and a simplified approach using two and three body interaction potentials. In the case of Al_5 they found a planar structure to be 0.2 eV lower in energy than the square pyramid structure when extensive electron correlation was included in the calculations. For Al_{13} an icosahedron was found to have the lowest energy structure, but this was only 0.3 eV more stable than a D_{3h} (hcp) structure and 1.2 eV more stable than an O_h (fcc) configuration. A planar star shaped structure was only 2.5 eV above the lowest energy structure. It should be noted, however, that the cohesive energy of the lowest energy form of Al_{13} found in these calculations is less than 50% of the measured total cohesive energy of this cluster (see Fig. 10).

Raghavachari [21] has recently performed ab initio calculations on a number of neutral and charged aluminum clusters: Al_2–Al_8, Al_2^+–Al_8^+, and Al_{13}^-. Both Al_3 and Al_3^+ were found to be triangular. For Al_3^+ a triplet state was found to be slightly lower in energy than the low spin states. Al_4 and Al_4^+ were found to be rhombuses, in agreement with previous work. For Al_5 a capped rhombus and a planar structure were found to have very similar energies, but for Al_5^+ the capped rhombus was the favored structure. Al_6^+ was found to have a capped pentagon structure, but for Al_6 two structures were very close in energy: a triplet distorted octahedron, and a singlet species with an open structure. Al_7^+ was found to have a particularly interesting structure. It is a distorted capped octahedron. The significance of this structure is that all atoms lie approximately the same distance from the center of mass of the cluster as if they are distributed over the surface of a sphere. Al_7^+ has 20 valence electrons and has a closed shell configuration according to the electronic shell model, so a spherical structure is expected. Raghavachari found that the electronic structure of this cluster could be related to the electronic shell model and found that there was a significant gap between the filled and unfilled orbitals as predicted by the electronic shell model. For clusters with up to seven atoms there is still a significant gap between the *s* and *p* levels and the bonding arises mainly from the *p* orbitals. The bonding is mainly covalent in nature, which accounts for the preference for rather open structures in these small clusters. Al_{13}^- is the next largest cluster expected to have a closed shell configuration. Calculations for this cluster suggest that it has an icosahedral structure, and that this structure is considerably more stable than fcc (cubo-octahedron) or hcp fragments. Again the electronic structure of this cluster could be related to the shell model and there was a significant gap between the filled and unfilled orbitals. The gap between the *s* and *p* bands has disappeared in Al_{13}^- and clearly the bonding in this cluster is more metallic since the close-packed structures are favored.

4.2.4 Gallium, Indium, and Thallium Clusters

Considerably less is known about clusters of these elements. Thallium is toxic, and it appears that little work has been done on clusters of this element except for a few studies of the dimer [52]. Several groups have investigated the distribution of gallium clusters and indium clusters produced by a liquid metal ion source [53]. In the case of gallium it appears that there are significant discontinuities in the cluster ion relative abundances at $n = 7$–8 and $n = 13$–14. *King* and *Ross* [54] have investigated the spontaneous unimolecular dissociation of sputtered aluminum, gallium, and indium cluster ions. The relative amount of unimolecular dissociation for clusters of these three elements showed roughly the same trends. The results are shown in Fig. 12. The similarity between the relative amounts of unimolecular dissociation for the different elements should not be taken to indicate that the dissociation energies are similar. According to the evaporative ensemble model, which is described in Chapter 2.7, the fraction of cluster *n* which dissociates in these experiments depends on the ratio of the dissociation energies of the *n* and $n + 1$ clusters, but it is virtually independent of the absolute magnitude of the dissociation energies. The cohesive energies of the bulk elements drop monotonically moving down group 13 of the periodic table, and it is likely that the cluster dissociation energies show the same trends. There are minima in the relative amounts of

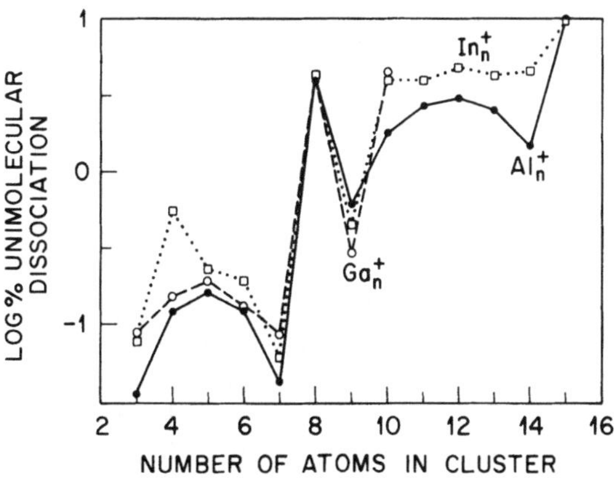

Fig. 12. Plot of the logarithm of the percentage of unimolecular dissociation versus cluster size for sputtered aluminum, gallium, and indium clusters. Adapted from Ref. [54]

unimolecular dissociation at M_7^+, M_9^+, and M_{14}^+, and sharp increases at M_8^+ and M_{15}^+. From these results it might appear that clusters with 7, 9, and 14 atoms are particularly stable. However, in the case of aluminum clusters there has been no other evidence to indicate that there is anything special about the nine atom cluster. The main products from the unimolecular dissociation of these clusters are M_{n-1}^+ and M^+. For aluminum the results are consistent with those described above. Al^+ dominates for $n = 4$–7 and Al_{n-1}^+ is the main product observed for larger clusters. Similar results were obtained for gallium clusters and indium clusters, though for these clusters M_{n-2}^+ was also observed as a minor product. These results suggest that the trends in the ionization energies for the clusters of these elements are similar: the ionization energies initially rise above the atomic value and then drop below it at around $n = 8$. On the otherhand, *Broyer* and coworkers [55] have measured ionization energies of indium clusters up to $n = 30$ using electron impact ionization. They found that the ionization energies initially drop below the atomic value, rise above it at $n = 7$–8, continue to increase up to In_{14}, and finally drop below the atomic value at $n = 18$–19. Thus the trend predicted by these electron impact measurements is almost the reverse of that predicted from the dissociation experiments. While the absolute values of ionization potentials determined from fragmentation patterns are questionable, if only In_{n-1}^+ is observed as a product in the dissociation of In_n^+ for $n > 7$ the conclusion that the ionization energies lie below the atomic value seems almost irrefutable. The origin of the problem with the electron impact measurements is not known, but the measurement of ionization energies by electron impact is known to be plagued with problems.

The group 13 elements are one component of the III–V materials. These materials, such as GaAs, are important semiconductors. There have been a few studies of the properties of III–V clusters. *Morse* and coworkers [56] have investigated the spectroscopy of jet cooled GaAs molecule and measured its bondlength. Surprisingly, the bondlength in the GaAs molecule is slightly longer than in bulk GaAs. In BN (the only other III–V molecule for which detailed spectroscopic data is available) and dimers of group 14 (C_2, Si_2, Sn_2, and Pb_2) the bondlength in the molecule is shorter than in the bulk. The contraction that occurs on going to the bulk in GaAs may arise from the increased ionic contribution to the bonding in the solid. Balasubramanian [57] has performed theoretical calculations on the GaAs molecule and several larger clusters. He predicted an $X\,^3\Sigma^-$ ground state for GaAs in agreement with the experimental work of *Morse* and coworkers [56]. *Smalley* and coworkers [58] have studied the photodissociation of GaAs clusters containing up to 30 atoms. Gallium and arsenic have close to the same atomic mass, and gallium has two isotopes so it is difficult to resolve the various Ga_xAs_y components of the peak in the mass spectrum corresponding to a particular total number of atoms. Both positively and negatively charged GaAs clusters appear to dissociate by sequential loss of atoms (as found for metal clusters) rather than by loss of "magic fragments" such as $n = 6$ and 10 observed for silicon and germanium clusters [18]. *Reents* [59] has investigated the reactions of small GaAs clusters with HCl. HCl etches the clusters according to the reaction

$$Ga_xAs_y^- + HCl \rightarrow Ga_{x-1}As_yH^- + GaCl \ . \qquad (10)$$

These studies also showed that at least two different structural forms (that could be differentiated kinetically) exist for a number of the small clusters.

Kolenbrander and *Mandich* [60] have reported a number of studies of InP clusters. The mass of indium is sufficiently different from that of phosphorus to allow complete separation of the various In_xP_y species in the mass spectrum, at least for moderately sized clusters. Kolenbrander and Mandich have recorded photodissociation spectra of small InP clusters in the visible region of the spectrum and also in the near infrared. Using RRKM simulations of the photodissociation data they have derived dissociation energies of $\sim$ 2.0–2.4 eV for In_5P_y–In_9P_y ($y = 1$–5). In order to record absorption spectra in the near infrared region of the spectrum, two photon dissociation spectroscopy was employed. Here the first laser was tuned through the near infrared region of the spectrum and the second laser was fixed, but with a photon energy sufficiently small that both photons are required to cause dissociation. The stoichiometric clusters (In_3P_3–In_6P_6) appear to show a continuum absorption with a threshold at around 1.3–1.5 eV. This is quite close to the bulk band gap of 1.18 eV. Most of the off-stoichiometric clusters also showed an apparent continuum absorption threshold. They also found discrete absorptions with transition energies considerably less than the bulk band gap for many of the clusters. Since the higher energy continuum absorption appears to exist for both the stoichiometric and off-stoichiometric clusters this feature was assigned to a $\sigma \rightarrow \sigma^*$ transition involving excitation of electrons in the In–P bonds. It was suggested that the absorptions with transition energies below the bulk band gap arise from $\sigma \rightarrow \sigma^*$ transitions in In–In bonds or transitions involving dangling bonds.

4.2.5 Conclusions

This chapter has summarized our current knowledge of the properties of clusters of group 13 in the periodic table. It should be apparent from reading this chapter that we know a considerable amount about the properties of aluminum clusters, but even here the picture is far from complete. As described in Chapter 4.1 the simple electronic shell model does a remarkably good job of describing the size dependent properties of alkali metal clusters. This simple model does not do so well for aluminum clusters. Many properties such as ionization energies, dissociation energies, electron affinities, and chemical reactivity show evidence for shell closings in small aluminum clusters. The polarizabilities, on the otherhand, are somewhat below the jellium model predictions (though this may simply reflect the suppressed response of the *s* band electrons). Theoretical calculations on small aluminum clusters suggest that the additional stability provided by the closed shell configuration forces the clusters to conform to the electronic shell model close to the shell closings. Away from the shell closings the electronic structure departs from the predictions of the shell model. This may explain why the shell model does not appear to work very well for the larger clusters, because the extra stability provided by a closed shell decreases with increasing cluster size. Another important issue with the s^2p^1 elements is the cluster size regime where the s and p bands overlap. Theoretical calculations suggest that overlap first occurs for clusters with < 13 atoms, but photoelectron spectra indicate that the s and p bands have not completely merged for clusters with 32 atoms. Chemically aluminum clusters appear to show similar reactivity trends to the bulk. Though often it appears that the clusters are less reactive than the bulk surfaces.

Much less is known about clusters of the other members of group 13. In the case of boron clusters there is considerable evidence that there is something special about the 13 atom cluster, which from the solid state chemistry of boron we would expect to be icosahedral or cubo-octahedral, but there is not yet any direct structural evidence. Work has just started on III–V clusters. The additional complications arising in two component clusters are formidable and it is clear that it will be sometime before we have a good understanding of clusters of these important materials.

References

1. L. Hanley, S.L. Anderson: J. Phys. Chem. **91**, 5161 (1987); L. Hanley, J.L. Whitten, S.L. Anderson: J. Phys. Chem. **92**, 5803 (1988); L. Hanley, J.L. Whitten, S.L. Anderson: J. Phys. Chem. **94**, 2218 (1990)
2. L. Hanley, S.L. Anderson: J. Chem. Phys. **89**, 2848 (1988)
3. S.A. Ruatta, L. Hanley, S.L. Anderson: J. Chem. Phys. **91**, 226 (1989)
4. P.A. Hintz, S.A. Ruatta, S.L. Anderson: J. Chem. Phys. **92**, 292 (1990)
5. R.A. Whiteside: Ph.D. Thesis, Carnegie Mellon University, Pittsburgh, 1981; J. Koutecky, G. Pacchioni, G.H. Jeung, E.C. Hass: Surf. Sci. **156**, 651 (1985); B. Sykja, S. Lunell: Surf. Sci. **141**, 199 (1984)
6. J. Berkowitz, W.A. Chupka: J. Chem. Phys. **40**, 2735 (1964)
7. P.J. Bruna, J.S. Wright: J. Chem. Phys. **91**, 1126 (1989)
8. M.F. Jarrold, J.E. Bower, J.S. Kraus: J. Chem. Phys. **86**, 3876 (1987)
9. W. Begemann, K.H. Meiwes-Broer, H.O. Lutz: Phys. Rev. Lett. **56**, 2248 (1986)
10. W. Begemann, S. Dreihofer, K.H. Meiwes-Broer, H.O. Lutz: in "The Physics and Chemistry of Small Clusters", ed. by P. Jena, B.K. Rao, S.N. Khanna (Plenum, New York, 1987)
11. L. Hanley, S.A. Ruatta, S.L. Anderson: J. Chem. Phys. **87**, 260 (1987)
12. M. Broyer, G. Delacretaz, P. Fayet, P. Labastie, N. Guoquan, W.A. Saunders, R.L. Whetten, J.-P. Wolf, L. Woste: in "Physics and Chemistry of Small Clusters", ed. by P. Jena, B.K. Rao, S.N. Khanna (Plenum, New York, 1987)
13. U. Ray, M.F. Jarrold, J.E. Bower, J.S. Kraus: Chem. Phys. Lett. **159**, 221 (1989)
14. U. Ray, M.F. Jarrold, J.E. Bower, J.S. Kraus: J. Chem. Phys. **91**, 2912 (1989)
15. P.J. Brucat, L.-S. Zheng, C.L. Pettiette, S. Yang, R.E. Smalley: J. Chem. Phys. **84**, 3078 (1986)
16. S.K. Loh, D.A. Hales, L. Lian, P.B. Armentrout: J. Chem. Phys. **90**, 5466 (1989); S.K. Loh, L. Lian, P.B. Armentrout: J. Amer. Chem. Soc. **111**, 3167 (1989)

17. M.E. Geusic, T.J. McIlrath, M.F. Jarrold, L.A. Bloomfield, R.R. Freeman, W.L. Brown: J. Chem. Phys. **84**, 2421 (1986); M.E. Geusic, M.F. Jarrold, T.J. McIlrath, R.R.Freeman, W.L. Brown: J. Chem. Phys. **86**, 3862 (1987)
18. L.A. Bloomfield, R.R. Freeman, W.L. Brown: Phys. Rev. Lett. **54**, 2246 (1985); Q.L. Zhang, Y. Liu, R.F. Curl, F.K. Tittel, R.E. Smalley: J. Chem. Phys. **88**, 1670 (1988); M.F. Jarrold, J.E. Bower: J. Phys. Chem. **92**, 5702 (1988)
19. J.L. Martins, R. Car, J. Buttet: Surf. Sci. **106**, 265 (1981); D.E. Beck: Solid State Comm. **49**, 381 (1984); W. Ekardt: Phys. Rev. B **29**, 1558 (1984); W.D. Knight, K. Clemenger, W.A. de Heer, W.A. Saunders, M.Y. Chou, M.L. Cohen: Phys. Rev. Lett. **52**, 2141 (1984); M.L. Cohen, M.Y. Chou, W.D. Knight, W.A. de Heer: J. Phys. Chem. **91**, 3141 (1987); W.A. de Heer, W.D. Knight, M.Y. Chou, M.L. Cohen: Solid State Phys. **40**, 93 (1987)
20. M.Y. Chou, M.L. Cohen: Phys. Lett. A **113**, 420 (1986)
21. K. Raghavachari: to be published
22. M.F. Jarrold, J.E. Bower: J. Chem. Phys. **87**, 1610 (1987)
23. D.M. Cox, D.J. Trevor, R.L. Whetten, A. Kaldor: J. Phys. Chem. **92**, 421 (1988)
24. K.E. Schriver, J.L. Persson, E.C. Honea, R.L. Whetten: Phys. Rev. Lett. **64**, 2539 (1990)
25. D.M. Wood: Phys. Rev. Lett. **46**, 749 (1981)
26. M.M. Kappes, M. Schar, P. Radi, E. Schumacher: J. Chem. Phys. **84**, 1863 (1986)
27. K.K. Sunhil, K.D. Jordan: J. Phys. Chem. **92**, 2774 (1988); C.W. Bauschlicher, L.A. Barnes, P.R. Taylor: J. Phys. Chem. **93**, 2932 (1989)
28. J.E. Harrington, J.C. Weisshaar: J. Chem. Phys. **93**, 854 (1990); also see, K. Fuke, S. Nonose, N. Kikuchi, K. Kaya: Z. Phys. D **12**, 571 (1989)
29. T.H. Upton: Phys. Rev. Lett. **56**, 2168 (1986); T.H. Upton: J. Phys. Chem. **90**, 754 (1986); T.H. Upton: J. Chem. Phys. **86**, 7054 (1987)
30. G. Gantefor, K.H. Meiwes-Broer, H.O. Lutz: Phys. Rev. A **37**, 2716 (1988); G. Gantefor, M. Gausa, K.H. Meiwes-Broer, H.O. Lutz: Z. Phys. D **9**, 253 (1988)
31. K.J. Taylor, C.L. Pettiette, M.J. Craycraft, O. Chesnovsky, R.E. Smalley: Chem. Phys. Lett. **152**, 347 (1988)
32. D.M. Cox, D.J. Trevor, R.L. Whetten, E.A. Rohlfing, A. Kaldor: J. Chem. Phys. **84**, 4651 (1986)
33. W.A. de Heer, P. Milani, A. Chatelain: Phys. Rev. Lett. **63**, 2834 (1989)
34. Z. Fu, G.W. Lemire, Y.M. Hamrick, S. Taylor, J.-C. Shui, M.D. Morse: J. Chem. Phys. **88**, 3524 (1988)
35. W.A. de Heer, K. Selby, V. Kresin, J. Masui, M. Vollmer, A. Chatelain, W.D. Knight: Phys. Rev. Lett. **59**, 1805 (1987)
36. W.A. Saunders, P. Fayet, L. Woste: Phys. Rev. A **39**, 4400 (1989)
37. G. Pacchioni, J. Koutecky: Ber Bunsenges. Phys. Chem. **88**, 242 (1984)
38. L.G.M. Pettersson, C.W. Bauschlicher, T. Halicioglu: J. Chem. Phys. **87**, 2205 (1987)
39. G. Pacchioni, P. Fantucci: in *Physics and Chemistry of Small Clusters*, ed. by P. Jena, B.K. Rao, S.N. Khanna (Plenum, New York, 1987)
40. A.R. Miedema: Faraday Symp. Chem. Soc. **14**, 136 (1980); A.R. Miedema, K.A. Gingerich: J. Phys. B **12**, 2081 (1979)
41. T.H. Upton, D.M. Cox, A. Kaldor: in *Physics and Chemistry of Small Clusters*, ed. by P. Jena, B.K. Rao, S.N. Khanna (Plenum, New York, 1987)
42. M.F. Jarrold, J.E. Bower: J. Amer. Chem. Soc. **110**, 70 (1988)
43. M.F. Jarrold, J.E. Bower: Chem. Phys. Lett. **144**, 311 (1988)
44. L. Hanley, S.L. Anderson: Chem. Phys. Lett. **129**, 429 (1986); S.A. Ruatta, L. Hanley, S.L. Anderson: Chem. Phys. Lett. **137**, 5 (1987); S.A. Ruatta, S.L. Anderson: J. Chem. Phys. **89**, 273 (1988)
45. M.F. Jarrold, J.E. Bower: J. Chem. Phys. **87**, 5728 (1987)
46. M.F. Jarrold, J.E. Bower: J. Amer. Chem. Soc. **110**, 6706 (1988)
47. R.E. Leuchtner, A.C. Harms, A.W. Castleman: J. Chem. Phys. **91**, 2753 (1989)
48. A.C. Harms, R.E. Leuchtner, S.W. Sigsworth, A.W. Castleman: J. Amer. Chem. Soc. **112**, 5673 (1990)
49. M.F. Jarrold, J.E. Bower: Chem. Phys. Lett. **149**, 433 (1988)

50. G. Pacchioni, D. Plavsic, J. Koutecky: Ber Bunsenges. Phys. Chem. **87**, 503 (1983); J. Koutecky, G. Pacchioni, G.H. Jeung, E.C. Hass: Surf. Sci. **156**, 650 (1985)
51. C.W. Bauschlicher, L.G.M. Pettersson: J. Chem. Phys. **84**, 2226 (1986); H. Patridge, C.W. Bauschlicher: J. Chem. Phys. **84**, 6507 (1986); C.W. Bauschlicher, L.G.M. Pettersson: J. Chem. Phys. **87**, 2198 (1987)
52. J. Drowart, R.E. Honig: J. Phys. Chem. **61**, 980 (1957); D.S. Ginter, M.L. Ginter, K.K. Innes: J. Chem. Phys. **69**, 480 (1965)
53. B. Wilkens: Nuc. Instr. Meth. A **236**, 340 (1985); D.L. Barr: J. Vac. Sci. Technol. B **5**, 184 (1987); F.G. Rudenauer, W. Steiger, H. Studnicka, P. Pollinger: Int. J. Mass Spectrom. Ion Processes **77**, 63 (1987)
54. F.L. King, M.M. Ross: Chem. Phys. Lett. **164**, 131 (1989)
55. D. Rayane, P. Melinon, B. Cabaud, A. Hoareau, B. Tribollet, M. Broyer: J. Chem. Phys. **90**, 3295 (1989)
56. G.W. Lemire, G.A. Bishea, S.A. Heidecke, M.D. Morse: J. Chem. Phys. **92**, 121 (1990)
57. K. Balasubramanian: Chem. Rev. **90**, 93 (1990); K. Balasubramanian: J. Mol. Spectrosc. **139**, 405 (1990)
58. Y. Liu, Q.-L. Zhang, F.K. Tittel, R.F. Curl, R.E. Smalley: J. Chem. Phys. **85**, 7434 (1986); Q.-L. Zhang, Y. Liu, R.F. Curl, F.K. Tittel, R.E. Smalley: J. Chem. Phys. **88**, 1670 (1988)
59. W.D. Reents: J. Chem. Phys. **90**, 4258 (1989)
60. K.D. Kolenbrander, M.L. Mandich: J. Chem. Phys. **90**, 5884 (1989); K.D. Kolenbrander, M.L. Mandich: J. Chem. Phys. **92**, 4759 (1990)

4.3 Transition Metal Clusters: Physical Properties

M.F. Jarrold

4.3.1 Introduction

Using the strict definition that transition elements contain partly filled *d* or *f* shells there are 56 transition elements. Thus well over half the elements in the periodic table are transition elements. Nearly all of them are hard, strong, high boiling point metals that are good conductors of heat and electricity. Transition metal cluster compounds, such as $Mn_2(CO)_{10}$ and $Au_{11}I_3(PR_3)_7$, have been studied for decades by inorganic chemists. Their physical and chemical properties have been investigated and the structures of many of them have been determined. The big difference between these clusters and the bare clusters which are the main topic of this book is that cluster compounds are surrounded by ligands and it is possible to generate macroscopic quantities of them. In recent years the inorganic chemists have developed techniques to make larger and larger cluster compounds. Perhaps the best known example is the $Au_{55}(PPh_3)_{12}Cl_6$ cluster [1]. Recently a cluster compound containing 561 palladium atoms has been prepared. This chapter will focus on the properties of bare transition metal clusters. We will be concerned almost exclusively with open *d* shell transition metals for the simple reason that very little work has been performed on clusters of the rare earth elements. A quick glance at the literature on transition metal clusters reveals that a large fraction of the studies have been concerned with investigating their chemical properties. Most catalysts are transition metals, and one motivation for these chemical studies is to obtain a detailed understanding of catalysis. The chemical studies are sufficiently numerous to warrant detailed discussion in a separate chapter. This chapter will focus on the physical properties.

4.3.2 Electronic Configuration and Bonding

The presence of the unfilled *d* shell in transition metal elements has a number of important consequences. For the isolated atoms there are often many low lying terms which arise because of the different ways of arranging the electrons in the unfilled *d* shell, and because the $(n + 1)s$ shell is almost isoenergetic with the *nd* shell (thus the energies of the $(n+1)s^2\,nd^m$ and $(n+1)s^1\,nd^{m+1}$ configurations are not very different). The large number of low lying excited electronic states are no doubt partly responsible for the important chemical properties of the transition metals. However, while the electronic states of the isolated atoms are understood, combining just two atoms to generate a dimer can result in a situation that is so complicated that normal ideas about the electronic structure of molecules may need to be abandoned. When the two atoms are brought together, each term of atom A must be combined with all terms of atom B, each combination resulting in many different combinations of orbital angular momenta and spin angular momenta. *Spain* and *Morse* [2] have enumerated the number of electronic states for V_2 and Ni_2 using Hunds rules. Within 1 eV above the ground state atoms there are ~ 1000 for Ni_2 and ~ 1900 for V_2. While it is not known how many of these states are bound, the label of diatomic metal seems appropriate for these species.

For many clusters the situation may not be as bad as implied above. For example, ab initio calculations for Fe_2 suggest that there are 112 electronic states within 0.5 eV of the ground state [3]. However, only two low lying states were found in the photoelectron spectrum of Fe_2^- [4]. While some states may not be accessible via a one electron photodetachment process when correlation effects are small, the disagreement between experiment and theory is striking and serves to illustrate the enormous problems facing the theorists that are attempting to perform calculations on transition metal clusters. More recent calculations indicate a much lower state density [5]. *Lineberger* and coworkers [4] have suggested that the photoelectron spectrum of Fe_2^- is consistent with a $(4s\sigma_g)^2\,(4s\sigma_u^*)^1\,(3d)^{13}$ ground state which correlates with one ground state $(4s^2\,3d^6)$ Fe atom and one excited state $(4s^1\,3d^7)$ Fe atom. The bonding for the configuration mentioned above arises from the 4*s* electrons and the bond order is formally 1/2. However, it is expected that there will also be significant additional 3*d* contribution to the bonding.

While our understanding of the chemical bonding in small transition metal clusters is primitive, *Morse* [6] has outlined a number of important factors: the size of the $(n+1)s$ and *nd* orbitals and the $(n+1)s^2\,nd^m \rightarrow (n+1)s^1\,nd^{m+1}$ promotion energy. It is well known that the tendency of transition metals to form multiple *nd* bonds is determined by the relative size of the $(n+1)s$ and *nd* orbitals. The $(n+1)s$ orbitals are larger than the *nd* orbitals so the $(n+1)s$ orbitals are expected to make a larger contribution to the bonding (the smaller *nd* orbitals do not overlap as well). The *nd* orbitals contract moving to the right in the periodic table and expand moving down the periodic table. Thus the *nd* contribution to the bonding is expected to be more important for the second and third transition series than the first. The importance of the $(n+1)s^2\,nd^m \rightarrow (n+1)s^1\,nd^{m+1}$ promotion energy can be demonstrated by considering Mn_2. Mn has a $4s^2 3d^5$ configuration. Due to the stability of the half filled *d* shell the $4s^2\,3d^5 \rightarrow 4s^1\,3d^6$ promotion energy is large (2.14 eV) and the ground state of Mn_2 is $(4s\sigma_g)^2\,(4s\sigma_u^*)^2\,(3d)^{10}$. Thus the formal bond order is zero. Mn_2 is only bound by 0.43 eV [7] and is often called a van der Waals molecule (though this is obviously an over-simplification). The dissociation energy of

Mn_2^+ is 1.39 eV [8]. Mn_2^+ with a $(4s\sigma_g)^2(4s\sigma_u^*)^1(3d)^{10}$ configuration has a formal bond order of 1/2. From the dissociation energies of Mn_2 and Mn_2^+ this 1/2 bond appears to be worth around 1 eV. Thus, it is clear that unlike Fe_2 discussed above (where the promotion energy is 0.87 eV), Mn_2 cannot recoup the 2.14 eV promotion energy from increased bonding, and so it has a $(4s\sigma_g)^2(4s\sigma_u^*)^2(3d)^{10}$ ground state instead of $(4s\sigma_g)^2(4s\sigma_u^*)^1(3d)^{11}$. The promotion energies for the other transition metals are generally much smaller and there is a delicate balance between the cost of promotion and the energy gained from increased bonding. The nature and relative contribution of the *nd* orbitals to the bonding is far from understood. Copper has a $4s^1 3d^{10}$ configuration and substantial $3d$ contraction occurs. A $3d^{10}3d^{10}4s\sigma_g^2\ {}^1\Sigma_g^+$ ground state is expected and oberved for Cu_2. However, even for Cu_2 it is necessary to consider the $3d$ electrons in order to account for the observed physical properties (D_0^0, r_e, and ω_e). Correlation of the $3d$ electrons increases the calculated bond energy of Cu_2 by 0.8 eV [9]. Correlation reduces the *d–d* repulsion and allows the atoms to move closer together forming a strong $4s\sigma_g^2$ bond.

4.3.3 Mass Spectra and Magic Numbers

As has been described in the preceding chapters the electronic shell model provides a useful way to understand the electronic structure of the clusters of some main group metals. The shell model provides a simple explanation for the occurrence of magic numbers in the mass spectra of alkali metal clusters. The simplicity of the shell model and its success accounting for the physical properties of alkali metal clusters has spawned further studies of these clusters and encouraged (rather tenuous) comparison to nuclear physics. However, the alkali metal clusters are special. With a few exceptions, which will be discussed in more detail below, mass spectra of transition metal clusters do not show magic numbers which can be accounted for by the electronic shell model. A characteristic of the chemistry of transition metals is their ability to adopt several oxidation states. Thus it is difficult to count electrons for transition metal clusters. The exceptions are the group 11 elements, Cu, Ag, and Au. These elements have $(n+1)s^1 nd^{10}$ electronic configurations and mass spectra of the clusters of these elements often show features which can be attributed to shell closings. The reason for this is simple, the *nd* shells are full and contracted so that the bonding arises mainly from the $(n+1)s$ orbitals.

4.3.4 Ionization Energies

Ionization energies provide important clues about the evolution of bulk properties and are directly relevant to chemical behavior because they provide a measure of the ability to donate electrons in a chemical reaction. Early studies of the ionization energies of alkali metal clusters by *Schumacher* and coworkers showed that ionization energies of these clusters approached the bulk work function as predicted by the simple spherical metal droplet model [10]. As described in preceding chapters this model predicts that the ionization energies are given by

$$\mathrm{IE}_n = \mathrm{WF} + \frac{\alpha e^2}{r}\,, \tag{1}$$

where WF is the bulk work function and r is the cluster radius. There has been considerable discussion over whether the correct value for α in this expression is 3/8 or 1/2. It appears that the correct value is 1/2 [11].

Smalley and coworkers [12] have bracketed the ionization energies of copper clusters containing up to 25 atoms using fixed frequency lasers. They found substantial odd-even oscillations in the ionization energies. The small even numbered clusters have ionization energies considerably larger than small clusters with an odd number of atoms. Ionization energies of V_n [13], Fe_n [14], Ni_n [14], and Nb_n [15] have been measured by *Kaldor* and coworkers using tunable lasers to determine the photoionization thresholds (except for Ni_n where the ionization energies were bracketed with fixed frequency lasers). These studies were limited to clusters with less than 30 atoms. More recently *Knickelbein, Riley,* and coworkers have reported more extensive measurements of the ionization energies of Ni_n [17], Fe_n [18], Co_n [18], and Nb_n [19]. These measurements were performed for clusters with up to 100 atoms. They used both tunable lasers to determine photoionization thresholds and chemical probes [20]. Adsorption of ammonia lowers the ionization energy of Fe_n, Co_n, and Ni_n so by measuring the minimum number of ammonia molecules required for ionization by the various excimer laser lines it is possible to extrapolate and determine the ionization energy of the bare cluster. There is good agreement between the measurements of Kaldor and coworkers and those of Knickelbein, Riley, and coworkers for the smaller clusters. The ionization energies determined by Knickelbein, Riley, and coworkers are generally slightly (~ 0.1 eV) lower than those determined by Kaldor and coworkers. This small difference probably arises because the clusters in the experiments of Knickelbein, Riley, and coworkers were colder and so the thermal tail on the ionization efficiency curves is smaller. The same general trends in the ionization energies were found for Ni_n, Fe_n, Co_n, and Nb_n. There are large variations in the ionization energies of the smaller clusters. The variations diminish with increasing cluster size, and the ionization energies for clusters with $n > 25$ vary smoothly with size. Figure 1 shows results for Fe_n and Co_n clusters. The solid line shows the prediction of the simple spherical metal droplet model. The measured ionization energies are considerably smaller than the predictions of this model. Furthermore the model predicts that the ionization energies should decrease by ~ 0.4 eV on going from $n = 25$ to $n = 100$ but the measured ionization energies change by very little over this cluster size regime. The work functions shown in Fig. 1 for bulk iron

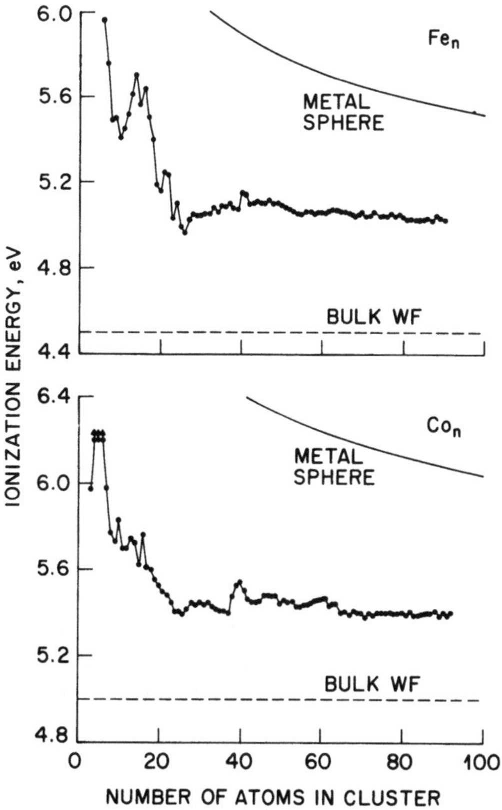

Fig. 1. Ionization energies measured for Fe_n and Co_n from Ref. [18]. The dashed line shows the work function of the polycrystalline bulk surface, and the solid line shows the prediction of the simple spherical metal droplet model described in the text

and bulk cobalt are for polycrystalline surfaces. Slightly different (by up to ~ 0.4 eV) work functions are obtained for single crystal surfaces, showing that surface structure influences the work function. Knickelbein, Riley and coworkers have suggested that the deviations of the measured ionization energies from the predictions of the simple spherical metal droplet model may indicate that the clusters have different structures than the bulk. Alternatively, the deviations may reflect the electronic structure of the clusters and in particular the development of the *d* band.

Niobium clusters provide one of the few examples where structural isomers have been observed for *metal* clusters. Isomers have been observed in the chemical reactions of both neutral Nb_n cluters and charged Nb_n^+ clusters [21]. In these chemical studies one isomer reacts away rapidly leaving behind a less reactive isomer. *Knickelbein* and *Yang* [19] have used the different chemical reactivity of the isomers to record ionization energies of both isomers. They found that the reactive isomers of Nb_9 and Nb_{12} had lower ionization energies than the unreactive isomers. For Nb_{10} the reactive and unreactive forms apparently have similar ionization energies.

4.3.5 Dissociation and Dissociation Energies

There have been several studies of the photodissociation and collision induced dissociation of transition metal cluster ions [22–30]. These experiments were performed with cluster ions because it is possible to size select charged clusters before fragmenting them. In most studies of cluster dissociation it is assumed that, regardless of the means of excitation, dissociation occurs by statistical unimolecular dissociation on the ground electronic state potential surface to give preferentially the lowest energy products. The high density of electronic states expected for transition metal clusters guarantees that there will be rapid internal conversion from excited electronic states to the ground state. Unlike clusters of covalently bonded materials such as carbon and silicon, which tend to dissociate by loss of particularly stable fragments, the transition metal clusters dissociate mainly by loss of single atoms (and in a few cases dimers). Loss of a dimer is favored on energetic grounds when the energy required to remove a second atom from the cluster is less than the dissociation energy of the dimer ($D^0_{n-1} < D^0_2$). However, since loss of a dimer is a more complicated process than loss of an atom it is possible that in some cases there is an activation barrier associated with this process, and dimer loss will not be observed even though it is favored on energetic grounds. Dimer loss is the lowest energy dissociation pathway for Nb_4^+ [27] and small odd-numbered copper [28] and silver [23, 24] cluster ions.

Measuring dissociation energies is considerably more difficult than determining what the fragments are. Knudsen cell mass spectrometry has been employed to measure dissociation energies for nearly all the transition metal dimers, and a few larger clusters [7]. *Morse* [6] has recently critically reviewed the dissociation energies measured by this approach. The dissociation energies are determined from measurements of equilibrium constants for the process

$$nM(\text{solid or liquid}) \rightleftarrows M_n(\text{gas}) \ . \tag{2}$$

High temperatures are required to obtain vapor pressures large enough to measure equilibrium constants. There are two methods for determining dissociation energies from the equilibrium constant measurements: the second law and third law methods (see Ref. [6] for a description). The third law method requires information on the physical properties of the clusters and in particular the low lying excited electronic states which are populated at the high temperatures required to perform the equilibrium constant measurements. Consequently, dissociation energies determined by this approach are not always reliable.

A more direct method for determining dissociation energies is to start, with cold clusters, excite them with a known amount of energy, and note whether dissociation occurs. This approach was first used by *Armentrout* and coworkers to measure the dissociation energy of Mn_2^+ [29]. They used collision induced dissociation. Subsequent experiments employing photodissociation with measurement of the kinetic energies of the fragments suggested that the collision induced dissociation value for the dissociation energy of Mn_2^+ was too low because the clusters contained excess internal energy [8]. This has been confirmed by more recent collision induced dissociation measurements [30]. *Spain* and *Morse* [2] have described a method to directly measure dissociation energies of neutral transition metal dimers using resonant two photon ionization. The dissociation energies can be determined with spectroscopic accuracy by noting when the resonant two photon ionization signal vanishes (or when the lifetimes drop dramatically) in the region of the dissociation threshold due to rapid internal conversion and dissociation. As described by Spain and Morse this method will only work if there is a high density of electronic states close to the dissociation threshold. They have used this method to determine the dissociation energies of V_2 (2.752 eV), VNi (2.100 eV), Ni_2 (2.068 eV), NiPt (2.798 eV), and Pt_2 (3.14 eV). It is not yet clear whether it will be possible to extend this approach to larger transition metal clusters.

Both collision induced dissociation and photodissociation have been used to determine the dissociation energies of larger cluster ions. *Armentrout* and coworkers have used collision induced dissociation to measure dissociation energies for Fe_n^+ [26] and Nb_n^+ [27]. The results are summarized in Fig. 2. The arrows on the right hand side show the bulk cohesive energies per atom. The dissociation energies measured for small Nb_n^+ clusters are much larger than those measured for small Fe_n^+ clusters, and this is consistent with the trends observed for the bulk metals. The main disadvantage of collision induced dissociation is that collisional activation transfers a broad range on energies into the clusters, and it is necessary to model the threshold region to determine the dissociation energy. On the otherhand, photoexcitation deposits a precisely known amount of energy into the cluster. However, efforts to obtain information on dissociation energies from photodissociation are hampered by the difficulty in distinguishing between one photon and multiple photon processes. It is difficult to distinguish between these processes from studies of the fluence

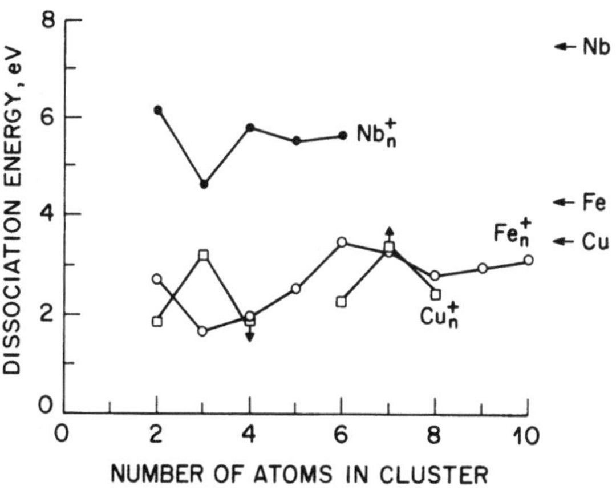

Fig. 2. Dissociation energies of Fe_n^+ [26], Nb_n^+ [27], Cu_n^+ [28]. The arrows on the right hand side show the bulk cohesive energy per atom

dependence because two photon processes can appear one photon if one of the transitions is saturated. For example, the dissociation energy determined from photodissociation of Fe_2^+ is in excellent agreement with the collision induced dissociation value [23]. But for small Nb_n^+ clusters photodissociation appears to be one photon (a linear influence dependence) at 2.33 eV photon energy [23], while the collision induced dissociation results indicate the photodissociation with 2.33 eV photons should be a multiphoton process.

For clusters with $n > 5$ another complication arises. The energy becomes distributed among the internal degrees of freedom and it takes a finite time for dissociation to occur. With increasing cluster size more energy is required to cause dissociation within a given time period. As described in Chapter 2.7, the lifetimes of excited clusters can be estimated from theory. However, the theoretical methods employed have not been generalized to a situation where there is a extremely large density of electronic states, a situation not encountered with normal molecules. *Armentrout* and coworkers have used the standard RRKM method in their simulations of collision induced dissociation thresholds [26, 27]. An alternative approach to measure dissociation energies is to directly determine the lifetimes for dissociation, and then use theory to model the energy dependence of the measured lifetimes. This approach was used by *Jarrold* and *Creegan* [28] in their studies of the dissociation of copper cluster ions. The results of these studies are shown in Fig. 2 along with those for Fe_n^+ and Nb_n^+. The dissociation energies for Cu_n^+ show evidence of substantial odd-even oscillations which is a characteristic of many of the physical properties of small Cu, Ag, and Au clusters. Clusters of Cu, Ag, and Au with an even number of electrons are found to be particularly stable and have high ionization energies. This observation has been rationalized by a bonding model in which the unpaired electron in the clusters with an odd number of electrons goes into a nonbonding or slightly antibonding orbital.

4.3.6 Magnetic Properties

Several transition metals have important magnetic properties, and most compounds of transition metals are paramagnetic. So far there have only been a couple of studies of the magnetic properties of bare transition metal clusters [31, 32] in the gas phase. *Cox* and coworkers were the first to report studies of the magnetic properties [31]. They used pulsed laser vaporization coupled with a Stern–Gerlach magnet and spatially resolved time of flight mass spectrometry to investigate the magnetic properties of small Fe_n and Fe_nO_m clusters. From detailed studies of the profile of the deflected beams they determined magnetic moments of 6.5 μ_B for Fe_2, 8.1 μ_B for Fe_3, and 6.5 μ_B for Fe_2O. These detailed measurements were not performed for larger clusters but the qualitative magnetic behavior of Fe_{4-14}, $Fe_{3-7}O$, and $Fe_{2-7}O_2$ was explored by measuring on-axis depletion factors when the magnetic field was turned on. From these

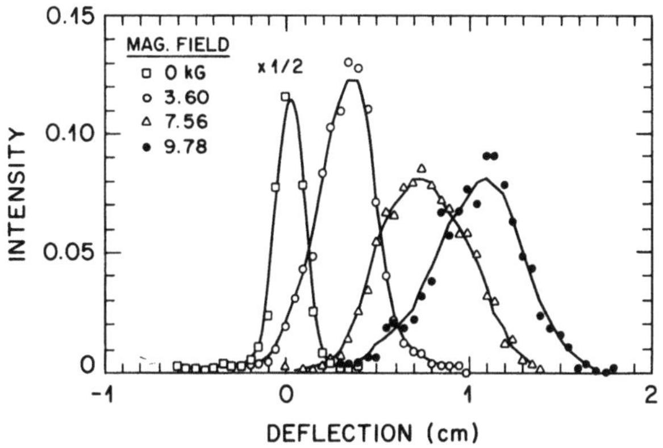

Fig. 3. Beam profiles measured for Fe_n ($n = 120$–140) passed through a Stern–Gerlach magnet. Profiles measured with a number of different magnetic field strengths are shown. Note that deflection occurs only in one direction. Adapted from Ref. [32]

results they deduced that the clusters had a constant spin per atom of $> 2.2\ \mu_B$. A different picture has emerged from the more recent work of *de Heer* and coworkers [32]. Using a similar experimental approach they have investigated the magnetic properties of iron clusters containing from 15 to 650 atoms. They found that the clusters deflect uniquely in the direction of the increasing magnetic field. Figure 3 shows beam profiles measured for clusters with 120–140 atoms for several different magnetic field strengths. In contrast, if an Fe atom (with a $^5D_4$ ground state) is passed through a Stern–Gerlach magnet the beam will be resolved into 9 M_J substates (± 4, ± 3, ± 2, $\pm 1, 0$) symmetrically dispersed around the undeflected $M_J = 0$ substate. Deflection only towards the high field is observed for macroscopic iron particles. The observation that deflection is only towards the high field for small clusters indicates that spin relaxation occurs much more rapidly than the flight time through the inhomogeneous magnetic field. The relaxation probably arises from coupling of the spin and rotational angular momentum. There have not yet been any studies of the magnetic properties of rare earth clusters, these studies would be particularly intriguing because of the interesting magnetic properties displayed by these elements.

4.3.7 Optical Spectroscopy

There have been numerous spectroscopic studies of transition metal dimers. Most of the earlier work was performed using either ovens (which make the spectra congested and difficult to interpret) or matrix isolation (where there is a problem of assigning the carrier of a particular transition). The matrix approach still continues to make important contributions but ovens have been superceded by laser vaporization and supersonic jets which cool the clusters down and result in simple spectra. *Morse* has recently reviewed the large amount of work that has been performed on both the homonuclear and heteronuclear transition metal dimers [6]. With increasing cluster size the spectroscopic information that is available drops off very sharply. In the gas phase, detailed spectroscopic studies (using for example resonant two photon ionization) have only been performed for three transitional metal trimers: Cu_3 [33], Ni_3 [34], and Ag_3 [35]; and one larger cluster: Cu_4^+ [36]. Even for most of the trimers mentioned above, only a single electronic transition has been identified. Of the three trimers, Cu_3 has been the most widely studied and the assignment of the 540 nm band system first observed by Smalley and coworkers in 1983 has been controversial. The ground electronic state of Cu_3 is now accepted to be a Jahn–Teller distorted $X\,^2E'$ state. On the ground state potential surface there are three equivalent C_{2v} (isosceles triangle) species with a barrier to pseudorotation of $\sim 100\ \mathrm{cm}^{-1}$. In pseudorotation the geometry changes from one of the equivalent isosceles triangle species to another. The Jahn–Tellar stabilization energy (the energy gained by distorting from an equilateral triangle geometry to an isosceles triangle) is $\sim 300\ \mathrm{cm}^{-1}$. Clearly Cu_3 is a very floppy species.

The only larger transition metal cluster for which detailed spectroscopic information is available in the gas phase is Cu_4^+ [36]. The spectrum for this cluster recorded over the entire visible region of the spectrum using photodissociation spectroscopy is shown in Fig. 4. Ten electronic transitions can be identified and most of them show resolved vibrational structure. Detailed assignment of the spectrum has not yet been performed. However, Jarrold and Creegan suggested that the group of transitions centered around 650 nm were due to $d \rightarrow s$ transitions, since this group of transitions appears to occur in a similar region of the spectrum as the $d \rightarrow s$ interband transition in bulk copper (650 nm). Recent calculations by *Balasubramanian* [37] support this assignment. The ground state structure determined by Balasubramanian for Cu_4^+ is a rhombus.

Spectroscopic information is available for a wider range of clusters trapped in rare gas matrices. A number of different spectroscopic techniques can be

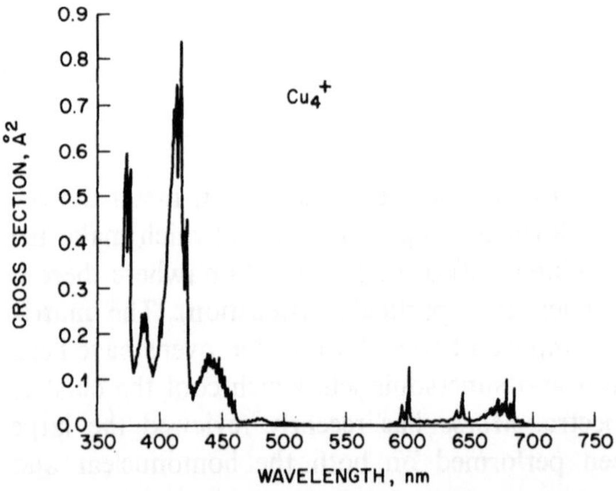

Fig. 4. Photodissociation spectrum recorded for Cu_4^+. Adapted from Ref. [36]

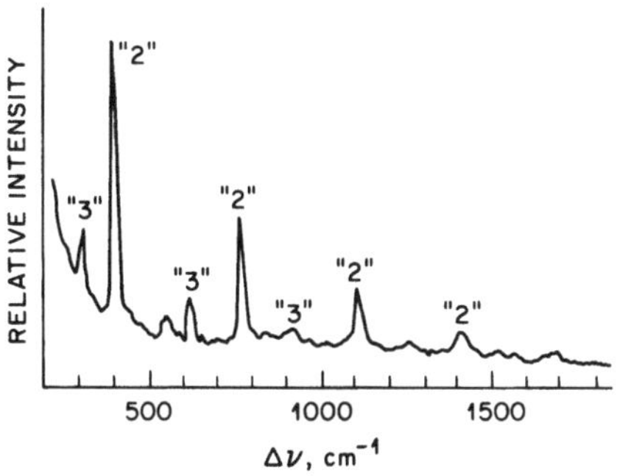

Fig. 5. Resonance Raman spectrum recorded for Cr containing argon matrix excited with 514.5 nm laser light. The features labelled "2" have been assigned to the dimer and those labelled "3" to the symmetric stretch of the trimer. Adapted from Ref. [40]

applied to the trapped clusters: UV and visible absorption spectroscopy, laser induced fluorescence, Raman spectroscopy, infrared absorption spectroscopy, and electron spin resonance. In some cases more than one technique has been employed, and this helps to resolve some of the problems associated with identifying the carrier of a particular spectroscopic feature. However, it is still difficult to unambiguously assign spectral features to a particular cluster size for clusters with more than three atoms. Electron spin resonance is unique in this regard because it is possible to determine the size of the cluster from the spectrum, if the spectrum can be recorded with high signal to noise ratio. The electron spin resonance technique will be discussed in more detail below. UV-visible absorption spectra have been assigned to trimers of Cr, Co, Ni, Cu, Mo, Rh, Ag, and Au, as well as a few larger clusters [38]. Raman or resonance Raman studies have been performed on Sc_3, Cr_3, Ni_3, Cu_3, and Ag_3 [39, 40]. Figure 5 shows a portion of the resonance Raman spectrum recorded for a Cr containing argon matrix excited with 514.5 nm laser light [40]. The features in the spectrum labelled "2" have been assigned to the dimer, and those labelled "3" to the trimer. The trimer progression with $\omega_e = 313\ cm^{-1}$ was assigned to a symmetric stretch vibration in C_{2v} symmetry. Two other lines at 123 cm^{-1} and 226 cm^{-1} observed under nonresonance conditions were assigned to the bending and asymmetric stretching vibrations of Cr_3.

4.3.8 Electron Affinities and Photoelectron Spectroscopy

Photodetachment studies of negative ions provide information on the electron affinity (the analog of the ionization energy of a neutral species). If the photoelectrons are energy analyzed then information can be obtained on the electronic structure of the *neutral* cluster in the geometry of the anion. Photoelectron spectroscopy thus provides a powerful tool to investigate the electronic structure of metal clusters. Early photodetachment studies focussed on determining the electron affinities by bracketing with fixed frequency lasers and investigating the competition between photodetachment and photodissociation

[24, 25]. Photodetachment appears to dominate for most metal cluster ions. The exception is silver cluster anions. For these clusters both photodissociation and photodetachment occurs. Photodetachment from copper cluster anions has been extensively studied [25, 41–43]. Electron affinities determined by bracketing with fixed frequency lasers (and in some cases scanning the threshold region with a tunable dye laser) show substantial odd-even oscillations and evidence of features which can be attributed to shell closings [25]. The electron affinities increase with cluster size. For a simple spherical metal droplet the electron affinities are expected to increase linearly with $1/r$ (r is the cluster radius). Photoelectron spectroscopy of copper cluster anions has been investigated by *Lineberger* and coworkers [41] and *Smalley* and coworkers [42, 43]. The photoelectron spectra of these clusters show features which can be attributed to the developing 4*s* and 3*d* bands. Figure 6 shows a plot of the electron affinities (onsets of the 4*s* bands) and the onset of the 3*d* bands determined from the photoelectron spectra of Cu_n^- (n = 1–410) as a function of cluster size. The black solid bars on the left hand side of this figure show the range of work functions and 3*d* band onsets observed for Cu(100), Cu(110), and Cu(111). It appears that the energy between the onsets of the 4*s* and 3*d* bands approaches the bulk separation for quite small clusters. However, even for a cluster with 410 atoms the width of the feature in the photoelectron spectrum attributed to the 3*d* band is much narrower than the 3*d* band of bulk copper. *Meiwes-Broer*, *Lutz*, and coworkers have measured photoelectron spectra for a number of different transition metal clusters (Ni_n^-, Ag_n^-, and Pd_n^-) [44, 45]. The photoelectron spectra of Ag_n^- show similar trends to those measured for Cu_n^-. For both Cu_n^- and Ag_n^- the electron affinities are in reasonable agreement with the predictions of the simple spherical metal droplet model. However, the electron affinities for Ni_n^- and Pd_n^- are considerably smaller than the predictions of this model. As already described above, the ionization energies of open *d* shell transition metals (Fe_n, Co_n, Ni_n, and Nb_n) are also lower than predicted by the metal droplet model. Meiwes-Broer, Lutz, and coworkers have attributed this difference to

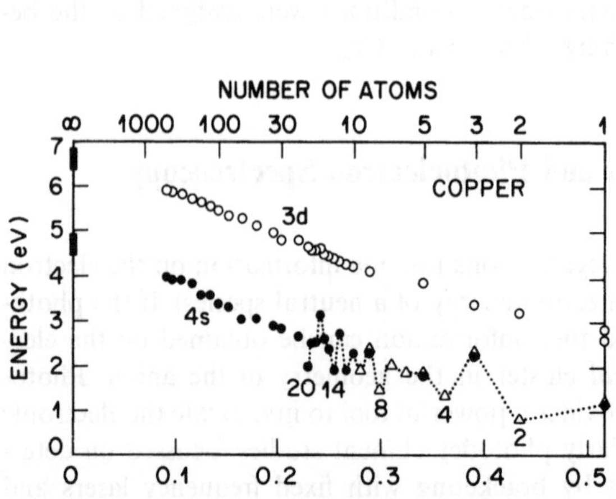

Fig. 6. Electron affinities (4*s* band onsets) and 3*d* band onsets determined from photoelectron spectra of size selected copper cluster anions with 1–410 atoms. The black solid bars on the left side of the figure show the 4*s* and 3*d* onsets determined for Cu(100), Cu(110), and Cu(111). Adapted from Ref. [43]

differences in the electronic structure. Ionization of copper and silver clusters involves delocalized valence electrons having mainly *s* character. But ionization of the open shell transition metal clusters involves more localized *d* electrons. Photoelectron spectra of the open *d* shell transition metal elements are much more complicated than those observed for Cu_n^- and Ag_n^-. Lineberger and coworkers have measured photoelectron spectra of a number of small transition metal clusters using relatively high resolution (up to 40 cm^{-1}). For Fe_2^-, Co_2^-, and Cu_2^- the photoelectron spectra showed resolved vibrational structure [4, 41, 46]. Spectra recorded for Ni_3^-, Pd_3^-, and Pt_3^- also showed some vibrational features (except for Ni_3^-) [47]. However, the high electronic state density, with many overlapping bands, made interpretation difficult. Zero kinetic energy photoelectron spectroscopy provides a way of achieving $\sim 3\ cm^{-1}$ resolution and several groups are currently attempting to extend this approach to metal cluster anions. In addition to the work described above on clusters in the gas phase, several groups have recorded photoelectron spectra for deposited clusters, probing both the valence bands and the core levels. The results of these studies are described in a separate chapter.

4.3.9 Geometric Structure

In the study of all types of clusters the largest challenge is determining the geometric structure. Most of the structural information available on transition metal clusters has been obtained from electron spin resonance studies of matrix isolated clusters. As noted above, electron spin resonance can provide information on the number of atoms in the cluster. However, this information is obtained from the number of lines in the spectrum and since the relative intensity of the weakest line can be several orders of magnitude smaller than the most intense feature in the spectrum it is necessary to record the electron spin resonance spectrum with high signal to noise ratio. Electron spin resonance spectra have been assigned to Sc_3, Y_3, Cr_4, Mn_5, Cu_3, Ag_3, Au_3, Cu_2Ag, Cu_5, $CuAg_4$, Cu_2Ag_3, Ag_7, and Sc_{13} [48, 49]. The assignment of the remarkable spectrum (consisting of over sixty equally spaced hyperfine lines) observed by *Welter* and coworkers to Sc_{13} is based mainly on the expected stability of this species [49].

Another approach to determining information on gas phase clusters is to use chemical probes. This approach has been used mainly by *Riley* and coworkers [50]. They have proposed structures for a number of small iron clusters based on their chemical behavior with a number of different reagents. A concern with these studies is that the structural information deduced may reflect the structure of the products rather than the bare cluster. In recent work equilibrium constants for the adsorption of water of hydrogenated cobalt clusters have been measured and show magic numbers in striking agreement with those expected for icosahedral packing [51].

The most direct way to obtain structural information on larger metal clusters is to employ one of the diffraction based techniques such as EXAFS (extended x-ray absorption fine structure) or electron diffraction. Electron diffraction studies have been performed on both deposited clusters and gas phase clusters, EXAFS studies have been limited to deposited clusters. *Stein* and coworkers have used electron diffraction to investigate the structures of Ag_n, using an oven and supersonic expansion with an inert carrier to generate the clusters [52]. For small Ag_n clusters (< 80 Å diameter) a partially liquid or amorphous model was proposed. *Montano* and coworkers [53] have used EXAFS to study silver clusters with a narrow size distribution isolated in argon. The coordination number of the smallest clusters investigated (25 Å diameter) was close to that of the fcc structure of the bulk.

4.3.10 Summary

This chapter has provided an overview of work performed investigating the physical properties of transition metal clusters. The transition metal clusters are much more difficult to understand than the alkali metal clusters where the electronic shell model provides a simple picture of the electronic structure. However, while they provide interesting models for studies of electron confinement, alkali metals have few commercial applications. Transition metals, on the otherhand, are extremely important commercially and a detailed understanding of the chemical bonding and properties of small transition metal clusters may have important consequences in catalysis and materials.

References

1. G. Schmid, R. Pfeil, R. Boese, F. Bandermann, S. Meyer, G. Calis, J. van der Velden: Chem. Ber. **114**, 3634 (1981)
2. E.M. Spain, M.D. Morse: Int. J. Mass Spectrum. Ion. Proc. **102**, 183 (1990)
3. I. Shim, K.A. Gingerich: J. Chem. Phys. **77**, 2490 (1982)
4. D.G. Leopold, W.C. Lineberger: J. Amer. Chem. Soc. **108**, 1379 (1986); D.G. Leopold, J. Almlof, W.C. Lineberger, P.R. Taylor: J. Chem. Phys. **88**, 3780 (1988)
5. M. Tomonari, H. Tatewaki: J. Chem. Phys. **88**, 1828 (1988)
6. M.D. Morse: Chem. Rev. **86**, 1049 (1986)
7. K.A. Gingerich: Faraday Symp. Chem. Soc. **14**, 109 (1980)
8. M.F. Jarrold, A.J. Illies, M.T. Bowers: J. Amer. Chem. Soc. **107**, 7339 (1985)
9. K. Raghavachari, K.K. Sunhil, K.D. Jordon: J. Chem. Phys. **83**, 4633 (1985)
10. M.M. Kappes, M. Schar, P. Radi, E. Schumacher: J. Chem. Phys. **84**, 1863 (1986)
11. G. Makov, A. Nitzan, L.E. Brus: J. Chem. Phys. **88**, 5076 (1988)
12. D.E. Powers, S.G. Hansen, M.E. Geusic, D.L. Michalopoulos, R.E. Smalley: J. Chem. Phys. **78**, 2866 (1983)
13. A. Kaldor, D.M. Cox, D.J. Trevor, M.R. Zakin: in *Metal Clusters*, ed. by F. Trager, G. zu Putlitz (Springer-Verlag, Berlin, 1986)

14. E.A. Rohlfing, D.M. Cox, A. Kaldor: Chem. Phys. Lett. **99**, 161 (1983); E.A. Rohlfing, D.M. Cox, A. Kaldor, K.H. Johnson: J. Chem. Phys. **81**, 3846 (1984)
15. E.A. Rohlfing, D.M. Cox, A. Kaldor: J. Phys. Chem. **88**, 4497 (1984)
16. R.L. Whetten, M.R. Zakin, D.M. Cox, D.J. Trevor, A. Kaldor: J. Chem. Phys. **85**, 1697 (1986)
17. M.B. Knickelbein, S. Yang, S.J. Riley: J. Chem. Phys. **93**, 94 (1990)
18. S. Yang, M.B. Knickelbein: J. Chem. Phys. **93**, 1533 (1990)
19. M.B. Knickelbein, S. Yang: J. Chem. Phys. **93**, 1476 (1990); **93**, 5760 (1990)
20. E.K. Parks, T.D. Klots, S.J. Riley: J. Chem. Phys. **92**, 3813 (1990)
21. M.R. Zakin, R.O. Brickman, D.M. Cox, A. Kadlor: J. Chem. Phys. **88**, 3555 (1988); Y. Hamrick, S. Taylor, G.W. Lemire, Z.-W. Fu: J.-C Shui, M.D. Morse: J. Chem. Phys. **88**, 4095 (1988); Y.M. Hamrick, M.D. Morse: J. Phys. Chem. **93**, 6494 (1989); J.L. Elkind, F.D. Weiss: J.M. Alford, R.T. Laaksonen, R.E. Smalley: J. Chem. Phys. **88**, 5215 (1988)
22. W. Begemann, S. Dreihofer, K.H. Meiwes-Broer, H.O. Lutz: in *Metal Clusters* ed. by F. Trager, G. zu Putlitz (Springer-Verlag, Berlin, 1986)
23. P.J. Brucat, L.-S. Zheng, C.L. Pettiette, S. Yang, R.E. Smalley: J. Chem. Phys. **84**, 3078 (1986)
24. P. Fayet, L. Woste: in *Metal Clusters*, ed. by F. Trager, G. zu Putlitz (Springer-Verlag, Berlin, 1986)
25. L.-S. Zheng, C.M. Karner, P.J. Brucat, S.H. Yang, C.L. Pettiette, M.J. Craycraft, R.E. Smalley: J. Chem. Phys. **85**, 1681 (1986)
26. S.K. Loh, D.A. Hales, L. Lian, P.B. Armentrout: J. Chem. Phys. **90**, 5466 (1989)
27. S.K. Loh, L. Lian, P.B. Armentrout: J. Amer. Chem. Soc. **111**, 3167 (1989)
28. M.F. Jarrold, K.M. Creegan: Int. J. Mass Spectrum. Ion Phys. **102**, 161 (1990)
29. K. Ervin, S.K. Loh, N. Aristov, P.B. Armentrout: J. Chem. Phys. **87**, 3593 (1983)
30. P.B. Armentrout: Proc. SPIE **620**, 38 (1986)
31. D.M. Cox, D.J. Trevor, R.L. Whetten, E.A. Rohlfing, A. Kaldor: Phys. Rev. B, **32**, 7290 (1985)
32. W.A. de Heer, P. Milani, A. Chatelain: Phys. Rev. Lett. **63**, 2834 (1989)
33. M.D. Morse, J.B. Hopkins, P.R.R. Langridge-Smith, R.E. Smalley: J. Chem. Phys. **79**, 5316 (1983); E.A. Rohlfing, J.J. Valentini: Chem. Phys. Lett. **126**, 113 (1986); W.H. Crumley: J.S. Hayden, J.L. Gole: J. Chem. Phys. **84**, 5250 (1986)
34. J.R. Woodward, S.H. Cobb, J.L. Gole: J. Chem. Phys. **92**, 1404 (1988)
35. P.Y. Cheng, M.A. Duncan: Chem. Phys. Lett. **152**, 341 (1988)
36. M.F. Jarrold, K.M. Creegan: Chem. Phys. Lett. **166**, 116 (1990)
37. K. Balasubramanian, K.K. Das: J. Chem. Phys. **94**, 2923 (1991)
38. M. Moskovits, J.E. Hulse: J. Chem. Phys. **66**, 3988 (1977); M. Moskovits, J.E. Hülse: J. Chem. Phys. **67**, 4271 (1977); W.E. Klotzbucher, G.A. Ozin: J. Amer. Chem. Soc. **100**, 2262 (1978); S.A. Mitchell, G.A. Ozin: J. Amer. Chem. Soc. **100**, 6776 (1978); G.A. Ozin, H. Huber: Inorg. Chem. **17**, 155 (1978); G.A. Ozin, A. Hanlan: Inorg. Chem. **18**, 1781 (1979); W. Schulze, H.U. Becker, H. Abe: Chem. Phys. **35**, 177 (1978); W.E. Klotzbucher, G.A. Ozin: Inorg. Chem. **18**, 2101 (1979); W.E. Klotzbucher, G.A. Ozin: Inorg. Chem. **19**, 3776 (1980); W.E. Klotzbucher, G.A. Ozin: Inorg. Chem. **19**, 3767 (1980); S.A. Mitchell, G.A. Ozin: J. Phys. Chem. **88**, 1425 (1984)
39. W. Schulze, H.U. Becker, R. Minkwitz, K. Manzel: Chem. Phys. Lett. **55**, 59 (1978); D.P. DiLella, M. Moskovits: J. Chem. Phys. **72**, 2267 (1980); D.P. DiLella, K.V. Taylor, M. Moskovits: J. Phys. Chem. **87**, 524 (1983); M. Moskovits, D.P. DiLella, W. Limm: J. Chem. Phys. **80**, 626 (1984); M. Moskovits: Chem. Phys. Lett. **118**, 111 (1985)
40. D.P. DiLella, W. Limm, R.H. Lipson, M. Moskovits, K.V. Taylor: J. Chem. Phys. **77**, 5263 (1982)
41. D.G. Leopold: J. Ho, W.C. Lineberger: J. Chem. Phys. **86**, 1715 (1987)
42. C.L. Pettiette, S.H. Yang, M.J. Craycraft: J. Conceicao, R.T. Laaksonen, O. Cheshnovsky, R.E. Smalley: J. Chem. Phys. **88**, 5377 (1988)
43. O. Cheshnovsky, K.J. Taylor: J. Conceicao, R.E. Smalley: Phys. Rev. Lett. **64**, 1785 (1990)
44. G. Gantefor, M. Gausa, K.-H. Meiwes-Broer, H.O. Lutz: Faraday Discuss. Chem. Soc. **86**, 197 (1988)
45. G. Gantefor, M. Gaussa, K.H. Meiwes-Broer, H.O. Lutz: J. Chem. Soc. Faraday Trans. **86**, 2483 (1990)
46. D.G. Leopold, W.C. Lineberger: J. Chem. Phys. **85**, 51 (1986)

47. K.M. Ervin: J. Ho, W.C. Lineberger: J. Chem. Phys. **89**, 4514 (1988)
48. R.J. Van Zee, C.A. Baumann, W. Weltner: J. Chem. Phys. **76**, 5636 (1982); C.A. Baumann, R.J. Van Zee, W. Weltner: J. Chem. Phys. **78**, 190 (1983); R.J. Van Zee, C.A. Baumann, W. Weltner: J. Chem. Phys. **82**, 3912 (1985); J.A. Howard, R. Sutcliffe, B. Mile: Surf. Sci. **156**, 214 (1985)
49. L.B. Knight, R.W. Woodward, R.J. Van Zee, W. Weltner: J. Chem. Phys. **79**, 5820 (1983)
50. K. Parks, B.H. Weiller, P.S. Bechthold, W.F. Hoffman, G.C. Nieman, L.G. Pobo, S.J. Riley: J. Chem. Phys. **88**, 1622 (1988); E.K. Parks, G.C. Nieman, L.G. Pobo, S.J. Riley: J. Chem. Phys. **88**, 6260 (1988)
51. T.D. Klots, B.J. Winter, E.K. Parks, S.J. Riley: J. Chem. Phys. **92**, 2110 (1990)
52. B.G. De Boer, G.D. Stein: Surf. Sci. **106**, 84 (1981)
53. P.A. Montano, W. Schulze, B. Tesche, G.K. Shenoy, T.I. Morrison: Phys. Rev. B **30**, 672 (1984)

4.4 Carbon Clusters

E.E.B. Campbell

4.4.1 Introduction

Carbon clusters fall into the category of strongly bound (non van der Waals) clusters and certainly constitute one of the most active and exciting areas of present day cluster research as witnessed by the large number of recent reviews [1–7] and press articles [8] on the subject. Carbon nucleates to form clusters far more readily than any other element in the periodic table (including such refactory elements as tungsten and tantalum), as can be seen from considering how easily soot is formed in "everyday life" (see Section 4.4.3.2 for a possible mechanism for soot formation). Aside from its great importance in combustion processes, carbon is also one of the most abundant elements in the universe (after hydrogen, helium and oxygen) and, in the present theories of stellar nucleosynthesis of the elements, is the first stable element that can be synthesized in fusion reactions involving the hydrogen and helium produced in the original "big bang" [9]. Carbon rich red giant stars are known to emit enormous amounts of carbon dust into the interstellar medium and indeed the main constituent of interstellar dust is thought to be carbon grains. The largest currently known molecules to have been definitely identified in space are based on long carbon chains (e.g. $HC_{11}N$) [5]. However, it is not only the technological and astrophysical applications which make the study of carbon clusters so interesting, the pure chemical and physical properties of the clusters, especially C_n with $n \geq 24$, the fullerenes, is proving to be a fascinating field of research.

Carbon clusters are normally formed using a laser vaporization source of the type described in Chapter 3.1.3. The second harmonic of a pulsed Nd YAG laser and a graphite substrate with a pulsed He beam to cool the laser produced plasma are the usual choices for the source with a UV laser for photoionization of the neutral clusters, although other combinations are possible. A typical mass spectrum is shown in Fig. 1 with the mass spectrum of another Group IV element, silicon, produced with the same type of source shown for comparison. Both spectra have a bimodal structure but the details are very different. The most apparent difference is the lack of carbon clusters with odd numbers of atoms beyond about $n = 24$ but the "magic numbers" i.e. the prominent mass peaks in the small mass range are also very different. (Under certain conditions it is possible to detect the odd numbered large carbon clusters [12] but they are

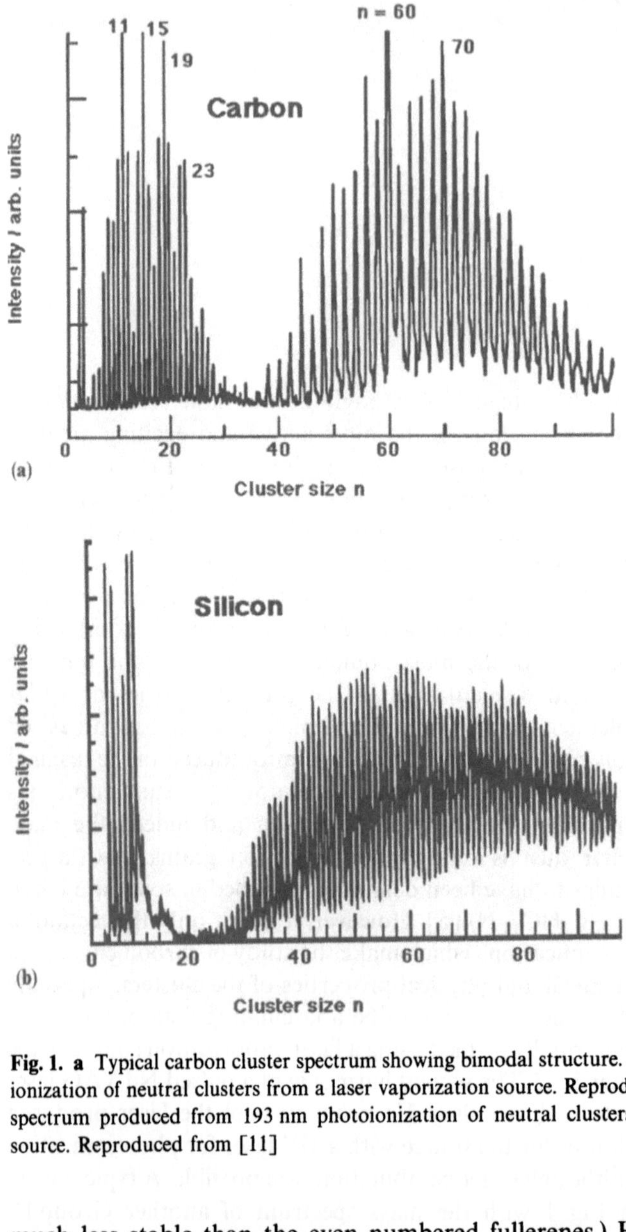

Fig. 1. **a** Typical carbon cluster spectrum showing bimodal structure. Produced by 193 nm photoionization of neutral clusters from a laser vaporization source. Reproduced from [10]. **b** Si cluster spectrum produced from 193 nm photoionization of neutral clusters from a laser vaporization source. Reproduced from [11]

much less stable than the even numbered fullerenes.) Hybridization of *s* and *p* orbitals occurs much more readily in carbon than in the other Group IV elements and π bonds are much stronger. Both these effects play an important role in determining the cluster geometries and stabilities and lead to the detail in the mass spectrum displayed in Fig. 1.

The structure and properties of the small carbon clusters C_n with $n \leq 24$ will be discussed in detail in Section 4.4.2 to be followed in Section 4.4.3 with a discussion of the large clusters $n \geq 24$ known as the fullerenes (there is also evidence that $n = 20$ fullerene, dodecahedrene, may exist [13]. The very special, highly symmetric cluster C_{60} known as Buckministerfullerene[1] (often shortened to "Buckyball"), which is apparent as a "magic number" in Fig. 1 (a), has become the centre of a flurry of research activity since a simple method was recently found to produce it in macroscopic quantities [15]. In Section 4.4.4 the unique properties of this fascinating cluster will be described and an attempt will be made to convey the excitement felt by many research groups now carrying out research on "Buckyballs" worldwide.

4.4.2 The Small Clusters

4.4.2.1 Mass Spectra and "Magic Numbers"

The first report of observations of positively charged carbon clusters up to $n = 15$, produced in a high frequency discharge between graphite electrodes and detected in a mass spectrometer, was made by *Otto Hahn* and his group in the early 1940s [16, 17]. The same method was then used some 16 years later by *Hintenberger* and coworkers for a somewhat more detailed investigation [18, 19]. The relative intensities of the positively charged ions detected in their experiment is shown in Fig. 2(a). An odd-even periodicity in the intensity up to $n = 10$ is clearly seen followed by a periodicity of 4 carbon atoms. A similar mass distribution was obtained in investigations of positive carbon clusters produced directly (i.e. without subsequent ionization) by laser vaporization of a thin carbon foil without a carrier gas [20] as illustrated in Fig. 2(b). The typical laser vaporization carbon cluster sources using a rare gas pulse to cool the products produce positive ion mass spectra like that shown in Fig. 2(c) which shows considerable enhancement for the "magic numbers" $n = 7$, 11, 15, 19 and 23 which are, to a much lesser extent, also prominent in Figs. 2(a) and (b).

Neutral cluster distributions are probed mainly by UV photoionization of the neutral species produced in the laser vaporization sources. A very similar mass distribution with the same "magic numbers" as in Fig. 2(c) is produced as shown in Fig. 3 although the cluster distribution in the C_1 to C_7 range depends strongly on experimental conditions. The similarity is perhaps not surprising since the small carbon clusters all have ionization potentials greater than 7 eV [21] which means that the laser ionization is at least a two-photon process and is known to lead to considerable photofragmentation [10]. (The ArF excimer laser which is the one most commonly used for the ionization has a wavelength of 193 nm i.e. a photon energy of 6.4 eV.)

---

[1] The name Buckminsterfullerene was chosen in honour of the American architect (mathematician, engineer, poet. . .) Buckminster Fuller whose geodesic domes were the inspiration for the correct guess at the special, hollow structure of C_{60} in 1985 [14] (see section 4.4.3.1).

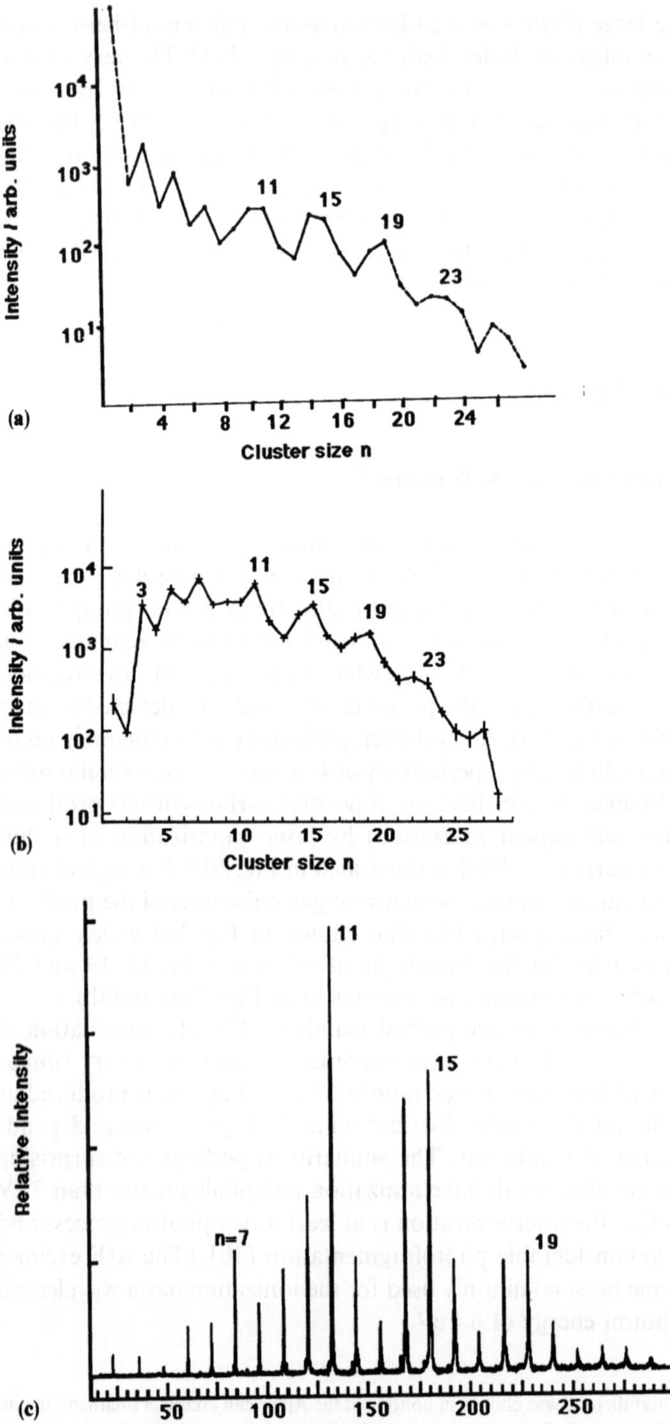
Intensity / arb. units
10⁴
10³
10²
10¹
11
15
19
23
(a)
4
8
12
16
20
24
Cluster size n
Intensity / arb. units
10⁴
10³
10²
10¹
3
11
15
19
23
(b)
5
10
15
20
25
Cluster size n
Relative Intensity
n=7
11
15
19
(c)
50
100
150
200
250
Mass / u

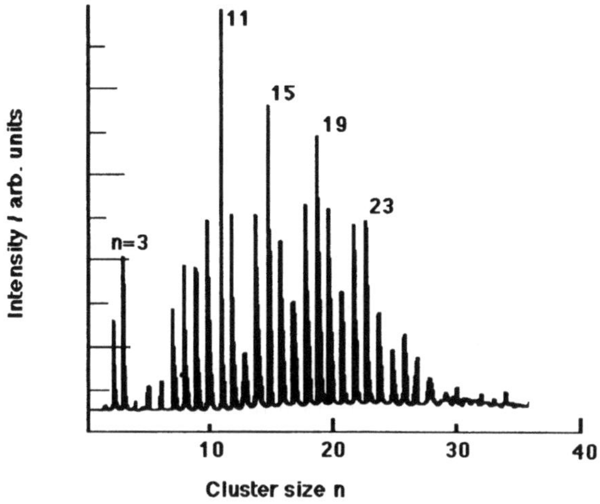

Fig. 3. Mass spectrum produced by UV photoionization of neutral clusters from a laser vaporization source (193 nm, 1.6 mJ). Reproduced from [10]

Detailed investigations of photofragmentation of mass selected positively charged clusters [22] show that the dominant fragmentation process is loss of a neutral C_3, while for some of the larger cluster ions the loss of a C_5 unit is also an important fragmentation channel. The same decay processes were observed metastably for the hot ions produced in the high frequency discharge source [18] and are accounted for by the high stability of the C_3 (and C_5) species.

Early molecular orbital theory calculations carried out for the neutral carbon clusters [23] predicted that the clusters form chains up to C_9 and rings for larger clusters. This can be qualitatively understood by considering that a covalently bonded cluster will always attempt to minimise the number of dangling bonds. Carbon is noted for its ability to form multiple bonds by which means the number of necessary dangling bonds can be reduced considerably. The second and third bonds are not as strong as the first but, in a finite system the multiply bonded structures will be favored, if the alternative is an increased number of dangling bonds. For the smallest clusters, the multiply bonded chain is the structure containing the smallest number of dangling bonds, e.g. C_3 is known to be linear[2] [24, 25]. Bending of a chain into a ring removes the dangling bonds at the ends of the chain but strain is introduced because the chain is *sp* bonded. For this reason formation of a ring is only favorable for larger clusters. Later, increasingly sophisticated calculations [e.g. 27–31] have

Fig. 2. Positively charged carbon clusters produced directly in the source showing typical "magic numbers". **a** From a high frequency graphite discharge. Reproduced from [18]. **b** From laser vaporization of a carbon foil, without carrier gas. Reproduced from [20]. **c** From laser vaporization of graphite, with carrier gas. Reproduced from [41]

[2] C_3 is, strictly speaking, quasi-linear with a very anharmonic and low bending frequency [26].

confirmed the crossover from chains to rings at about $n = 10$ although there are some indications that at low temperatures the favored form of the small even clusters, $n = 4, 6, 8$, should be the ring structure [28, 31]. The relative stabilities of the clusters can be inferred from simple bonding considerations [23, 27]. In the chain clusters all intermediate atoms contribute 2 electrons to the σ-bonding system to form bonds with each adjacent atom and the end atoms use 3 electrons to form one σ bond and an unshared pair. There are thus $(n-1)\sigma$ bonds containing $2n-2$ electrons and two lone pairs, each with 2 electrons, leaving $2(n-1)$ electrons to occupy the π manifold. The orbitals of the π levels are doubly degenerate and can accommodate four electrons each and, since clusters with fully occupied HOMOs (highest occupied molecular orbital) are the most stable, this implies that neutral, linear clusters with odd numbers of atoms will be more stable and hence more abundant than those with even numbers of atoms. In the ring structures the two π bonding planes (parallel and perpendicular to the ring) each have a nondegenerate totally symmetric level as the lowest level, followed by doubly degenerate levels. Since a ring cluster has $2n$ σ electrons and $2n$ π electrons (no lone pairs), the most stable rings with fully occupied HOMOs are for $n = 4k + 2$, with k an integer. (This is in analogy to Hückels rule for cyclic hydrocarbons where the carbon atoms contribute one π electron each and π bonding occurs only in the plane perpendicular to the ring).

The relative stabilities and the change from chain structures to ring structures predicted theoretically are consistent with results from very nice experiments carried out by the *Smalley* group [32] in which the ultraviolet photoelectron spectra (PES) of negative cluster ions formed in a laser vaporisation source were measured. The observed photodetachment onsets provide a measure of the electron affinities of the corresponding neutral clusters. Here one would expect that the even numbered chain clusters which are open shell species will have higher electron affinities than the closed shell odd numbered clusters where the next electron has to go into the next highest shell. Similarly the rings with $n = 4k + 2$ should have low electron affinities. This is exactly what is seen experimentally as shown in Fig. 4 where the measured electron affinity is plotted against cluster size.

The observed structure in the mass spectra of the positively charged clusters produced directly in the source (Fig. 2) cannot be understood so easily by a simple electron counting model. A model for gas phase cluster aggregation including the simultaneous growth of neutral and charged species with electronic structure calculations to determine the kinetic parameters of the model [27] can reproduce the "magic numbers" for the ring structures very successfully when assuming temperatures $T_{av} \leq 500$ K. This indicates that significant cooling occurs in the source. A calculated mass spectrum is shown in Fig. 5 which should be compared with the experimental spectra in Fig. 2. As the temperature is increased the chain structures become more dominant for the larger ions. The mass distribution for $n < 10$ is extremely sensitive to experimental conditions unlike the larger clusters, probably since the small clusters are formed when the system is far from equilibrium and are more sensitive to the source conditions.

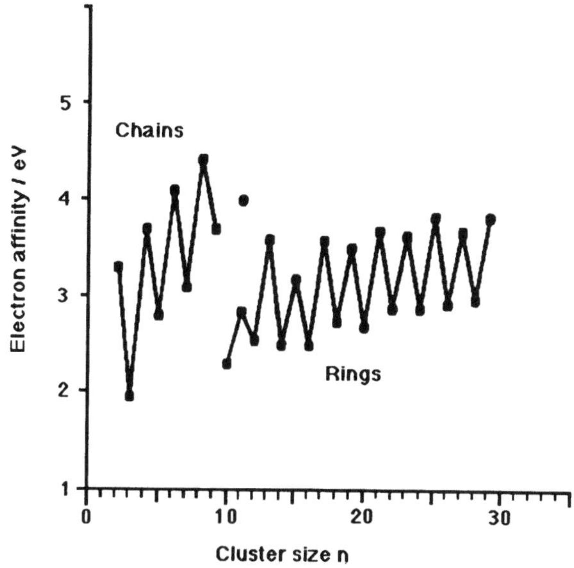

Fig. 4. Electron affinity of neutral carbon clusters as measured by the observed photodetachment thresholds of the negative cluster ions. Reproduced from [32]

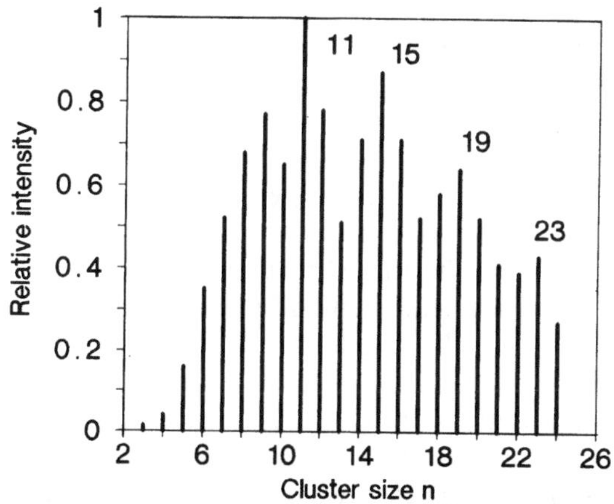

Fig. 5. Calculated relative abundances for positively charged clusters. Compare with Fig. 2. Data from [27]

The mass spectrum observed on photoionization of the neutral clusters is, as mentioned above, strongly dependent on the photoionization process. Two main effects will contribute: Firstly, the ionization potential of neutral clusters with $4k + 3$ atoms (i.e. $n = 11, 15, \ldots$) will be lower than that for the neutral "magic numbers", $n = 10, 14, \ldots$, since the last pair of π electrons occupies an antibonding orbital. Secondly, photofragmentation of the highly abundant clusters with $n = 4(k + 1) + 2$ will, due to C_3 ejection, lead predominantly to positively charged clusters with $n = 4k + 3$. Both effects yield for different reasons the same "magic numbers" in the mass spectra of the photoionized ring clusters as observed in the mass spectra of positively charged ring clusters produced directly in the source.

4.4.2.2 Experimental Structural Determination

High resolution optical spectroscopy can be used to determine the structure of small neutral clusters. Optical spectroscopy of C_3 dates back to 1881 [33] when a unique group of spectral lines at 4050 Å was detected in the spectrum of a comet. However, it took another 70 years before these lines could be attributed to C_3 [34]. A very thorough analysis of the vibronic transitions carried out in 1965 [24] showed that they were due to a linear, symmetric molecule. More recently, observation of vibrational-rotational lines of C_3 in the circumstellar spectrum of a carbon star [35] followed by high resolution diode laser absorption spectroscopy of C_3 [36] and C_5 [37] have confirmed the theoretically predicted linear, chain structures. Practically simultaneously with the laboratory measurements, C_5 was also detected in the infrared spectrum of a carbon star [38].

Direct geometrical images of molecules or clusters can be obtained using the Coulomb Explosion Imaging (CEI) technique [39] and this method has been applied to the small carbon clusters C_3 [25] and C_4 [40]. The technique takes advantage of the large Coulomb repulsion between the nuclei in a cluster when the atoms are stripped of their electrons. By measuring the relative momenta of the atomic fragments, one can in principle trace the trajectories backward in time through the Coulomb interaction to deduce the initial spatial positions of the fragments before the explosion. Since the timescale for electron stripping is very short compared to the characteristic vibrational and rotational motions in the cluster, the Coulomb explosion of each individual cluster gives a "snapshot" of the relative positions of the nuclei within that cluster. The experiments confirmed the linear form for C_3 [25] and showed, in contrast to the PES results Fig. 4, that many of the C_4 clusters have the ring structure [40]. The latest theoretical calculations using the Car–Parrinello method [31] do in fact show that the lowest energy structure for C_4 is the ring form but at temperatures greater than about 250 K the chain isomer is favored for entropy reasons. However, if, due to the preparation conditions, the cluster is trapped in the ring form and kept at fairly low temperatures it is likely to remain stable since the transition to the chain structure has a very high energy barrier. Thus, differences between the PES results shown in Fig. 4 which detect only the chain isomer for C_4 and the CEI measurements are most likely due to very different source conditions. (One likely possibility is that the method used to produce the negative clusters in [32] discriminated against the low electron affinity species [39].)

4.4.2.3 Reactivity

A number of studies of the reactivity of mass selected, positively charged carbon clusters ($n < 20$), with diatomic molecules [41, 42], small hydrocarbons [43] and HCN [44], have been carried out. Reaction is seen to occur predominantly at the ends of the linear cluster ions for collisions with D_2 and O_2 with the main reaction products being C_nD^+ and C_nO^+ respectively. The results provide convincing evidence for the presence of both the chain and ring forms of C_7^+ in the cluster beam [41] (later studies also indicate that both isomers could be present for C_3^+ to C_6^+ [42]). As the complexity of the collision partner increases, the types of reactions undergone by the carbon clusters become more complex. The predominant product in reactions between the small linear clusters C_n^+ and e.g. the unsaturated hydrocarbon C_2H_2 is $C_{n+2}H^+$ leading to growth of the cluster [43]. C_2H_2 is also the first molecule investigated which has been shown to react with the ring clusters. The only product ion formed results from an association reaction, i.e., incorporation of the neutral reactant within the cluster ion without the loss of a neutral fragment: $C_n^+ + C_2H_2 \rightarrow C_{n+2}H_2$ [43]. The product is then stabilized by photon emission. Similar association reactions are observed for the ring clusters in reactions with HCN [44] to form C_nHCN^+, however, with the exception of $n = 7$, these adduct ions are not stable and will easily lose the HCN with just a few electron volts of excitation energy. The small linear clusters ($n < 6$), on the other hand, stabilize after the association reaction by eliminating a fragment of the HCN to form predominantly C_nCN^+. The larger linear clusters ($n = 6, 8, 9$) can accommodate the excess energy long enough for the adduct to stabilize, possibly by emitting IR photons. The C_nHCN^+ product ions did not react further in contrast to the C_nCN^+ indicating that, after insertion of one of the C_n^+ dangling bonds into the H–CN bond, the H migrates to the other end of the chain to give the structure HC_nCN^+, thus blocking the second dangling bond site [44]. Such species can also be formed when graphite is vaporized in the presence of NH_3 and CH_3CN [45]. Investigations of these reaction processes are certainly of great relevance for understanding the mechanism of formation of such long chain HC_nN^+ molecular ions in space. Incidentally, it was the interest in the spectroscopy of such molecules that led to the discovery of the fullerenes which will be the subject of the rest of this chapter.

4.4.3 The Fullerenes: C_n with $n \geq 24$

4.4.3.1 The Discovery of the Fullerenes

In 1985 Harry Kroto from Sussex University in England visited Texas with the idea of simulating carbon star chemistry with the help of Bob Curl and Rick Smalley's laser vaporization cluster source. Similar sources had already been used by groups at Exxon Research and Engineering [10] and AT&T Bell Labs [46] to produce carbon clusters. Figure 1(a) shows a mass spectrum produced by the Exxon group. They made the interesting observation that only even numbered clusters were present for n larger than about 30 which was falsely interpreted as being due to a "carbyne phase" of carbon with the clusters consisting of cross-linked acetylene like linear chains $(-C{\equiv}C-)_n$. Although C_{60}^+ was seen to be prominent in the spectra no additional mention or interpretation of this fact was given in the text.

Initially Kroto and Smalley were interested mainly in the small clusters but as the work progressed they became more and more fascinated by the C_{60} peak which, under certain source conditions, could be made to fully dominate the mass spectrum as shown in Fig. 6. After much discussion, inspired guess work, beer drinking and playing with pieces of paper and sticky tape (see [5–7] for interesting insights into this period of carbon cluster research!) the controversial but intuitively very attractive truncated icosahedron structure for C_{60}, shown in Fig. 7, was suggested and it was christened with the name "Buckminsterfullerene" [14] in recognition of the inspiration obtained from considering the structures of Buckminster Fuller's extremely sturdy but incredibly light geodesic domes.

The carbon clusters discussed in Section 4.4.2 start out as one-dimensional *sp*-hybridized chains but as they grow tend to form two-dimensional *sp*-hybridized rings. An sp^3-hybridized array of carbons is three-dimensional (e.g. diamond) but has a surface with attendant (destabilizing) dangling bonds. Similarly an sp^2-hybridized planar fragment of a graphite sheet has a reactive edge with dangling bonds. In C_{60}, however, the incorporation of pentagons into the hexagonal graphite planes allows the sp^2-hybridized structure to bend around and form a cage, thus removing any dangling bonds and being further stabilized due to the overlap of adjacent π-electrons [e.g. 47]. This kind of structure had, in

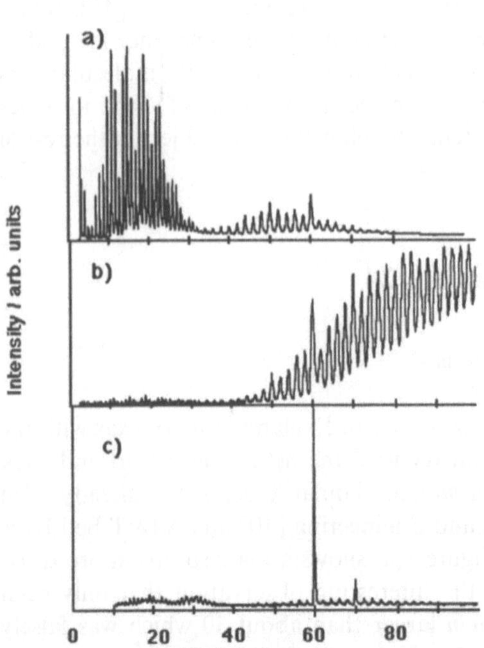

Fig. 6. Mass spectra of carbon cluster distributions in a supersonic beam produced by laser vaporization in the presence of a carrier gas under conditions of increasing extent of clustering (**a**–**c**). Reproduced from [5]

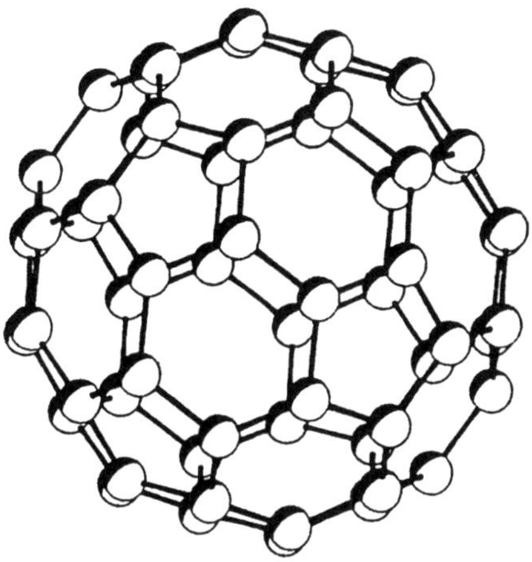

Fig. 7. Truncated icosahedral (soccerball) structure of C_{60}

fact, been predicted on a number of occasions before the *Smalley/Kroto* experiments were carried out [48–51]. Simple geometrical considerations, outlined below, show that it is in principle possible to envisage similar cage structures for all of the even carbon clusters with $n \geq 20$, with the exception of $n = 22$. Odd-numbered clusters will always have at least one atom with a remaining dangling bond and will thus be very reactive and unstable.

Each carbon atom is bonded to 3 others so that the number v of atoms (or vertices in the cage) can be related to the number e of bonds (edges) as follows

$$2e = 3v \ .$$

Similarly if f_n is the number of n-sided rings (faces) we have

$$2e = \sum_n n f_n \ .$$

Euler's theorem for convex polyhedra gives the further condition

$$v + \sum_n f_n = e + 2 \ .$$

If we restrict the rings to be pentagons and hexagons, which have the largest π-electron stabilization [47], and eliminate v and e we obtain

$$5f_5 + 6f_6 = 3\left(\frac{1}{2}(5f_5 + 6f_6) + 2 - f_5 - f_6\right) \Rightarrow f_5 = 12$$

i.e. we need 12 pentagons to produce a cage, with the number of hexagons given by

$$f_6 = \frac{v - 20}{2} \ .$$

The cages should be stable if the curvature related strain is symmetrically (geodesically) distributed and if the pentagons are isolated as much as possible

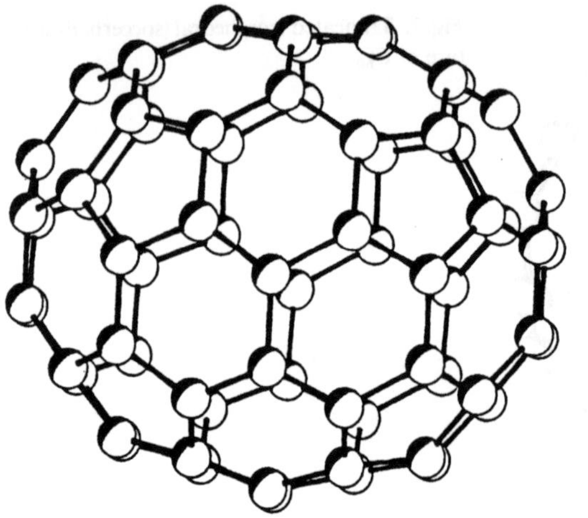

Fig. 8. C_{70}

by the hexagons to avoid the inherent instability of fused-pentagon configurations [52, 47]. The unique stability of C_{60} can thus be understood since it is the only 5/6-ring cage for which all the atoms are equivalent (i.e. the strain is perfectly distributed) and it is the smallest fullerene for which all pentagons can be isolated. The next fullerene which is able to avoid abutting pentagons is C_{70}, shown in Fig. 8, which is also observed experimentally to be particularly stable (Fig. 6(c)).

Support for the truncated icosahedral structure of C_{60} was obtained from PES measurements on C_{60}^- [53]. The PES pattern shown in Fig. 9 is characteristic of a cluster having a closed shell electronic structure for the neutral species with a large HOMO-LUMO gap (1.9 eV), in accordance with theoretical predictions for the truncated icosahedron [e.g. 5.4]. The extra electron added to form the negative ion must lie alone in the LUMO and it is this electron that is detached at threshold. The next electrons (and, for a cluster of this size there will generally be many such electrons) lie in what corresponds to the HOMO of the neutral cluster. The PES pattern thus consists of a small initial feature followed by the HOMO-LUMO gap followed by a much larger feature. These PES results provided evidence that C_{50}, C_{60} and C_{70} all had closed shell electronic structures.

One question which immediately springs to mind is why should such highly symmetric (and thus low entropy) clusters form under the violent, high temperature conditions of the laser vaporization source. One of the first suggestions was that the laser pulse removed pieces of the graphite intact from the surface which then rearranged to eliminate the dangling bonds, thus forming the cage structures [14]. This interpretation was finally ruled out when investigations of UV laser ablation of aromatic polymers (especialy polyimide; C:H:N:O = 22:10.5:2) produced practically identical mass spectra for the large mass range ($> 400\ u$) [55, 56]. The fact that pure, even numbered carbon clusters were produced from such a source [55] was somewhat astonishing and additional

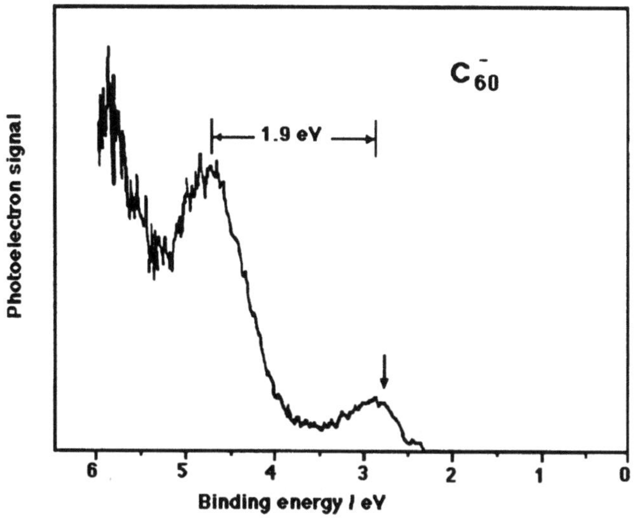

Fig. 9. UV photoelectron spectrum of C_{60}^- taken with an ArF excimer laser (6.4 eV). Reproduced from [53]

evidence that these species were particularly stable and unreactive. Another suggestion was that these clusters were by-products on the way to soot formation [57].

4.4.3.2 Fullerenes and Soot Formation

Careful examination of the data presented in Fig. 6 showed that C_{60} was not increasing much in intensity in the supersonic beam on going from Fig. 6(a) through (b) to (c) but the other clusters were decreasing [57]. In other words C_{60} is a survivor of some processes occuring in the cluster source. Simple gas kinetic mechanisms have been used to model the growth of the carbon clusters from carbon atoms and small carbon radicals present in the laser produced plasma [58]. Good agreement with experiment [55] indicates that this is a reasonable model. The energetically most favorable form of small carbon radicals at high temperature in a carbon vapor is probably the linear chain (see Section 4.4.2.1 and e.g. [23]). As these chains grow via aggregation in the dense plasma they will reach a length e.g. 20–30 carbon atoms where they are likely to occasionally form polycyclic aromatic rings or, in other words, graphitic sheets [59]. These structures have the advantage of increasing the average coordination number of the carbon atoms but the disadvantage of having a large number of unsatisfied dangling bonds at the edges. For small sheets it will, however, be energetically favorable for the sheets to bend and start to close due to the bond energy released by eliminating the reactive edge [60]. If the carbon atom density is high enough the cluster will continue to grow at the reactive edge as indicated

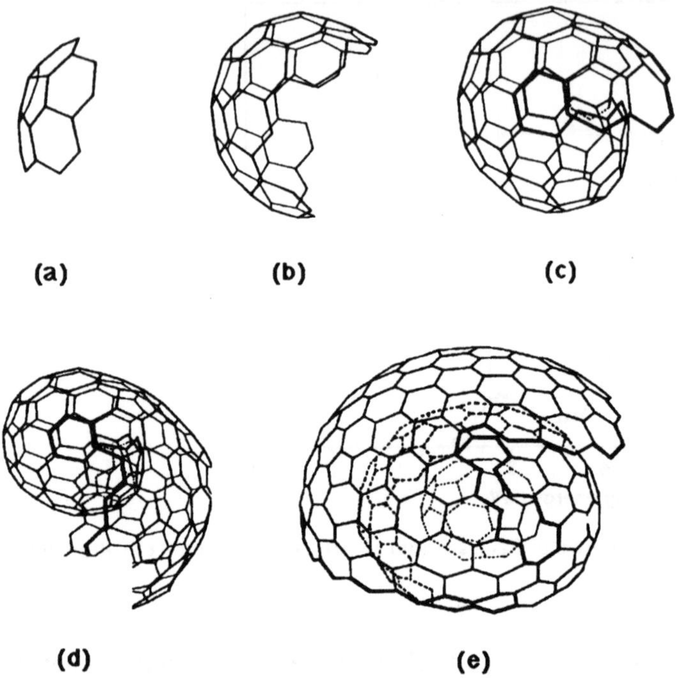

Fig. 10. Icospiral nucleation scheme proposed as a possible mechanism for soot formation. Reproduced from [5, 6]

in Fig. 10, overrunning and burying the opposite edge as it curls. Once this burying has occurred there is no ready way to terminate the growth process and large spiral structures of the kind shown in Fig. 10(e) will form, eventually leading to the formation of macroscopic soot particles [60]. If, at stage (c) in Fig. 10 accidental closure of the network occurs with the correct disposition of pentagons, C_{60} and the other fullerenes will be formed. The particularly stable fullerenes, predominantly C_{60} and C_{70}, will be unable to grow further after closure has occurred and will be left behind as the others continue to grow into giant particles. Further evidence for this scenario was obtained in 1987 when the fullerenes were detected in sooting hydrocarbon–oxygen flames [61].

4.4.3.3 Fullerene Fragmentation and Photophysics

In contrast to the small carbon clusters which predominantly fragment to give neutral C_3, the fullerenes fragment by losing C_2 – a consequence of the high stability of the even numbered clusters. Rate constants for unimolecular decay of positively charged fullerenes produced directly in a laser vaporization source without any carrier gas to aid cooling have been reported using two different experimental techniques. The first measurements involved fullerene ions produced from laser vaporization of graphite with fragmentation on a microsecond timescale being monitored in a double-focusing, reverse-geometry mass spectrometer [62]. The second approach used a polyimide ablation source for the fullerene ions and a reflectron time of flight mass spectrometer to detect the fragmentation [63]. A comparison of the reported rate constants is given in Fig. 11. Although the variations in rate constant with cluster mass were similar, the absolute value was approximately a factor of 5 smaller for the measurements with the time of flight mass spectrometer. This discrepancy was originally attributed entirely to differing source conditions but has since been shown to be due mainly to interferences occurring in the double-focusing mass spectrometer [64, 65] which led to the rate constants being overestimated by a factor of 2 to 3 [66]. The fullerenes from the polyimide ablation source [63] are, however, still "colder" than those from the graphite vaporization source [62, 66] i.e. they still have a somewhat smaller rate constant for fragmentation. This can be explained by considering that polyimide ablation also produces a large amount of small neutral molecules which, in the supersonic expansion from the polyimide surface [67], will act as a carrier gas thus producing some cooling of the carbon clusters. This leads to the interesting observation, shown in Fig. 12, that if the metastable fragmentation of carbon clusters with a particular average velocity is investigated as a function of the fluence of the ablating laser, the rate constant is seen to decrease as the laser fluence is increased. This is due to the increased gas pressure and thus more efficient cooling during the expansion from the polymer surface [63].

Estimates of the binding energy of C_2 to C_{58}^+, C_{60}^+ and C_{62}^+ have been made by measuring the kinetic energy release distributions and using statistical phase space theory to model the experimental data [68, 69]. The results indicate that C_2 is bound to C_{58}^+ and C_{60}^+ by about 4.6 eV, to C_{62}^+ by only about 3 eV. A model

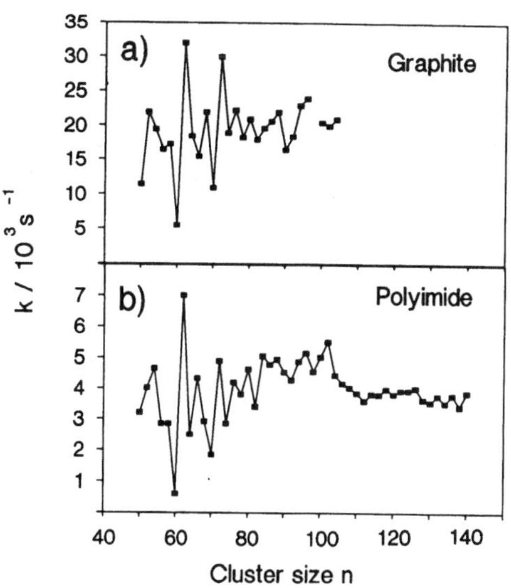

Fig. 11. Comparison of rate constants for C_2 metastable fragmentation. **a** Values obtained by *Radi* et al. from graphite [62]. **b** Values obtained from laser ablation of polyimide [63]

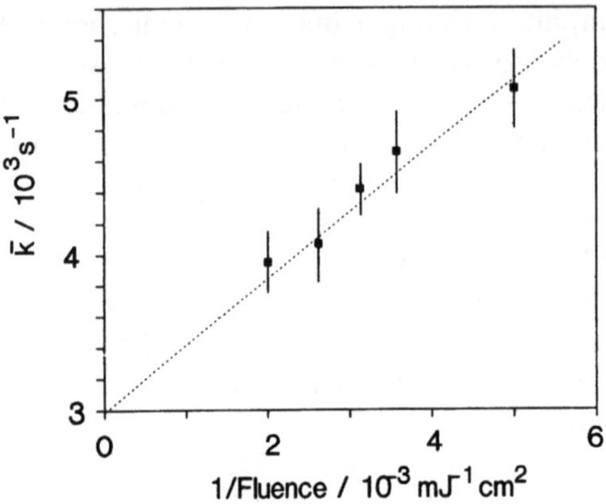

Fig. 12. Measured average rate constant for C_2 loss from $70 < n < 100$, produced from laser ablation of graphite, as a function of ablating laser fluence for ions of the same velocity [63]

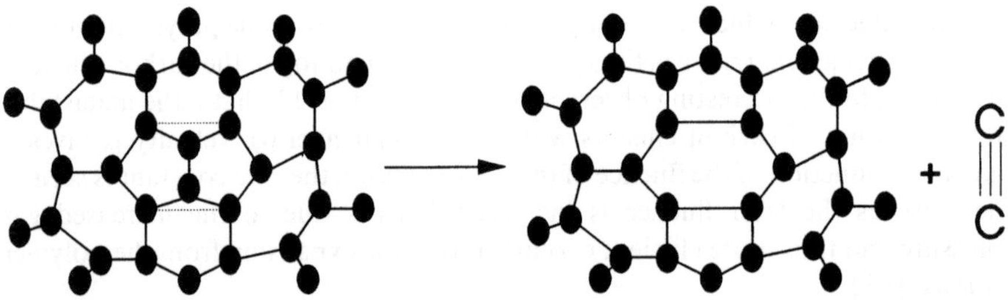

Fig. 13. Proposed mechanism for C_2 loss from fullerenes conserving the number of pentagons. Reproduced from [70]

for the C_2 loss mechanism has been put forward by *O'Brien* et al. [70]. Figure 13 shows how a C_n^+ shell with a fused five-membered ring system could lose C_2 and rearrange to give a C_{n-2}^+ shell. The fused pentagons become a hexagon and the two hexagons above and below the fused pentagons become pentagons thus keeping the number of pentagons constant, as required for a closed shell structure, with the net loss of a hexagon. Such a mechanism requires that rapid surface geometry reorganisation can take place in order to bring two pentagons together in the larger clusters. A model for this pentagon rearrangement has been suggested [71] which may well be relevant for the "hot" clusters investigated in these experiments.

Photofragmentation experiments [70, 72] produce even numbered fragments from the fullerenes down to the 30 atom range. Figure 14 shows the fragments produced on irradiating a mass selected C_{60}^+ with 15 mJcm^{-2} of 193 nm excimer laser light. As the laser fluence is increased the smaller fragments grow in sequentially. This C_2 loss continues until C_{32}^+ is reached. It is not yet certain whether fragmentation occurs by successive C_2 losses or by single step elimination of even units. When C_{32}^+ is reached it does not fragment by

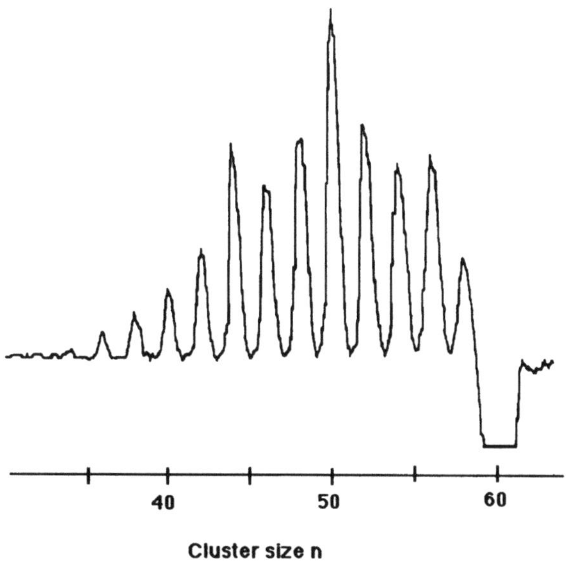

Fig. 14. C_{60}^+ mass selected and fragmented with $15\,mJ\,cm^{-2}$ ArF excimer light (193 nm). Data without laser subtracted from data with laser resulting in depletion of parent ions. Reproduced from [70]

losing a C_2 but breaks up into a number of smaller fragments all less than C_{20} [70]. Experiments in a Fourier transform ion cyclotron resonance (FT-ICR) apparatus, where the ions could be stored and exposed to many laser pulses over an extended time period, showed that when the fluence of the laser was very low ($< 0.5\,mJcm^{-2}$) no fragmentation of C_{60}^+ was observed even although the clusters were exposed to the laser for 30 seconds at a repetition rate of 50 Hz [1]. A rough estimate shows that the clusters must have absorbed about 300 laser photons in this time [1], each with an energy of 6.4 eV. The lack of fragmentation has been explained by suggesting that C_{60}^+ undergoes efficient radiative cooling [1] in contrast to the situation with high fluence when the cluster can absorb enough photons on a nanosecond timescale for fragmentation to occur, as described above.

The whole question of energy storage and conversion within C_{60} and the other fullerenes is a very topical one. Thermionic emission has been observed to occur from neutral fullerenes [73], giant ($n = 150$–600) positively charged fullerenes [72] and negatively charged C_{60}^- [74] on a timescale of microseconds. Figure 15 displays the time dependence of the ion signal obtained on irradiating neutral C_{60} and C_{70} with 308 nm (4 eV) photons. The different time constants observed were attributed to absorption of different numbers of photons [73]. These results, although at first sight surprising, can be at least qualitatively explained by using a simple model for thermionic emission [75]. An aspect of the measurements which is not yet understood but is the subject of current research, is the complex interplay and competition which must exist between electronic excitation, thermionic emission, electron-phonon coupling and metastable fragmentation of the fullerenes.

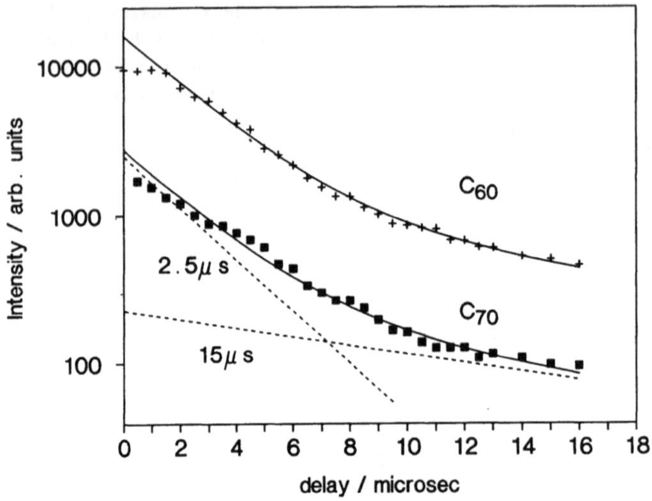

Fig. 15. Ion intensity as a function of delay between ionizing laser and ion production. The full lines are fits to two exponential decays with different lifetimes. The dashed lines show these two exponential decays for the C_{70} ionization. Taken from [73]

4.4.3.4 Metal-Containing Fullerenes

If you think you have a hollow cage-like cluster one obvious thing to try to do is to put something inside it. In 1985 laser vaporization of $LaCl_3$ impregnated graphite produced a prominant mass peak at a mass corresponding to $C_{60}La^+$ which had high photophysical stability [76]. Later, the *Smalley* group performed extensive measurements in the magnetic trap of an FT-ICR mass spectrometer to obtain convincing evidence that metal atoms were caught inside the cage [77]. The only clusters in the 300–1500 u mass range which survived the injection and thermalization cycles used to fill the ICR cell were the even numbered bare carbon clusters and the same clusters with a single metal atom attached. The stored clusters were then photofragmented with an ArF excimer laser (6.4 eV). Figure 16 shows the photofragmentation pattern of $C_{60}K^+$. The primary photoprocess was observed to be loss of C_2, as for the bare carbon clusters, and not loss of the metal (the C_n^+ seen in the spectrum are fragments from C_{64}^+ injected into the trap with $C_{60}K^+$ to provide a calibration). The cage can, of course, only hold the metal if there is enough room inside. From Fig. 16 it would seem that the smallest cluster which can contain potassium is $n = 44$. Similarly the smallest cluster to contain Cs was seen to be $n = 48$.

The method of producing these metal-fullerene complexes has recently been improved by carrying out the laser vaporization in an oven at 1200 °C [78]. The products can be deposited on a Cu substrate, removed from the vacuum chamber and transported in air without being destroyed. LaC_{82} was found to be especially stable when exposed to air which may be related to a closed shell

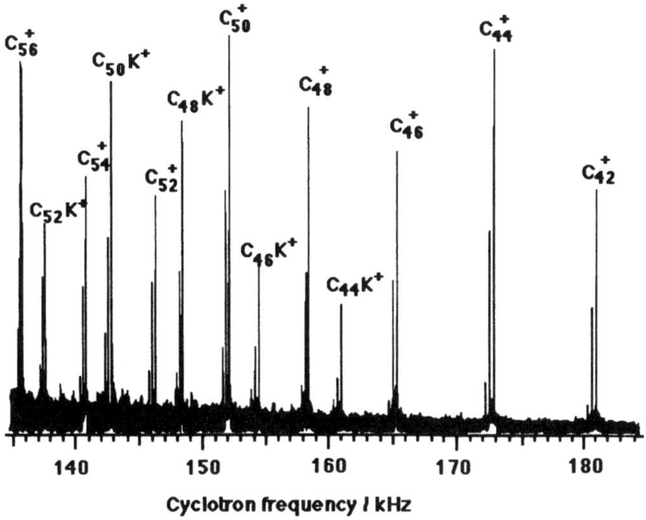

Fig. 16. Photofragmentation pattern of $C_{60}K^+$ detected by FT-ICR mass spectrometry. The bare C_n^+ clusters are fragments from C_{64}^+ injected into the ICR trap along with $C_{60}K^+$ providing an internal calibration. Reproduced from [77]

electronic structure [78]. There is even evidence to show that it is possible to get up to three La atoms inside clusters with $n \geq 88$. On the other hand, K_3C_{60} only has one potassium inside with two on the outside (it has been suggested that such compounds be written as $K_2(K@C_{60})$ [78]). Fullerene compounds involving the replacement of a carbon atom in the cage with boron such as $C_{59}B$ and $K@C_{59}B$ can also be readily formed.

4.4.4 The "Buckyball" Era

In September 1990 *Krätschmer* et al. published a paper in "Nature" [15] which led to a veritable explosion in C_{60} research activity with many groups from many different fields rushing to play with the carbon footballs. The paper reported a surprisingly simple method of producing and purifying C_{60}. The soot formed when graphite electrodes were evaporated in a He atmosphere was dissolved in benzene (or toluene). When this solution crystallised out, the crystals (given the name fullerite) were found to consist of a C_{60}/C_{70} mixture (in a ratio of about 3:1) with less than 2% of larger fullerenes present [79]. In the year that has elapsed since the publication of this paper a number of improvements in the formation process have been reported with a production of more than 10 g per day of C_{60} now possible [59, 80], the C_{60}/C_{70} mixture making up about 10–15% of the "raw soot". Chromatographic techniques can then be used to separate the fullerenes [81]. By using different solvents it is also possible to extract significant amounts of the large fullerenes up to about C_{300} from the soot [82]. Pure C_{60} and C_{70} are now available commercially using coal instead of graphite in the production [83], which produces a similar percentage yield but is more economically viable. The most likely means of production for eventually producing industrially relevant quantities of C_{60} would seem at the present time to be benzene flames which have been shown to produce 3 g of C_{60}/C_{70} for every kg of benzene burnt [84].

The availability of relatively large amounts of pure C_{60} made a detailed structural determination possible. The first method applied was that of ^{13}C NMR [81, 85, 86] which showed a single peak for C_{60} at room temperature confirming that all the carbon atoms were identical as expected for the truncated icosahedron structure. The results of the measurement on chromatographically separated C_{60} are shown in Fig. 17(a). Figure 17(b) displays the equivalent measurement for pure C_{70} giving 5 lines with intensity ratios expected from the proposed structure shown in Fig. 8. Structural studies of crystalline C_{60} indicate that at room temperature the C_{60} clusters are orientationally disordered and the crystal structure is a face-centered cubic (fcc) configuration of the C_{60} spheres [87], see Fig. 18(a). Below 249 K the molecules become orientationally ordered and a simple cubic lattice results (Fig. 18(b)) [88]. Atomic force microscopy confirmed the room temperature fcc crystal structure [89].

All of the above measurements were highly supportive of the truncated icosahedron structure but did not prove it nor provide information on the exact positions of the carbon atoms within C_{60}. This was first done when an osmyl group was attached to C_{60} (forming $C_{60}(OsO_4)$(4-tert-butylpyridine)) in order to break its pseudospherical symmetry and fix the position of the carbon framework relative to the osmyl unit. By using X-ray crystallography [90] it was

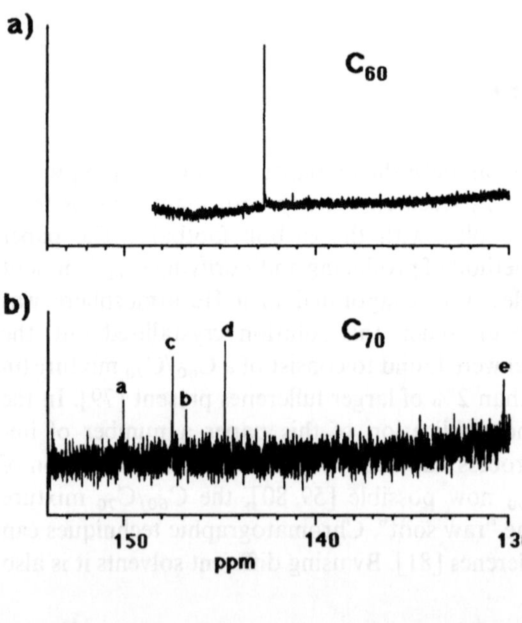

Fig. 17. ^{13}C NMR spectra (a) C_{60} showing 1 line proving that all 60 atoms have the same chemical environment (b) C_{70} showing the 5 lines expected from the structure shown in Fig. 8. Reproduced from [81]

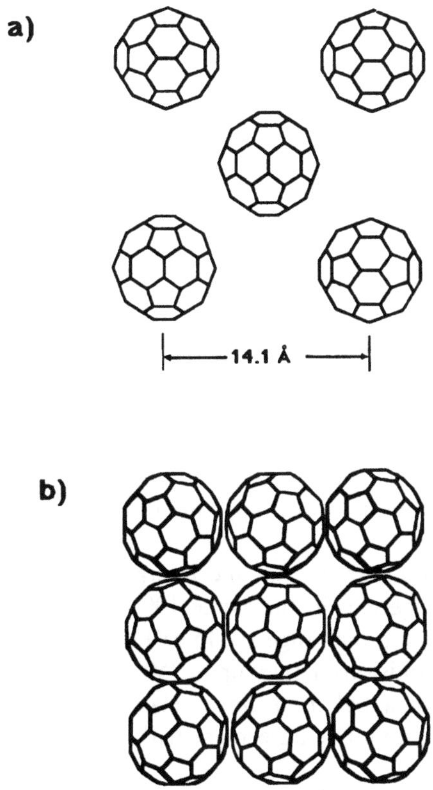

Fig. 18. **a** Room temperature fcc crystal structure of C_{60}. **b** Orientationally ordered simple cubic crystal structure formed at temperatures less than 249 K

then finally possible to confirm the "soccer ball" arrangement of the carbon atoms and give values for the two different C–C bond lengths; 1.432(5) Å for bonds between 5- and 6-membered rings and 1.388(9) Å for bonds between two 6-membered rings, in agreement with theoretical predictions [4].

Cluster scientists have thus been presented with the means of producing intense, mass-selected beams of neutral (or charged) clusters with well defined structures. The work is only just beginning but already a wealth of information both theoretical and experimental has been collected on these fascinating species. This is neither the proper place nor time to give a review of the state of "buckyball" research, however, some of the experimental investigations that have been carried out on the purified gas phase fullerenes in recent months will be briefly mentioned to give a flavour of the intense activity in this field.

a) The ionization potential of C_{60} has been determined to be 7.58 eV using single photon ionization with synchrotron radiation [91] in close agreement with values obtained from PES of solid C_{60} [92].
b) Threshold photodetachment of cold C_{60}^- has determined the electron affinity to be 2.65 eV [74]. In the course of these measurements evidence was obtained for the occurrence of electron-"phonon" scattering in "hot"

C_{60}^- which, along with thermionic emission [73] (see Section 4.4.3.3 above), is very much a "bulk" phenomenon.

c) A giant plasmon resonance above the ionization threshold [93] has been observed, in close agreement with a theoretical prediction [94] from linear response theory, indicating that the collective oscillation of the 240 electrons plays a significant role in the electronic structure of the fullerenes. There is still a great deal of work to do before the electronic structure of the fullerenes can be fully understood, however, it would seem that an understanding of these species could provide important insights into the transition between the molecular and solid state descriptions of matter.

d) Preliminary investigations of the electronic spectra of supersonically cooled C_{60} and C_{70} using resonant two photon ionization spectroscopy [95] in the near UV and visible region have produced spectra with many sharp, clear features. Although the spectra can not yet be understood, they will certainly help to gain a detailed understanding of the electronically excited states, including vibronic couplings, when combined with appropriate theoretical treatments.

e) Collisions of C_{60}^+ with a graphite surface show that the interaction is extremely inelastic (with initial kinetic energies in the range 100–500 eV the final kinetic energy of the C_{60}^+ is always in the range 10–15 eV) but that the cluster rebounds essentially intact for collision energies up to 250 eV [96–98]. A simple, intuitive, classical model of a ball bouncing inelastically from a surface can give surprisingly good qualitative agreement with the experimental results [97].

f) It is possible for a He atom to be captured and trapped inside the C_{60}^+ cage in collisions of a few keV energy [99, 100]. Measurements of the energy threshold for capture [101] are in reasonably good agreement with estimates of the barrier for a He atom to pass through a C_6 ring [102].

A great deal has been speculated about the possible technological applications of C_{60}. Of particular interest is the fact that C_{60} becomes superconducting when doped with alkalis [103] (and perhaps also I_2). The alkali atoms sit between the C_{60} clusters in the fcc crystal lattice (see Fig. 18(a)). It has the highest transition temperature T_c of any organic superconductor and at the moment it is impossible to predict where the maximum possible T_c will lie (at the time of writing the highest published T_c was 33 K for C_{60} doped with a Cs/Rb mixture [104] with reliable reports of 43 K for Rb/Tl mixtures and rumours of 70 K with I_2). Another property with possible applications is the extremely large infrared nonlinear optical response of C_{60} [105]. "Nanometer ballbearings" has been suggested [106] as a possible application for $C_{60}F_{60}$ which has recently been produced in the laboratory [101] and should be very inert and a superb lubricant. There has also been considerable speculation that the discovery of the fullerenes could play as important a role in organic chemistry as the discovery of the ring structure of benzene [106], opening up the possibility of three-dimensional aromatic chemistry.

Whether C_{60} will prove to be the great technological wonder of present speculations or not is something that only time will tell. In any case it is certain to remain a fascinating object of research for chemists and physicists for many years to come.

4.4.5 Recent Developments

The developments in the field of fullerene research since this chapter was written are far to numerous to mention briefly here. There is now even a journal devoted entirely to the fullerenes "Fullerene Science and Technology" (Marcel Dekker, New York). The interested reader is referred to the books and review articles listed in ref. 108–113.

Acknowledgements. The author would like to thank Ingolf Hertel and Harry Kroto for their careful and critical reading of the manuscript and their suggestions for improvements.

References

1. R.E. Smalley: in *Atomic and Molecular clusters*, ed. by E.R. Bernstein (Elsevier Science, 1990) 1
2. R.F. Curl, R.E. Smalley: Scientific American, October (1991) 54
3. H.W. Kroto: Pure Appl. Chem. **62**, 407 (1990)
4. W. Weltner, R.J. van Zee: Chem. Rev. **89**, 1713 (1989)
5. H.W. Kroto: Science **242**, 1139 (1988)
6. R.F. Curl, R.E. Smalley: Science **242**, 1017 (1988)
7. R.E. Smalley: The Sciences **31**, 22 (1991)
8. e.g. New York Times Dec. 25, 13 (1990); Wall St. Journal Apr. 2, 29 (1991); Newsweek Apr. 29, 68 (1991); Time May 6, 66 (1991)
9. R. Kippenhahn, A. Weigert: *Stellar Structure and Evolution* (Springer Verlag, 1991)
10. E.A. Rohlfing, D.M. Cox, A. Kaldor: J. Chem. Phys. **81**, 3322 (1984)
11. D.J. Trevor, D.M. Cox, K.C. Reichmann, R.O. Brickman, A. Kaldor: J. Phys. Chem. **91**, 2598 (1987)
12. E.A. Rohlfing: J. Chem. Phys. **93**, 7851 (1990)
13. H.W. Kroto: private communication
14. H.W. Kroto, J.R. Heath, S.C. O'Brien, R.F. Curl, R.E. Smalley: Nature **318**, 162 (1985)
15. W. Krätschmer, L.D. Lamb, K. Fostiropoulos, D.R. Huffman: Nature **347**, 354 (1990)
16. O. Hahn, F. Strassman, J. Mattauch, H. Ewald: Naturwiss. **30**, 541 (1942)
17. J. Mattauch, H. Ewald, O. Hahn, F. Strassman: Z. Phys. **20**, 598 (1943)
18. E. Dornenburg, H. Hintenberger: Z. Naturforsch. **14A**, 765 (1959)
19. E. Dornenburg, H. Hintenberger: Z. Naturforsch. **16A**, 532 (1961)
20. N. Fürstenau, F. Hillenkamp: Int. J. Mass Spec. Ion Phys. **37**, 135 (1981)
21. S.B.H. Bach, J.R. Eyler: J. Chem. Phys. **92**, 358 (1990)
22. M.E. Geusic, M.F. Jarrold, T.J. McIlrath, R.R. Freeman, W.L. Brown: J. Chem. Phys. **86**, 3862 (1987)
23. K.S. Pitzer, E. Clementi: J. Am. Chem. Soc. **81**, 4477 (1959)

24. L. Gausset, G. Herzberg, A. Lagerqvist, B. Rosen: Astrophys. J. **142**, 45 (1965)
25. I. Plesser, Z. Vagar, R. Naaman: Phys. Rev. Lett. **56**, 1559 (1986)
26. W.P. Kraemer, P.R. Bunker, M. Yoshimine: J. Mol. Spec. **107**, 191 (1984)
27. J. Bernholc, J.C. Phillips: J. Chem. Phys. **85**, 3258 (1986)
28. K. Raghavachari, J.S. Binkley: J. Chem. Phys. **87**, 2191 (1987)
29. A.K. Ray: J. Phys. B **20**, 5233 (1987)
30. P. Joyes, M. Leleyter: J. Physique **45**, 1681 (1984)
31. W. Andreoni, D. Scharf, P. Giannozzi: Chem. Phys. Lett. **173**, 449 (1990)
32. S. Yang, K.J. Taylor, M.J. Craycraft, J. Conceicao, C.L. Pettiette, O. Cheshnovsky, R.E. Smalley: Chem. Phys. Lett. **144**, 431 (1988)
33. W. Huggins: Proc. Roy. Soc. (London) **33**, 1 (1882)
34. A.E. Douglas: Astrophys. J. **114**, 466 (1951)
35. K.W. Hinkle, J.J. Keady, P.F. Bernath: Science **242**, 1319 (1988)
36. K. Matsumura, H. Kanamori, K. Kawaguchi, E. Hirota: J. Chem. Phys. **89**, 3491 (1988)
37. J.R. Heath, A.L. Cooksy, M.H.W. Gruebele, C.A. Schmuttenmaer, R.J. Saykally: Science **244**, 564 (1989)
38. P.F. Bernath, K.H. Hinkle, J.J. Keady: Science **244**, 562 (1989)
39. Z. Vagar, R. Naaman, E.P. Kanter: Science **244**, 426 (1989)
40. M. Algranati, H. Feldman, D. Kella, E. Malkin, E. Miklazky, R. Naaman, Z. Vagar, J. Zajfman: J. Chem. Phys. **90**, 4617 (1989)
41. S.W. McElvany, B.I. Dunlap, A. O'Keefe: J. Chem. Phys. **86**, 715 (1986)
42. P.A. Hintz, M.B. Sowa, S.L. Anderson: Chem. Phys. Lett. **177**, 146 (1991)
43. S.W. McElvany: J. Chem. Phys. **89**, 2063 (1988)
44. D.C. Parent, S.W. McElvany: J. Am. Chem. Soc. **111**, 2393 (1989)
45. H.W. Kroto, J.R. Heath, S.C. O'Brien, R.F. Curl, R.E. Smalley: Astrophys. J. **314**, 352 (1987); J.R. Heath, Q. Zhang, S.C. O'Brien, R.F. Curl, H.W. Kroto, R.E. Smalley: J. Am. Chem. Soc. **109**, 359 (1987)
46. L.A. Bloomfield, M.E. Geusic, R.R. Freeman, W.L. Brown: Chem. Phys. Lett. **121**, 33 (1985)
47. T.G. Schmalz, W.A. Seitz, D.J. Klein, G.E. Hite: J. Am. Chem. Soc. **110**, 1113 (1988)
48. D.E.H. Jones: New Scientist **32**, 245 (1966)
49. E. Osawa: Kagaku (Kyoto) **25**, 854 (1970)
50. D.A. Bochvar, E.G. Galipern: Dokl. Akad. Nauk. SSSR **209**, 610 (1973)
51. R.A. Davidson: Theor. Chim. Acta **58**, 193 (1981)
52. H.W. Kroto: Nature **329**, 529 (1987)
53. S. Yang, C.L. Pettiette, J. Conceicao, O. Cheshnovsky, R.E. Smalley: Chem. Phys. Lett. **139**, 233 (1987)
54. S. Larsson, A. Volosov, A. Rosen: Chem. Phys. Lett. **137**, 501 (1987)
55. E.E.B. Campbell, G. Ulmer, B. Hasselberger, H.-G. Busmann, I.V. Hertel: J. Chem. Phys. **93**, 6900 (1990)
56. W.R. Creasy, J.T. Brenna: J. Chem. Phys. **92**, 2269 (1990)
57. Q.L. Zhang, S.C. O'Brien, J.R. Heath, Y. Liu, R.F. Curl, H.W. Kroto, R.E. Smalley: J. Phys. Chem. **90**, 525 (1986)
58. W.R. Creasy: J. Chem. Phys. **92**, 7223 (1990)
59. R.E. Haufler, Y. Chai, L.P.F. Chibante, J. Conceicao, C. Jin, L.-S. Wang, S. Maruyama, R.E. Smalley: Mat. Res. Soc. Proc. **206**, 627 (1991)
60. H.W. Kroto, K.G. McKay: Nature **331**, 328 (1988)
61. P.H. Gerhardt, S. Löffler, K.H. Homann: Chem. Phys. Lett. **137**, 306 (1987)
62. P.P. Radi, M.T. Hsu, J. Brodbelt-Lustig, M. Rincon, M.T. Bowers: J. Chem. Phys. **92**, 4817 (1990)
63. E.E.B. Campbell, G. Ulmer, H.-G. Busmann, I.V. Hertel: Chem. Phys. Lett. **175**, 505 (1990)
64. D. Schröder, D. Sülzle: J. Chem. Phys. **94**, 6933 (1991)
65. T. Drewello, K.-D. Asmus, J. Stach, R. Herzschuh, M. Kao, C.S. Foote: J. Phys. Chem. **95**, 10554 (1991)
66. M.T. Bowers, P.P. Radi, M.T. Hsu: J. Chem. Phys. **94**, 6934 (1991)

67. G. Ulmer, B. Hasselberger, H.-G. Busmann, E.E.B. Campbell: Appl. Surf. Sci. **46**, 272 (1990)
68. P.P. Radi, M.T. Hsu, M.E. Rincon, P.R. Kemper, M.T. Bowers: Chem. Phys. Lett. **174**, 223 (1990)
69. C. Lifshitz, M. Iraqi, T. Peres, J.E. Fischer: Int. J. Mass Spec. Ion Phys. **107**, 565 (1991)
70. S.C. O'Brien, J.R. Heath, R.F. Curl, R.E. Smalley: J. Chem. Phys. **88**, 220 (1988)
71. A.J. Stone, D.J. Wales: Chem. Phys. Lett. **128**, 501 (1986)
72. S. Maruyama, M.Y. Lee, R.E. Haufler, Y. Chai, R.E. Smalley: Z. Phys. D **19**, 409 (1991)
73. E.E.B. Campbell, G. Ulmer, I.V. Hertel: Phys. Rev. Lett. **67**, 1986 (1991)
74. L.-S. Wang, J. Conceicao, C. Jin, R.E. Smalley: Chem. Phys. Lett. **182**, 5 (1991)
75. C.E. Klots: Chem. Phys. Lett. **186**, 73 (1991)
76. J.R. Heath, S.C. O'Brien, Q. Zhang, Y. Liu, R.F. Curl, H.W. Kroto, F.K. Tittel, R.E. Smalley: J. Am. Chem. Soc. **107**, 7779 (1985)
77. F.D. Weiss, J.L. Elkind, S.C. O'Brien, R.F. Curl, R.E. Smalley: J. Am. Chem. Soc. **110**, 4464 (1988)
78. Y. Chai, T. Guo, C. Jin, R.E. Haufler, L.P. Chibante, J. Fure, L. Wang, J.M. Alford, R.E. Smalley: J. Phys. Chem. **95**, 7654 (1991)
79. G. Ulmer, E.E.B. Campbell, R. Kühnle, H.-G. Busmann, I.V. Hertel: Chem. Phys. Lett. **182**, 114 (1991)
80. A.S. Koch, K.C. Khemani, F. Wudl: J. Org. Chem. **56**, 4543 (1991)
81. R. Taylor, J.P. Hare, A.K. Abdul-Sada, H.W. Kroto: J. Chem. Soc. Chem. Comm. 1423 (1990)
82. D.H. Parker, P. Wurz, K. Chatterjee, K.R. Lykke, J.E. Hunt, M.J. Pellin, J.C. Hemminger, D.M. Gruen, L.M. Stock: J. Am. Chem. Soc., **113**, 7499 (1991)
83. L.S.K. Pang, A.M. Vassallo, M.A. Wilson: Nature **352**, 480 (1991)
84. J.B. Howard, J.T. McKinnon, Y. Makarovsky, A. Lafleur, M.E. Johnson: Nature **352**, 139 (1991)
85. R.D. Johnson, G. Meijer, D.S. Bethune: J. Am. Chem. Soc. **112**, 8983 (1990)
86. H. Ajie, M.M. Alvarez, S.J. Anz, R.D. Beck, F. Diederich, K. Fostiropoulos, D.R. Huffman, W. Krätschmer, Y. Rubin, K.E. Schriver, D. Sensharma, R.L. Whetten: J. Phys. Chem. **94**, 8630 (1990)
87. J.E. Fischer, P.A. Heiney, A.R. McGhie, W.J. Romanow, A.M. Denenstein, J.P. McCauley Jr., A.B. Smith III: Science **252**, 1288 (1991)
88. W.I.F. David, R.M. Ibberson, J.C. Matthewman, K. Prassides, T.J.S. Dennis, J.P. Hare, H.W. Kroto, R. Taylor, D.R.M. Walton: Nature **353**, 147 (1991)
89. E.J. Snyder, M.S. Anderson, W.M. Tong, R.S. Williams, S.J. Anz, M.M. Alvarez, Y. Rubin, F.N. Diederich, R.L. Whetten: Science **253**, 171 (1991)
90. J.M. Hawkins, A. Meyer, T.A. Lewis, S. Loren, F.J. Hollander: Science **252**, 312 (1991)
91. J. de Vries, H. Steger, B. Kamke, C. Menzel, B. Weisser, W. Kamke, I.V. Hertel: Chem. Phys. Lett., **188**, 159 (1992)
92. D.L. Lichtenberger, K.W. Nebesny, C.D. Ray, D.R. Huffman, L.D. Lamb: Chem. Phys. Lett. **176**, 203 (1991)
93. I.V. Hertel, H. Steger, J. de Vries, B. Weisser, C. Menzel, B. Kamke, W. Kamke: Phys. Rev. Lett. **68**, 784 (1992)
94. G.F. Bertsch, A. Bulgac, D. Tomanek, Y. Wang: Phys. Rev. Lett. **67**, 2690 (1991)
95. R.E. Haufler, Y. Chai, L.P.F. Chibante, H.R. Fraelich, R.B. Weisman, R.F. Curl, R.E. Smalley: J. Chem. Phys. **95**, 2197 (1991)
96. R.D. Beck, P. St. John, M.M. Alvarez, F. Diederich, R.L. Whetten: J. Phys. Chem. **95**, 8402 (1991)
97. H.-G. Busmann, T. Lill, I.V. Hertel: Chem. Phys. Lett. **187**, 459 (1991)
98. R.C. Mowrey, D.W. Brenner, B.I. Dunlap, J.W. Mintmire, C.T. White: J. Phys. Chem. **95**, 7138 (1991)
99. T. Weiske, D.K. Bohme, J. Hrusak, W. Krätschmer, H. Schwarz: Angew. Chem. Int. ed. Engl. **30**, 884 (1991)
100. M.M. Ross, J.H. Callahan: J. Phys. Chem. **95**, 5720 (1991)
101. E.E.B. Campbell, A. Hielscher, R. Ehlich, J.M.A. Frazao, I.V. Hertel: Z. Phys. D. **23**, 1 (1992)

102. J. Hrusak, D.K. Bohme, T. Weiske, H. Schwarz: Chem. Phys. Lett. **193**, 97 (1992)
103. A.F. Hebard, M.J. Rosseinsky, R.C. Haddon, D.W. Murphy, S.H. Glarum, T.T.M. Palstra, A.P. Ramirez, A.R. Kortan: Nature **350**, 600 (1991)
104. K. Tanigaki, T.W. Ebbesen, S. Saito, J. Mizuki, J.S. Tsai, Y. Kubo, S. Kuroshima: Nature **352**, 222 (1991)
105. W.J. Blau, H.J. Byrne, D.J. Cardin, T.J. Dennis, J.P. Hare, H.W. Kroto, R. Taylor, D.R.M. Walton: Phys. Rev. Lett. **67**, 1423 (1991)
106. New Scientist **1776** (1991) 25
107. J.H. Holloway, E.G. Hope, R. Taylor, G.J. Langley, A.G. Avent, T.J. Dennis, J.P. Hare, H.W. Kroto, D.R.M. Walton: J. Chem. Soc. Chem. Comm. **14**, 966 (1991)
108. MRS Boston meeting on Buckyballs, Symp. Proc. **206** (1991)
109. "Fullerenes: Synthesis, Properties and Chemistry of Large Carbon Clusters", eds. G.S. Hammond, V.J. Kuck, ACS Symp. Series **481** (ACS, Washington D.C., 1992)
110. "Electrical, Optical and Magnetic Properties of Organic Solid State Materials, Part II", eds. L.Y. Chiang, A.F. Garito, D.J. Sandman, MRS Symposium Series **247** (1992)
111. "The Fullerenes", eds. H.W. Kroto, J.E. Fischer, D.E. Cox (Pergamon Press, 1993)
112. "The Chemistry of Fullerenes", R. Taylor, D.R.M. Walton, Nature, **363**, 685 (1993)
113. "The Physics and Chemistry of Fullerenes", ed. K. Prassides, NATO ARW Proc. (Plenum Press, 1993)

4.5 Oxides and Halides of Alkali Metals

T.P. Martin

4.5.1 Introduction

The alkali halides have long served as model materials in solid state physics because of the simplicity of the bonding and of sample preparation. The same properties make these materials ideally suited to represent a class of clusters. Many properties of alkali halide crystals, e.g., stucture, vibrational frequencies, and even ionization energies, can be described with a classical electrostatic model. The same model parameters used to describe a crystal can also be used to describe alkali halide molecules. In spite of this, the properties of an NaCl molecule and the corresponding crystal are quite different. The molecule is a much more tightly bonded entity. The interatomic distance in the molecule is much smaller than in the crystal. The vibrational frequencies are much higher. The study of alkali halide clusters will allow us to observe the transition from molecule to crystal [1].

The elements cesium and oxygen combine to form a bewildering array of crystalline compounds (Cs_7O, Cs_4O, Cs_3O_{11}, Cs_2O_2, Cs_2O_3, CsO_2, CsO_3). The structures of the suboxides Cs_7O, Cs_4O, and Cs_3O_{11} are particularly interesting and unusual. Crystallographic investigations indicate that these compounds are composed of $Cs_{11}O_3$ clusters having strongly ionic character [2]. However, the bonding between the clusters is metallic. The most immediate support of this bonding model is found by comparing the interatomic distances in the suboxides with the corresponding distances in the normal oxides and pure metals. The metal-oxygen distance in the suboxide clusters is nearly the sum of the ionic radii, while the metal-metal distances between clusters are nearly those found in a pure alkali metal. Cs_2O is a semiconductor with a band gap of about 2.0 eV [3]. Cs_2O_2 appears to be an insulator containing O_2^{2-} anions [4]. The superoxides form because oxygen possesses the capability of appearing also as O_2^-. Clusters composed of varying amounts of cesium and oxygen also represent three distinct classes of materials. Merely by changing the oxygen pressure in the cluster condensation chamber we can study metal clusters, semiconductor clusters, semiconductor clusters doped with donors, or insulator clusters [5].

4.5.2 Interatomic Forces

A realistic total energy of covalently-bonded clusters can be obtained only with the most sophisticated quantum mechanical many-body methods. The total energy of clusters with purely ionic bonding can be calculated much more simply. It appears to be possible to define a classical two-body interaction potential which is independent of cluster size and symmetry. In its simplest form the interaction energy among the ions in a cluster can be expressed with only two terms: the electrostatic Coulomb energy and a repulsive Born–Mayer energy:

$$V_{ij} = \frac{q_i q_j}{r_{ij}} + A_{ij}\exp(-r_{ij}/\rho) \ . \tag{1}$$

The ions carry a charge q_i and are separated by a distance r_{ij}. The total energy is evaluated as the sum of all interaction pairs in a cluster. Starting from an initial configuration the energy is lowered by variation of the position coordinates until a local minimum is achieved. This procedure is carried out for several initial configurations. The end configuration having the lowest energy is assumed to be the most stable one. For alkali halide molecules a second model has been introduced by *Rittner* [6]. This model allows each ion to assume a dipole moment due to the fact that the electron shells are polarized by the electric field of the other ions in the cluster. The following terms have to be added to the rigid-ion potential: a monopole-dipole interaction,

$$V^{MD} = -\frac{q_i}{r_{ij}^3}(\mu_j r_{ij}) - \frac{q_j}{r_{ij}^3}(\mu_i r_{ij}) \ , \tag{2}$$

where U_j is the dipole moment on atom j, a dipole-dipole interaction,

$$V^{DD} = \frac{3}{r_{ij}^5}(\mu_j r_{ij})(\mu_i r_{ij}) + \frac{\mu_i \mu_j}{r_{ij}^3} \ , \tag{3}$$

and the self-energy of a dipole

$$V^D = \frac{\mu^2}{2\alpha} \ , \tag{4}$$

where α is the polarizability of an ion. The Rittner model can be improved by remembering that the other electrons do not move rigidly with the core electrons. Due to the deformation of the electron shells of the ions, the distance r_{ij} in the Born–Mayer repulsion term has to be replaced by an effective distance:

$$r_{ij}^{\mathrm{eff}} = r_{ij} + \frac{\mu_i}{Q_i} - \frac{\mu_j}{Q_j} \ . \tag{5}$$

In addition, the deformation tends to increase the dipole moment on positive ions and decrease that on negative ions. However, this effect can be accounted for by introducing an effective polarizability. This potential has been applied to alkali halide monomers and clusters by *Welch* et al. [7, 8]. They have determined the ion specific parameters α and Q from data on all of the monomers containing this ion. By introducing the dipole moments, the energy of all configurations can be lowered. This is not a trivial result because the self-energy of the dipoles gives a positive contribution to the total energy of the cluster. At the same time the symmetry of the structures is lowered. For example, a square, formed by two cations and two anions is deformed into a rhombus because the polarizabilities of the larger ions are much greater than those of the smaller ones. In order to demonstrate the effect of polarizability, calculations have been made for four materials containing cations and anions with all combinations of small and large polarizabilities.

4.5.3 Neutral Clusters at Zero Temperature

The most stable configurations of $(MX)_n$ clusters are shown in Fig. 1. The energy minima were located in the following way. Each ion in the cluster is assigned an initial position. The resulting cluster configuration will not be a reasonable one because the initial $3N$ coordinates are chosen using a random number generator. From this starting point a minimization routine is initiated. It is of great importance that this routine be not only fast, but also precise. Speed is essential in order to allow minimization from a large number of starting configurations. Only in this way can the multidimensional surface be adequately searched. Precision is essential to avoid the danger of falsely identifying a saddle point as a true minimum. Therefore, the routine first performed a course minimization using the steepest descents method which was followed by a refinement using a so-called Davisdon–Fletcher–Powell method [9]. Starting from a variety of initial configurations, the energy of a given minimum was reproducible to six digits and the vibrational frequencies to four digits. The calculations have been carried out for NaCl, NaI, CsF and CsI [10]. The form of the most stable configuration is usually model and material dependent. The differences between the two models go beyond distortions of a cubic structure caused by the polarizability. In some cases the two models produce totally different structures. This happens at cluster sizes: $n = 3, 4, 5, 11, 13, 16, 18$. Most of the configurations are built out of hexagonal or cubic elements. The most stable structures of CsF clusters turn out to be rock-salt-like configurations. For $n = 3, 6, 9, 12, 15, 18$ a column of hexagonal rings or cube-like structures are possible. CsF and CsI. both prefer the cube-like structure. The calculation gives no hint that they will crystallize into two different crystal structures. The NaCl clusters $n = 6, 9, 12, 15$ prefer the hexagonal structures in spite of the fact that NaCl crystallizes info an fcc lattice. For $n = 18$, however, a fcc-like form is preferred. NaI also prefers hexagonal elements for $n = 3, 6, 8, 9, 10, 11, 14, 15, 18$, but for $n = 12$ the truncated octahedron is more stable than a column of hexagonal rings.

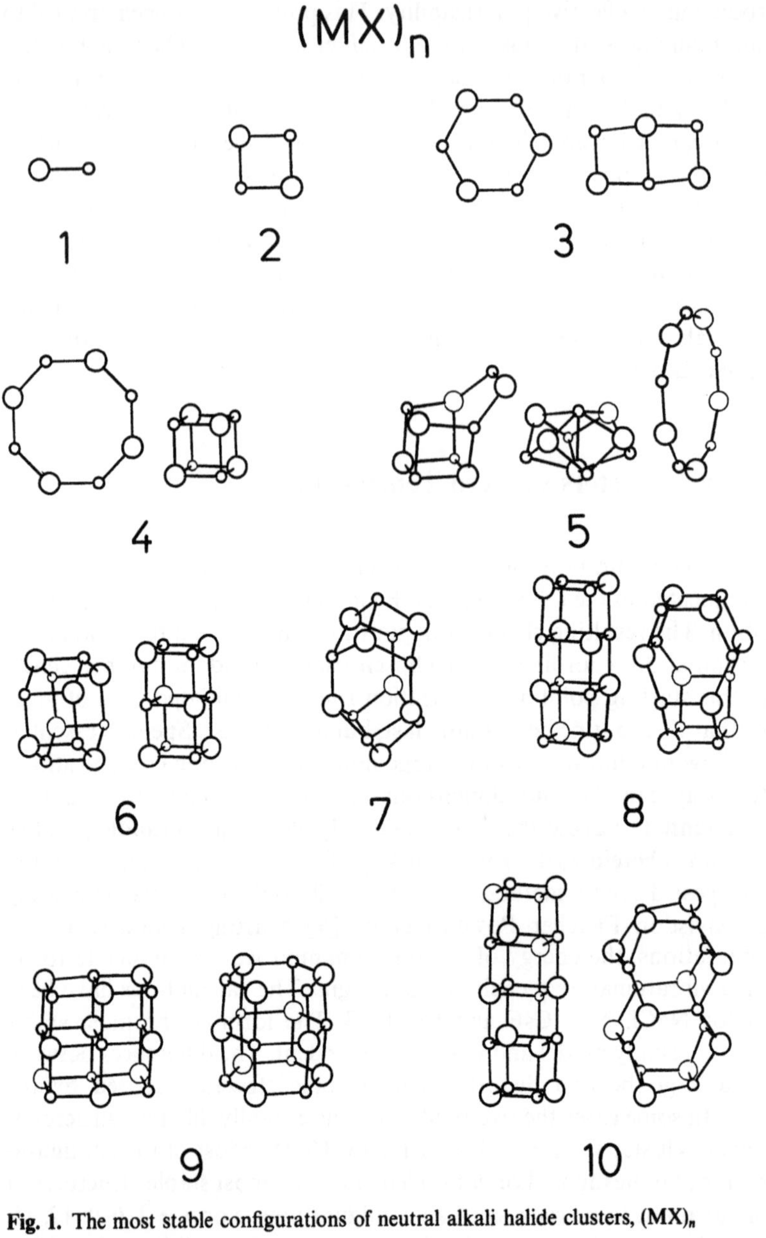

Fig. 1. The most stable configurations of neutral alkali halide clusters, $(MX)_n$

The charge of an oxygen ion is not uniquely defined. It depends on the oxygen ion's environment. It is thought to be about − 1.8 in the compound $Cs_{11}O_3$. However, this compound provides the oxygen atoms with a cesium-rich environment. The oxygen charge in our clusters is chosen to be − 1.6 and the charge on cesium to be + 0.8. The remaining parameters A and β have been fitted to the vibrational frequency of the antisymmetric mode of matrix isolated Cs_2O molecules [11] and to the interatomic distance in Cs_2O crystals. The values used are $A = 6.2 \times 10^{-9}$ ergs and $\beta = 3.22\,\text{Å}^{-1}$. The most stable form of Cs_6O_3 is shown at the top of Fig. 2 [12]. It has 35.34 eV of binding energy. However, it was found only twice in 1000 minimizations from random starting configurations. The "chain" isomer with 35.24 eV binding energy, although less stable, has a significantly larger catchment area. It was found 712 times in 1000 minimizations. Qualitatively, this result indicates that the chain will tend to become more stable with increasing temperature. The relation between catchment area and entropy will be discussed in more detail in the later section. We

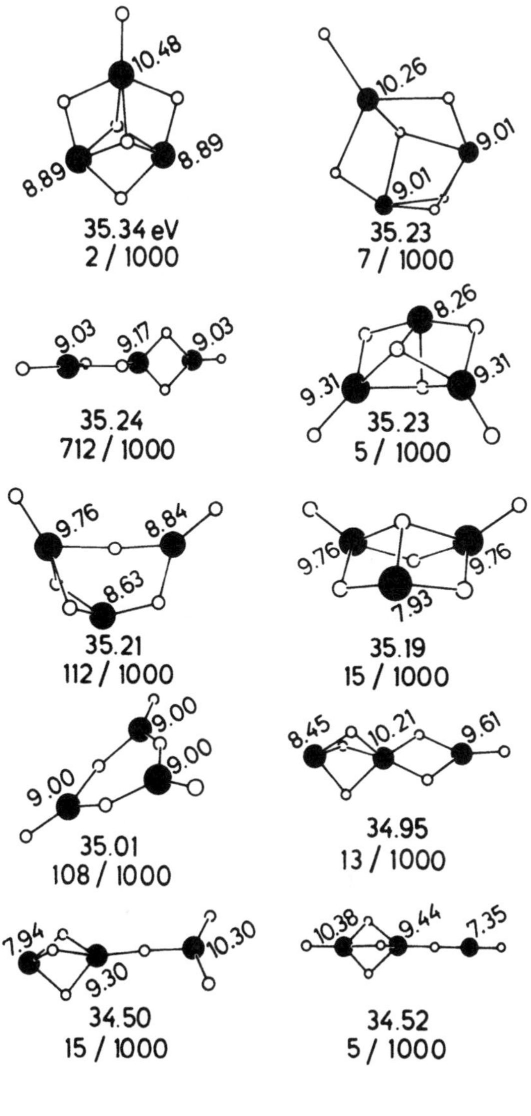

Fig. 2. Calculated stable isomers of Cs_6O_3. The total binding energy and the number of times each isomer is obtained after minimization from 1000 random starting configurations has been indicated. Also indicated is the electrostatic energy at each oxygen ion

have found 11 stable isomers of $(Cs_2O)_3$, 18 stable isomers of $(Cs_2O)_4$, and more than 60 isomers of $(Cs_2O)_5$.

4.5.4 Charged Clusters at Zero Temperature

Clusters having the compositions $M(MX)_n^+$ have been detected by means of a mass spectrometer. For certain values of *n*, the mass peaks are very intense indicating aggregates of enhanced stability. Figure 3 shows the most stable

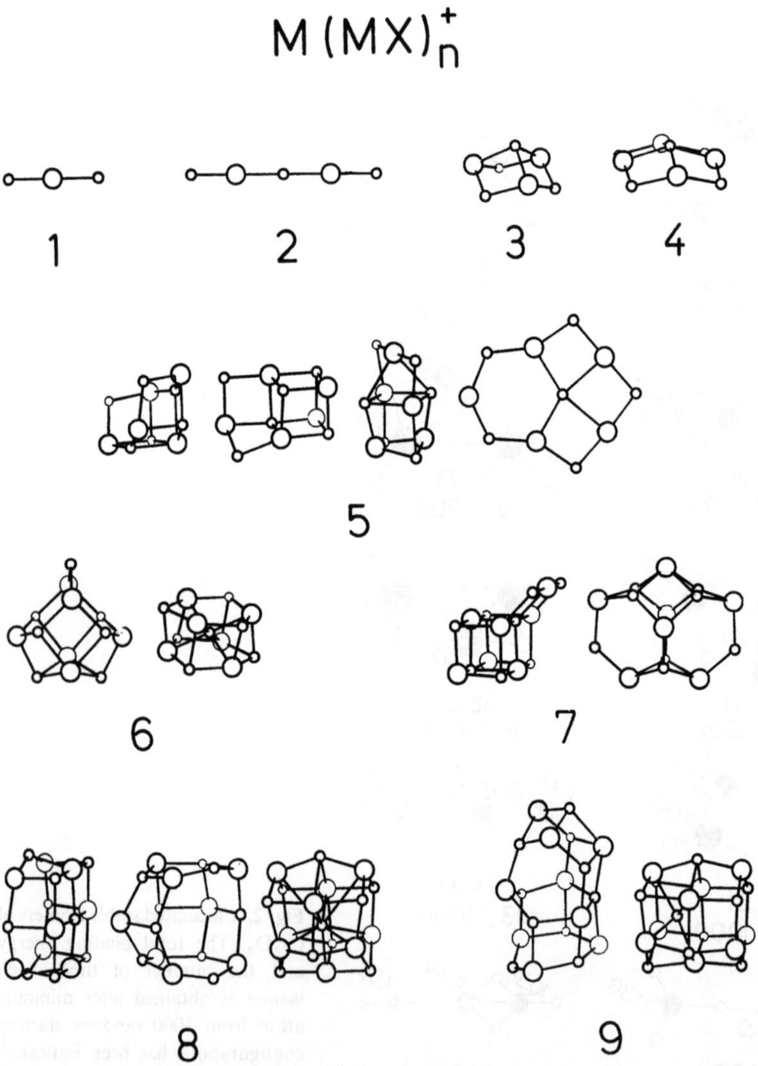

Fig. 3. The most stable configurations of positively charged alkali halide clusters, $M(MX)_n^+$

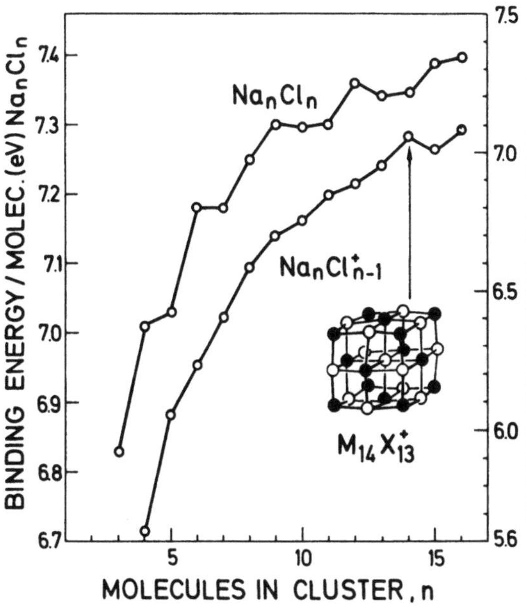

Fig. 4. Energy per molecule of the most stable forms of $(NaCl)_n$ and $Na(NaCl)_n^+$ clusters

configurations of $M(MX)_n^+$ clusters with $n = 1$ to 8. Even for the smallest clusters the most stable configuration is model dependent. If a dipole-monopole interaction is included, the linear form of Na_2Cl^+ and Na_2I^+ is unstable. For $M_3X_2^+$ clusters the rigid-ion model leads to linear forms independent of the material. The symmetry of the $M_4X_3^+$ clusters is always the same, a distorted six-atom ring with one ion over its center. This cluster might form when a neutral eight-ion cube loses one anion. The $M_5X_4^+$ cluster has in no case a planar structure. The center ion may therefore tunnel through the plane of the four anions. For larger clusters there exists a great variety of different forms. Therefore, it is not easy to find the absolute minimum on the energy surface. There is one exception, the cluster $M_{14}X_{13}^+$ – well known from mass spectra. This cluster prefers to assume the shape of a cube with three ions on one edge. However, the $Cs_{14}F_{13}^+$ cluster has not the full symmetry when calculated with the polarizable-ion model: the central ion is displaced from the center of the cube. For this size a cluster with CsCl structure is also conceivable. It can be thought of as a cube highly distorted along the body diagonal. But this configuration is found to be unstable when the position coordinates are varied.

Figure 4 shows the energies per molecule of the most stable structures of $(NaCl)_n$ and $Na(NaCl)_n^+$ clusters. The sizes with relatively low energy are $n = 4$, 6, 9, 13 (rigid-ion model) and $n = 3$, 6, 9, 13 (polarizable-ion model). Mass spectra [13–23] show that clusters with $n = 6, 9, 13$ have enhanced stability. The calculated forms of these clusters are all cube-like.

4.5.5 Catchment Area and Free Energy

Another important characteristic of a total energy surface is often called the "catchment" region. Imagine rain falling onto the surface. Every raindrop, no matter where it hits the surface, will eventually find its way to a minimum. That area of the surface which collects rain drops for a specific minimum is said to define a catchment area. All of configuration space can be uniquely divided up into a set of such areas. In this way, each point in configuration space can be said to belong to a given minimum (except, of course, those points defining the boundary line between catchment areas). In the above discussion the concept of catchment area proved useful in a qualitative description of a total energy surface. We will now attempt to use the concept in a more quantitative way. Catchment area has a particularly simple meaning for the case of a microcanonical ensemble. In such an ensemble each point in configuration space is equally probable, therefore, the probability of the cluster having a given configuration is simply proportional to the length of a constant energy contour in the corresponding catchment area. This interpretation must be only slightly modified when considering the canonical ensemble. In this case the probability of reaching a given point in phase space must be weighted by $\exp(-H/KT)$. Specifically, the probability that a cluster occupies catchment region a relative to the probability that it occupies region b is just,

$$P_a/P_b = \int_a \exp(-H/KT)dq^3\,dp^3 \Big/ \int_b \exp(-H/KT)dq^3\,dp^3 \ . \tag{6}$$

For the purposes of this discussion we will assume that the integration over momentum coordinates is independent of the configuration since both configurations have the same mass. However, the configurations will not, in general, have the same moment of inertia. The integration over configuration coordinates is easily carried out if it is further assumed that each catchment region has a parabolic form. This harmonic assumption is certainly valid at low temperatures.

$$V_a = E_a + \sum_i \omega_a^2(i)\, q_a^2(i)/2m \ , \tag{7}$$

where $\omega_a(i)$ is the ith vibrational frequency in the parabolic catchment region belonging to minimum a. The corresponding displacement amplitude has been denoted with $q_a(i)$. Using this expression the integral in Eq. (6) is easily evaluated to give:

$$P_a/P_b = \exp[(E_b - E_a)/KT]\prod_i \omega_b(i)/\omega_a(i) \ . \tag{8}$$

Here we have used the classical expression for the energy of a harmonic oscillator in order to retain a simple geometrical description of the probabilities.

The quantum mechanical result is obtained if each ω in the product in Eq. (8) is replaced with a corresponding $1 - \exp(-h\omega/KT)$. Notice that the relative probability of a cluster being in a given minimum is dependent on the depth of the minimum (through the Boltzmann factor) and on the extent of the catchment region (through the product of inverse frequencies). Just one very low frequency vibration can, at non-zero temperature, stabilize an energetically unfavorable configuration [1, 24]. Now it is possible to state and understand a rather unusual observation. *Cluster configurations corresponding to shallow minima tend to become more favorable at high temperatures. Cluster configurations corresponding to deep minima are less likely to be found at high temperatures.* The reason for this behavior is the following: energetically unfavorable clusters sitting in shallow minima usually have open structures which possess incipient instabilities. Such clusters resist only weakly certain types of deformations, i.e. they have low frequency vibrations. The entropy generating low frequency vibrations tend to make the configuration more favorable as the temperature increases. In principal, that configuration with the lowest frequencies will always become the most favorable if the temperature is made high enough. We will now present a specific example to illustrate this point. Consider the NaCl tetramer. The tetramer has two stable forms at low temperature, the cube and the ring as shown in Fig. 5. The total binding energy of the cube is slightly higher at low temperature. However, the ring has eight modes of vibration with frequencies below 100 cm^{-1}, the lowest being 29 cm^{-1}. Clearly, the ring will tend to be stabilized at high temperature by its low frequency vibrations. The question remains: Is this temperature dependent contribution to the free energy large

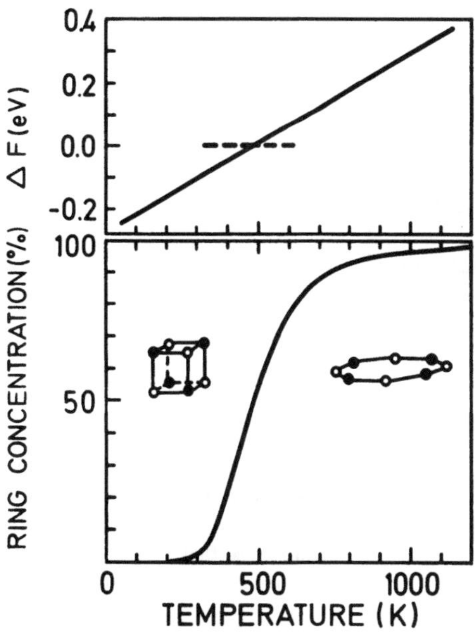

Fig. 5. Calculated free energy difference between a cubic and a ring-shaped $(NaCl)_4$ cluster (top). Temperature dependence of the relative concentration of ring-shaped tetramers (bottom) [5]

enough to compensate for the 0.27 eV difference in the potential energy? The difference in the free energy of the cube and ring is plotted in Fig. 5. Clearly, the compensation occurs at about 500 K. The ring concentration as a function of temperature is also shown in Fig. 5. Below 500 K the cube tetramer dominates, above 500 K the ring tetramer.

4.5.6 Atomic Vibrations

Having established the equilibrium conformations of the clusters, we are now in a position to calculate the vibrational frequencies. It should be emphasized that this can be done without introducing any new parameters. The $3N \times 3N$ second derivatives of the total energy with respect to atomic displacements can be evaluated numerically. These derivatives can be used to define a dynamical matrix which, when diagonalized, gives the frequencies and eigenvectors of the modes of vibration. The calculated frequencies for $(NaCl)_n$ are shown in Fig. 6. The monomer frequency is very high, consistent with the small interatomic spacing. The frequencies of the dimer and larger polymers are considerably

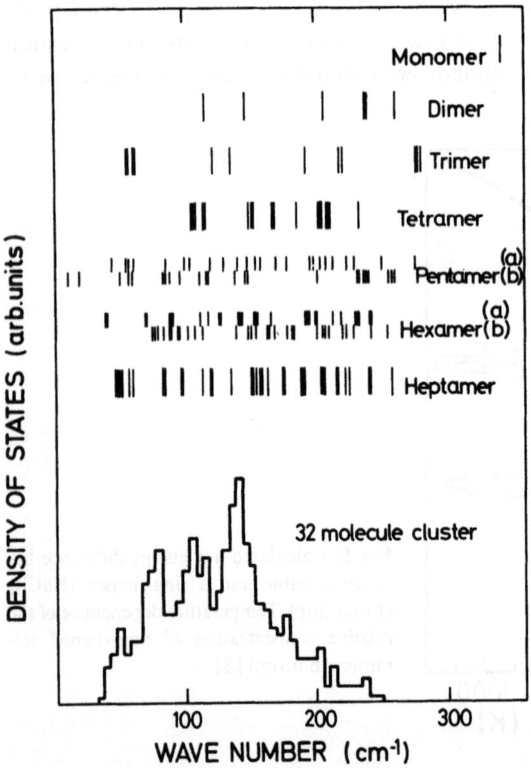

Fig. 6. Calculated vibrational frequencies of NaCl clusters. Degeneracies have been artificially removed so that all degrees of freedom are represented [5]

lower and, in fact, nicely span the phonon density of states of crystalline NaCl. The frequencies in Fig. 6 were plotted without degeneracy. This is partly a graphical device to better show concentrations of vibrational modes and partly due to the fact that the symmetry of the cluster is lowered due to the finite steps used in approaching the equilibrium configuration. For example, a tetramer that is a perfect cube would have two nondegenerate A_1 modes, two doubly degenerate E modes, 3 triply degenerate T_2 modes, and one triply degenerate T_1 mode. The assignment of these modes is clear from the groupings of frequencies. The high energy form, b, of the pentamer has two very low frequency modes indicating its incipient instability. It would be pleasing to see the crystalline vibrational density of states develop as the clusters increase in size. However, for the size of clusters considered in Fig. 6, this simply does not happen. The density of states turns out to be more dependent on the shape of the cluster than on its size. In order to see this convergence, the vibrational frequencies of much larger polymers containing 32 and 72 molecules were also calculated. However, in these cases a specific regular atomic configuration was assumed stable. A determination of the configuration by minimizing the energy would have required excessively large amounts of computer time. The vibrational eigenvalues and eigenvectors were determined also using the rigid ion model according to the method of *Kellerman* [25]. The force constant matrix is separated into a Coulomb term ϕ_c and a short-range term ϕ_r. An element of ϕ_c can be written

$$\phi_c(l\alpha, l'\alpha') = \frac{Z_l Z_{l'}}{|(r(l, l')|^3} \left[-\delta_{\alpha\alpha'} + \frac{3r_\alpha(l, l') r_{\alpha'}(l, l')}{|r(l, l')|^2} \right], \tag{9}$$

where $r_\alpha(l, l')$ is the α component of the vector between the equilibrium position of atom l and atom l'. The elements of ϕ_r are assumed to be zero, unless they correspond to nearest neighbor atoms, and then:

$$\phi_r(l\alpha, l', \alpha') = \frac{Z_l Z_{l'}}{4|r(l, l')|^3} \left[B\delta_{\alpha\alpha'} + \frac{(A - B) r_\alpha(l, l')\, r_{\alpha'}(l, l')}{|r(l, l')|^2} \right]. \tag{10}$$

The parameters Z, A and B were determined from the Madelung constant, the reststrahl frequency and the compressibility of crystalline NaCl. The dashed histogram in Fig. 7 shows the density of states of bulk NaCl calculated assuming cyclic boundary conditions. A sampling of 1000 points in the Brillouin zone was made. In order to smooth out the histograms, we have assumed that a mode of vibration contributes not only to the frequency interval in which its eigenfrequency falls, but also to one frequency interval on each side. The density of states demonstrates two distinct peaks corresponding to critical points at the Brillouin zone edge. Using the same parameters, we have calculated the density of states of microcrystals containing 64, Fig. 6, and 144, Fig. 7, atoms. The 64-atom microcrystal was cube shaped with four atoms along each edge. The 144-atom microcrystal was rectangular with edge ratio 4 to 6 to 6. These histograms are both quite similar to the density of states of the bulk material.

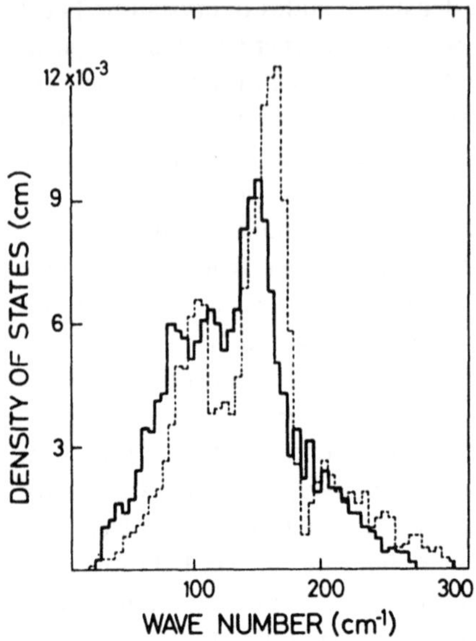

Fig. 7. Calculated density of vibrational states of NaCl microcrystals with atomic dimensions $4 \times 6 \times 6$ (solid line) and of NaCl using cyclic boundary conditions (dashed line) [5]

This similarity is surprising, but not implausible. The peaks in the density of states assuming cyclic boundary conditions correspond to short wavelength modes. It is just these modes which are best reproduced in a microcrystal. The low frequency, long wavelength modes are, of course absent in the microcrystal.

4.5.7 Photoionization of Cs–O Clusters

Does the ionization energy of an atom transform continuously into the work function of a solid with increasing cluster size? Early experiments indicated that the transition should not be smooth [26]. In addition, calculations [27, 28] predicted that clusters having closed electronic shells with orbital angular momentum ℓ should be very hard to ionize; open shell configurations should give up their electrons easily. These predictions were based on the assumption that clusters have spherical symmetry, an assumption which could not be supported by structure calculations. There the matter stood until 1984, when mass spectra of Na clusters were published showing unusually strong mass peaks for clusters containing 8, 20, 40, 58, 92 valence electrons, i.e., closed-shell configurations [29]. In the meantime, mass spectra have been published by many groups, apparently confirming the conclusion that closed-shell clusters have an enhanced stability. More direct evidence concerning the electronic structure can be obtained by examining the ionization energy of clusters [30–36]. Photoionization experiments on alkali metal clusters have been carried out successfully using several types of light sources; tunable pulsed dye lasers and Hg lamp radiation monochromatized by gratings. We have chosen to use a continuous dye laser because it provides a highly monochromatic beam of light without producing high peak power levels which could promote multiphoton ionization and fragmentation. However, a continuous laser also has a distinct disadvantage. Usable dyes exist only for wavelengths less than 3 eV and frequency doubling is less efficient than with a pulsed laser. Since we are limited to studying photoionization thresholds below 3 eV, the choice of the materials to be investigated is also severely limited. Cs has the lowest ionization energy of all elements, 4.0 eV. As we shall see, Cs compound clusters have even lower ionization thresholds. Photoionization spectra that consist of one sharp step can be straightforwardly interpreted. However, in practice the spectra are usually much more complicated, Fig. 8. For the otpimist this means that the spectra are very rich in information.

Ion counting experiments have a very large dynamical range, usable signals vary from 10 cps to 10^5. Therefore, data must be presented on a logarithmic scale if information is not to be lost, even though such a presentation tends to suppress detailed but reproducible structure. It is not exceptional to find that the photoionization spectra have two or more well-defined steps, Fig. 8. Possible explanations are:

1) Different electrons within the cluster require different energies for ionization.
2) The cluster has several isometric forms, each with its own threshold.
3) The cluster ion observed can also be produced as a fragment resulting from ionization of a larger cluster. Clearly, the interpretation of photoionization threshold spectra is far from straightforward. Assumptions must be made concerning, the temperature of the cluster, the electronic-vibrational coupling (Franck–Condon factors), the form of the vibration-free photoionization efficiency curve, and the extent and the temperature dependence of isomerization and fragmentation. All of these factors have been discussed in detailed reports on these measurements. For this qualitative presentation we will merely interpret breaks in a semilogarithmic plot of the photoionization spectra as defining ionization energies. This assumption, while leading to consistently high values, does not effect the qualitative discussion which will now be presented. The results of the experiments are summarized in Fig. 9. Here, the ionization energies of Cs_nO_m are plotted as a function of n and m. No experimental information could be obtained above about 3.0 eV. All points plotted at this limiting value correspond to thresholds outside the experimentally accessible spectral region.

In discussing these results it is useful to speak of four types of cluster; "metal-rich" clusters with composition from Cs_n to $Cs_{2n+2}O_n$, "doped oxide" clusters containing just one excess Cs atom $Cs(Cs_2O)_n$, "oxide" clusters with the composition $(Cs_2O)_n$, and "oxygen-rich" clusters containing more oxygen than oxide clusters. Using this terminology it is now possible to formulate several experimental observations. Three of these observations are qualitative:

1) The ionization energies of metal-rich clusters *decrease* with increasing oxygen content.

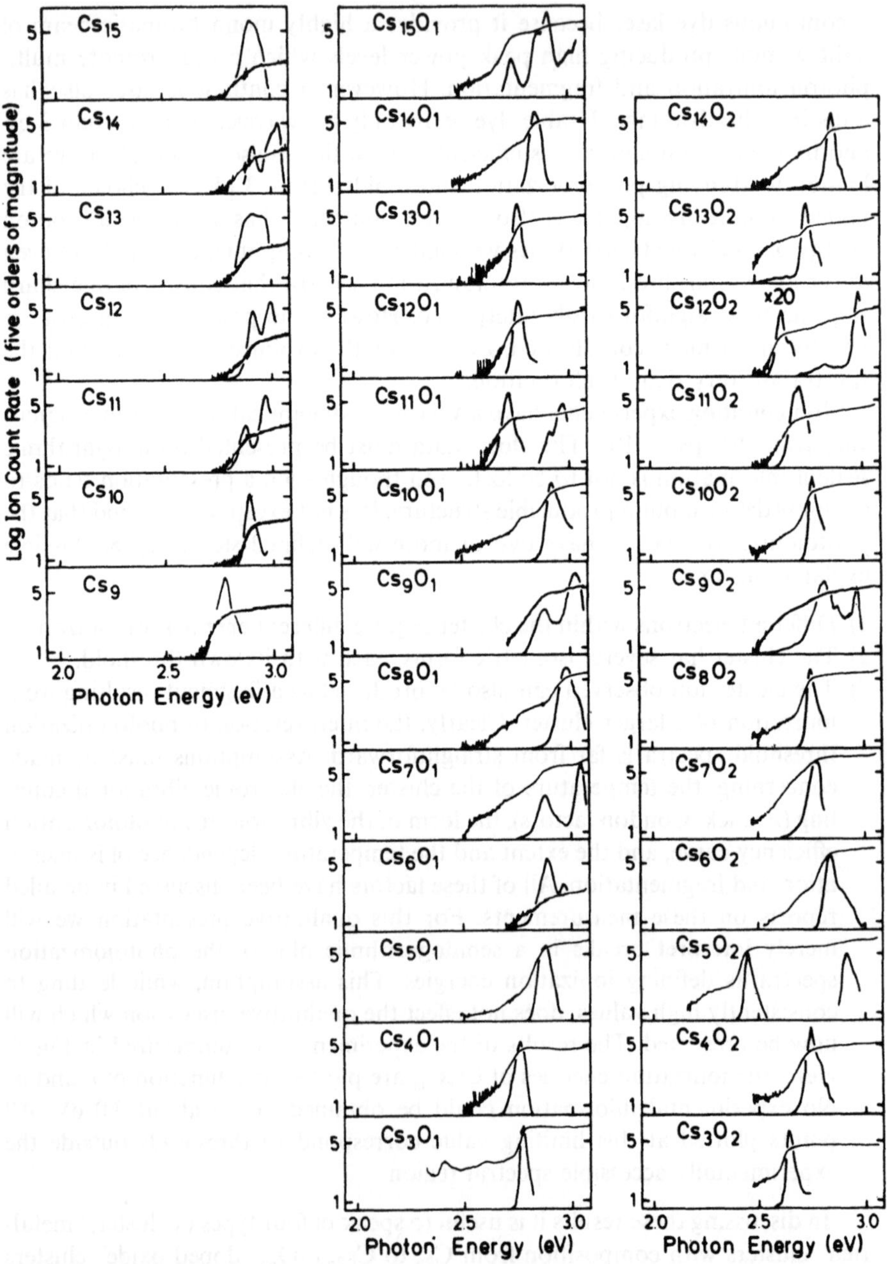

Fig. 8. Laser frequency dependence of the ionization energy of Cs–O Clusters. The derivative spectra are also shown [35]

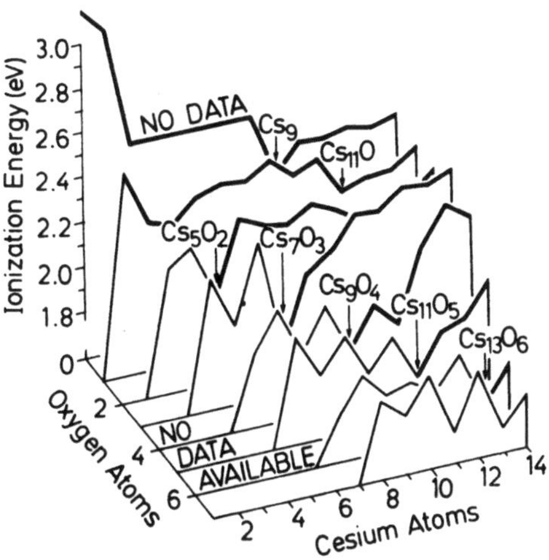

Fig. 9. Energy of the break-point in the photoionization spectra as a function of *n* and *m* for Cs_nO_m clusters [37]

2) "Doped oxide" clusters have unusually low ionization energies.
3) The ionization energies of oxygen-rich clusters Cs_nO_m alternate; high for even values of *n*, low for odd values of *n*.

   In addition to these general observations several specific results are worth mentioning.
4) Cs_9, Cs_4O, $Cs_{11}O$ and have particularly low ionization energies.
5) Cs_2O, $Cs_{10}O$ have particularly high ionization energies. Each of these results will now be discussed in turn. Will the ionization energy of a Cs cluster increase or decrease when an oxygen atom is added? Intuitively one might think that a metal should be easier to ionize than an oxide. Figure 10 clearly shows that this is not true for Cs. In fact, oxidized Cs has perhaps the lowest work function of any known material, a property utilized for decades in the construction of photocathodes sensitive in the infrared. Many models have been presented over the years to explain this effect on solid surfaces. Perhaps the effect can be made plausible in a cluster in the following way. Among the alkali metal oxides, the Cs–O bond has the highest ionicity. An electron removed from a Cs *s*-orbital will be given an additional push by the nearby negative oxygen ion. Clearly, realistic ab initio calculations for Cs_nO clusters are needed to clarify this issue.

"Doped oxide" clusters containing one extra cesium atom have very low ionization energies. Optical transmission measurements on thin films of Cs_2O indicate that the normal oxide is a semiconductor with a band gap of about 2.0 eV. If the material is prepared under conditions which result in an excess of cesium, measurements of the activation energy of electrical conductivity indicate the presence of donor states 0.25 eV below the conduction band. In analogy with the bulk properties, the excess cesium in an oxide cluster may act as a donor state well above the highest occupied orbital (top of the valence band) of the normal oxide, resulting in an appreciable lowering of the ionization energy (work function).

The third effect listed above can also be interpreted as being due to doping by an excess Cs atom. However, in this case, the doped material is not a simple oxide but a mixture of normal and peroxides. For example, Cs_8O_5 can be thought of as composed three different building units; Cs^+, O^{2-}, and O_2^{2-}. Using this scheme Cs_8O_5 becomes $(Cs_2O)_3(Cs_2O_2)$, i.e., a mixed oxide with no excess cesium atoms. On the other hand, Cs_9O_5 can be written as $Cs(Cs_2O)_3(Cs_2O_2)$, i.e., this cluster is doped with an excess cesium atom and has, therefore, a low ionization energy. This results in the even-odd alternation in the ionization energy of oxide-rich clusters. Even-odd alternations have been observed previously in the mass spectra of metal clusters and metal compound clusters. The most obvious explanation is that a paired electron is difficult to remove.

Both Cs_9 and $Cs_{11}O$ have unusually low ionization energies. Both clusters contain nine free electrons. As explained in the introduction to this section, both theory and experiment give support to the view that the ninth electron in a pure, spherical metal cluster opens a new angular momentum shell and that this electron is weakly bonded. Here we see that apparently such considerations qualitatively explain the electronic properties of not only pure alkali metal clusters, but also of clusters containing an oxygen atom.

4.5.8 Recent Developments

This contribution has described some of the early work done on oxides and halides of alkali metals. However, during the past four years the field has undergone a transformation. Exciting discoveries have been made which were undreamed of at the time the manuscript was prepared. Space limitations require me to give only a few references which will introduce the reader to some of these fascinating developments. R.D. Beck, P. St. John, M.L. Homer, R.L. Whetten: Science **253**, 879 (1991). Xiuling Li, R.D. Beck, R.L. Whetten: Phys. Rev. Lett. **60**, 3420 (1992). N.G. Phillips, C.W.S. Conover, L.A. Bloomfield: J. Chem. Phys. **94**, 4980 (1991). Y.A. Yang, L.A. Bloomfield, C. Jin, L.S. Wang, R.E. Smalley: J. Chem. Phys. **96**, 2453 (1992).

References

1. For a review of early work see: T.P. Martin: Phys. Rep. **95**, 167 (1983)
2. A. Simon: in *Structure and Bonding*, Vol. 36 (Springer-Verlag, Berlin, 1979)
3. J.E. Davy: J. Appl. Phys. **28**, 1031 (1957)
4. H. Föppl: Z. Anorg. Allg. Chem. **297**, 12 (1957)
5. T.P. Martin: J. Chem. Phys. **67**, 5207 (1977); **69**, 2036 (1978); **72**, 3506 (1980)
6. E.S. Rittner: J. Chem. Phys. **19**, 1030 (1951)

7. D.O. Welch, O.W. Lazareth, G.J. Dienes: J. Chem. Phys. **64**, 835 (1976)
8. D.O. Welch, O.W. Lazareth, G.J. Dienes, R.D. Hatcher: J. Chem. Phys. **68**, 2159 (1978)
9. J. Stoer: *Einführung in die Numerische Mathematik I*; 3rd ed. (Springer, Berlin, Heidelberg, 1979)
10. J. Diefenbach, T.P. Martin: J. Chem. Phys. **83**, 4585 (1985); **83**, 2238 (1985)
11. R.C. Spiker, L. Andrews: J. Chem. Phys. **59**, 1851 (1973)
12. T.P. Martin, B. Wassermann: J. Chem. Phys. **90**, 5108 (1989)
13. J.E. Campana, T.M. Barlak, R.J. Colton, J.J. DeCorpo, J.R. Wyatt, B.I. Dunlap: Phys. Rev. Lett. **47**, 1046 (1981)
14. K. Sattler, J. Mühlbach, O. Echt, P. Pfau, E. Recknagel: ibid. **47**, 160 (1981)
15. T.M. Barlak, J.E. Campana, J.R. Wyatt, R.J. Colton: J. Phys. Chem. **87**, 3441 (1983)
16. W. Ens, R. Beavis, K.G. Standing: Phys. Rev. Lett. **50**, 27 (1983)
17. R. Viswanathan, K. Hilpert: Ber. Bunsenges. Phys. Chem. **88**, 125 (1984)
18. R. Pflaum, P. Pfau, K. Sattler, E. Recknagel: Surf. Sci. **156**, 165 (1985).
19. R. Pflaum, K. Sattler, E. Recknagel: Chem. Phys. Lett. **138**, 8 (1987)
20. C.W.S. Conover, Y.A. Yang, L.A. Bloomfield: Phys. Rev. B **38**, 3517 (1988)
21. E.C. Honea, M.L. Homer, P. Labastie, R.L. Whetten: Phys. Rev. Lett. **63**, 394 (1989)
22. Y.A. Yang, C.W.S. Conover, L.A. Bloomfield: Chem. Phys. Lett. **158**, 279 (1989)
23. H.J. Hwang, D.K. Sensharama, M.A. El-Sayed: Chem. Phys. Lett. **60**, 243 (1989); J. Phys. Chem. **93**, 5012 (1989)
24. J. Jortner, D. Scharf, U. Landman: *Elemental and Molecular Clusters*, Springer Ser. Mat. Science, Vol. 6, Eds. G. Benedek, T.P. Martin, G. Pacchioni (Springer, Berlin, 1988) p. 148; U. Landman, D. Scharf, J. Jortner: Phys. Rev. Letts. **54**, 1860 (1985)
25. E.W. Kellermann: Philos. Trans. R. Soc. Lond. **239**, 513 (1940)
26. A. Herrmann, S. Leutwyler, E. Schumacher, L. Wöste: Helv. Chim. Acta **61**, 453 (1978)
27. J.L. Martins, R. Car, J. Buttet: Surf. Sci. **106**, 265 (1981)
28. W. Ekardt: Ber. Bunsenges. Phys. Chem. **88**, 289 (1984)
29. W.D. Knight, K. Clemenger, W.A. de Heer, W.A. Saunders, M.Y. Chou, M.L. Cohen: Phys. Rev. Lett. **52**, 2141 (1984)
30. M.M. Kappes, R.W. Kunz, E. Schumacher: Chem. Phys. Lett. **91**, 413 (1982)
31. K.I. Peterson, P.D. Dao, R.W. Farley, A.W. Castleman, jr.: J. Chem. Phys. **80**, 1780 (1984)
32. C. Brechignac, Ph. Cahuzac, J.Ph. Roux: Chem. Phys. Lett. **127**, 445 (1986)
33. W.A. Saunders, K. Clemenger, W.A. de Heer, W.D. Knight: Phys. Rev. B. **32**, 1366 (1985)
34. K.L. Peterson, P.D. Dao, A.W. Castleman, jr.: J. Chem. Phys. **79**, 777 (1983)
35. H.G. Limberger, T.P. Martin: J. Chem. Phys. **90**, 2979 (1989)
36. T. Bergmann, T.P. Martin: J. Chem. Phys. **90**, 2848 (1989)
37. T.P. Martin, H.G. Limberger, in Ref. 24.

4.6 Rare Gas Clusters

H. Haberland

The rare gases play a special role in the periodic table of elements. Their valence electron shells are filled, and the electronic charge density is spherically symmetrical. At sufficiently low temperatures the rare gases from crystals, which have been well studied. The zero point vibrations are so large for helium that pressure is needed for solidification. From this one can deduce that free He clusters are always liquid, as they are in a vacuum environment.

Clusters consisting of one kind of atoms will be treated nearly exclusively. Also, in view of what has been said in the Introduction to this book, trimers will not be considered as clusters in this chapter.

The interaction between two rare gas atoms is of the "van der Waals" type. The most popular, albeit not very accurate, interaction potential is that proposed by Lennard–Jones:

$$V_{LJ}(R) = \varepsilon \cdot [(R/R_0)^{-12} - 2(R/R_0)^{-6}] \ , \qquad (4.6.1)$$

where ε is the depth of the potential well which is located at the internuclear distance $R = R_0$. The interaction in the ground state can be described as that of soft, weakly attractive balls. The first term in the square brackets is due to the repulsion of the filled orbitals. The R^{-12} behavior is too steep, an exponential dependence like in a Morse potential would be more appropriate. The attractive long range R^{-6} part results from the induced dipole – induced dipole interaction. The potential between the rare gas atoms is very known very well. Accurate piecewise analytical potential have been determined by a fit to all available experimental data [1, 2].

The first electronic excitation lies in the 10 eV (for xenon) to 20 eV (for helium) range. Rare gas crystals, liquids, and gases have therefore a colorless, transparent appearance. The dimer and the bulk have been intensively studied. The information relevant to cluster studies will be briefly reviewed now. Figure 1 shows very schematically the dimer potential drawn for the case of argon. The well depths ε of the neutral dimer is about 12 meV. The well depth of the excited and the ionized states are much larger. The physical origin of this behavior is explained below in 4.6.1 and 4.6.2. The internuclear distance and the energies of the electronic levels do not change very much from the dimer to the bulk. The potential diagram of Fig. 1 has therefore often been used to explain bulk phenomena.

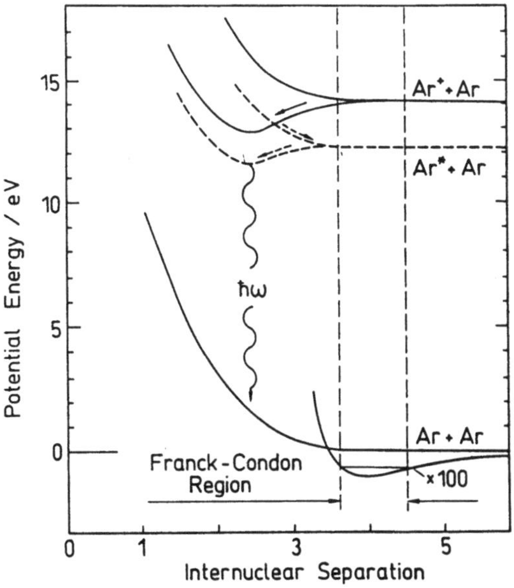

Fig 1. Dimer rare gas potential drawn for the case of argon. As the energies and internuclear distances do not change much in going from the dimer to the bulk, this diagram has often been used to explain cluster and bulk phenomena. Electronic relaxation in the cluster, solid or liquid can be described by the arrows. Only the $\Sigma_{g/u}$ potentials are drawn for clarity. Note that the internuclear distance of the potential minimum shrinks by about 30% upon excitation or ionization. An electronically excited neutral cluster can decay by vibrational relaxation only to the bottom of the excited state potential (drawn with a dashed line). Relaxation to the ground electronic state happens by photo emission

Electronic excitation is a process fast compared to heavy particle motion. This is the physical foundation of the separation of electronic and nuclear motion as exemplified by the success of the Born–Oppenheimer approximation, from which the Franck–Condon principle can be derived. Its classical version is: the electronic transition is so fast, that the atoms do not move during an electronic transition. The quantum-mechanical version says: the electronic transition probability is proportional to the square of the overlap of the vibrational wavefunctions in the initial and final state. The extension of the vibrational wavefunction of the ground state is indicated as the "Franck–Condon" region in Fig. 1. Outside the dashed vertical lines the vibrational wavefunction becomes very small, effectively zero. According to the Franck–Condon principle the upper state can therefore only be reached within these geometric limits.

Figure 1 has also been used extensively to discuss electronic excitation in the rare gas solids. An ion, say Ar^+, interacts with its neighbors via the potential marked $Ar^+ + Ar$. It can follow the solid arrow to form an Ar_2^+, if the kinetic energy lost in the R-coordinate of Fig. 1 is taken up by the surrounding atoms as vibrational energy. This is not possible in the dimer, but only in a cluster or in the solid. The vibrational energy is dissipated, and in the language of solid state physics the positive charge has self-trapped on an Ar_2^+ center or defect. If the ion can catch an electron, which happens to wander nearby it can follow the dashed line to a neutral excited state. This process is possible in the condensed phase, where the rare gas might be frozen to some metal support which is supplying the electrons. Thus the neutralization process is possible for the bulk but not for the cluster, where there are no free electrons. If the neutralization happens near a surface an electronically excited atom can be ejected. Alternatively a self-trapped molecular exciton Ar_2^* can be formed. (In solid state language a bound electron-hole pair is called an exciton). The fast nonradiative decay is no longer possible for the bound Ar_2^*, as this needs potential curves crossings. The Ar_2^* can only decay via UV-photon emission, as indicated by the curly line. As the lower potential is strongly repulsive, a well-known broad continuum is emitted. The shape of this continuum does not change much, if a gas, liquid or solid is excited, indicating that the same mechanism is operating in all three cases, and indicating also that the potentials are very similar indeed. All these processes are well studied, and the details are of course much more complicated than discussed here, see Ref. [3] for a comprehensive review.

Figure 2 shows a photoemission process in the band scheme of a rare gas solid, say Xe. The valence band, composed of the six Xe 5*p* orbitals, is completely filled. The lowest possible excitation are those from the valence band to the exciton states. These cannot be classified as $5p^5 7s$, $5p^5 8s$ etc., but must be considered as bound electron-hole states, whose energy converges to the bottom of the empty conduction bands. There the electron is free to move through the crystal. In the exciton state the positive and negative charge have to move together.

There is one important difference between the energy scales of Fig. 1 and Fig. 2. The total potential energy, i.e. the sum of all electronic energies is plotted in Fig. 1. In Fig. 2 single particle energies are plotted, wherever possible. This

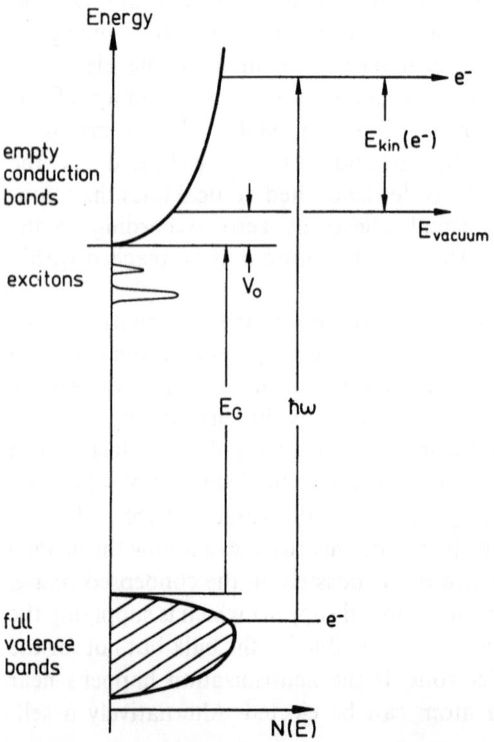

Fig. 2. A photoemission process described in the band structure picture of condensed Kr or Xe gases. For the lighter rare gases, the value of V_0 is negative, i.e. the bottom of the conduction band lies higher than the vacuum level (E_{vacuum})

cannot be done for the exciton, which consists of a electron and a hole moving together. Here the energy of this new entity, the exciton, is treated as one particle, and its effective single particle energy can be included in Fig. 2.

The gap energy E_G can be measured as the threshold for photoconductivity. For this experiment solid xenon, for example, is frozen between two electrically isolated plates in a ultra-high-vacuum system. Light from a synchrotron is passed through the solid xenon. If the photon energies reaches E_G a current can be measured between the two plates. The photon has transferred a quasifree electron into the empty conduction band, and the electron can now carry an electric current through this perfect isolator.

For krypton and xenon an additional energy V_0 is necessary to eject an electron from the bottom of the conduction band into the vacuum. The energy of the "vacuum level" E_{vac} is given by $E_{vac} = E_G + V_0$, where E_{vac} is per definition the energy of an electron with zero kinetic energy and a large distance away from the surface. For He and Ne the value of V_0 is positive, an electron in the conduction band of He or Ne gains the energy V_0 on leaving the cluster. By measuring kinetic energies of the emitted photoelectrons one can obtain very detailed information on the band structure. For solid Ar, Kr, or Xe the V_0 values are + 0.4 eV, − 0.3 eV, and − 0.4 eV, respectively. One can thus expect therefore, that negatively charged Kr and Xe clusters can be stable (see Section 4.6.5). For He "bubble states" are observed, which consist of an electron which has expelled the neighboring atoms from its surrounding. More details are discussed in the comprehensive review [3].

4.6.1 Neutral Rare Gas Clusters in the Ground Electronic State

Nearly the only information on neutral, ground state clusters is derived from numerical simulations on a computer, and from electron diffraction. In the latter case the fast electron takes a 10^{-16} s "snapshot" of the cluster. Only electrons which have been scattered elastically will interfere coherently to form a diffraction pattern, inelastic events contribute to a background only. As explained in Chapter 3.3 electron diffraction can give information on the diameter, the structure, and temperature of the cluster. The interpretation and modelling of the diffraction patterns involves quite heavy a theoretical effort. A potential is assumed, nearly always the Lennard-Jones potential of Eq. (4.6.1). For rare gases the three body (Axilrod–Teller) contribution can often safely be neglected [1]. The interaction potential has to be summed over all pairs of atoms. To find the absolute minimum is a non-trivial task even on a very fast computer. It is always possible to find a local minimum, or many of them. Safe for the smallest clusters it is often impossible to prove that the absolute minimum has been reached, the calculation takes just too long. A 13 atom cluster interacting via a Lennard-Jones potential has at least 988 local minima, and this number increases very fast with N. A softer potential, e.g. the Morse potential, or the well known rare-gas potentials [1, 2], have a much reduced number of local minima. An influence of the chosen potential can therefore be expected at least for the thermodynamic properties.

Growing of Rare Gas Clusters and Crystals. Figure 3 shows a calculated growth sequence. The trimer, $N = 3$, has the geometry of an equilateral triangle; $N = 4$ corresponds to a tetrahedron, $N = 5$ to a triangular bipyramid. For $N = 7$ one

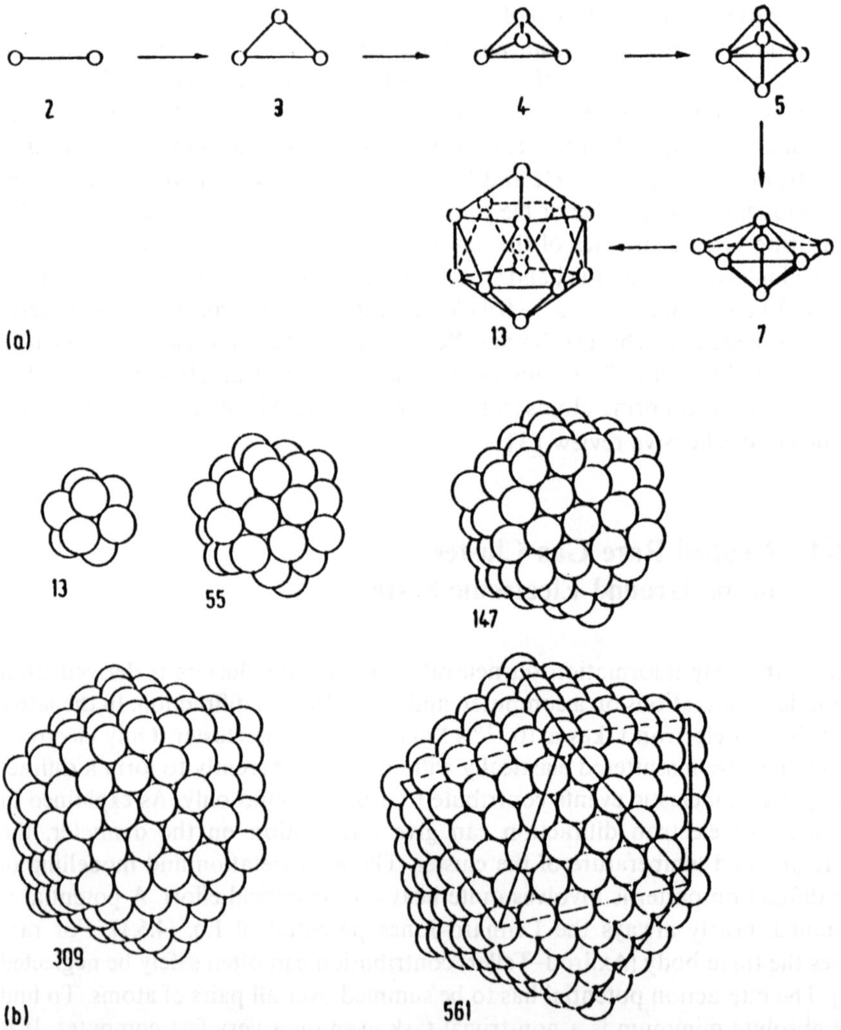

Fig. 3. Growth sequence of small and large neutral rare gas clusters. Five-fold symmetry, forbidden in the bulk, appears for the first time for $N = 7$. The icosahedra $N = 13, 55, 147, 309$, etc. all have five-fold symmetry. For small clusters this symmetry gives a lower energy than the face-centered-cubic structure of bulk rare gas solids

gets a pentagonal (five fold) symmetry, a pentagonal bipyramid consisting of a ring of five atoms, with one atom above and below it. Pentagonal symmetry is forbidden for the standard infinite lattices of solid state physics. It cannot lead to complete space-filling without distortions. But this symmetry is ubiquitous in cluster physics. For $N = 13$ one has the smallest member of the famous icosahedral geometry, an interior atom with two pentagonal caps. The clusters grow by addition of atoms to their sides, until the next larger icosahedron has been built, having $N = 55$, 147, 309, 561, ... The details of this growing sequence have been discussed in detail for neutral [4] and ionized [5] rare gas clusters. The number of nearest neighbors in an icosahedron is larger than in a small piece from a crystalline lattice. This is the reason for its stability. The icosahedron can be constructed from 20 slightly distorted fcc units whose faces are (111) planes and which share a common vertex. The interatomic spacing is not uniform in an icosahedron. This leads to a built-up of mechanical stresses, so that for some large N the fcc lattice becomes favored. From electron diffraction one obtains $800 \leq N \leq 3500$ [6, 7], while theoretically much larger N values are calculated [8]. The discrepancy might be due to a temperature effect, the calculations are done at $T = 0$ K, while an evaporative ensemble with $T \approx 38 \pm 4$ K is used in the experiments [9].

Helium. The mass of the helium atom is so small and the He–He interaction so weak that quantum effects dominate at low temperature. The more abundant ^{4}He isotope shows Bose-condensation and both isotopes superfluidity in the bulk. It is generally assumed, that these properties persist also in clusters. Experimental proof is so far lacking. The knowledge up to mid-1988 has been reviewed by *Toennies* [10], newer references can be found in [11].

4.6.2 Potentials for Excited and Ionized Rare Gas Dimers and Clusters

The binding energy of a rare gas dimer changes dramatically upon excitation or ionization, as can be seen from Fig. 1. The simplest explanation is given by the molecular orbital diagram of Fig. 4. It is drawn for the case of helium but applies equally for the heavier rare gases. The only complication is that one has to consider *p*- instead of *s*-orbitals.

Helium has a $1s^2$ electronic structure. From the $1s$ atomic orbitals one can construct two molecular orbitals (MO): a bonding σ_g and an antibonding σ_u MO. The indices g and u describe the symmetry of the wavefunction with respect to interchange of the nuclei. The electronic wavefunction having σ_g symmetry does not change sign on interchanging the nuclei. This puts a lot of electron density between the two nuclei, giving a strong bonding contribution. The reverse is true for the σ_u orbital. The method is explained in many textbooks of physical chemistry. Each σ orbital can accommodate up to two electrons of

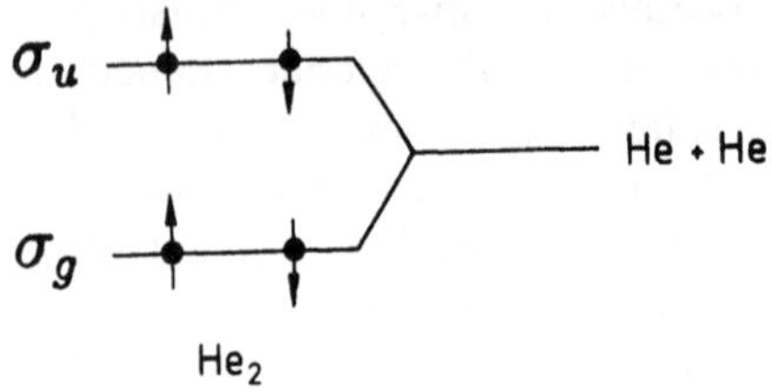

Fig. 4. Molecular orbital diagram for the neutral He_2 molecule. There are two electrons in the bonding σ_g and two electrons in the antibonding σ_u molecular orbital. The bonding and antibonding effects cancel each other nearly, so that the well depth of the He_2 interaction potential is only about 1 meV. The removal of an electron from the antibonding σ_u orbital leads to a strong chemical bond for He_2^+ molecular ion, which has a well depth of about 2.5 eV. This is an increase of more than 3 orders of magnitude. The effect is similar for the heavier rare gases. The only difference is that p type orbitals have to be used instead of the s orbitals of helium

opposite spin. For He_2 one has thus 2 binding and two antibinding electrons, giving a near cancellation of the binding and antibinding effects. This explains on a qualitative basis the low binding energy of He_2 and the heavier rare gas neutral dimers. But the molecular ion, He_2^+, has 2 bonding and only 1 antibonding electron. The large binding energy of He_2^+ and of the heavier rare gas dimer ions can thus be understood by this simple picture.

Figure 1 shows that not only the potentials for the dimer ions possess a deep minimum, but also those for the neutral excited states. To a good approximation one can write the electronic configuration of these states as a dimer ion (e.g. Ar_2^+) surrounded by an outer electron: $Ar_2^* = (Ar_2^+)e^-$. This makes the strong interaction in these states plausible. For an excited or ionized dimer consisting of two different atoms, say ArXe*, the interaction is much weaker, as the molecule is no longer symmetric with respect to the interchange of the nuclei. Consequently there is no $\sigma_u - \sigma_g$ splitting.

Charge Localization. The large binding energy of the homonuclear dimer ions is responsible for the charge localization in positively charged rare gas cluster ions. The cluster can lower its total energy upon charge localization. The internuclear distance of the potential minimum shrinks by about 30% upon ionization or excitation (compare Fig. 1). The dimer or trimer ions thus have a much smaller interatomic distance than the dimers or trimers in the neutral, ground electronic state. In other words, a large geometric change of the neighboring atoms is necessary for a movement of the charge. In more physical terms: the overlap (or hopping) integral of the charged site to the neighboring neutral one is reduced to a very small value. In the language of condensed matter science the charge has self localized.

It was originally conjectured that the charge would localize on a dimer ion [12]. This agrees with a calculation for neon [13], i.e. an Ne_n^+ ion has an electronic structure like $Ne_2^+Ne_{n-2}$. More than 97% of the charge is shared just between the two atoms. The dimer and trimer ion have nearly the same energy for helium [14], while for the heavier rare gases charge localization on a trimer or tetramer ion is preferred [15].

4.6.3 Experiments with Neutral Rare Gas Clusters

Intense beams of neutral rare gas clusters can easily be generated by a supersonic expansion. Only He poses some experimental problems as the nozzle has to be cooled to about 10 K. The experiments performed to date can be classified according to their time scale, as shown in Table 1. Some experiments interrogate the cluster on so fast a time scale, that the atoms do not have time to move. These are classified as very fast in Table 1. Fast and slow is meant with respect to a vibrational period in the cluster.

Among the seven experiments discussed below for neutral clusters, only the first three probe the property of the neutral cluster. In 4.6.3.1 an experiment to measure the vibrational spectrum and in 4.6.3.2, 4.6.3.3 the electronic spectra of neutral rare gases is discussed. In the other four experiments the cluster is ionized, i.e. the experiment changes its charge state. In 4.6.3.4 and 4.6.3.5 the cluster ions are measured in order to understand mass spectra, structure of the cluster ions, and fragmentation. The photon energies necessary to eject an electron are discussed in 4.6.3.6, while in 4.6.3.7 the energy distribution of the emitted photo electrons is measured. With one exception, all experiments have been performed on not mass selected beams.

Table 1. Order of magnitude of the time scales involved in the different experiments. Fast and slow is meant with respect to a vibrational period in the cluster, which is a little less than 1 ps

Time scale		Experiment	Type	
very fast (electronic processes, atoms do not move)	10^{-16} to 10^{-14} s	electron diffraction photoabsorption photoelectron spectroscopy electron scattering		
comparable slow	10^{-13} to 10^{-12} s $\leq 10^{-8}$ s	atom-cluster collisions fluorescence	N	no
very fast to very slow	10^{-12} to 10^{-4} s	mass spectra by electron impact or photo ionization electron impact mass spectra	N	yes
slow to very slow	10^{-7} to 10^{-4}	metastable decay		
very fast to very slow	10^{-12} to 10^{-4} s	photofragmentation	I	yes

yes (no) = experiment with (not) mass selected clusters; N (I) = experiment with neutral (ionized) cluster; e^- = electron.

4.6.3.1 Inelastic He–Ar_n Collisions, the Phonon Spectrum of a Free Cluster

Unclustered helium atom beams from supersonic expansions can have very sharp velocity distributions ($\delta v/v \approx 10^{-2}$). It is a standard tool of surface science to scatter these beams from a surface. Vibrational modes (phonons) can be excited and deexcited. The phonon energy can be determined by measuring the velocity distribution of the scattered beam. *Buck* and coworkers have applied this method to He + Ar_n collisions. The first vibrational spectrum of a free cluster could thus be measured [16]. Figure 5 shows a result, which agrees well with theoretical expectations.

4.6.3.2 Photoabsorption and Fluorescence

The first optical absorptions of the rare gas atoms, molecules, clusters and solids/liquids lie far in the vacuum ultra-violet. So synchrotron radiation is convenient to be used for these experiments. The radiation is monochromatized and focused into an unselected cluster beam. The process can be written as:

$$h\nu_1(\text{from synchrotron}) + Ar_n \rightarrow Ar_n^* \,. \tag{4.6.2}$$

The star (*) denotes electronic excitation. Part of the electronic excitation is converted to vibrational energy, as indicated by the dashed arrow in Fig. 1. This

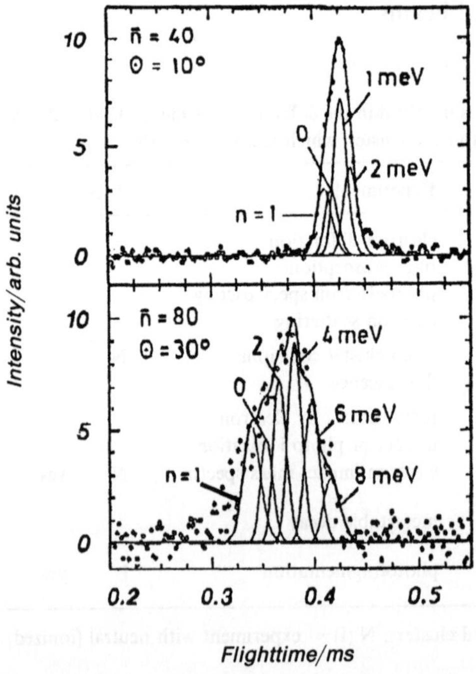

Fig. 5. Measured time-of-flight spectrum of monoenergetic He atoms scattered from Ar clusters with mean cluster sizes of 40 and 80 atoms, respectively and two different scattering angles. The curve marked "1" corresponds to scattering from Ar atoms which are also present in the cluster beam. The curves marked 0, 2, . . . correspond to a transfer of that amount of kinetic energy in meV from the He atom to vibrational motion of the cluster

heats the cluster and leads to ejection of atoms:

$$Ar_n^* \rightarrow Ar_m^* + (n - m)Ar \ . \qquad (4.6.3)$$

The potential curves of the ground and excited states do not cross as a function of the internuclear distance in the energy range available to the cluster. Conversion of electronic to vibrational energy is effective only at potential curve crossings. So the cluster can relax down only to the minimum of the electronically excited state. There conversion of electronic into vibrational energy stops. After about 10^{-8} s (or longer for a metastable decay) a photon is emitted:

$$Ar_m^* \rightarrow Ar_{m-k} + k\,Ar + h\nu_2 \ . \qquad (4.6.4)$$

The electronic excitation in the above two equations can be either localized or delocalized. The localized state has a lower energy but a potential barrier sometimes impedes its formation. If localization occurs it is generally assumed that it happens on a dimer, e.g. the Ar_m^* of the two equations above would have an electronic structure like $(Ar_2^*)Ar_{m-2}$.

Two different kind of experiments have been performed, mainly by the DESY group [17, 18, 19]:

1. Measurement of the total fluorescence yield. All photons $h\nu_2$ of Eq. (4.6.4) are collected undispersed, while $h\nu_1$ is scanned. If "dark" relaxation channels (those which do not give rise to a photon, i.e. ionization) are of minor importance the photon yield is proportional to the photoabsorption cross section of the neutral cluster. Below the ionization energy this condition should be well fulfilled.
2. Measurement of the spectrally resolved fluorescence. For a fixed photon energy $h\nu_1$ the fluorescence $h\nu_2$ is recorded spectrally dispersed. This can and has also been done using electron impact for excitation [20, 21].

Figure 6 shows spectra for krypton [17]. The data span the range from the atom, over small and large clusters, all the way to the solid. The atom displays two sharp lines, which are due to the spin-orbit split $4p^6 \rightarrow 4p^5 5s$ transition. With increasing cluster size the transition broadens, shifts a bit, and splits into several maxima. In the bulk the sharp lines are interpreted as being due to the excitation of "excitons". An electron is excited, and a positive charge (a hole in solid state language) remains on the atom. Electron and hole attract each other via the Coulomb attraction, forming a series of states similar to that of the hydrogen atom. The data in Fig. 6 are just the lowest member of this series. It is evident from the figure, that the overall shift with increasing cluster size is visible but not a very strong one. The solid state aspects of excitons are discussed in great detail in Ref. [3] and the cluster aspects in Ref. [17, 19].

In order to record the spectrum of Fig. 7 a fraction of 10^{-4} xenon was diluted in argon. Thus the clusters consist mainly of argon. Some of them contained a Xe atom. From a comparison of experimental and theoretical data, the three peaks are identified as belonging to Xe atoms I) sitting on top of the cluster, II) inside the surface, and III) inside the cluster. A detailed calculation

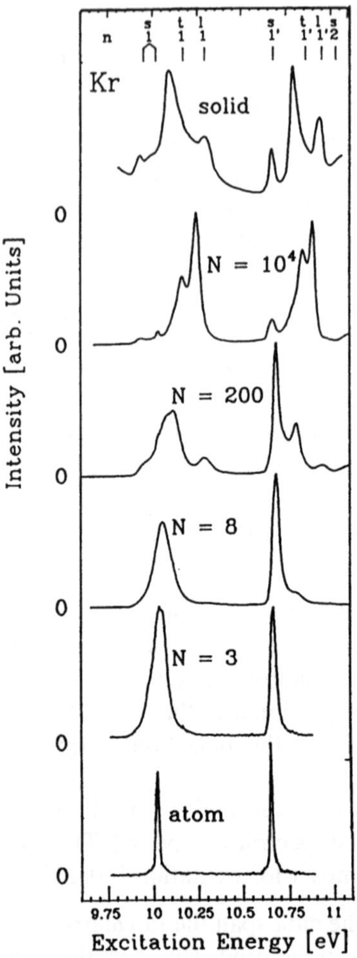

Fig. 6. Fluorescence excitation spectra of krypton clusters as a function of the mean cluster size N. The main splitting is due to spin-orbit interaction of the electron with the hole. The finer details are due to surface (s) and two kinds of volume excitons (l = longitudinal, t = transverse). The excitation does not change dramatically in going from the atom to the bulk. The total absorption cross section is proportional to the number of atoms in the cluster

has been performed for the energy relaxation pathways for these mixed rare gas clusters [22]. The agreement between theory and experiment is good.

4.6.3.3 Inelastic Electron Scattering

A first result on inelastic electron scattering has appeared [23]. An electron beam is energy selected to a half width of 30 meV and scattered from a not mass selected Ar and Kr cluster beam. The energy of the scattered electrons is analysed. It is planned in a future extension to measure the electrons in coincidence with the ions produced. So far rather high electron energies are used and the scattered beam is measured in the forward direction. In this case of small momentum transfer electron excitation is similar to photoionization, and the results are indeed similar to those discussed above.

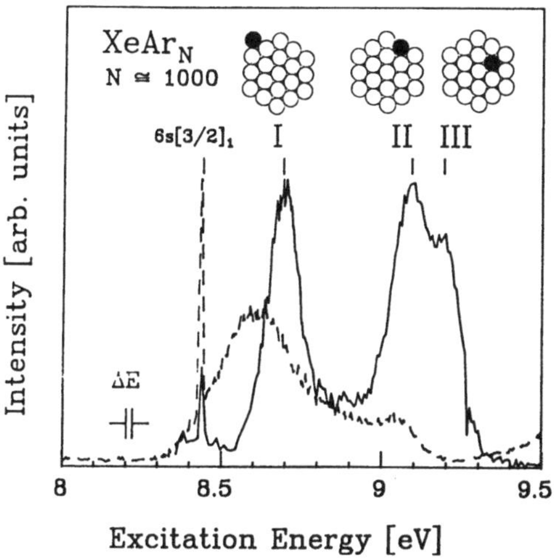

Fig. 7. Spectrum of $XeAr_N$, with $N \approx 1000$. The three bands are attributed to excitations of Xe atoms on top of the cluster surface (I), inside the cluster surface (II), and in the interior of the cluster (III). The three different positions of the Xe atoms (solid circle) are illustrated above the spectra

4.6.3.4 Mass Spectra

One of the easiest experiments in cluster science is to record a mass spectrum. However it is no longer easy to obtain the resolving power to record a spectrum like that shown in Fig. 8 [24]. How can these spectra be interpreted, what kind of information is contained in them? Note that the intensity of the cluster peaks is not smoothly varying with cluster size. There are several intensity anomalies, peaks with higher intensity, or peaks where a strong drop in intensity occurs. These have become known as "magic numbers", in analogy to the magic numbers of nuclear physics. Today, there is nothing magic about these numbers, their physical origin is well understood: they result from an interplay of the size dependence of the binding energy of the cluster ions with the fragmentation processes after ionization. Many authors prefer to speak of intensity anomalies instead of magic numbers.

The positive charge in a rare gas cluster is not delocalized like in, say Na_n^+. Localization on a dimer, trimer, or tetramer ion is energetically favored, as discussed above. The Ar_n^+ clusters of Fig. 8 thus have an electronic structure like $(Ar_x^+)Ar_{n-x}$ with $x = 3$ or 4. This will be discussed in more detail below (see 4.6.4.1). There seems to be common agreement that neutral atoms group themselves in icosahedral shells around this molecular ion [24]. The number of atoms in an icosahedron of i shells is given as:

$$n(i) = 1 + \sum_{k=1}^{i} (10k^2 + 2) \; . \tag{4.6.5}$$

Here $(10k^2 + 2)$ is the number of atoms in the kth shell. Icosahedral shell closings thus occur at $n = 13, 55, 147, 309, 561 \ldots$ Either a large intensity or a sharp drop in intensity is seen at or very near these numbers. The structure between these numbers is attributed to subshell closures, which are obtained by

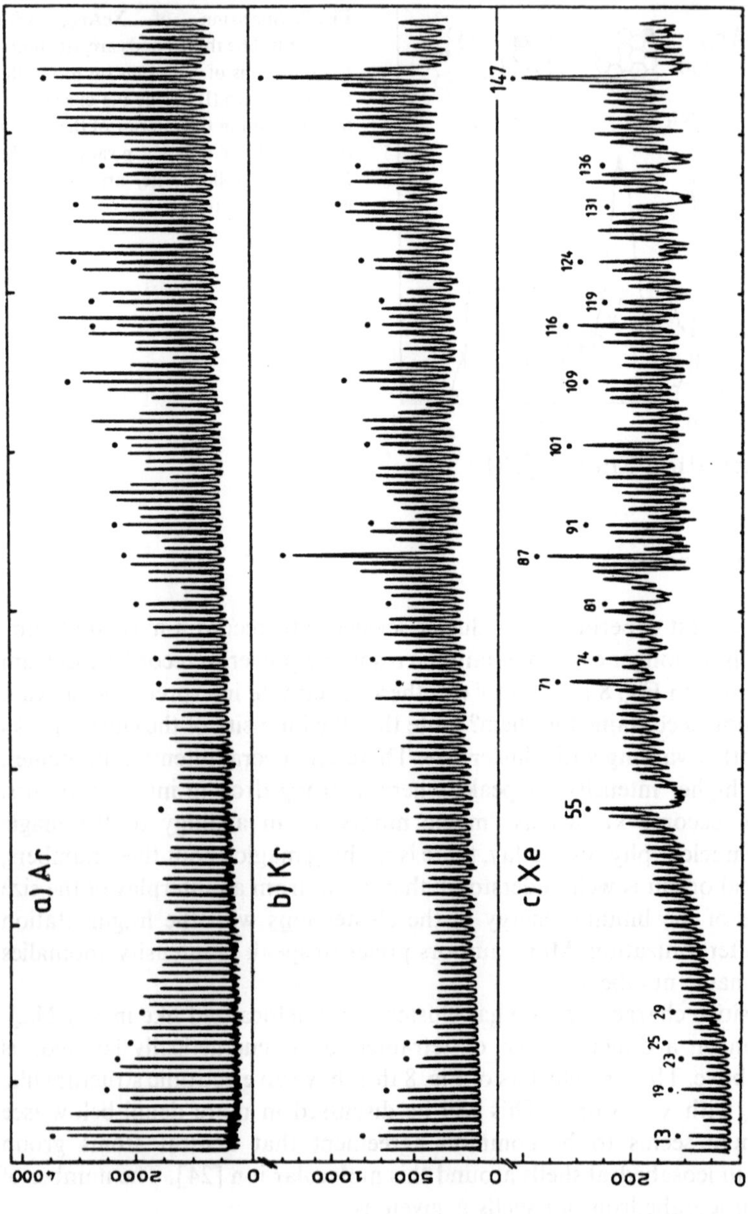

Fig. 8. Mass spectra of positively charged Ar, Kr, and Xe clusters. The Xe-intensity anomalies are labelled with dots, and the number of atoms per clusters given. The intensity anomalies are less pronounced for Ar, and Kr

peeling off atoms from the faces of the closed shell icosahedra. A satisfactory agreement is obtained using these purely geometric arguments [4, 24].

As explained in detail in Chapter 3, clusters from a supersonic expansion are not cold. They leave the expansion zone "boiling hot", and cool by evaporation. The clusters one uses in the experiment still have enough internal energy to evaporate another 2 to 3 atoms. With this "evaporative ensemble" the experiments are performed. For the rare gases the ionization process deposits about 1 to 2 eV in the cluster, which leads to further heating and evaporation. It is thus difficult to reconstruct from a mass spectrum of the cluster ions the size distribution of the neutral clusters. A simulation has appeared of the evaporation processes after ionization and the evolution of the mass spectra [235]. A statistical distribution of the internal energies has been assumed in this calculation. Very new experiments (see 4.6.4.2 below and Ref. [26]) point to the fact, that the ionized core is not strongly coupled to the neutral atoms. Thus the statistical assumption is probably not valid, especially not for long decay times. Nevertheless Ref. [25] gives a very instructive example of the development of the structures in a mass spectrum, and how the variations in binding energy of the cluster ions manifest themselves.

Although it is presently not possible to reconstruct the mass spectrum of the neutral clusters from that of the ionized ones, it is possible to estimate the average number of atoms evaporated [27, 28]. This is an important number for all the experiments on neutral rare gas clusters, as it allows to calculate from the mean of the ionized distribution the mean of the neutral one.

4.6.3.5 Mass Spectra from Mass Selected Clusters

Small neutral clusters can be mass selected using the method explained in Section 3.4. If a beam of neutral Ar_5 is ionized by electrons of 70 eV kinetic energy, only a very small signal is observed on Ar_3^+, Ar_4^+, or Ar_5^+. The dimer ion Ar_2^+ has the highest intensity. These experiments [29] were the first direct experimental proof of the dramatic fragmentation of rare gas clusters, which had been predicted earlier based on the electronic properties of the rare gases [12]. The result that Ar_2^+ is the dominant fragment channel has an interesting counterpart in that Ar_2^* is the dominant fragment channel, when small argon clusters are photoexcited [30].

4.6.3.6 Thresholds for Photoionization

The threshold for photoionization has been measured by two groups [31, 32]. Monochromatized radiation from a synchrotron is used to ionize the clusters. The emitted photoelectrons are passed through a special analyzer, which transmits only electrons of near zero kinetic energy. These are measured in coincidence with the cluster ions. Thus only those cluster ions are recorded which have emitted an electron of near zero kinetic energy. In this way the energy deposited in the cluster is exactly known. The method is named TPEPICO for Threshold Photo Electron Photo Ion COincidence. The process whose threshold one wants to study can be written as:

$$h\nu(\text{from synchrotron}) + Ar_n \rightarrow Ar_n^+ + e^-(E_{kin} = 0) \ . \tag{4.6.6}$$

Note that the excited and ionized cluster are supposed to contain the same number n of atoms, which is at first surprising given the discussion above. The energy of the ionizing radiation is decreased, until the signal vanishes. Ionization energies are obtained from a careful study of the threshold region. Some results are collected in Fig. 9. One sees a large jump from the monomer to the dimer, which has been rationalized above (see discussion of Figs. 1 and 4). Afterwards the ionization energies (IE) decrease much more slowly. The lower limit of the experimental IEs reach nearly the bulk value already for Ar_{20}, while for Kr_{20} it is still about one half of an eV away. Note that the IE of metals converge much slower.

It is at first sight surprising from the discussion given so far, that no fragmentation occurs and that an adiabatic ionization energy can be measured by this method. According to Fig. 1 there is no finite Franck–Condon factor connecting the minima of the Ar_2 and Ar_2^+ potential. (The Franck–Condon principle is discussed in the context of Fig. 1). A direct transition by a photon between the two minima is impossible. Very probably a highly excited Rydberg state of neutral Ar_n is first excited, which then makes an autoionizing transition.

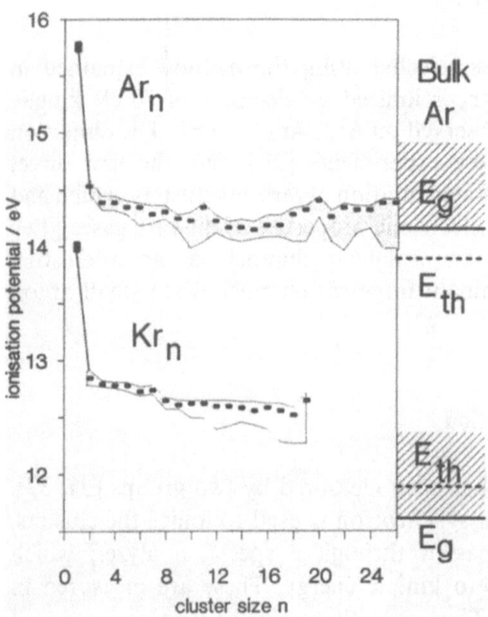

Fig. 9. Experimental ionization thresholds for Ar_n and Kr_n. The bulk value is E_{th}. The shaded areas give the experimental error limits

The actual process studied is thus:

$$h\nu(\text{from synchrotron}) + \mathrm{Ar}_n \rightarrow \mathrm{Ar}_n^* \rightarrow \mathrm{Ar}_n^+ + e^-(E_{\mathrm{kin}} = 0)\ . \tag{4.6.7}$$

The star (*) denotes electronic excitation. After a vertical transition to Ar_n^* the atoms can start to move, and the system can cross from an excited state to an ionized one.

In Ref. [31, 32] the fragmentation just above the threshold was studied in detail. Only the first 100 meV above the threshold fragmentation can be avoided. In another synchrotron experiment an electron from the core levels of argon was photo ejected and the ensuing strong fragmentation processes studied [33].

4.6.3.7 Photoelectron Spectroscopy

The photoelectron spectra (PES) of Ar, Kr, and Xe clusters have been measured [34, 35]. The process is:

$$h\nu + \mathrm{Ar}_n \rightarrow \mathrm{Ar}_n^+ + e^-\ . \tag{4.6.8}$$

Now the photon energy is so high that the electron is directly transferred to the continuum (see Fig. 2). The kinetic energy of the emitted electrons is measured. The electron leaves the cluster so fast, that the atoms do not move on this time scale. The kinetic energy of the electron carries therefore information on the Ar_n^+ in the geometry of the neutral cluster. Once the electron has left, the cluster is "electronically relaxed" (i.e. the electrons of the cluster have adjusted themselves to the new force field); but it is not "geometrically relaxed" as the atoms have had no time to adjust their geometric positions to the new force field.

The photo electron spectra have been calculated for some highly symmetric neutral structures (Ar_3, Ar_7, and Ar_{13}), and compared to the experimental data [34]. The authors of [34] identify these as "ionization chromophores". Ar_{13}^+ is the ionization chromophore for large clusters and the solid. This is not surprising. The long range Ar^+–Ar interaction goes as $\alpha e^2/2R^4$, where α is the polarizability of the neutral Ar, e the electronic charge, and R the internuclear distance. This interaction is effectively zero beyond one atomic layer around an ion, which make together 13 atoms.

One should always bear in mind that this ionization chromophore corresponds to the geometrically unrelaxed state, and has nothing in common with the geometrically relaxed chromophore discussed below. The time scale for an electronic movement is about 10^{-16} s. After 10^{-13} to 10^{-12} s after the electron has left the cluster the atoms start to move in the new force field; note that this is a very long time on the electronic time scale. Due to the 1 to 2 eV deposited in the cluster during the ionization, the cluster heats up, boils off a lot of atoms, and the charge finally localizes on the trimer or tetramer ion as discussed below (see 4.6.4.1).

4.6.4 Experiments with Positively Charged Rare Gas Clusters

The photoinduced processes in positively charged clusters will be discussed now. All the experiments are done mass selectively. Note, that when the cluster interacts with the photon, it has undergone already the following processes: The cluster was formed in a supersonic expansion. It leaves the condensation zone very hot and cools by evaporation. It is ionized by electron impact. This deposits another 1 to 2 eV of energy into it. The cluster evaporates more atoms. The times for evaporation of the last few atoms are very long, so that one works usually with cluster ions having an internal energy sufficient to evaporate several atoms.

4.6.4.1 Photoabsorption

The neutral rare gases are transparent in the visible part of the spectrum. The localized charge on the other hand has a strong absorption spectrum in this range, as well as in the near IR and UV. This had been first confirmed experimentally by the *Lineberger* group [36]. Newer data and references to other work can be found in [37]. The calculated structures are very floppy, many geometries with very similar energies exist, so that Fig. 10, which shows a rigid sphere and stick model of the $n = 13$ and 19 cluster ions, is a bit misleading. It is still very difficult to impossible to cool a cluster or cluster ion down to a very low temperature.

For the heavier rare gases the trimer and tetramer ions are linear due to the orientational characteristics of the missing *p*-electron. The density of the globelike grids on the central atoms in Fig. 10 is proportional to the calculated ground state charge density, which is about 0.25:0.5:0.25 for the trimer, and 0.1:0.4:0.4:0.1 for the tetramer. For the tetramer 10% of the positive charge resides on each of the outer two atoms, the inner ones carry 40% each. This structure is not so different from the dimer ion, which would have a 0.5:0.5 charge distribution. The charged atoms are surrounded by neutral atoms, which are polarized by the charge nearby. The neutral atoms place themselves in rings or "crowns" around the chromophore, whose spectroscopy will now be discussed. It allows an easy distinction between a charge localization on a dimer or larger unit, as their optical spectra are very different. Assuming an Ar trimer ion core, the process to measure the cross section can be written as:

$$h\nu + (Ar_3^+)Ar_n \rightarrow (Ar_3^{+*})Ar_n \rightarrow [(Ar_3^+)Ar_n]^{\#}$$
$$\rightarrow (Ar_3^+)Ar_m + (n-m)Ar \ . \qquad (4.6.9)$$

The photon induces an electronic transition in the chromophore. The electronically excited state is indicated by a star (*), vibrational energy by #. The electronic energy relaxes fast into vibrational energy, which in turn leads to

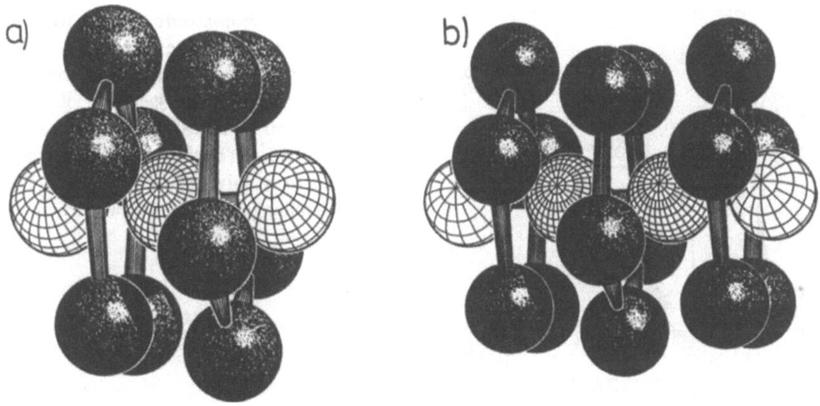

Fig. 10. Stick and sphere model of a trimer ion core of a $n = 13$ cluster (top), and a tetramer ion core in a $n = 19$ cluster. The linear charged chromophores are surrounded by polarized atoms, which are arranged in rings or "crowns" around the chromophore

ejection of atoms. Experimentally an ion, say $(Ar_3^+)Ar_n$, is selected by a mass spectrometer, irradiated by photons from a laser, and the decrease of the intensity of the originally selected clusters measured. From this data the cross section for photoabsorption can be obtained, as explained in the experimental section.

Figure 11 shows the optical absorption spectrum of Xe_3^+ and Xe_{19}^+. Two peaks are observed for the trimer, and three peaks for the larger cluster ions. The two peaks of the trimer spectrum are due to spin-orbit splitting. Indeed for Ar_3^+ the spin-orbit splitting is much smaller, and only one peak has been observed [36, 37, 38, and many references therein]. Upon addition of atoms, the trimer peaks develop into the two peaks at the high energy side of the $n = 19$ spectrum. This shift is just due to the rings of neutral atoms surrounding the trimer core. The polarization interaction between the neutral and charged atoms is different in the ground and excited states [13, 14, 15]. This shifts the energy of the two states differently, and it is this difference which is probed by the photons. The third peak of Xe_{19}^+ has been tentatively explained as being due to a transition in a tetramer ion core [37]. If this is true has to wait for a theoretical calculation.

Assuming that the infrared peak of Xe_{19}^+ corresponds indeed to a tetramer ion, one has thus two structures for this ion: $(Xe_3^+)\,Xe_{16}$ or $(Xe_4^+)\,Xe_{15}$. One can ask, whether these are two well separated geometric and electronic structures? Or can the cluster go over from one to the other one easily? The experimental answer is, that the isomerization between trimer and tetramer ion core is faster than 10 ns [37]. This is probably due to the internal excitation in the cluster. An experiment to measure this isomerization rate as a function of the cluster's internal energy would be very interesting.

The charged dimer, Ar_2^+, has a weak visible and a strong UV absorption. The latter has also been observed in small ($n \geq 3$) clusters [39]. It had been

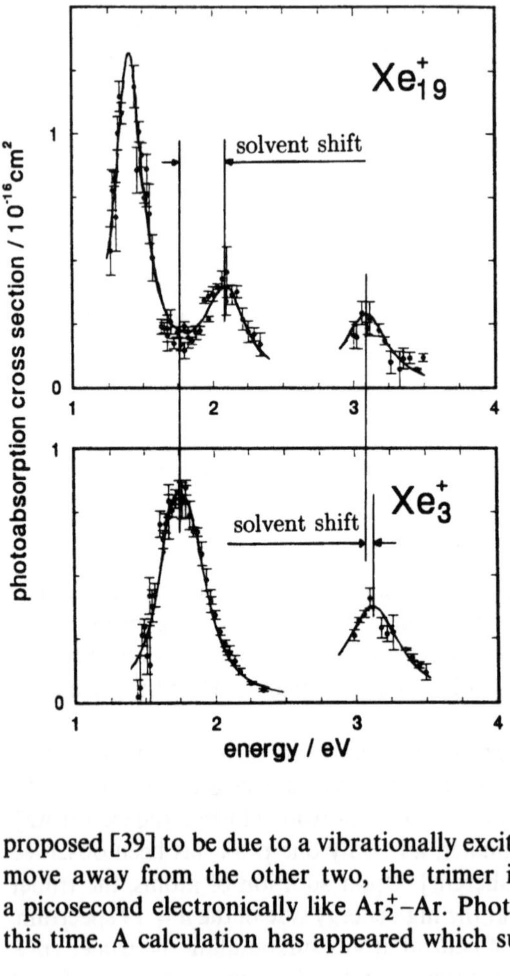

Fig. 11. Photoabsorption spectrum of Xe_3^+ and Xe_{19}^+. The two peaks of the trimer spectrum evolve into the two small peaks of the spectrum of Xe_{19}^+, whose structure can thus be written as $(Xe_3^+)Xe_{16}$. The third peak is probably due to a solvated tetramer ion. The interchange between the two structures is faster than 10 ns

proposed [39] to be due to a vibrationally excited Ar_3^+. If one outer atom would move away from the other two, the trimer ion would look for fractions of a picosecond electronically like Ar_2^+–Ar. Photoabsorption is fast compared to this time. A calculation has appeared which supports this idea [40].

4.6.4.2 Photofragmentation

Figure 12 illustrates fragmentation after photoexcitation. An Ar_{81}^+ was selected and irradiated by 2 eV-photons. At a low photon flux, only the fragment peaks around 56 ± 5 are observed. Thus 2 eV of energy deposited in this cluster leads to an ejection of about 25 atoms. If the laser fluence is increased, the fragments themselves can absorb a photon. They fragment down to the masses around 35 ± 4. A third photon gives fragments around $n = 18$.

A large amount of fragmentation spectra have been analyzed, in order to decide whether the fragmentation is statistical or not [41]. If cluster Ar_{n+1}^+ has an internal energy $E^*(n+1)$ the cluster Ar_n^+ produced by an ejection of a single atom has an energy:

$$E^*(n) = E^*(n+1) - D(n+1) - \varepsilon(n+1) \ , \qquad (4.6.10)$$

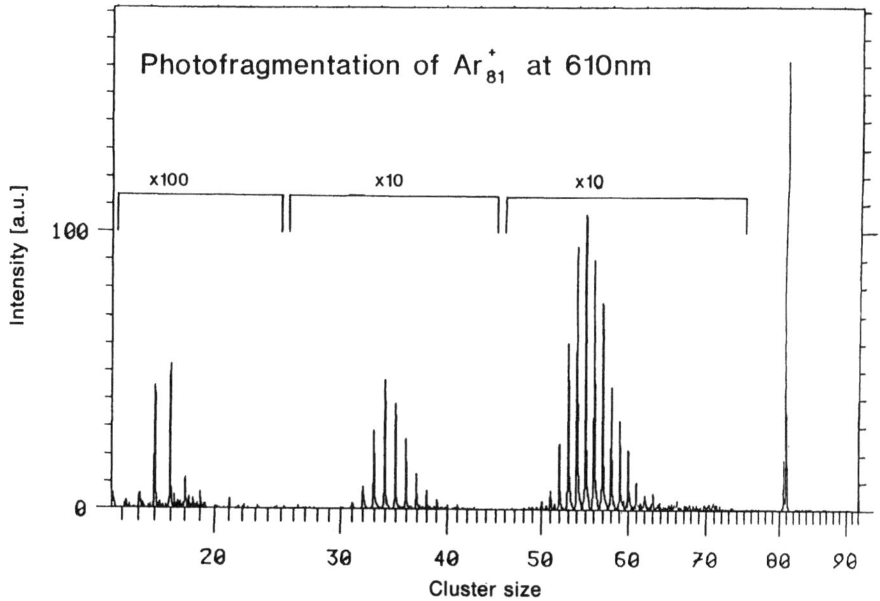

Fig. 12. Photofragmentation mass spectrum. Absorption of one photon by Ar_{81}^{+} leads to the masses around 56. These fragments can absorb another photon leading in turn to the mass peaks around 35, which are fragmented by a third photon to the mass peaks around 18. A statistical simulation of this spectrum was not successful due to nonstatistical effects at very short and very long times

where $D(n+1)$ is the dissociation energy needed to separate one atom from Ar_{n+1}^{*}, and $\varepsilon(n+1)$ is the kinetic energy of the recoiling products. For each ejected atom the cluster looses an amount of $D+\varepsilon$ of internal energy. Thus:

$$\sum_{n\approx 56}^{81} D(n) + \varepsilon(n) \approx 2\ \text{eV} \ . \tag{4.6.11}$$

The equation is only approximately correct as one has a distribution of final sizes, and also the different internal energies of the first and last fragmentation step have not been taken into account. But this can be done in a theoretical analysis. More difficult is the question which kinetic energy to choose. From the contribution by *M. Jarrold* (Chapter 2.7 of this book) one learns that the theoretical predictions for ε differ by a factor of two. A lot of photofragmentation data for argon and xenon cluster ions have been analysed in detail. The result is, that a simple statistical interpretation seems not to be possible [41]. This can have two reasons:

1. The first steps of the energy relaxation are not statistical, e.g. the photon induces a direct ejection of an atom.
2. The vibrational spacing of the trimer/tetramer ions is very much larger than that of their neutral surrounding. Because of this mismatch the energy might not be statistically distributed at all, especially at late times.

Probably both mechanisms are operative. The kinetic energies of the photo emitted atoms have been first measured by *Stace* et al. [42]. The results have been extended to larger clusters and up to Ar_{33}^+, nonstatistical photo desorption has been seen [26]. Also very probably conjecture 2 holds, too. Without it the wide mass distributions around the fragments, e.g. 56 ± 5, seems not to be explainable [41]. From photoemission data from condensed rare gases a similar result is known: the relaxation of the molecular exciton is incomplete before fluorescence occurs [21].

4.6.4.3 Metastable Decay

Metastable decay has been discussed by *Märk* and *Echt* in Chapter 5.6 and will not be treated here.

4.6.5 Experiments with Negatively Charged Rare Gas Clusters

For bulk krypton and xenon the bottom of the conduction band lies lower than the vacuum level (see Fig. 1). Thus an electron can propagate through a Kr or Xe crystal without having enough energy to leave it. This property can be expected to persist in sufficiently large Kr and Xe clusters. Indeed Xe_n^-, $n \geq 6$ have been observed, as explained in more detail in Chapter 5.5. Even the negatively charged atom, Xe^- might be stable [43].

4.6.6 Summary

The geometric arrangement of the atoms in rare gas clusters are given by packing arguments. The densest structure having most nearest neighbors is energetically favored. This leads to the observed and calculated icosahedral growth sequence. For charged clusters, the positive charge localizes on 2, 3 or 4 atoms. The size of these charged chromophores depends on the rare gas, the cluster size, and possibly on the cluster's temperature. An extra negative charge for Kr and Xe is delocalized. Photo, electron, and mass spectroscopy give a consistent, though still crude picture of the development from the atom and dimer, over the cluster to the condensed phase.

References

1. J.A. Barker: Interatomic potentials for inert gases from experimental data. In *Rare Gase Solids Vol.* 1. Academic Press, London, 1977
2. R.A. Aziz: Inert gases. In *Springer Series in Chemical Physics* **34**. Springer, Berlin, 1984

3. N. Schwentner, E.-E. Koch, J. Jortner: *Electronic Excitations in Condensed Rare Gases*. Springer Tracts in Modern Physics. Springer-Verlag, Berlin, 1985
4. J.A. Northby: J. Chem. 6166 (1987)
5. O. Echt, O. Kandler, T. Leisner, W. Miechle, E. Recknagel: J. Chem. Soc. Faraday Soc. 2411 (1190)
6. J. Farges, M.F. de Feraudy, B. Raoult, G. Torchet: J. Chem. Phys. **78**, 5067 (1983)
7. J.W. Lee, G.D. Stein: J. Phys. Chem. **91**, 2450 (1987)
8. B.W. van der Waal: J. Chem Phys. **90**, 3407 (1989)
9. J. Farges, M.F. de Feraudy, B. Raoult, G. Torchet: Surf. Sci. 95 (1981)
10. J.P. Toennies: Helium clusters. *Proc. Int. School "Enrico Fermi", Course CVII, Varenna 1988*, ed. by G. Scoles, 597, 1988
11. H. Buchenau, J.P. Toennies, J.A. Northby: J. Chem. Phys. **95**, 8134 (1991)
12. H. Haberland: Surf. Sci. **156**, 305 (1985)
13. M. Fieber, A.M.G. Ding, P.J. Kuntz: Z. Phys. **D23**, 171 (1992)
14. V. Staemmler: Z. Phys. **D16**, 219 (1990)
15. M. Amarouche, G. Durand, J.P. Malrieux: J. Chem. Phys. 88, 1010 (1988) and references therein
16. U. Buck, R. Krohne, J. Siebers: In *Nuclear Physics Concepts in Atomic Cluster Physics*. ed. by R. Schmidt, H.O. Lutz, R. Dreizler. Springer Verlag, Berlin, 1992
17. J. Stapelfeldt, J. Wörmer, T. Möller: Phys. Rev. Lett. **62**, 98 (1989)
18. M. Lengen, M. Joppien, R. Müller, J. Wörmer, T. Möller: Phys. Rev. Lett. **68**, 2362 (1992)
19. T. Möller: *Progress and Application of Synchroton Radiation to Molecules and Clusters* ed. by A. Ding. Cambridge University Press
20. E.T. Verkhotseva, E.A. Bondarenko, Yu.S. Dornin: Chem. Phys. Lett. **140**, 181 (1987)
21. F. Coletti, J.M. Debever, G. Zimmerer: J. Chem. Phys. **83**, 49 (1985)
22. D. Scharf, J. Jortner, U. Landman: J. Chem. Phys. **88**, 495 (1988)
23. A. Burose, C. Becker, A. Ding: Symposium on Atomic and Surface Physics **90** (1990)
24. W. Miehle, O. Kandler, T. Leisner, O. Echt: J. Chem. Phys. **91**, 5940 (1989)
25. R. Casero, J.M. Soler: J. Chem. Phys. **95**, 2927 (1991)
26. B.v. Issendorff, H. Haberland: unpublished results
27. J. Wörmer, M. Joppien, T. Möller: Chem. Phys. Lett. **182**, 632 (1991)
28. R. Müller, M. Joppien, J. Wörmer, T. Möller: Rev. Sci. Instrum., submitted
29. U. Buck, H. Meier: J. Chem. Phys. **84**, 4854 (1986)
30. M. Joppien, F. Groetelüschen, T. Kloiber, M. Lengen, T. Möller, J. Wörmer, G. Zimmerer, J. Keto, M. Kykta, M.C. Castex: Journal of Luminescence **48 & 49**, 601 (1991)
31. G. Ganteför, G. Bröker, E. Holub-Krabbe, A. Ding: J. Chem. Phys. **91**, 7972 (1989)
32. W. Kamke, J. de Vries, J. Krauss, E. Kaiser, B. Kamke, I.V. Hertel: Z. Phys. **D14**, 339 (1989)
33. E. Rühl, H.W. Jochims, C. Schmale, E. Biller, A.P. Hitchcock, H. Baumgärtel: Chem. Phys. Lett. **178**, 558 (1991)
34. F. Carnovale, J.B. Peel, R.G. Rothwell, J. Valldorf: J. Chem. Phys. **90**, 1452 (1989)
35. F. Carnovale, J.B. Peel, R.G. Rothwell: J. Chem. Phys. **95**, 1473 (1991)
36. N.E. Levinger, D. Ray, W.C. Lineberger: J. Chem. Phys. **89**, 5654 (1988)
37. H. Haberland, B.v. Issendorff, Th. Kolar, H. Kornmeier, Ch. Ludewigt, A. Risch: Phys. Rev. Lett. **67**, 3290 (1991)
38. H. Haberland, B.v. Issendroff, H. Kornmeier, W. Orilk, T. Kolar, C. Ludewigt, T. Reiners, A. Risch: eds: P. Jena, R.N. Rao, S.N. Khanna, 1992. Physics and Chemistry of Finite Systems: From Clusters to Crystals, Vol. II, p. 943, NATO ASI Series, Vol. 374, Kluwer Academic Publishers, Dordrecht
39. M.J. Deluca, M. Johnson: Chem. Phys. Lett. **162**, 445 (1989)
40. M.T. Bowers, W.E. Palke, K. Robins, C. Roehl, S. Walsh: Chem. Phys. Lett. **180**, 235 (1991)
41. M. Schmidt, H. Haberland: unpublished results
42. J.A. Smith, N.G. Gott, J. Winkel, R. Hallet, C.R. Woodward, A.J. Stace, B.J. Whitaker: J. Chem. Phys, **97**, 397 (1992)
43. H. Haberland, T. Kolar, T. Reiners: Phys. Rev. Lett. **63**, 1219 (1989)

4.7 Neutral Molecular Clusters

U. Buck

4.7.1 Introduction

In this chapter clusters of neutral molecules are treated which are bound by weak interactions. In case of stable molecules with closed shells the only forces holding them together are van der Waals or dispersion forces and, if they have permanent moments, also induction forces are present. For systems which contain H-atoms and electronegative elements such as O, N, or F-atoms, in addition a sort of charge transfer occurs and the wellknown hydrogen bonding is found. Typical examples of the former case are $(CO_2)_n$, $(SF_6)_n$, and $(C_6H_6)_n$ with binding energies per bond smaller than 100 meV and examples of the latter case are $(HF)_n$, $(H_2O)_n$, and $(CH_3OH)_n$ with binding energies smaller than 300 meV [1, 2]. The binding energy is about an order of magnitude larger than that of rare gas clusters which have been treated in Chapter 4.6. The additional binding forces and the molecular character thus change the behavior of these clusters in detail, but some of the general trends remain.

As example we discuss the transition to the solid as was investigated in electron diffraction experiments in connection with theoretical calculations [3–5]. For Ar_n it was demonstrated that icosahedral packing [6, 7] is preferred in small clusters for $n = 20$ to 50 and that the crystalline face centered cubic structure occurs at very large cluster sizes around $n = 800$ [3, 4]. Similar results were found for hydrogen bonded water clusters. Only very large clusters with $n = 400$ exhibit the diagonal cubic structure of bulk ice [5] which is, however, not the usual hexagonal form of ice. In addition, small clusters around $n = 20$ did not show the clathrate structure which consists of a regular dodecahedron with one H_2O molecule at each of the 20 vertices and one additional molecule in the center. Best agreement with electron diffraction data is observed for a model of amorphous ice clusters which are composed of a network of distorted rings with three to six hydrogen bonded H_2O molecules [4].

A further characteristic feature of neutral clusters is their behavior following ionization. Rare gas clusters are known for their strong fragmentation when they are ionized [8, 9]. The reason is that the small ionic clusters of these species which are formed after charge localization are much more tightly bound ($\sim$1 eV) and are usually shifted in their equilibrium distances to smaller values compared with the neutral clusters. In a Franck–Condon like vertical transition from the neutral cluster into the ionic ground state, the ionic cluster is highly vibrationally excited which leads to subsequent evaporation of single atoms out of the cluster (for details see Chapter 4.6). These features are also present in molecular clusters, however, the resulting fragmentation pattern can be strongly modified by the possibility of fast chemical reactions of the formed molecular ions with its partner molecules in the cluster. Examples are presented in [10] and [11]. A further change occurs for aromatic molecules. Since in this case the electrons are delocalized, the neutral and the ionic configuration do not differ very much from each other. In the vertical transition during the ionization process, the vibrational ground state can be reached so that, at least, at the threshold, fragmentation can be avoided [12].

A further important point is the role which these molecular clusters play as solvents in chemical reaction dynamics. Studies of electronically and vibrationally excited states energetics and dynamics of these molecular complexes establish the connection between the level structure and the relaxation path of an isolated molecule and a solvent perturbed molecule which manifests itself in the observables of spectral line shifts and linewidths. In this way basic information on solvent perturbations is provided as investigated at a microscopic point of view.

All these arguments make it worthwhile to treat neutral, weakly bound molecular clusters in a separate chapter. We will restrict ourselves to homogeneous species, since the heterogeneous ones and the ions are discussed in Vol. II Chapters 2.1 to 2.5 and results about stronger bound clusters can be found in Chapter 4.5. In the discussion of the experimental tools the emphasis will be laid on methods which make sure that the experiments are carried out with neutral clusters selected according to their size. Because of the fragmentation problem, simple mass spectrometry does not reflect the size distribution of the neutral clusters. Therefore, mainly spectroscopic and scattering methods will be applied to reach this goal. We start the chapter with a short description of typical cluster structures calculated by theoretical means. Then the experimental methods are introduced and illustrated with a few characteristic examples starting with electronic excitation. Finally vibrational spectroscopy is discussed with application to high resolution data and infrared photodissociation of clusters which are size selected in a scattering experiment.

4.7.2 Structure Calculations

Theoretical calculations of the structure are very important tools for the interpretation of the data, because, in general, complete experimental information on cluster structure like in solid state or molecular physics is not yet available. For the structure calculations which are based on procedures of finding the minimum of the total energy, a reliable interaction potential has to be used. The potential is usually written as

$$V = V_{\text{rep}} + V_{\text{dis}} + V_{\text{elec}} + V_{\text{ind}} \,. \tag{1}$$

V_{rep} is the repulsive first order contribution to the electronic exchange energy. The distance and orientation dependence can be approximated by a sum of site-site interactions at i and j on different molecules,

$$V_{\text{rep}} = \sum_{i,j} A_{ij} \exp(-B_{ij} R_{ij}) \,. \tag{2}$$

It is usually obtained from ab initio calculations at the Hartree–Fock self consistent field (SCF) level or further approximative schemes like the test-particle method of *Ahlrichs* and coworkers [13] or empirical methods [14] in which the parameters are fitted to experimental data. The second term represents the attractive long ranged dispersion or van der Waals forces which are obtained in second order perturbation theory and which are often truncated after the first instantaneous dipole-dipole term

$$V_{\text{dis}} = -\sum_{i,j} C_{ij}/R_{ij}^6 \cdot F_{ij}(R_{ij}) \,. \tag{3}$$

A damping function $F_{ij}(R_{ij})$ reduces this term at small distances [15]. This contribution is calculated by ab initio methods or taken from semiempirical formulas using the known polarizabilities [13]. The third term is the first-order contribution to the electrostatic energy

$$V_{\text{elec}} = \frac{1}{4\pi\varepsilon_0} \sum_{i,j} q_i q_j / R_{ij} \tag{4}$$

which is written as the sum of interactions between point charges q_i. ε_0 is the dielectric constant. The site charges can be determined by calculating or measuring the electrostatic multipole potential which is then reproduced by the point charge model. The last term is caused by the induction of multipole moments in the polarizable molecule j by the charge distribution of the other molecules at different sites i, j and vice versa. This is the only term which is nonadditive and which causes in the case of hydrogen bonded systems the well known cooperative effect, an increase of bonding energy per molecule with increasing cluster size for small clusters [16, 17]. For van der Waals systems it is small and very often the pairwise additive model is also used for hydrogen bonded systems. To account for the nonadditive contributions, the parameters of the model are adjusted accordingly to high quality calculations or measurements [18].

As example, results for methanol will be discussed which is a typical case for a linear hydrogen bond similar to water. This is demonstrated for the dimer in Fig. 1a. The bond takes place between the H-atom of the donor molecule and the O-atom of the acceptor molecule. The structure of the α-phase of solid

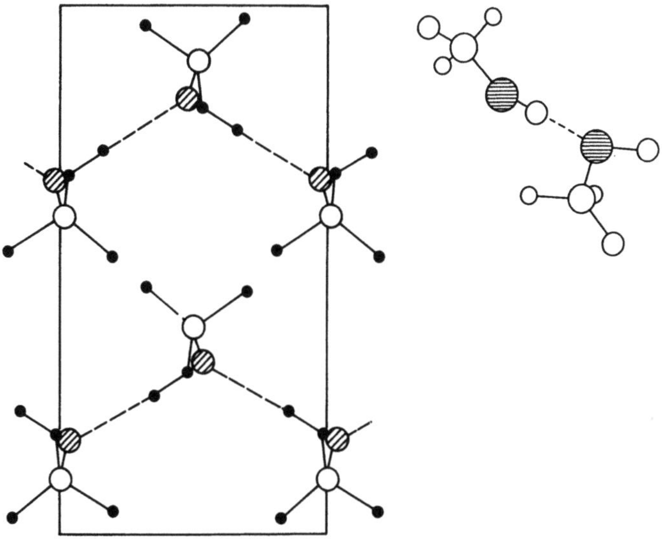

(a)

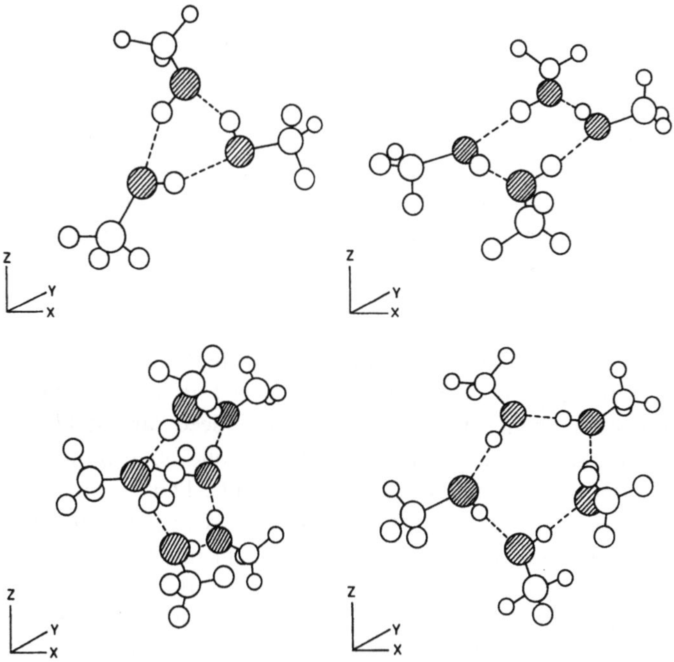

(b)

Fig. 1. **a** Crystal structure of α-methanol showing the projection of the unit cell after [19] and the dimer structure [20]. **b** Calculated minimum energy configurations for the methanol trimer (C_{3h}), tetramer (C_{4h}), pentamer (C_1) and hexamer (C_2) after [20] methanol is also shown in Fig. 1a as is derived from neutron powder diffraction measurements [19]. The molecular geometry is found to be very similar to that in the gas phase and the crystal is formed by infinite hydrogen bonded chains of molecules with adjacent chains pointing in opposite direction and coordination number two. Now the interesting question arises how clusters will behave with a finite number of molecules. The results obtained for the empirical potential of [18] are shown in Fig. 1b [20]. The potential OPLS (optimized potential for liquid simulations) consists of a simple Lennard-Jones potential for the terms of Eqs. (2) and (3) and a Coulomb term (Eq. (4)) with parameters fitted to data of the complex and the liquid. The structures are obtained by starting with randomly chosen geometries and minimizing the total energy. In [20] the results were compared with those of two other empirical potentials and SCF calculations [21] and it was found that, aside from minor differences, they all exhibit the same characteristic features. For the trimer, tetramer and pentamer rings are the preferred structure. Trimer and tetramer are exactly planar and the pentamer is almost planar with a simple alternation of the CH_3 groups pointing up and down. The energy which is necessary to remove one molecule increases from 28.5 kJ/mol for the dimer over 44.9 kJ/mol for the trimer to 51.4 kJ/mol for the tetramer and then decreases again, a clear manifestation that at least part of the cooperative effect is included. The next stable isomers for the trimer and tetramer with much lower energies are chains which have indeed a larger energy per bond, for example 31.4 kJ/mol compared with 24.5 kJ/mol for the ring trimer. But the arrangement with one more bond leads to a larger global binding energy and is therefore preferred. The question arises up to what size a cluster can grow forming ring structures. The results for the hexamer demonstrate that the most stable structure is still a ring which is, however, not planar anymore. In addition, the number of stable isomers increases rapidly and for the second stable configuration a ring of five molecules with an attached monomer is found. In a similar study it was found that ring structures continue up to $n = 11$ and then several combined rings appear [22]. It is interesting to compare these results with those obtained for the simulation of liquids using the same potential function [18]. The percentage of molecules participating in two hydrogen bonds, an indication of a chain or ring structure is indeed largest (73%). However, there is also a significant percentage (19%) in one bond indicating chain ends and three bonds (8%) indicating branching points. Thus we conclude that the ring structure of small clusters which is not found in the condensed phase is a consequence of their finite number. Larger clusters approach the behavior of the liquid with *ring/chain* mixtures of finite length. Similar calculations have been carried out for a number of systems, partly with a much better theoretical foundation based on ab initio calculations of different types for $(HCN)_n$ [23], $(NH_3)_n$ [24], $(C_2H_2)_n$ [25] and $(C_2H_4)_n$ [26]. It is noted that all these calculations of the minimum structure correspond to $T = 0$. For any realistic system Monte Carlo or Molecular Dynamics simulations have to be performed in order to account for temperature effects. Another possibility is to calculate the free energy [22, 26].

4.7.3 Electronic Spectroscopy

Optical spectroscopy by laser induced fluorescence is one of the most successful tools to investigate the structure and dynamics of molecules. So it is not surprising that it was also applied to weak complexes [27, 28]. The detection method, however, has one drawback when applied to clusters, it is not size-specific. Therefore only the variant of resonant two-photon ionization (R2PI) combined with mass spectrometric detection proved to be the adequate method for studying clusters [29–31]. The first step in the process is resonant with an electronically excited state, the second step is generally non-resonant, leading into the ionization continuum. A diagram of the R2PI scheme is shown in Fig. 2. For the first step, tunable laser light is required which is easily available. Thus mainly aromatic molecules are investigated [32]. Very often the experiments are carried out with two photons of one color. In such a case there is always the danger that fragmentation occurs because of the uncontrolled excess energy of the second photon and the size specificity is partly lost. A way out of this problem is the use of two photons with different frequencies (2CR2PI) [33–36] so that the energy of the second photon can be chosen to lie below the dissociation threshold of the cluster ion. The procedure implies that the potential curves of the neutral and the ionic cluster are quite similar as is drawn in Fig. 2 so that no intrinsic excess energy is available. Fortunately most of the aromatic molecules show such a behavior.

A further means to identify different cluster sizes are characteristic spectral features such as line shifts provided they are free of accidental coincidences. The

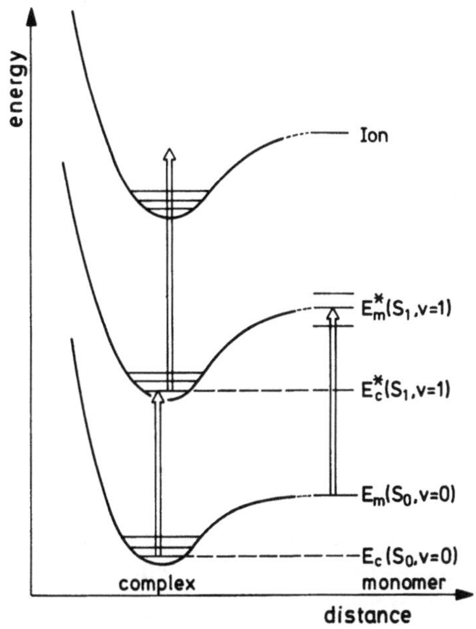

Fig. 2. Schematic diagram of the resonant two-photon ionization of a cluster with energies E_c in one dissociating coordinate. The energies of the monomer E_m in the ground (S_0) and the electronically excited state (S_1) with the vibrational quantum number v are indicated (after [32])

line shift occurs, because the binding energy of the complex in the ground state $D_c = E_m - E_c$ is different from that in the excited state $D_c^* = E_m^* - E_c^*$. For the uncomplexed transition of the monomers we have $\hbar\omega_0 = E_m^* - E_m$ with the monomer energies in the excited and ground state, respectively. According to Fig. 2 the transition for the complex is given by

$$\hbar\omega = E_m^* - D_c^* - (E_m - D_c) = E_m^* - E_m + (D_c - D_c^*) = \hbar\omega_0 + \Delta\hbar\omega \ . \quad (5)$$

The frequency shift $\Delta\hbar\omega$ is expressed by the difference $D_c - D_c^*$. If the two potentials of the complex have identical shapes, this term and thus the net shift would all be zero. If the binding energy in the excited state is larger (smaller), a red (blue) shift occurs.

An example of a state-of-the-art experiment using this technique is the investigation of benzene clusters [37]. They are generated in a supersonic expansion through a pulsed nozzle of a 100 µm orifice with 5 bar He backing pressure. The 2CR2PI-spectra are obtained by two frequency doubled pulsed dye lasers. The first one is used to scan the resonant S_1 state around 262 nm and the wavelength of the second laser was set that it would not ionize the monomer benzene molecule but the larger complexes which have lower ionization potentials. The ions are detected with a reflectron time-of-flight mass spectrometer. The results for the trimer and the tetramer recorded simultaneously are shown in Fig. 3. The trimer spectrum shows the $v = 0$ to $v' = 0$ transition on the left side shifted by $-122.5\ \mathrm{cm}^{-1}$ to the red and a splitting of $1.9\ \mathrm{cm}^{-1}$. The peaks which appear further red shifted are dissociation products of the tetramer. This is easily visualized by comparison with the tetramer spectrum which shows the 0–0 transition at $-161.6\ \mathrm{cm}^{-1}$ splitted by $2.4\ \mathrm{cm}^{-1}$. The peak in the middle at $-146.4\ \mathrm{cm}^{-1}$ is again a dissociation product, now from the pentamer. In spite of using two colors for the ionization, fragmentation could not be avoided. Because of low intensities, the laser intensity and frequency was increased and the amount of excess energy above the threshold for ionization was so large that dissociation occurred. Fortunately, the line shifts are so pronounced for these clusters that it is no problem to identify the different contributions. To overcome this problem experimentally, one can separate the extraction of the ions out of the ionization region so that the dissociation products with higher kinetic energy leave the region from which ions can get to the detector [38].

We note that the red shift increases up to the tetramer and then starts to decrease again. Similar measurements have been reported for $(C_6D_6)_n$ clusters up to $n = 20$ [39]. In these experiments, however, R2PI with one color is used for detection so that the results are strongly influenced by fragmentation. Nevertheless, the same trends in the line shift are observed which reach about $-150\ \mathrm{cm}^{-1}$ for $n = 20$, a value which is still far from the value of the solid at $-262\ \mathrm{cm}^{-1}$. This result is interpreted by taking a simple growth sequence for sphere packing [6].

The 0–0 transitions of the dimer, trimer and tetramer show a very interesting line splitting which the authors attribute to the coupling of identical excited partners in the cluster and thus call it exciton splitting. Using first order

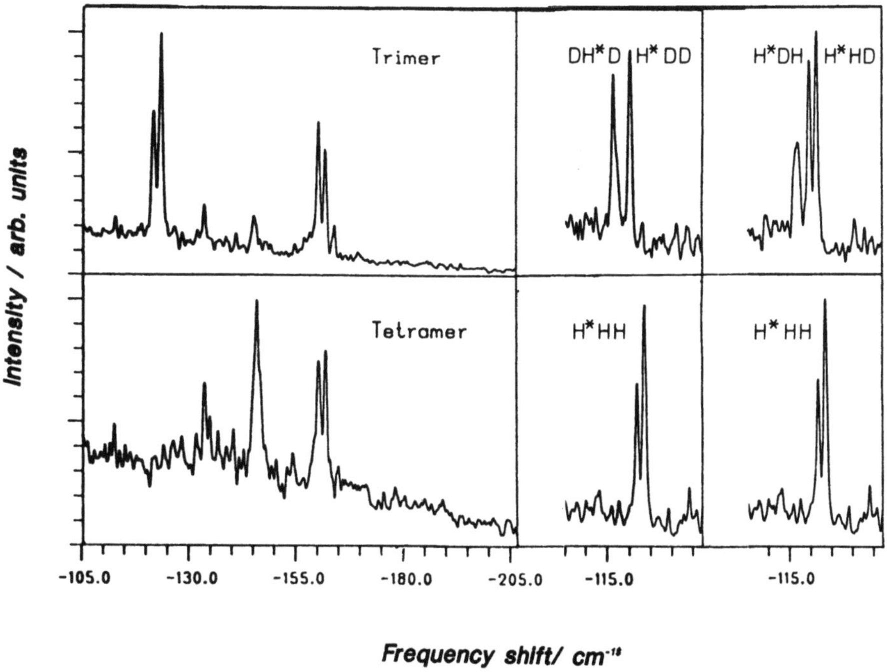

Fig. 3. Two colour resonant two-photon ionization spectra of benzene clusters at the 0-0 transition. Left panels: Trimer and tetramer spectrum with the fragmentation products of the tetramer and pentamer. Right panels: Isotopically substituted trimer spectra (H protonated, D deuterated) showing exciton splitting (after [37]).The lower right panels show the trimer spectrum from the upper left panel for comparison

degenerate perturbation theory, the energy difference is given by $\Delta E = 2\beta$ [40] where ΔE is the observed line splitting and β the interaction matrix element. To explore this behavior in more detail, the authors measured the spectra of isotopically substituted trimers in which one (marked by HHD) or two benzene molecules (marked by HDD) are replaced by perdeuterated benzene. The laser frequency is tuned to excite only the protonated ring (H). The results are displayed in the right panels of Fig. 3. In the case in which two identical partners are left for the excitation one split and one single line is observed and both are slightly shifted from each other. This is a clear indication that two different isomers of the type H*HD and H*DH are present. Only the arrangement with two neighbored protonated molecules (H*HD) shows the exciton splitting. The spectrum with two perdeuterated molecules exhibits as expected no exciton splittings. However, the existence of the two slightly shifted lines is again an indication of two non-equivalent positions in the trimer: one in the middle DH*D and one at the end H*DD. These examples clearly demonstrate that the line shift and the exciton splitting of isotopically substituted molecules gives detailed information on the structure of the clusters.

The structure which the authors found to be most compatible with their experimental results (only two equivalent sites in the trimer) is the zig-zag structure. This structure which is also present in the solid and in the dimer is found in total energy calculations for the trimer with an energy of -22.1 kJ/mol [41]. Other calculations obtained a cyclic type of arrangement with a minimum energy of -31.1 kJ/mol [42, 43], which is also supported by infrared photodissociation experiments. A probable explanation why the zig-zag structure is preferred in the described experiment could be a high cluster temperature which favors chains over rings [22].

4.7.4 Vibrational Spectroscopy

Vibrational spectroscopy has become a very popular method to investigate weakly bound complexes [44–46]. With the advent of tunable laser sources, mainly two methods are used in connection with molecular beams: the direct absorption technique with pulsed jets and slit expansions [46] and the optothermal detection method which is based on the use of a liquid helium cooled bolometer to measure the laser induced change in the molecular beam energy resulting from the vibrational excitation of the molecules [47]. When the excited state is stable for a time that is long with respect to the molecular flight time to the detector, a "positive" signal is observed. For weak complexes, the excitation energy of one vibrational quantum of a monomer in the cluster exceeds the bonding energy of the cluster so that dissociation occurs before the detector is reached and a "negative" bolometer signal is measured. Of course, the method of beam depletion can also be utilized with mass spectrometer detection [48–51].

What is measured in the depletion experiment is the fraction of dissociating molecules P_{diss} which is given by

$$P_{diss} = 1 - \exp[-\sigma(\omega) \cdot F/\hbar\omega] \tag{6}$$

in which $\sigma(\omega)$ is the dissociation cross section, F is the laser fluence (energy/area) and $\hbar\omega$ the photon energy. In principle, these dissociation spectra contain the following information:

1. The line shift $\Delta\hbar\omega$, which is caused by the interaction of the excited oscillator with the surrounding molecules and which therefore gives infomation about the structure of the cluster (see Eq. (5)),
2. the linewidth Γ which, if homogeneously broadened, contains information about the lifetime and, therefore, about the dynamical coupling of the intramolecular vibrational mode to the intermolecular cluster modes,
3. the dissociation cross section which is related directly to the absorption and decay process.

For a more quantitative description of the latter two points, the nature of this process is illustrated in Fig. 4. The monomer with the internal vibrational

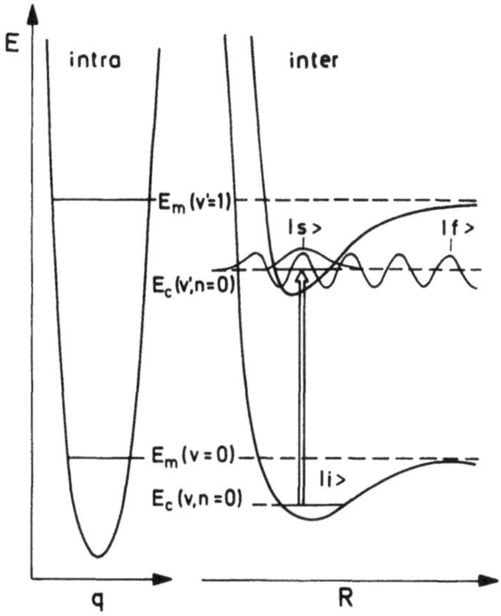

Fig. 4. Schematic diagram of vibrational predissociation of molecular clusters which are first excited from $|i\rangle$ to the bound state $|s\rangle$ and then $|s\rangle$ is coupled to the continuum state $|f\rangle$ after [81]

coordinate q and the intramolecular potential $U_m(q)$ is excited within the cluster from the ground state $v = 0$ to the excited state $v' = 1$. The corresponding intermolecular interaction potentials $V(q, R)$ of the complex with the stretching coordinate R and the quantum number n are shown on the right hand part of Fig. 4. Because of the excitation the potential is shifted upwards and, generally, slightly modified. The dissociation proceeds in two steps. First the cluster is excited from the zeroth-order state $|\Psi_i\rangle = |v = 0, n = 0\rangle$ to $|\Psi_s\rangle = |v = 1, n = 0\rangle$ and then the bound state $|\Psi_s\rangle$ is coupled to the continuum of the ground state $|\Psi_f\rangle = |v = 0, E_c\rangle$. In the framework of time dependent first order perturbation theory for the electric dipole transition and the Golden Rule for the decay rate Γ the cross section is given by [52]

$$\sigma(i \rightarrow f, \omega) = \frac{\pi}{\varepsilon_0}\frac{\omega}{c}|\langle \Psi_s|\boldsymbol{\mu}\cdot\mathbf{e}|\Psi_i\rangle|^2 \frac{1}{\pi}\frac{\Gamma}{(E - E_s)^2 + \Gamma^2} \tag{7}$$

with

$$\Gamma = \pi|\langle \Psi_s|V(q, R)|\Psi_f\rangle|^2 \; . \tag{8}$$

Here, $\boldsymbol{\mu}$ is the transition dipole moment, $\mathbf{e}$ the polarization vector of the photon, ε_0 the dielectric constant and c the velocity of light. This expression reflects the two steps and predicts a Lorentzian lineshape. The predissociation rate scales as the square of the strength of the coupling function $V(q, R)$ and the overlap of the bound-continuum wavefunctions. The latter point leads to the famous momentum gap laws [53, 54] which say that the most facile dissociation will be the one that leads to fragments with small translational energy. Otherwise the overlap of the two wavefunctions is small.

A simple approach for the calculation of vibrational line shifts [55] in solvents can be adapted to treat molecular clusters: As a starting point both the intra- and intermolecular potentials are expanded in a Taylor series in dimensionless normal coordinates [56,57] of the isolated molecule

$$U_m(q) = \frac{1}{2}\sum_i \omega_i q_i^2 + \frac{1}{6}\sum_{ijk} \phi_{ijk} q_i q_j q_k + \cdots \tag{9}$$

$$V(q, R) = V_0(R) + \sum_i \frac{\partial V}{\partial q_i} q_i + \frac{1}{2}\sum_{ij} \frac{\partial^2 V}{\partial q_i \partial q_j} q_i q_j + \cdots \tag{10}$$

The first equation corresponds to the conventional normal mode approach including cubic anharmonicities. Note that for symmetric molecules some of the cubic force constants ϕ_{ijk} have to vanish. The second equation describes the intermolecular interaction in terms of the same coordinates. Using non-degenerate perturbation theory one obtains for the shift of the vibrational excitation frequency ($v = 0 \rightarrow v = 1$)

$$\Delta\hbar\omega = \frac{1}{2}\frac{\partial^2 V}{\partial q_i^2} - \sum_j \frac{\phi_{iij}}{\omega_j}\frac{\partial V}{\partial q_j} . \tag{11}$$

The first term, which is obtained in first order, represents the change of the force constant, while the second term obtained in second order represents the effect of a force ($-\partial V/\partial q_j$) which shifts the equilibrium value of the normal coordinate. Thus a steeper potential ($\partial^2 V/\partial q_i^2 > 0$) causes a blue shift, while an elongation of the bond ($\partial V/q_i < 0$) leads to a red shift, if the anharmonic force constant is negative.

Most of the experimental results published so far deal with binary complexes [46]. The reason is that the experimental methods are not size-selective. This is obvious for the bolometer. But because of strong fragmentation problems also the mass spectrometer does not help very much. Therefore many experiments are carried out under such source conditions that only dimers are present. In addition, the spectrum, if rotationally resolved, is used as fingerprint for the cluster size. Such a procedure gets to a natural limit, if the cluster size increases. Thus only very few examples are available in which clusters other than dimers have been unambiguously identified. As an example we will discuss $(HCN)_3$ [58]. Results of similar quality have been obtained for $(C_2H_2)_4$ [59].

In the case of the hydrogen cyanide trimer an additional feature is in favor of an easy identification of the spectra. The line shifts of this strongly hydrogen bonded system is quite different for the different cluster sizes as was predicted by theoretical calculations [60]. The experiment was carried out with the optothermal detection technique after the excitation of the CH stretching mode with a *F*-center laser under high resolution conditions [58]. The authors observed three well seperated bands with simple *P*- and *R*-branch structures.The missing *Q*-branch is a clear indication of a linear structure. Two of them were rotationally resolved so that lifetimes could be extracted. The results are presented in Table 1 and the underlying structure is shown in Fig. 5. For the ν_1-mode, which is nearly free, a very small line shift of $-4.7\ cm^{-1}$ compared with the monomer

$$\overset{r_1}{\text{H–C}} \equiv \text{N} \cdots \overset{r_2}{\text{H–C}} \equiv \text{N} \cdots \overset{r_3}{\text{H–C}} \equiv \text{N}$$

Fig. 5. Structure of the linear hydrogen cyanide trimer (after [60])

Table 1. Observed line positions and dissociation lifetimes for the $(HCN)_3$

	Linear			Cyclic
	ν_1	ν_2[a]	ν_3[a]	
Line positions/cm^{-1}	3306.8	3231.1	3212.9	3273.5
Lifetimes/ns	> 140	(0.01)[b]	2.8	–

[a] Antisym. and sym. linear combination of the r_2 and r_3 motions of Fig. 5.
[b] Single lines not resolved, fit to the band contour.

frequency of 3311.5 cm^{-1} is observed. In addition, the linewidth is instrumentally limited which places a lower limit of 140 ns on the predissociation lifetime according to the relation $\tau = (2\pi\Delta\nu)^{-1}$. In contrast, the two H-bonded stretches, ν_2 and ν_3, are shifted by -80 cm^{-1} and -99 cm^{-1} and have lifetimes of 10 ps and 2.8 ns. Only in the latter case a clearly resolved structure results, while in the former case the number is derived from a fit to an unresolved band contour. Thus the process of vibrational predissociation is extremely mode-selective. As expected the coupling is much more effective for a mode in the bond, the promotor, than for the free one, the spectator. A further band which exhibits no gap in the center originated from an oblate symmetric top and was identified with that of a cyclic trimer. In this case only one band appears, since from the three frequencies two are degenerate and one is infrared inactive. The shift of -21 cm^{-1} lies between the two extreme cases of the linear configuration. The fact that the signals observed for the two isomers are comparable, suggests that their binding energies are quite similar. This, indeed, is confirmed by calculations which give both in the SCF-approximation [60] and including correlation [61] very similar binding energies for the linear and cyclic conformers. The calculated red shifts further increase up to -144 cm^{-1} for the linear pentamer and reach finally -170 cm^{-1} for the solid which is also composed of linear chains. In that respect the rings are typical structures of small systems which get their stabilization energy only from the additional bond. In the case of the trimer, the ring formation leads to such a large strain that also the linear arrangement has nearly the same binding energy.

4.7.5 Infrared Photodissociation of Size Selected Clusters

Given all the problems which arise from the non-existence of a monosized cluster source, it is highly desirable to carry out spectroscopic measurements with clusters which are selected according to their size. We have developed such

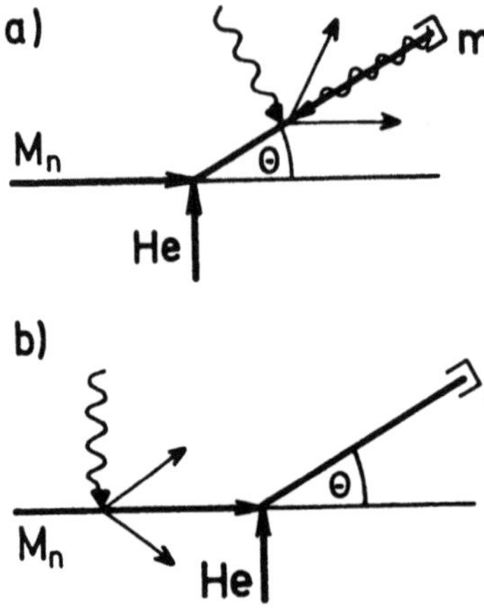

Fig. 6. Schematic experimental arrangements for the photodissociation of size-selected clusters: **a** laser interaction with collisional excited clusters after the scattering process, **b** laser interaction with cold clusters before the scattering process

a method which is described in detail in Chapter 3.10. The different clusters sizes are separated from each other by momentum transfer in a scattering experiment with He atoms [62]. This method can easily be combined with the IR-photodissociation depletion technique decribed in the last section. In practice, two different arrangements are used which are shown in Fig. 6.

In version (a) the cluster beam is first scattered from He for size selection. By measuring the intensity at different angles and masses, different cluster sizes are selected. Then the cluster is dissociated by the infrared radiation either of a pulsed, line-tunable CO_2 laser perpendicular to the beam or by a cw laser collinear with the scattered beam. The dissociation is measured by monitoring the decrease of intensity as a function of laser wavelength and fluence. Because of the scattering process with He, a certain amount of energy is transferred into internal energy of the cluster so that photodissociation takes place with internally excited, "warm" clusters. In the complementary arrangement (b), the laser-molecular beam interaction takes place before the scattering center where the clusters are still cold. Then the cluster beam is dispersed by the He beam and the cluster specific detection is obtained as in the first case. Although the laser interacts with all clusters in the beam, only the dissociation of the cluster size to which the detector is adjusted is measured. This will only cause problems, if a large cluster which is a dissociation product does not leave the beam, reaches the scattering region and causes a positive detector signal. Thus method (a) is better suited for measuring larger clusters, while method (b) is superior for resolving structure in smaller clusters, since the clusters are colder. The scattering method allows to measure the photodissociation not only as a function of cluster size but also for one size as a function of the internal excitation. Systems studied so far in arrangement (a) are C_2H_4 [63, 64, 26], CH_3OH [20, 65], N_2H_4 [66], CH_3CN [67], and CH_3NH_2 [68], those studied in arrangement (b) include C_2H_4 [69], SF_6 [70], CH_3OH [71], NH_3 [72] and C_6H_6 [43]. Surveys of this work can be found in [11, 73–75].

Table 2. Observed line positions in cm^{-1} for various clusters

	Mode	$n = 1$	$n = 2$	$n = 3$	$n = 4$	$n = 5$	$n = 6$
CH_3OH	CO stretch	1034	1027	1041	1044	1048	1040
			1052				1052
N_2H_4	NH_2 wag	937	979	992	1025	1025	1026
			985	1023	1046	1049	1049
N_2H_4	NN stretch	1098	1082	1088	1090	1090	1094

In what follows, results of methanol and hydrazine clusters will be presented as case studies. Assuming that the observed dissociation spectrum of each cluster is a single homogeneously broadened line, we use Eq. (6) and (7) to fit the measured spectra. If more than one peak is found, the exponential function of Eq. (6) is replaced by a sum over exponentials. The results for the peak frequencies are given in Table 2.

Methanol. Methanol clusters have been thoroughly investigated by using the scattering method for size selection both for cold clusters and pulsed lasers [71] (method (b) of Fig. 6) up to $n = 4$ and using internally excited clusters and cw lasers [20] (method (a) of Fig. 6) up to $n = 8$. In both cases a few percent of CH_3OH is seeded in He and Ne as carrier gases that allows to vary the amount of internal excitation within each of the methods. The detection occurs at the protonated ions $(CH_3OH)_{n-1}H^+$ so that a good mass separation against the next cluster size is achieved.

The dimer spectrum after the excitation of the CO-stretching mode is shown in Fig. 7. The spectra are characterized by a two peak structure with one peak shifted by 7 cm^{-1} to the red and one peak shifted by 18 cm^{-1} to the blue compared with the gas phase frequency. The linewidth depends strongly on the internal excitation of the cluster: It varies on the average from 4.0 cm^{-1} for cold dimers seeded in Ne, up to 32 cm^{-1} for collisionally excited dimers with an average transferred energy of $\Delta E = 33$ meV. A probable reason for this behavior is the inhomogeneity of the linewidth so that the measured profile is only the band contour of many rotational lines. The more initial states are excited, the broader is the final distribution.

The explanation for the line shift is found in the calculated structure of the dimer which is shown in Fig. 1. The excited CO-stretching mode is in a non-equivalent position with respect to the hydrogen bond. In the acceptor the O-atom participates directly in the bond, while this is not the case in the donor. The calculation of the line shift [76] based on Eq. (11) and using the potential model of [18], gives a red shift of 4 cm^{-1} for the acceptor and a blue shift of 23 cm^{-1} for the donor, in surprisingly good agreement with experiment. The red shift of the acceptor originates mainly from the elongation of the C–O distance which is caused by the attractive hydrogen bond. The blue shift of the donor results from a larger force constant and the coupling to the O–H mode which squeezes the C–O distance.

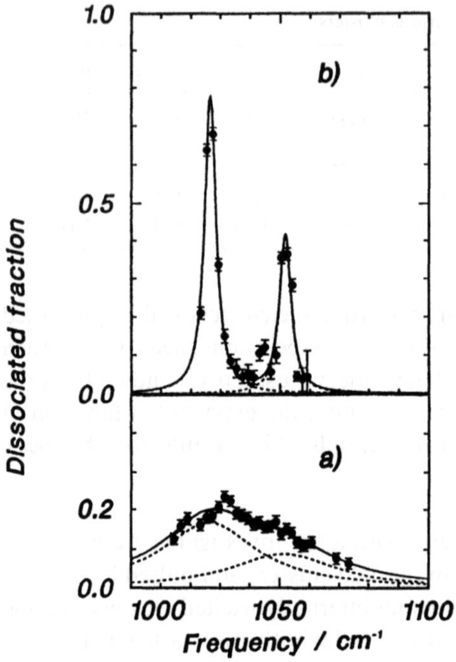

Fig. 7. IR-photodissociation spectra of methanol dimers taken in the two arrangements of Fig. 6 **a** collisional excited [20], **b** cold dimers [71]

For the next larger clusters, the trimer, tetramer and pentamer, only one peak is found shifted to the blue by $+7\,\text{cm}^{-1}$, $+11\,\text{cm}^{-1}$, and $+15\,\text{cm}^{-1}$, respectively, as is illustrated in Fig. 8. These values are the averaged results of two series of experiments with CH_3OH seeded in He and Ne [20]. This behavior changes again with the hexamer. Now a double peak structure appears with one peak shifted by only $+6\,\text{cm}^{-1}$ and one peak shifted further to the blue by $+18\,\text{cm}^{-1}$ compared with the pentamer. The double structure continues then up to $n = 8$.

In the structure calculations planar or nearly planar rings are found for the trimer, tetramer and pentamer with all the molecules in equivalent positions with respect to the CO-excitation (see Fig. 1). Each molecule acts as donor as well as acceptor which explains the frequency peaks between those of the dimer. The fact, that only one peak occurs can be rationalized for the planar trimer which has C_{3h} symmetry. The three CO-stretching modes lie in the plane. One mode, the total symmetric combination of the three CO-stretches is infrared inactive and the other two infrared active modes are degenerate, as in the case of the HCN-trimer. For the hexamer a drastic structural change is observed which is also displayed in Fig. 1. The ring is now strongly distorted and also a series of isomers appear. We attribute the two peaks to two non-equivalent positions of the excited mode. Further experiments and in particular calculations will provide a complete explanation of this problem. In any case we find a strong correlation between the line shift and the structure of the methanol clusters. The comparison with spectra obtained without size selection [77, 78] clearly shows

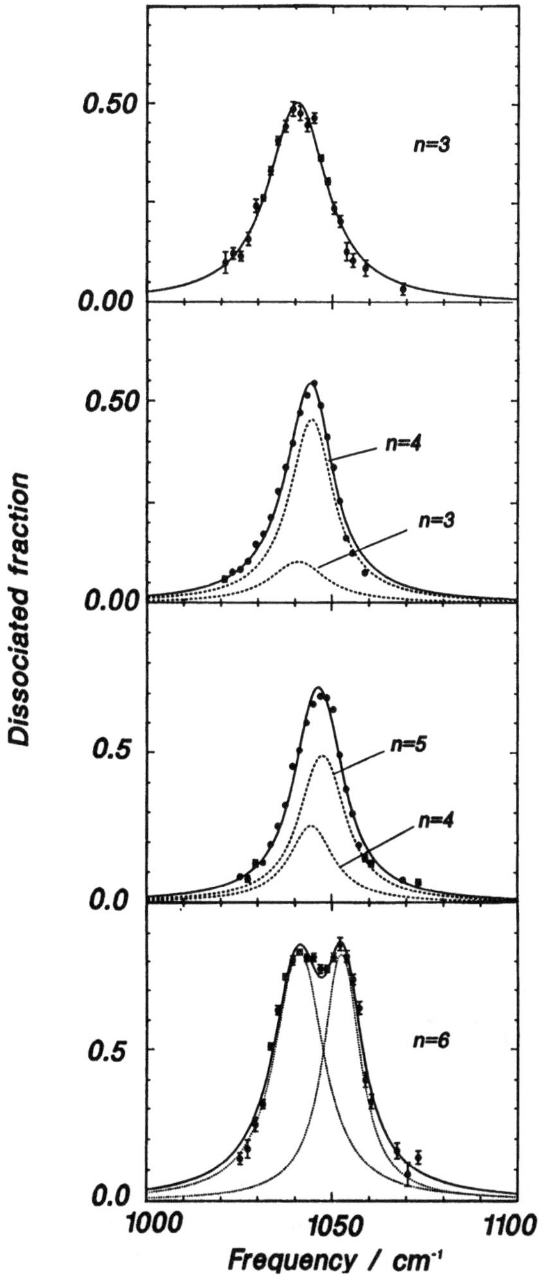

Fig. 8. IR-photodissociation spectra of methanol clusters for $n = 3$, 4, 5 and 6 [20]. The dashed curves are fits of Lorentzian profiles to the data according to the measured cluster composition

that the details of the present results are washed out because of contributions of fragments from different cluster sizes.

Hydrazine. $(N_2H_4)_n$ clusters are generated in a 5.8% mixture in He and are also detected at their protonated ions $(N_2H_4)_{n-1}H^+$. The photodissociation spectra for size-selected clusters for $n = 2$ to $n = 6$ are shown in Fig. 9 [66]. In the range

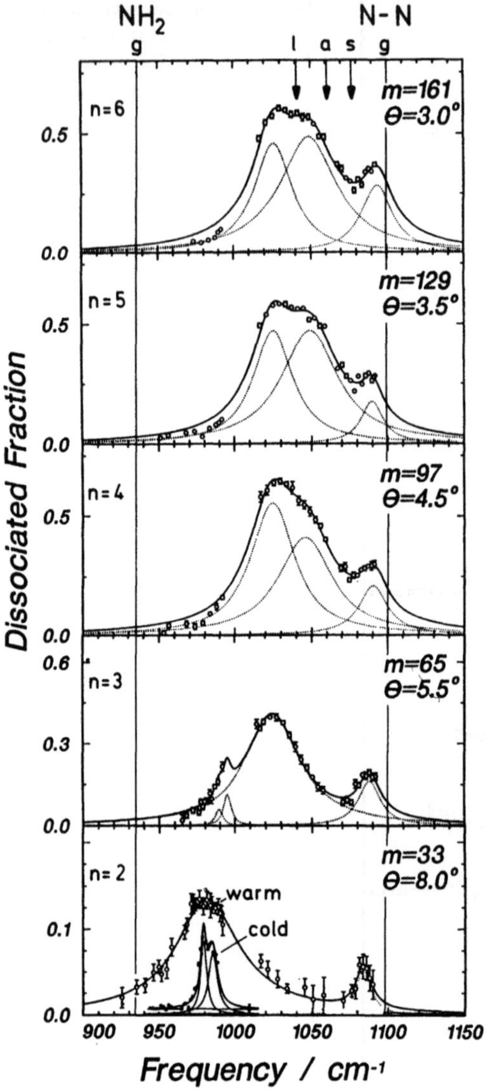

Fig. 9. IR-photodissociation spectra of hydrazine clusters for $n = 2, 3, 4, 5$, and 6. The dotted lines are fits of Lorentzian profiles to the data. The straight solid lines indicate the gas (g) phase absorption frequency of the NH_2-wag and the NN-stretch mode. The arrows indicate the frequencies of the liquid (l), amorphous (a) and crystalline (s) solid [66]

of our laser, the antisymmetric NH_2-wagging (ν_{12} at 937 cm^{-1}) and the NN-stretching mode (ν_5 at 1098 cm^{-1}) are excited. The NN-stretching mode is characterized by one peak which is only slightly shifted to the red. The other mode, however, in which all the four hydrogen atoms move in phase parallel to the N–N axis shows a completely different behavior.

Already the dimer exhibits an average blue shift of 45 cm^{-1}. This trend is continued from the trimer, for which the main peak is observed blue shifted by 88 cm^{-1}, to the tetramer, pentamer, and hexamer for which the shift reaches 89 cm^{-1} and 112 cm^{-1}. Structure calculations of the dimer indicate that the two NH_2 molecules are bound by two non-linear hydrogen bonds. This is shown

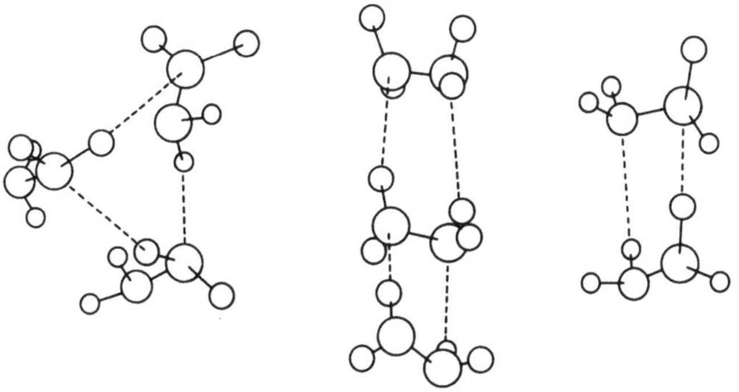

Fig. 10. Calculated minimum energy configurations of hydrazine dimers and trimers [74]

together with the trimer configurations in Fig. 10. Therefore, the motion of the H-atoms is hindered in one of the monomers but not in the other and a double peak structure is expected as was observed for the methanol dimer. This is indeed found in experiments with cold clusters for which a line splitting of 4 cm^{-1} has been observed. The spectrum of the trimer is attributed to two different isomers which are also displayed in Fig. 10. The smaller peaks close to the dimer peaks are probably caused by a chain-like structure consisting of two dimer configurations with two hydrogen bonds in each case. The expected double peak structure is not resolved in this incomplete spectrum of internally excited clusters. The larger peak is explained to be caused by the ring configuration with only one hydrogen bond per monomer. The larger clusters $n > 3$ exhibit also a peak at the same position as is found for the ring trimer, which is therefore attributed to a similar ring structure with one hydrogen bond per molecule. In addition, a second peak shifted to the blue develops, which is very close to the absorption frequency of the amorphous solid [79], but away from the one of the crystalline solid [80]. In summary, the NH_2 motion is heavily distorted and leads, similar to the results for the umbrella motion of NH_3 [72], to large blue shifts in the cluster spectra which, in turn, allow to derive detailed conclusions on the structure.

4.7.6 Summary

The field of weakly bound neutral molecular clusters has enjoyed, by virtue of several spectroscopic techniques, a respectable growth in the last 5 years. The main experimental tools are combinations of different spectroscopic methods with those of size selecting the clusters. Pure mass spectroscopic methods do not give valuable information on the neutral cluster structures and energetics, because strong fragmentation occurs during the ionization process and the measured intensities reflect very often only the stability of the cluster ions after a complicated fragmentation and evaporation process. Mainly three experimental methods gave valuable spectroscopic information on the clusters:

1. The high resolution rotationally revolved spectrum itself is used as indication for the cluster size. The method has found a widespread use in the infrared spectroscopy of dimers and small clusters.
2. Two-colour resonant two-photon ionization (2CR2PI) has been successfully applied to get size specific information for clusters containing aromatic molecules. These species have the additional advantage that when ionized near the threshold nearly no fragmentation occurs.
3. The combination of the size selective process of the scattering by a He-beam with the infrared photodissociation resulting in depletion is a further method to achieve the preset goal.

The cluster sizes investigated so far are small. The major effects of the cluster on the spectral properties of the monomer is (a) to shift the peaks and (b) to broaden the width. The former point gives information about the structure, while the latter one is responsible for the decay dynamics.

What did we learn so far from the experiments? The spectra exhibit a great variety of structures. The lines are strongly shifted to the red (CH-stretch in HCN), to the blue (NH_2-wag in N_2H_4) or undergo only small shifts (CO-stretch in CH_3OH, CH_2-wag in C_2H_4). We observe single lines, splitting because of non-equivalent positions, and exciton splitting. All these data are strongly correlated with the structure of the clusters. Together with calculations of the minimum energy configurations and data from electron diffraction studies the following general picture emerges. Very often the dimer structure is also found in the solid. With coordination number two these solids are infinite chains of "dimers". This behavior is found in linear hydrogen bonds (CH_3OH, HCN), non linear hydrogen bonds (N_2H_4), dipolar interactions (CH_3CN) and also in systems with π-electrons (C_6H_6). For smaller clusters such as trimers, tetramers and pentamers usually rings are preferred with very symmetric and often planar structures. The reason is simply that the number of bonds which is now available for these structures is larger compared with open chain-like arrangements. Then it does not matter that the stabilization energy per bond is usually smaller, since the ideal bonding configuration of the dimer has to be strained. In a few cases, e.g. for HCN and N_2H_4 trimers, both configurations are equally present in the experiments. In the first case the dimer configuration is linear and the strain is very large. In the second case one additional hydrogen bond is lost (see Fig. 10). For $n = 4$ and $n = 5$ usually rings are the most stable configuration. The larger number of molecules leads to less strained hydrogen bonds thus making the cyclic isomers the most stable ones. From the hexamer upwards, cyclic structures still dominate. However, now they are distorted and strong deformations start to appear. In addition, the number of isomers with configurations nearly as stable as the global cyclic structures gets larger with increasing cluster size. They often consist of combinations of rings and chains with branching points and look like the frozen structure of a liquid or better an amorphous solid. This trend is still present up to $n = 20$. The transition to the ordered solid structure usually occurs at much larger cluster sizes.

The decay dynamics of the vibrational predissociation studies in the infrared which are reflected in the homogeneously broaded linewidth of the spectra have indicated some fascinating nonstatistical behavior. The corresponding times reach from some 0.1 ms (ArHF) to 10 ps ($(HCN)_3$) and many of the measured features can be rationalized by the gap laws based on Eq. (8). Quite remarkable is the extreme mode selectivity of the process. The decay rate of different modes in the same cluster can differ by orders of magnitude depending whether the excitation takes place directly in the bond or away from it. Recently a strong correlation was found between the decay rate Γ and the square of the lineshift $(\Delta\hbar\omega)^2$ [45, 81]. Both effects arise from the same intramolecular–intermolecular coupling function $V(q, R)$. Therefore it is not so surprising that also the line shifts are mode selective as is demonstrated for N_2H_4 for which the excitation of the NH_2 wag leads to more than 100 cm^{-1} blue shift, while the NN stretch is red shifted by only 10 cm^{-1}.

Directions for future activities in the field are manifold. The extension to larger cluster sizes is highly desirable in order to test the behavior of clusters between $n = 10$ and $n = 20$. This will certainly cause some problems for the measurements of completely resolved spectra. However, the 2CR2PI-method has already been applied in this size region [32] and also the scattering method can be extended to larger clusters by changing the scattering partner and the velocities of the colliding particles. The problem of multiple isomeric configurations may be tackled by applying double-resonance techniques which allow to distinguish whether a line splitting is caused by non-equivalent positions in one cluster or by two different isomers. The elucidation of dynamical effects in the photodissociation experiments can be much improved by measuring not only the depletion but also the distributions of the products. Finally, we should mention that one of the most critical parameters in all these experiments, the temperature of the cluster, is not really known. Any new attempt for a realistic determination is welcome. As was already mentioned, the experimental work has to be accompanied by theoretical calculations of the cluster structure and simulations of their phase and their spectral features for finite temperatures. For larger clusters the rigid structure calculation at $T = 0$ have to be replaced by statistical mechanics computer simulation techniques, if possible with quantum or semiclassical methods [82].

4.7.7 Recent Developments

Substantial progress has been made in the interpretation of the line shift data presented in Section 4.7.5. The procedure is based on the ideas given in Section 4.7.4 and leads to a complete understanding of the spectra and the cluster structures involved [83]. For the methanol hexamer a new lowest energy configuration of S_6 symmetry has been found (see Fig. 1). The progress made in the calculation of complexes containing benzene molecules is discussed in [84]. Some of the future activities mentioned in the Summary have been realized in the meantime. These include the extension of the cluster separation method by atomic scattering to sizes $n = 13$ for $(CH_3CN)_n$ [85], double resonance experiments with size selected clusters for identifying different isomers [86, 87], and the detection of products in state-to-state IR-photodissociation experiments [88, 89].

Acknowledgement. I acknowledge with gratitude the contributions of many coworkers, as cited in the references. I am pleased to thank B. Schmidt, M. Winter, M. Hobein, and J. Siebers for their comments on this manuscript. Part of the work was supported by the Deutsche Forschungsgemeinschaft (SFB 93 and 357).

References

1. J. Jortner: In: Ber. Bunsenges. Phys. Chem. **88**, 188 (1984); J. Jortner, D. Scharf, U. Landmann: In: *Elemental and Molecular Clusters*, G. Benedek, T.P. Martin, G. Pacchioni (eds.), p. 148. Springer, Berlin, 1988
2. The complete issue of Chem. Rev. **88,** 813–988 (1988)
3. J. Farges, M.F. de Feraudy, B. Raoult, G. Torchet: J. Chem. Phys. **84**, 3491 (1986)
4. J. Farges, M.F. de Feraudy, B. Raoult, G. Torchet: Adv. Chem. Phys. **70**, Part 2, 45 (1988)
5. G. Stein: Physics Teacher **17**, 503 (1979)
6. M.R. Hoare: Adv. Chem. Phys. **40**, 49 (1979)
7. M.R. Hoare, J.A. McInnes: Adv. Phys. **32**, 791 (1983)
8. H. Haberland: Surf. Sci. **156**, 305 (1985)
9. O. Echt: In: *Elemental and Molecular Clusters*, G. Benedek, T.P. Martin, G. Pacchioni (eds.), p. 263. Springer, Berlin, 1988
10. U. Buck: J. Phys. Chem. **92**, 1023 (1988)
11. U. Buck: In: *The Chemical Physics of Atomic and Molecular Clusters* (Enrico Fermi School, Varenna), G. Scoles (ed.), p. 543. North Holland, Amsterdam, 1990
12. M. Kappes, S. Leutwyler: In: *Atomic and Molecular Beam Methods*, G. Scoles (ed.), p. 380. Oxford University Press, New York, 1988
13. H.J. Böhm, R. Ahlrichs: J. Chem. Phys. **77**, 2028 (1982)
14. J. Snir, R.A. Nemenoff, H.A. Scheraga: J. Phys. Chem. **82**, 2497 (1978)
15. R. Ahlrichs, R. Penco, G. Scoles: Chem. Phys. **19**, 119 (1977); K.T. Tang, J.P. Toennies: J. Chem. Phys. **80**, 3725 (1984)
16. A. Karpfen, A. Beyer, P. Schuster: Chem. Phys. Lett. **102**, 289 (1983)
17. J. Detrich, G. Corongiu, E. Clementi: Chem. Phys. Lett. **112**, 426 (1984)
18. W.L. Jorgensen: J. Phys. Chem. **80**, 1276 (1986)
19. B.H. Torrie, S.-X. Weng, B.M. Powell: Mol. Phys. **67**, 575 (1989)
20. U. Buck, X.J. Gu, Ch. Lauenstein, A. Rudolph: J. Chem. Phys. **92**, 6017 (1990)
21. L.A. Curtiss, M. Blander: Chem. Rev. **88**, 827 (1988)
22. T.P. Martin, T. Bergmann, B. Wassermann: In: *Large finite Systems*, J. Jortner, B. Pullmann, (eds.) Reidel, Dordrecht, 1987
23. M. Kofranek, A. Karpfen, H. Lischka: Chem. Phys. **113**, 53 (1987)
24. J.C. Greer, R. Ahlrichs, I.V. Hertel: Chem. Phys. **139**, 191 (1989)
25. R.G.A. Bone, R.D. Amos, N.C. Handy: J. Chem. Soc. Faraday Trans. **11**, 1931 (1990)

26. R. Ahlrichs, S. Brode, U. Buck, M. DeKieviet, Ch. Lauenstein, A. Rudolpf, B. Schmidt: Z. Phys. D **15**, 341 (1990)
27. R.E. Smalley, L. Wharton, D.H. Levy, D.H. Chandler: J. Chem. Phys. **68**, 2487 (1978).
28. A. Amirav, U. Even, J. Jortner: J. Phys. Chem. **86**, 3345 (1982)
29. S. Leutwyler, U. Even, I. Jortner: J. Chem. Phys. **79**, 5769 (1983)
30. U. Boesl, H.J. Neusser, E.W. Schlag: Chem. Phys. **55**, 193 (1981)
31. K. Rademann, B. Brutschy, H. Baumgärtel: Chem. Phys. **80**, 129 (1983)
32. S. Leutwyler, J. Bösinger: Chem. Rev. **90**, 489 (1990)
33. K.H. Fung, W. Henke, H.L. Selzle, E.W. Schlag: J. Phys. Chem. **85**, 3560 (1981)
34. J.B. Hopkins, D.E. Powers, R.E. Smalley: J. Phys. Chem. **85**, 3739 (1981)
35. U. Even, N. BenHorin, J. Jortner: Phys. Rev. Lett. **62**, 140 (1989)
36. R. Knochenmuss, S. Leutwyler: J. Chem. Phys. **92**, 4686 (1990)
37. K.O. Börnsen, S.H. Lin, H.L. Selzle, E.W. Schlag: J. Chem. Phys. **90**, 1299 (1989)
38. V. Beushausen: Dissertation, Göttingen, 1989; E.W. Bieske, M.W. Rainbird, A.E.W. Knight: J. Chem. Phys. **90**, 2068 (1989)
39. D.C. Easter, M.S. ElShall, M.Y. Hahn, R.L. Whetten: Chem. Phys. Lett. **157**, 277 (1989)
40. C. Cohen-Tannoudji, B. Diu, F. Laloë: *Quantum Mechanics*, Vol. 1, p. 405. John Wiley, New York, 1977
41. D.E. Williams: Acta Cryst. **A36**, 715 (1980)
42. B.W. van de Waal: Chem. Phys. Lett. **123**, 69 (1986)
43. A. de Meijere, F. Huisken: J. Chem. Phys. **92**, 5826 (1990)
44. Structure and Dynamics of Weakly Bound Molecular Complexes. A. Weber (ed.). NATO ASI Series C212. Reidel, Dordrecht, 1987
45. R.E. Miller: Science **240**, 447 (1988)
46. N.J. Nesbit: Chem. Rev. **88**, 843 (1988)
47. T.E. Gough, R.E. Miller, G. Scoles: J. Chem. Phys. **69**, 1588 (1978)
48. M.P. Casassa, D.S. Bomse, K.C. Janda: J. Chem. Phys. **74**, 5044 (1981)
49. M.F. Vernon, D.J. Krajnovich, H.S. Kwok, J.M. Lisy, Y.R. Shen, Y.T. Lee: J. Chem. Phys. **77**, 47 (1982)
50. M.A. Hoffbauer, K. Liu, C.F. Giese, W.R. Gentry: J. Chem. Phys. **78**, 5567 (1983)
51. D.W. Michael, J.M. Lisy: J. Chem. Phys. **85**, 2528 (1986)
52. J.A. Beswick: In: *Structure and Dynamics of Weakly Bound Molecular Complexes*. A. Weber (ed.), p. 563. NATO ASI Series C212. Reidel, Dordrecht, 1987
53. J. A. Beswick, J. Jortner: J. Chem. Phys. **68**, 2277 (1978)
54. G.E. Ewing: Chem. Phys. **29**, 253 (1978)
55. A.D. Buckingham: Trans. Faraday Soc. **56**, 763 (1960)
56. I.M. Mills: In: *Molecular Spectroscopy*, K.N. Rao, C.W. Matheus (eds), p. 115. Academic Press, New York, 1972
57. P.O. Westlund, R.M. Lynden-Bell: Mol. Phys. **60**, 1189 (1987)
58. K.W. Jucks, R.E. Miller: J. Chem. Phys. **88**, 2196 (1988)
59. G.W. Bryant, D.F. Eggers, R.O. Watts: J. Chem. Soc. Faraday Trans. 2 **84**, 1443 (1988)
60. M. Kofranek, A. Karpfen, H. Lischka: Chem. Phys. **113**, 53 (1987)
61. I.J. Kurnig, H. Lischka, A. Karpfen: J. Chem. Phys. **92**, 2469 (1990)
62. U. Buck, H. Meyer: J. Chem. Phys. **84**, 4854 (1986)
63. U. Buck, Ch. Lauenstein, A. Rudolph, B. Heijmen, S. Stolte, J. Reuss: Chem. Phys. Lett. **144**, 396 (1988)
64. U. Buck, F. Huisken, Ch. Lauenstein, H. Meyer, R. Sroka: J. Chem. Phys. **87**, 6276 (1987)
65. U. Buck, X.J. Gu, Ch. Lauenstein, A. Rudolph: J. Phys. Chem. **92**, 5561 (1988); J. Chem. Phys. **92**, 6017 (1990)
66. U. Buck, X.J. Gu, M. Hobein, Ch. Lauenstein: Chem. Phys. Lett. **163**, 455 (1988)
67. U. Buck, X.J. Gu, R. Krohne, Ch. Lauenstein: Chem. Phys. Lett. **174**, 247 (1990)
68. U. Buck, X.J. Gu, R. Krohne, Ch. Lauenstein, H. Linnartz, A. Rudolph: J. Chem. Phys. **94**, 23 (1991)
69. F. Huisken, T. Pertsch: J.Chem. Phys. **86**, 106 (1987)

70. F. Huisken, M. Stemmler: Chem. Phys. Lett. **132**, 351 (1989)
71. F. Huisken, M. Stemmler: Chem. Phys. Lett. **144**, 391 (1988)
72. F. Huisken, T. Pertsch: Chem. Phys. **126**, 213 (1988)
73. U. Buck: In *Dynamics of Polyatomic van der Waals Complexes*, N. Halberstadt, K.C. Janda (eds.), p. 43. NATO ASI Series. Plenum, New York, 1990
74. U. Buck, X.J. Gu, M. Hobein, Chr. Lauenstein, A. Rudolph.: J. Chem. Soc. Faraday Trans. **86**, 1923 (1990)
75. F. Huisken: Adv. Chem. Phys. **81**, 63 (1991)
76. U. Buck, B. Schmidt: J. Mol. Liq. **46**, 181 (1990)
77. M.A. Hoffbauer, C.F. Giese, W.R. Gentry: J. Phys. Chem. **88**, 181 (1984)
78. J. Crooks, A.J. Stace, B.J. Whitaker: J. Phys. Chem. **92**, 3554 (1988)
79. J.A. Roux, B.E. Wood: J. Opt. Soc. Amm. **73**, 1181 (1983)
80. P.A. Giguère, I.D. Liu: J. Chem. Phys. **20**, 136 (1952)
81. R.J. LeRoy, M.R. Davies, M.E. Lam: J. Phys. Chem. **95**, 2167 (1991)
82. R.O. Watts: In: *The Chemical Physics of Atomic and Molecular Clusters* (Enrico Fermi School, Varenna) G. Scoles, (ed.), p. 271. North Holland, Amsterdam, 1990
83. U. Buck, B. Schmidt: J. Chem. Phys. **98**, 9410 (1993)
84. P. Hobza, R. Zahradnik: Ber. Bunsenges. Phys. Chem. **96**, 1294 (1992)
85. U. Buck: Ber. Bunsenges. Phys. Chem. **96**, 1275 (1992)
86. W. Scherzer, H.L. Selzle, E.W. Schlag: Chem. Phys. Lett. **195**, 11 (1992)
87. U. Buck, M. Hobein: Z. Phys. D (1993), in press
88. E.J. Bohàc, R.E. Miller: J. Chem. Phys. **98**, 2604 (1993)
89. E.J. Bohac, M.D. Marshall, R.E. Miller: J. Chem. Phys. **97**, 4901 (1992)

Subject Index

Springer Series in Chemical Physics

Editors: Vitalii I. Goldanskii   Fritz P. Schäfer   J. Peter Toennies

1 **Atomic Spectra and Radiative Transitions**
By I. I. Sobelman
2 **Surface Crystallography by LEED**
Theory, Computation and Structural Results. By M. A. Van Hove, S. Y. Tong
3 **Advances in Laser Chemistry**
Editor: A. H. Zewail
4 **Picosecond Phenomena**
Editors: C. V. Shank, E. P. Ippen, S. L. Shapiro
5 **Laser Spectroscopy**
Basic Concepts and Instrumentation
By W. Demtröder   3rd Printing
6 **Laser-Induced Processes in Molecules**   Physics and Chemistry
Editors: K. L. Kompa, S. D. Smith
7 **Excitation of Atoms and Broadening of Spectral Lines**   By I. I. Sobelman, L. A. Vainshtein, E. A. Yukov
8 **Spin Exchange**
Principles and Applications in Chemistry and Biology
By Yu. N. Molin, K. M. Salikhov, K. I. Zamaraev
9 **Secondary Ion Mass Spectrometry SIMS II**   Editors: A. Benninghoven, C. A. Evans, Jr., R. A. Powell, R. Shimizu, H. A. Storms
10 **Lasers and Chemical Change**
By A. Ben-Shaul, Y. Haas, K. L. Kompa, R. D. Levine
11 **Liquid Crystals of One- and Two-Dimensional Order**
Editors: W. Helfrich, G. Heppke
12 **Gasdynamic Laser**   By S. A. Losev
13 **Atomic Many-Body Theory**
By I. Lindgren, J. Morrison
14 **Picosecond Phenomena II**
Editors: R. M. Hochstrasser, W. Kaiser, C. V. Shank
15 **Vibrational Spectroscopy of Adsorbates**   Editor: R. F. Willis
16 **Spectroscopy of Molecular Excitions**
By V. L. Broude, E. I. Rashba, E. F. Sheka
17 **Inelastic Particle-Surface Collisions**
Editors: E. Taglauer, W. Heiland
18 **Modelling of Chemical Reaction Systems**   Editors: K. H. Ebert, P. Deuflhard, W. Jäger
19 **Secondary Ion Mass Spectrometry SIMS III**
Editors: A. Benninghoven, J. Giber, J. László, M. Riedel, H. W. Werner
20 **Chemistry and Physics of Solid Surfaces IV**   Editors: R. Vanselow, R. Howe
21 **Dynamics of Gas-Surface Interaction**
Editors: G. Benedek, U. Valbusa
22 **Nonlinear Laser Chemistry**
Multiple-Photon Excitation
By V. S. Letokhov
23 **Picosecond Phenomena III**
Editors: K. B. Eisenthal, R. M. Hochstrasser, W. Kaiser, A. Laubereau
24 **Desorption Induced by Electronic Transitions DIET I**   Editors: N. H. Tolk, M. M. Traum, J. C. Tully, T. E. Madey
25 **Ion Formation from Organic Solids**
Editor: A. Benninghoven
26 **Semiclassical Theories of Molecular Scattering**   By B. C. Eu
27 **EXAFS and Near Edge Structures**
Editors: A. Bianconi, L. Incoccia, S. Stipcich
28 **Atoms in Strong Light Fields**
By N. B. Delone, V. P. Krainov
29 **Gas Flow in Nozzles**
By U. G. Pirumov, G. S. Roslyakov
30 **Theory of Slow Atomic Collisions**
By E. E. Nikitin, S. Ya. Umanskii
31 **Reference Data on Atoms, Molecules, and Ions**   By A. A. Radzig, B. M. Smirnov
32 **Adsorption Processes on Semiconductor and Dielectric Surfaces I**
By V. F. Kiselev, O. V. Krylov
33 **Surface Studies with Lasers**
Editors: F.R. Aussenegg, A. Leitner, M.E. Lippitsch
34 **Inert Gases**
Potentials, Dynamics, and Energy Transfer in Doped Crystals.   Editor: M. L. Klein
35 **Chemistry and Physics of Solid Surfaces V**   Editors: R. Vanselow, R. Howe
36 **Secondary Ion Mass Spectrometry, SIMS IV**   Editors: A. Benninghoven, J. Okano, R. Shimizu, H. W. Werner
37 **X-Ray Spectra and Chemical Binding**
By A. Meisel, G. Leonhardt, R. Szargan
38 **Ultrafast Phenomena IV**
By D. H. Auston, K. B. Eisenthal
39 **Laser Processing and Diagnostics**
Editor: D. Bäuerle

MIX
Papier aus verantwortungsvollen Quellen
Paper from responsible sources
FSC® C105338

If you have any concerns about our products,
you can contact us on
ProductSafety@springernature.com

In case Publisher is established outside the EU,
the EU authorized representative is:
Springer Nature Customer Service Center GmbH
Europaplatz 3, 69115 Heidelberg, Germany

Printed by Libri Plureos GmbH
in Hamburg, Germany